Crystal Plasticity at Micro- and Nano-scale Dimensions

Crystal Plasticity at Micro- and Nano-scale Dimensions

Editors

Ronald W. Armstrong
Wayne L. Elban

MDPI • Basel • Beijing • Wuhan • Barcelona • Belgrade • Manchester • Tokyo • Cluj • Tianjin

Editors
Ronald W. Armstrong
University of Maryland
USA

Wayne L. Elban
Loyola University Maryland
USA

Editorial Office
MDPI
St. Alban-Anlage 66
4052 Basel, Switzerland

This is a reprint of articles from the Special Issue published online in the open access journal *Crystals* (ISSN 2073-4352) (available at: https://www.mdpi.com/journal/crystals/special_issues/Crystal_Dimensions).

For citation purposes, cite each article independently as indicated on the article page online and as indicated below:

LastName, A.A.; LastName, B.B.; LastName, C.C. Article Title. *Journal Name* **Year**, *Volume Number*, Page Range.

ISBN 978-3-0365-0874-0 (Hbk)
ISBN 978-3-0365-0875-7 (PDF)

Cover image courtesy of Daniel Kiener.

© 2021 by the authors. Articles in this book are Open Access and distributed under the Creative Commons Attribution (CC BY) license, which allows users to download, copy and build upon published articles, as long as the author and publisher are properly credited, which ensures maximum dissemination and a wider impact of our publications.

The book as a whole is distributed by MDPI under the terms and conditions of the Creative Commons license CC BY-NC-ND.

Contents

About the Editors vii

Preface to "Crystal Plasticity at Micro- and Nano-scale Dimensions" ix

Ronald W. Armstrong and Wayne L. Elban
Crystal Strengths at Micro- and Nano-Scale Dimensions
Reprinted from: *Crystals* **2020**, *10*, 88, doi:10.3390/cryst10020088 1

Daniel Kiener, Jiwon Jeong, Markus Alfreider, Ruth Konetschnik and Sang Ho Oh
Prospects of Using Small Scale Testing to Examine Different Deformation Mechanisms in Nanoscale Single Crystals—A Case Study in Mg
Reprinted from: *Crystals* **2021**, *11*, 61, doi:10.3390/cryst11010061 11

Dongyue Xie, Binqiang Wei, Wenqian Wu and Jian Wang
Crystallographic Orientation Dependence of Mechanical Responses of FeCrAl Micropillars
Reprinted from: *Crystals* **2020**, *10*, 943, doi:10.3390/cryst10100943 27

Qian Lei, Jian Wang and Amit Misra
Mechanical Behavior of Al–Al$_2$Cu–Si and Al–Al$_2$Cu Eutectic Alloys
Reprinted from: *Crystals* **2021**, *11*, 194, doi:10.3390/cryst11020194 41

G. Sainath, Sunil Goyal and A. Nagesha
Plasticity through De-Twinning in Twinned BCC Nanowires
Reprinted from: *Crystals* **2020**, *10*, 366, doi:10.3390/cryst10050366 51

Xiaolei Chen, Thiebaud Richeton, Christian Motz and Stéphane Berbenni
Atomic Force Microscopy Study of Discrete Dislocation Pile-ups at Grain Boundaries in Bi-Crystalline Micro-Pillars
Reprinted from: *Crystals* **2020**, *10*, 411, doi:10.3390/cryst10050411 59

James C. M. Li, C. R. Feng and Bhakta B. Rath
Emission of Dislocations from Grain Boundaries and Its Role in Nanomaterials
Reprinted from: *Crystals* **2021**, *11*, 41, doi:10.3390/cryst11010041 79

Petr Šesták, Miroslav Černý and Jaroslav Pokluda
Extraordinary Response of H-Charged and H-Free Coherent Grain Boundaries in Nickel to Multiaxial Loading
Reprinted from: *Crystals* **2020**, *10*, 590, doi:10.3390/cryst10070590 91

Vladyslav Turlo and Timothy J. Rupert
Interdependent Linear Complexion Structure and Dislocation Mechanics in Fe-Ni
Reprinted from: *Crystals* **2020**, *10*, 1128, doi:10.3390/cryst10121128 103

Zhiyuan Yu, Xinmei Wang, Fuqian Yang, Zhufeng Yue and James C. M. Li
Review of γ' Rafting Behavior in Nickel-Based Superalloys: Crystal Plasticity and Phase-Field Simulation
Reprinted from: *Crystals* **2020**, *10*, 1095, doi:10.3390/cryst10121095 115

Jie Ding, Yifan Zhang, Tongjun Niu, Zhongxia Shang, Sichuang Xue, Bo Yang, Jin Li, Haiyan Wang and Xinghang Zhang
Thermal Stability of Nanocrystalline Gradient Inconel 718 Alloy
Reprinted from: *Crystals* **2021**, *11*, 53, doi:10.3390/cryst11010053 139

Hao Liu, Long Yu and Xiazi Xiao
Hardness-Depth Relationship with Temperature Effect for Single Crystals—A Theoretical Analysis
Reprinted from: *Crystals* **2020**, *10*, 112, doi:10.3390/cryst10020112 151

Roya Ermagan, Maxime Sauzay, Matthew H. Mecklenburg and Michael E. Kassner
Determination of Long-Range Internal Stresses in Cyclically Deformed Copper Single Crystals Using Convergent Beam Electron Diffraction
Reprinted from: *Crystals* **2020**, *10*, 1071, doi:10.3390/cryst10121071 161

Qian Qian Zhao, Brad L. Boyce and Ryan B. Sills
Micromechanics of Void Nucleation and Early Growth at Incoherent Precipitates: Lattice-Trapped and Dislocation-Mediated Delamination Modes
Reprinted from: *Crystals* **2021**, *11*, 45, doi:10.3390/cryst11010045 173

Kuntimaddi Sadananda, Ilaksh Adlakha, Kiran N. Solanki and A.K. Vasudevan
Analysis of the Crack Initiation and Growth in Crystalline Materials Using Discrete Dislocations and the Modified Kitagawa–Takahashi Diagram
Reprinted from: *Crystals* **2020**, *10*, 358, doi:10.3390/cryst10050358 193

Si Gao, Takuma Yoshimura, Wenqi Mao, Yu Bai, Wu Gong, Myeong-heom Park, Akinobu Shibata, Hiroki Adachi, Masugu Sato and Nobuhiro Tsuji
Tensile Deformation of Ultrafine-Grained Fe-Mn-Al-Ni-C Alloy Studied by In Situ Synchrotron Radiation X-ray Diffraction
Reprinted from: *Crystals* **2020**, *10*, 1115, doi:10.3390/cryst10121115 211

Andrey Pereverzev and Tommy Sewell
Elastic Coefficients of β-HMX as Functions of Pressure and Temperature from Molecular Dynamics
Reprinted from: *Crystals* **2020**, *10*, 1123, doi:10.3390/cryst10121123 227

Mohammad S. Dodaran, Jian Wang, Nima Shamsaei and Shuai Shao
Investigating the Interaction between Persistent Slip Bands and Surface Hard Coatings via Crystal Plasticity Simulations
Reprinted from: *Crystals* **2020**, *10*, 1012, doi:10.3390/cryst10111012 243

Jialin Liu, Xiaofeng Fan, Yunfeng Shi, David J. Singh and Weitao Zheng
The Effect of Strain Rate on the Deformation Processes of NC Gold with Small Grain Size
Reprinted from: *Crystals* **2020**, *10*, 858, doi:10.3390/cryst10100858 257

Xiazi Xiao, Hao Liu and Long Yu
On the Size Effect of Strain Rate Sensitivity and Activation Volume for Face-Centered Cubic Materials: A Scaling Law
Reprinted from: *Crystals* **2020**, *10*, 898, doi:10.3390/cryst10100898 273

Iyad Alabd Alhafez and Herbert M. Urbassek
Influence of the Rake Angle on Nanocutting of Fe Single Crystals: A Molecular-Dynamics Study
Reprinted from: *Crystals* **2020**, *10*, 516, doi:10.3390/cryst10060516 283

Haokun Deng, Thapanee Sarakonsri, Tao Huang, Aishui Yu and Katerina Aifantis
Transformation of SnS Nanocompisites to Sn and S Nanoparticles during Lithiation
Reprinted from: *Crystals* **2021**, *11*, 145, doi:10.3390/cryst11020145 299

About the Editors

Ronald W. Armstrong Professor Emeritus Ronald W. Armstrong, University of Maryland, College Park, obtained a Bachelor of Engineering Science (B.E.S.) degree from Johns Hopkins University in 1955 and a Doctor of Philosophy (Ph.D.) degree in metallurgical engineering from Carnegie Institute of Technology, now within Carnegie-Mellon University, in 1958. He spent a post-doctoral year at the Houldsworth School of Applied Science, Leeds University, UK, during 1958–1959, followed by a Westinghouse Research Laboratory appointment during 1959–1964 and was then at the Commonwealth Scientific and Industrial Research Organization (CSIRO), Division of Tribophysics, University of Melbourne, Australia, during 1964. Tenured academic positions were at Brown University, 1965–1968, and the University of Maryland, College Park, 1968–1999. His sabbatical leave periods have been at the Physics and Engineering Laboratory (PEL), Department of Science and Industrial Research (DSIR), Lower Hutt, NZ, 1974; Department of Metallurgy and Materials Science, University of Cambridge, UK, 1984; Department of Physics, Cavendish Laboratory, University of Cambridge, UK, 1991. From 2000 to 2003, he was a Senior Scientist in the Munitions Directorate, Eglin Air Force Base, FL. He has held temporary positions with the U.S. Office of Naval Research, London, UK, during 1982–1984 and 1991 as a liaison scientist and at other overseas and U.S. universities and government laboratories. His research experience has mainly dealt with the dislocation mechanics of plasticity and fracturing in single crystal and polycrystalline materials.

Wayne L. Elban received a BChE with distinction in 1969 and a Ph.D. in Applied Sciences: Metallurgy in 1977 from the University of Delaware and a MS in Engineering Materials in 1972 from the University of Maryland, College Park. From 1969 to 1985, he was a research engineer at the Naval Surface Warfare Center, White Oak Laboratory, Silver Spring, Maryland. Since 1985, he has been in the engineering department at Loyola College (now Loyola University Maryland). In 2020, he retired as Professor Emeritus. He has worked on a variety of projects, including development and characterization of energetic materials, metallographic examination and hardness testing of historic wrought iron and synthesis and characterization of organic clay hybrids.

Preface to "Crystal Plasticity at Micro- and Nano-scale Dimensions"

The advent of "nanotechnology" in the 1980s is often credited to the foresight expressed in one of Nobel Prize winner Richard Feynman's lectures given in late 1959, "There's Plenty of Room at the Bottom". The lecture featured the potential advantages in chemistry to be gained by the manipulation of atoms on an atomic scale, and that nano-scale mechanical devices would be designed and developed in future engineering applications. The lecture pre-staged the Nobel Prize-winning developments of atomic force microscopy and scanning tunneling microscopy. An important industrial impetus was provided by the development of microelectromechanical system (MEMS) devices and also later by the development of both nano-scale testing systems, such as nano-indentation hardness testing, and micro-scale specimen fabrication methods, such as the focused ion beam (FIB) manufacturing of micro-pillars and the production of nanoparticles and nano-wires.

Less attention has been given to the order-of-magnitude mechanical strength level advantage enabled at micro- to nanoscale crystal dimensions and within nano-polycrystalline microstructures, both of which are topics included in the present Crystals journal's Special Issue, now issued as an e-book. Such an achievement in the strength level has coincided with other recent advancements in the crystal dislocation mechanics theories that have recently been developed to explain these exceptional crystal strength properties. The advanced testing methods and accompanying model simulations of crystal properties provide an opportunity for quantitatively accounting for the character of individual dislocations as well as their interactions within small groups. The aim of the present Special Issue was to provide a welcome venue for contributed works on this rapidly developing subject.

We Guest Editors express our appreciation to the authors of the articles contained in the present e-book and to the approximately two times greater number of peer reviewers who have helped to present these final collected works in the best possible light. Such recognition is especially deserved because of all persons coping with the worldwide pandemic generated by the coronavirus. Dr. Stephen Walley at the Cavendish Laboratory, University of Cambridge, is thanked for providing helpful contacts with a number of prospective authors. A special note of gratitude is expressed to Ms. Dancy Yu for the very considerable assistance provided in every aspect of achieving both the initial Special Issue and the present e-book summary of results.

Ronald W. Armstrong, Wayne L. Elban
Editors

Editorial

Crystal Strengths at Micro- and Nano-Scale Dimensions

Ronald W. Armstrong [1,*] and Wayne L. Elban [2]

1. Department of Mechanical Engineering, University of Maryland, College Park, MD 20742, USA
2. Department of Engineering, Loyola University Maryland, Baltimore, MD 21210, USA; welban@loyola.edu
* Correspondence: rona@umd.edu

Received: 3 February 2020; Accepted: 3 February 2020; Published: 5 February 2020

Abstract: Higher strength levels, achieved for dimensionally-smaller micro- and nano-scale materials or material components, such as MEMS devices, are an important enabler of a broad range of present-day engineering devices and structures. Beyond such applications, there is an important effort to understand the dislocation mechanics basis for obtaining such improved strength properties. Four particular examples related to these issues are described in the present report: (1) a compilation of nano-indentation hardness measurements made on silicon crystals spanning nano- to micro-scale testing; (2) stress–strain measurements made on iron and steel materials at micro- to nano-crystal (grain size) dimensions; (3) assessment of small dislocation pile-ups relating to Griffith-type fracture stress vs. crack-size calculations for cleavage fracturing of α-iron; and (4) description of thermally-dependent strain rate sensitivities for grain size strengthening and weakening for macro- to micro- to nano-polycrystalline copper and nickel materials.

Keywords: crystal strength; micro-crystals; nano-crystals; nano-polycrystals; nano-wires; whiskers; pillars; dislocations; hardness; crystal size dependencies; fracture; strain rate sensitivity

1. Introduction

To underscore the relevance of the current topic, a number of example engineering-based references are provided for micro- and nano-crystal mechanical property measurements and the applications of them [1–5]. In brief, the topics relate to crystal size effects in manufacturing; strength measurements relating to MEMS applications; micro-forming of foils; nano-metric machining defects; and advanced machine tool manufacturing. Such observations involving extremely small dimensionally-dependent mechanical property measurements have followed a pioneering report by Brenner [6] of greatly-enhanced strength levels being achieved for smaller diameter "whisker" materials. This was later established for nano-grained steel wire [7], micro-pillar α-iron [8], and copper [9] materials. The strength levels have been attributed either to higher applied stresses being needed to nucleate individual dislocations within a dislocation-free environment or because of the need for internal stress concentrations to be produced by small dislocation pile-ups [10–12], for example, as established in the description of higher micro-hardness levels reached for nano-grained nickel material [13].

Here, we consider four example cases in more detail, spanning micro- and nano-scale strength levels and their connection with conventional crystal strength properties, as follows: (1) hardness-based nano-indentation load, P, values vs. corresponding respective contact diameters, d_i, for silicon crystals; (2) theoretical-limiting strength levels achieved for the mentioned ultrafine grain size iron and steel materials; (3) dislocation pile-up characterizations of strength levels at nano-scale dimensions relating to the Griffith-based theory of brittle cleavage fracturing; and (3) order-of-magnitude increases in the thermally-dependent strain rate sensitivity properties of nano-crystalline copper and nickel materials associated both with grain size strengthening and grain size-dependent weakening behaviors, in the latter case, connected most often with high temperature creep behavior.

2. Elastic, Plastic, and Cracking Aspects of Crystal Nano-Indentations

Figure 1 provides a background log/log compilation of indentation load, P, vs. surface-projected contact diameter, d, spanning a range from pioneering nano-indentation measurements to micro-scale indentation fracture mechanics test results obtained on silicon crystals, as identified in the top-left corner of the figure [14]. The figure was constructed to illustrate a d^3 dependence of load for elastic contact in pioneering nano-indentation measurements made by Pethica, Hutchings, and Oliver [15], as compared with an expected d^2 dependence for a constant material hardness; D is the actual or effective spherical tip diameter for each indentation. The $d_c^{3/2}$ dependence applies for crack tip-to-crack tip measurements across the enclosed indentations.

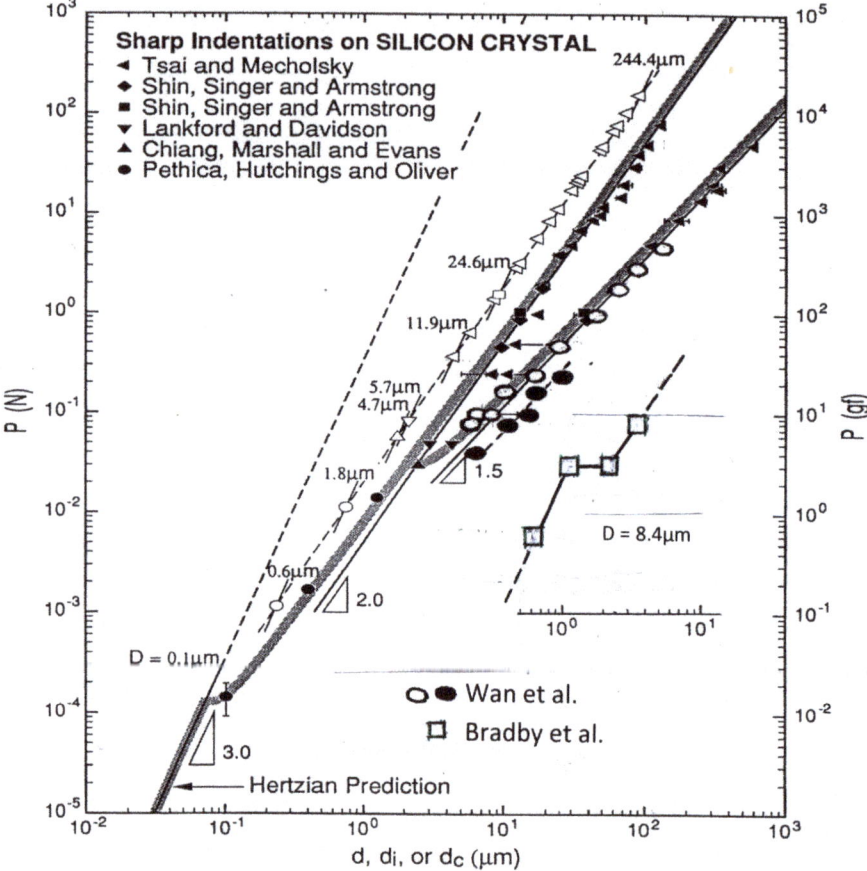

Figure 1. Load, P, vs. either elastic contact diameter, d_e, plastic diamond pyramid diagonal, d_d, or tip-to-tip crack length, d_c, for silicon crystal indentations [14], and including additional cracking [16] and inset elastic-plastic spherical Berkovich-type nano-indentation measurements [17].

In Figure 1, more recently added open- and closed-ellipsoidal points [16] have covered many of the earlier data points on the $d^{3/2}$ fracture mechanics dependence and, likewise, the inset open-square points [17] for both elastic, $d_e{}^3$, and plastic, $d_p{}^2$, Berkovich spherically-tipped nano-indentations would have significantly overlapped the previously established hardness dependence. Thus, these points that illustrate a significant "pop-in" behavior at initial plastic yield have been shifted on the abscissa scale for clarity. Close examination of the deviation from the labeled d^2 dependence in Figure 1 shows that an increasing hardness applies for a smaller indentation size. Calculations of the stress–strain behaviors both at the onset of plastic yielding and follow-on nano-indentation strain hardening behaviors have been reported very recently for NaCl, MgO, and copper crystals [18], tungsten crystals [19], ammonium perchlorate, and α-iron crystals [20]. Beyond the well-established determination of very high flow stress levels for initial plastic yielding, whether gradual or of pop-in type, the plastic strain hardening behavior has also been shown to be exceptionally high. The dimensionally smaller plastic deformation zones account for both higher values of initial yielding and subsequent flow on the basis of the smaller dislocation line lengths and their interactions.

3. Crystal (Grain) Size-Dependent Strengths

Brenner [6] found an inverse dependence of strength on the specimen wire diameter for α-iron and copper whiskers. The same dependence has been found for integrated circuit-connected measurements on gold micro-wires [21] as well as for thin copper wires [22] and other nickel nano-pillar [23] materials, also with consideration taken into account of the nano-polycrystal grain structures. The status of nano-polycrystal micro-pillar strength properties have been reviewed by Shahbeyk et al. [24]. Kiener et al. [25] have provided an analysis of the strength properties on the basis of the small volumes of the materials that were tested.

Figure 2 illustrates on an expanded abscissa scale a further connection of the size dependence described previously for a compilation of strength properties obtained on conventional and ultrafine crystal (grain) size of α-iron and steel materials [26]. In the figure, the continuous slightly curved line is an extrapolation of the Hall–Petch (H–P) inverse square root of grain size measured at ambient temperature for a number of steel materials. The filled-square points connected by a dashed line just below the extrapolated H–P dependence are pioneering measurements reported for patented eutectoid steel wire materials. The filled triangle points were also obtained more recently for eutectoid steel wire materials but at smaller effective grain sizes as is indicated as well for the filled circle points that were obtained for ball-milled α-iron material. The topmost filled diamond point is a latest measurement reported by Li et al. [7] for severely drawn eutectoid material.

Of particular importance in Figure 2 is a dot-dashed dislocation pile-up model description of the Hall–Petch dependence fitted to the reported H–P parameters at larger grain sizes and which dependence is shown to transition at ultrafine grain size from an $\ell^{-1/2}$ to an ℓ^{-1} dependence. Such size-dependent transition has been predicted for the variation in H–P measurements assessed earlier for copper materials [27]. The transition occurs when the dislocation pile-up length is sufficiently reduced such that only one dislocation loop, $n = 1.0$, is able to be produced within a restricted slip plane length to overcome the nano-polycrystal grain boundary resistance. On such a restricted dislocation number basis, the dislocation model description for an H–P dependence connects with the described strength dependence of whiskers, nano-wires, and micro-pillars on their specimen diameters.

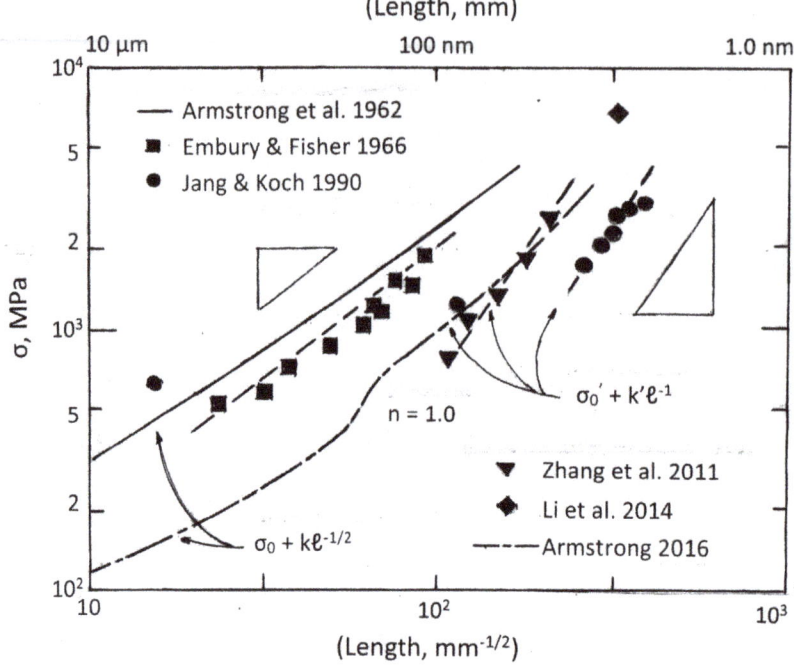

Figure 2. Strength of α-iron and steel materials on the basis of an inverse square root of grain size dependence that transitions to an inverse size dependence depending on the level of the dislocation pile-up stress intensity; an expanded version of compiled results is reported in reference [26].

4. Crystal Dislocation Pile-Ups at Small Dimensions

The reduction in numbers of dislocations in pile-ups within ultrafine crystal or grain size materials raises an issue of analogous pile-up lengths and crack sizes associated with brittle fracturing of conventional microstructures. Petch determined an inverse square root of grain size dependence for the cleavage fracture strength of a combination of α-iron and related steel materials with grain sizes larger than ten microns and also provided the now widely accepted explanation for the dependence in terms of the dislocation pile-up model [28]. Just afterwards, Stroh provided a follow-up calculation of the condition whereby the concentrated stress at the tip of the pile-up was the same as for a Griffith-type cleavage crack [29]. The similar relationship of crystal grain size and crack size dependencies for describing fracture strengths has been reviewed [30].

Figure 3 provides a comparison of the stress concentrations at a distance, r, ahead of two pile-ups and cracks for 5 and 49 dislocations in the pile-ups [12]. In the model figure description, τ_{23} and τ_∞ are the local shear stress and applied shear stress; A_i is dislocation position within the pile-up divided by the characteristic length, $a' = Gb/2\pi\tau_\infty$, in which G is the shear modulus and b is dislocation the Burgers vector. The $n = 5$ and $n = 49$ open circles point to the level of the stress ratios at the pile-up tips. The Griffith stress concentration follows an $r^{-1/2}$ dependence as expected and is approximately followed at small (r/a') by the forward pile-up stress concentration. At large (r/a'), the pile-up stress concentration is seen to follow an r^{-1} dependence as is sometimes employed to represent the pile-up as a multiple Burgers vector dislocation of strength, nb.

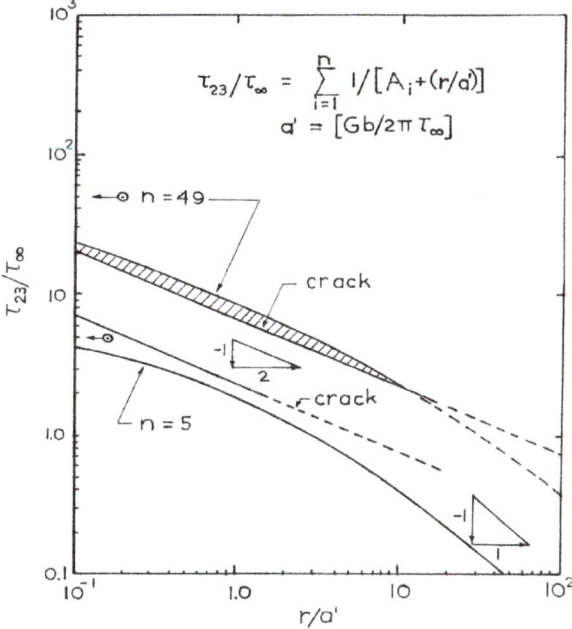

Figure 3. Comparison of the concentrated stresses ahead of dislocation pile-ups consisting of 5 or 49 dislocations and Griffith cracks of equal lengths [12].

The notable result shown in Figure 3 is that the smaller dislocation pile-up does not achieve a sufficient stress concentration to be modeled as a Griffith crack. On such theoretical basis, therefore, one might expect that cleavage would be more difficult to initiate within an ultrafine grain size material. In fact, experimental evidence for the greater difficulty of initiating cleavage in ultrafine grain size α-iron material has been reported by Hohenwarter and Pippan [31]. The result adds to the well-known experimental observation of lowering the ductile-to-brittle transition behavior of steel and related bcc metals through refinement of the polycrystal grain size, particularly via measurements obtained in Charpy impact testing [30].

5. Crystal Size-Dependent Strain Rate Sensitivity

A dislocation mechanics description of loading rate dependence based on thermally-activated dislocation motion has provided a valuable additional avenue for the investigation of crystal size effects at micro- to nano-scale sizes. An early model description of the behavior for conventional grain size hcp cadmium involved separation of the strain rate sensitivity property into grain volume and grain boundary components in the relation [32]:

$$(1/v^*) = (1/v_0^*) + (k_\varepsilon/2m_T \tau_C v_C^*)\ell^{-1/2} \tag{1}$$

The parameters in Equation (1) are an activation volume, $v^* = A^*b = k_B T[\{\Delta \ln(d\gamma/dt)\}/\Delta \tau\}]_T$, in which A^* is an activation area; k_B is Boltzmann's constant; T is temperature; $\Delta \ln(d\gamma/dt)$ is the imposed change in the plastic shear strain rate; $\Delta \tau$ is the accompanying change in shear stress; k_ε is the H–P stress intensity, m_T is the Taylor orientation factor; τ_C is the local shear stress at the grain boundary; and ℓ is the grain size. The first term, $(1/v_0^*)$, measures the rate-dependent contribution of the dislocation motion within the grain volume and the second term, involving the factor $(1/v_C^*)$, is a measure of the thermally-activated resistance to transmission of plastic flow across the grain boundary [33]. The product $(\tau_C v_C^*)$ is constant, and thus Equation (1) follows an H–P type dependence.

The current Figure 4 is an added-to version of a previous description of copper and nickel measurements in which a comparison was made with the Hall–Petch type grain size prediction given in Equation (1) [34]. The upper curve in the figure is for T = 195 K and the lower curve for 300 K. Transition from grain size strengthening to weakening is shown to occur in the smaller grain size regime at the "jump" in the filled-triangle values of ($1/v^*$). The unmistakable jump by an approximate order-of-magnitude leads to v^* values of near atomic dimensions that are generally determined for higher temperature (creep) deformations. However, additional open circle points shown in the figure, as obtained from measurements reported by Chen et al. [35], are seen to follow the H–P grain size strengthening prediction. The positions of these points near to the H–P dependence are significant in terms of the grain boundary structures being stable and continuing to resist slip transmission from within the grains. These same authors provided confirmation of H–P grain size strengthening via separately reported hardness measurements. In related research on nano-grained copper and nickel materials, Zhou et al. have given emphasis to the importance of establishing thermal stability of the grain boundary structures in such materials [36]. Otherwise, Figure 4 shows v^* to be an effective monitor for detecting the onset of grain size weakening, for example, in line with the determination of material creep behavior as described in a previous report by Armstrong et al. [37].

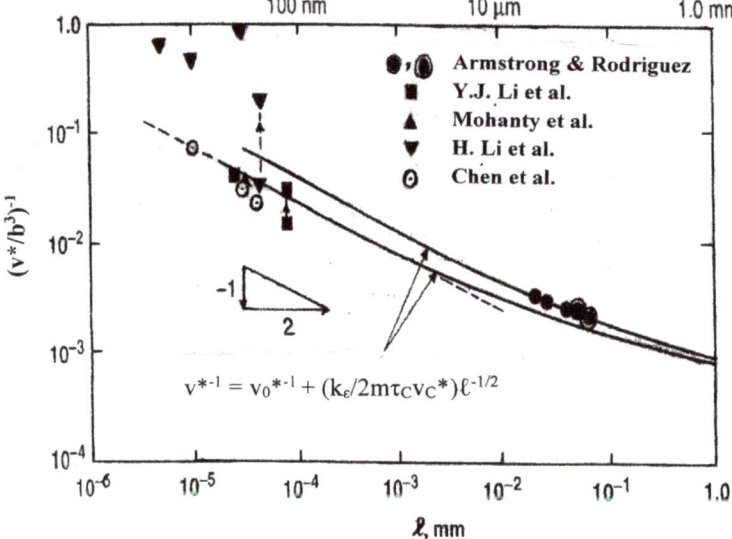

Figure 4. Strain rate sensitivity-based reciprocal activation volume measurements for copper and nickel materials spanning conventional and micro- to nano-scale grain sizes [34,35].

6. Discussion

The present description of experimental strength measurements involved examples, in turn, of the topics: localized deformations in micro- and nano-indentation tests; tensile or compression tests of whisker, micro-wire, and micro-pillar crystal or polycrystalline specimens; and stress–strain, fracturing, and deformation rate dependencies of micro-to nano-polycrystalline materials. In a number of cases, the micro- or nano-scale properties have been connected with (lower) strength levels normally achieved in conventional bulk materials.

An example of connection between nano-indentation testing incorporated in Figure 1 and technical alloy development has been given in a report on aluminum single crystals by Filippov and Koch [38]. The test method was employed to probe the material anisotropic elastic deformation behavior through testing the diamond pyramid hardness of (100), (110), and (111) crystal surfaces. The research relates to reference [20] in which a comparison was made between the experiment and simulation of nano-indentations in (111) and (100) α-iron crystal surfaces. At an opposite dimensional scale, an important civil engineering connection spanning nano- to macro-size dimensions was provided by a comparison of the ultrafine steel wire measurements described for the drawn micro-wire strengths presented in Figure 2 and the comprehensive description by Ono [39] of steel wire materials employed in historical and current transportation bridge constructions.

A novel example of exceptionally high strength measurements being made on silicon nano-particles, relating to the results shown in Figure 1, has been reported by Nowak et al. [40]. Near theoretical limiting strength levels were determined by compression of the particles between hardened steel platens. In another study, so-called length-scale 'architectured' polycrystalline copper micro-pillars have been fabricated with an achievement of surprisingly higher strength advantage [41]. Another useful study of micro-pillars has involved the measurement of solute effects in aluminum alloys [42]. Basic features of slip behavior have been reported for Fe-3% Si micro-pillar crystals [43]. Dendrite/nano-structured (high entropy alloy) titanium-based composite material has been investigated in static and dynamic compression tests [44]. The use of probing spherical nano-indentation hardness testing has been applied as far afield as in the evaluation of the elastic modulus and strain energy within the mineral, antigorite, to understand planetary tectonic plate subduction behavior [45], thus spanning the seemingly largest imaginable range in dimensions.

Finally, in discussion, the several examples presented of dislocation mechanics modeling of hardness, stress–strain, fracturing, and strain rate sensitivities of materials at smaller dimensions have focused on particular features accompanying the behavior of single or small groups of dislocations. The higher strength levels and exceptional strain hardening behaviors can be understood in a general manner in terms of the higher values of internal dislocation stresses accompanying smaller dislocation line lengths and smaller separations accompanying subsequent dislocation interactions. Stress levels approaching the theoretical strength are obtained [46], for example, as indicated in Figure 2, for the smallest effective grain size of heavily drawn eutectoid steel wire material [7].

7. Summary

This opening editorial report is presented for the current *Crystals* journal Special Issue on the topic of "Crystal Plasticity at Micro- and Nano-Scale Dimensions". A number of example references are provided to demonstrate current research interests in the topic. Four particular descriptions of micro- to nano-strength property evaluations are given and are compared with conventional material strength properties. The dislocation mechanics-based model descriptions of the strength properties are based on the nucleation of dislocations and their interactions at corresponding micro- to nano-scale dimensions.

Author Contributions: This editorial has been jointly written by both co-authors. All authors have read and agreed to the published version of the manuscript.

Acknowledgments: The authors express appreciation to Dancy Yu for editorial assistance.

Conflicts of Interest: The authors declare no conflict of interest.

References

1. Vollertsen, F.; Biermann, D.; Hansen, H.N.; Jawahir, I.S.; Kuzman, K. Size effects in manufacturing of metallic components. *CIRP Ann.–Manufact. Tech.* **2009**, *58*, 566–587. [CrossRef]
2. Asano, K.; Tang, H.; Chen, C.-Y.; Nagoshi, T.; Chang, T.-F.; Yamane, D.; Konishi, T.; Machida, K.; Masu, K.; Sone, M. Promoted bending strength in micro-cantilevers composed of nano-grained gold toward MEMS applications. *Microelect. Eng.* **2018**, *196*, 20–24. [CrossRef]
3. Cheng, C.; Wan, M.; Meng, B.; Zhao, R.; Han, W.P. Size effect on the yield behavior of metal foil under multiaxial stress states: Experimental investigation and modelling. *Int. J. Mech. Sci.* **2019**, *151*, 760–771. [CrossRef]

4. Zhang, L.; Zhao, H.; Dai, Y.; Yang, Y.; Du, X.; Tang, P.; Zhang, L. Molecular dynamics simulation of deformation accumulation in repeated nanometric cutting on single-crystal copper. *RSC Adv.* **2015**, *5*, 12678–12685. [CrossRef]
5. Vorotilo, S.; Loginov, P.; Mishnaevsky, L.; Siderenko, D.; Levashov, E. Nano-engineering of metallic alloys for machining tools: Multiscale computational and in situ TEM investigation of mechanisms. *Mater. Sci. Eng. A* **2019**, *739*, 480–490. [CrossRef]
6. Brenner, S.S. Properties of whiskers. In *Growth and Perfection of Crystals*; Doremus, R.H., Roberts, B.W., Turnbull, D., Eds.; John Wiley & Sons, Inc.: New York, NY, USA, 1958; pp. 157–190.
7. Li, Y.; Raabe, D.; Herbig, M.; Choi, P.-P.; Goto, S.; Kostka, A.; Yarita, H.; Borchers, C.; Kirchheim, R. Segregation stabilizes nano-crystalline bulk steel with near theoretical strength. *Phys. Rev. Lett.* **2014**, *113*, 106104. [CrossRef]
8. Rogne, B.R.S.; Thaulow, C. Strengthening mechanisms of iron micro-pillars. *Philos. Mag. A* **2015**, *95*, 1814–1828. [CrossRef]
9. Wheeler, J.M.; Kirchlechner, C.; Micha, J.-S.; Michler, J.; Kiener, D. The effect of size on the strength of FCC metals at elevated temperatures: Annealed copper. *Phil. Mag.* **2016**, *96*, 3379–3395. [CrossRef]
10. Armstrong, R.W.; Chou, Y.T.; Fisher, R.M.; Louat, N. The limiting grain size dependence of the strength of a polycrystalline aggregate. *Phil. Mag.* **1966**, *14*, 943–951. [CrossRef]
11. Li, J.C.M.; Liu, G.C.T. Circular dislocation pile-ups: I. Strength of ultrafine polycrystalline aggregates. *Phil. Mag.* **1967**, *15*, 1059–1063. [CrossRef]
12. Armstrong, R.W. Hall-Petch analysis for nano-polycrystals. In *Nano-Metals—Status and Perspective; 33rd Risö International Symposium on Materials Science*; Fæster, S., Hansen, N., Huang, X., Juul Jensen, D., Ralph, B., Eds.; Technical University of Denmark: Roskilde, Denmark, 2012; pp. 181–199.
13. Hughes, G.D.; Smith, S.D.; Pande, C.S.; Johnson, H.R.; Armstrong, R.W. Hall-Petch strengthening for the micro-hardness of twelve nanometer grain diameter electrodeposited nickel. *Scr. Metall.* **1986**, *20*, 93–97. [CrossRef]
14. Armstrong, R.W.; Mecholsky, J.J.; Shin, H.; Tsai, Y.L. Elasticity, plasticity and cracking at indentations in single crystal silicon. *J. Mater. Sci. Letts.* **1993**, *12*, 1274–1275. [CrossRef]
15. Pethica, J.B.; Hutchings, R.; Oliver, W.R. Hardness measurement at penetration depths as small as 20 nm. *Phil. Mag.* **1983**, *48*, 593–606. [CrossRef]
16. Wan, H.; Shen, Y.; Chen, Q.; Chen, Y. A plastic damage model for finite element analysis of cracking of silicon under indentation. *J. Mater. Res.* **2010**, *25*, 2226–2236. [CrossRef]
17. Bradby, J.E.; Williams, J.S.; Wong-Leung, J.; Swain, M.V.; Munroe, P. Mechanical deformation in silicon by micro-indentation. *J. Mater. Res.* **2001**, *16*, 1500–1507. [CrossRef]
18. Armstrong, R.W.; Elban, W.L. Exceptional crystal strain hardening determined over macro- to micro- to nano-size scales in continuous spherical indentation tests. *Mater. Sci. Eng. A* **2019**, *757*, 95–100. [CrossRef]
19. Armstrong, R.W.; Elban, W.L. Tungsten (111) crystal strain hardening in nano-indentations. *Int. J. Refract. Met. Hard Mater.* **2020**, *87*, 105140. [CrossRef]
20. Armstrong, R.W.; Elban, W.L. Dislocation reaction mechanism for enhanced strain hardening in crystal nano-indentations. *Crystals* **2020**, *10*, 9. [CrossRef]
21. Yang, H.K.; Cao, K.; Han, Y.; Wen, M.; Guo, J.M.; Tan, Z.L.; Lu, J.; Lu, L. The combined effects of grain and sample sizes on the mechanical properties and fracture modes of gold micro-wires. *J. Mater. Sci. Tech.* **2019**, *35*, 76–83. [CrossRef]
22. Guo, S.; He, Y.; Li, Z.; Lei, J.; Liu, D. Size and stress dependences in the tensile stress relaxation of thin copper wires at room temperature. *Int. J. Plast.* **2019**, *112*, 278–296. [CrossRef]
23. Yuan, L.; Xu, C.; Shivpuri, R.; Shan, D.; Guo, B. Size effect in the uniaxial compression of polycrystalline Ni nanopillars with small number of grains. *Metall. Mater. Trans. A* **2019**, *50A*, 4462–4479. [CrossRef]
24. Shahbeyk, S.; Voyiadjis, G.Z.; Habibi, V.; Astaneh, S.H.; Yaghoobi, M. Review of size effects during micro-pillar compression test: Experiments and atomistic simulations. *Crystals* **2019**, *9*, 591. [CrossRef]
25. Kiener, D.; Fritz, R.; Alfreider, M.; Leitner, A.; Pippan, R.; Maier-Kiener, V. Rate limiting deformation mechanisms of bcc metals in confined volumes. *Acta Mater.* **2019**, *166*, 687–701. [CrossRef]
26. Armstrong, R.W. Crystal engineering for mechanical strength at nano-scale dimensions. *Crystals* **2017**, *7*, 315. [CrossRef]

27. Armstrong, R.W.; Smith, T.R. Dislocation pile-up predictions for the strength properties of ultrafine grain size fcc metals. In *Processing and Properties of Nanocrystalline Materials*; Suryanarayana, C., Singh, J., Froes, F.H., Eds.; TMS-AIME: Warrendale, PA, USA, 1996; pp. 345–354.
28. Petch, N.J. The cleavage strength of polycrystals. *J. Iron Steel Inst.* **1953**, *174*, 25–28.
29. Stroh, A.N. The formation of cracks as a result of plastic flow. *Proc. R. Soc. (Lond.) A* **1954**, *223*, 404–414.
30. Armstrong, R.W. Material grain size and crack size influences on cleavage fracturing. *Phil. Trans. R. Soc. (Lond.) A* **2015**, *373*, 20140124. [CrossRef]
31. Hohenwarter, A.; Pippan, R. Anisotropic fracture behavior of ultrafine-grain iron. *Mater. Sci. Eng. A* **2010**, *527*, 2649–2656. [CrossRef]
32. Prasad, Y.V.R.K.; Armstrong, R.W. Polycrystal versus single-crystal strain rate sensitivity of cadmium. *Phil. Mag.* **1974**, *29*, 1421–1425. [CrossRef]
33. Armstrong, R.W.; Rodriguez, P. Flow stress/strain rate/grain size coupling for fcc nano-polycrystals. *Phil. Mag.* **2006**, *86*, 5787–5796. [CrossRef]
34. Armstrong, R.W. Size effects on material yield strength/deformation/fracturing properties. *J. Mater. Sci.* **2019**, 1–16. [CrossRef]
35. Chen, J.; Lu, L.; Lu, K. Hardness and strain rate sensitivity of nano-crystalline Cu. *Scr. Mater.* **2006**, *54*, 1913–1918. [CrossRef]
36. Zhou, X.; Li, X.Y.; Lu, K. Enhanced thermal stability of nano-grained metals below a critical grain size. *Science* **2018**, *360*, 526–530. [CrossRef] [PubMed]
37. Armstrong, R.W.; Conrad, H.; Nabarro, F.R.N. Meso-to-nano-scopic polycrystal/composite strengthening. In *Mechanical Properties of Nano-Structured Materials and Nano-Composites*; Ovid'ko, I., Pande, C.S., Krishnamoorti, R., Lavernia, E., Skandan, G., Eds.; Mater. Res. Soc.: Warrendale, PA, USA, 2004; pp. 69–77.
38. Filippov, P.; Koch, U. Nano-indentation of aluminum single crystals: Experimental study on influencing factors. *Materials* **2019**, *12*, 3688. [CrossRef] [PubMed]
39. Ono, K. Size effects of high strength steel wires. *Metals* **2019**, *9*, 240. [CrossRef]
40. Nowak, J.D.; Beaber, A.R.; Ugurlu, O.; Girshick, S.L.; Gerberich, W.W. Small size strength dependence on dislocation nucleation. *Scr. Mater.* **2010**, *62*, 819–822. [CrossRef]
41. Hou, X.D.; Krauss, S.; Merle, B. Additional strengthening in length-scaled architectured copper with ultrafine and coarse domains. *Scr. Mater.* **2019**, *165*, 55–59. [CrossRef]
42. Yan, S.; Zhou, H.; Qin, Q.H. Microstructure versus size: Nano/micro-scale deformation of solute-strengthening Al alloys via pillar compression tests. *Mater. Res. Lett.* **2019**, *7*, 53–59. [CrossRef]
43. Kheradmand, N.; Rogne, B.R.; Dumoulin, S.; Deng, Y.; Johnsen, R.; Barnoush, A. Small scale testing approach to reveal specific features of slip behavior in bcc metals. *Acta Mater.* **2019**, *174*, 142–152. [CrossRef]
44. Dong, J.L.; Wang, Z.; Han, L.N.; Tang, Y.L.; Qiao, J.W.; Shi, X.H.; Wang, Z.H. Novel *in-situ* Ti-based dendrite/nano-structured matrix composites with excellent mechanical performances upon dynamic compression. *J. Alloy Compounds* **2019**, *781*, 716–722. [CrossRef]
45. Hansen, L.N.; David, E.C.; Brantut, N.; Wallis, D. Insight into the micro-physics of antigorite deformation from spherical nano-indentation. *Phil. Trans. R. Soc. (Lond.) A* **2020**, *378*, 20190197.
46. Han, W.-Z.; Huang, L.; Ogata, S.; Kimizuka, H.; Yang, Z.-C.; Weinberger, C.; Li, Q.-J.; Liu, B.-Y.; Zhang, X.X.; Li, J.; et al. From "Smaller is stronger" to "Size-independent strength plateau": Towards measuring the ideal strength of iron. *Adv. Mater.* **2015**, *27*, 3385–3390. [CrossRef] [PubMed]

© 2020 by the authors. Licensee MDPI, Basel, Switzerland. This article is an open access article distributed under the terms and conditions of the Creative Commons Attribution (CC BY) license (http://creativecommons.org/licenses/by/4.0/).

Article

Prospects of Using Small Scale Testing to Examine Different Deformation Mechanisms in Nanoscale Single Crystals—A Case Study in Mg

Daniel Kiener [1,*], Jiwon Jeong [2,3], Markus Alfreider [1], Ruth Konetschnik [1] and Sang Ho Oh [2,4,*]

1. Department Materials Science, Chair of Materials Physics, Montanuniversität Leoben, Roseggerstrasse 12, 8700 Leoben, Austria; markus.alfreider@unileoben.ac.at (M.A.); r.konetschnik@gmail.com (R.K.)
2. Department of Materials Science and Engineering, Pohang University of Science and Technology (POSTECH), Pohang 37673, Korea; j.jeong@mpie.de
3. Department of Structure and Nano-/Micromechanics of Materials, Max-Planck Institut für Eisenforschung GmbH, 40237 Düsseldorf, Germany
4. Department of Energy Science, Sungkyunkwan University (SKKU), Suwon 16419, Korea
* Correspondence: daniel.kiener@unileoben.ac.at (D.K.); sanghooh@skku.edu (S.H.O.)

Citation: Kiener, D.; Jeong, J.; Alfreider, M.; Konetschnik, R.; Oh, S.H. Prospects of Using Small Scale Testing to Examine Different Deformation Mechanisms in Nanoscale Single Crystals—A Case Study in Mg. *Crystals* **2021**, *11*, 61. https://doi.org/10.3390/cryst11010061

Received: 12 November 2020
Accepted: 11 January 2021
Published: 14 January 2021

Publisher's Note: MDPI stays neutral with regard to jurisdictional claims in published maps and institutional affiliations.

Copyright: © 2021 by the authors. Licensee MDPI, Basel, Switzerland. This article is an open access article distributed under the terms and conditions of the Creative Commons Attribution (CC BY) license (https://creativecommons.org/licenses/by/4.0/).

Abstract: The advent of miniaturised testing techniques led to excessive studies on size effects in materials. Concomitantly, these techniques also offer the capability to thoroughly examine deformation mechanisms operative in small volumes, in particular when performed in-situ in electron microscopes. This opens the feasibility of a comprehensive assessment of plasticity by spatially arranging samples specifically with respect to the crystal unit cell of interest. In the present manuscript, we will showcase this less commonly utilised aspect of small-scale testing on the case of the hexagonal metal Mg, where, besides dislocation slip on different slip planes, twinning also exists as a possible deformation mechanism. While it is close to impossible to examine individual deformation mechanisms in macroscale tests, where local multiaxial stress states in polycrystalline structures will always favour multiple mechanisms of plasticity, we demonstrate that miniaturised uniaxial experiments conducted in-situ in the scanning electron microscope are ideally suited for a detailed assessment of specific processes.

Keywords: dislocation plasticity; twinning; miniaturised testing; in situ electron microscopy; magnesium

1. Introduction

Small scale experiments have provided numerous insights regarding the mechanical properties of confined volumes. This is largely owed to the fact that by testing small volumes, individual constituents of the microstructure can be examined. Since the first report on the intriguing mechanical properties of micron sized metallic objects [1], understanding related size effects was a prime driver of research in the micromechanics community [2,3]. Sparked by this initial discovery, the activities and interests continuously grew and spread into many different routes of research, which cannot be covered within the scope of this article. One of the first subsequent questions that received considerable interest regarded the effect of specific or multiple interfaces within a small probed volume [4–7], while another facet of experimental efforts concerned the deformation mechanisms in more complex crystal structures such as hexagonal-close packed metals (Mg [8,9] or Ti [10,11]).

The deformation of Mg is accommodated by deformation twinning (tensile twin, contraction twin) or dislocation slip (basal <a>, prismatic <a>, and pyramidal <c+a> slip). Each deformation mode has been investigated in order to understand the fundamentals of the anisotropic deformation behaviour in Mg. In the case of <a> basal slip, the deformation mechanism appears quite similar to that of face-centred cubic metals, because <a> dislocations are easily nucleated and glide on basal planes [12]. In the case of <c+a> pyramidal slip, however, the deformation behaviour is quite different compared to <a> basal slip, because <c+a> dislocations cannot easily glide on their slip plane due to the large Burgers vector and the high Peierls stress of pyramidal planes [8,9,13–15]. In the case of the $\{10\bar{1}2\}$ tensile twin, the twinning mechanism is still not well understood and under debate because of the extremely small twinning shear compared to the other twinning modes [16,17].

Thinking beyond the significant open questions that still exist for pure magnesium, one might be interested, for example, in aspects concerning the critical resolved shear stress of certain deformation mechanisms [12,18,19] or the influence of alloying on solid solution strengthening and precipitation hardening affecting the material twinnability [20,21].

With the increasing complexity of the research in question, it becomes more and more demanding to assess the single crystal properties for a fundamental understanding rather than just examining the bulk response of a new experimental alloy. This is where the strength of small scale testing should be embraced, as it allows comprehensive testing utilising only minimal amounts of material. Thus, the intention of this manuscript is to demonstrate the feasibility to test all fundamental deformation mechanisms of magnesium utilising only a small material volume. Due to the large difference in the activation energy of each deformation mode in Mg, strongly anisotropic deformation behaviours can be investigated depending on loading axes and conditions. While this allows not only the examination of the influence of crystal orientation, but also the impact of crystal size, strain rate and deformation temperature on the resulting mechanical response, we will limit this consideration to the fundamental modes of deformation, the related extensions towards the mentioned other variables are trivial but beyond the scope of this work.

As such, in the remainder we will detail a possible testing scheme using in-situ testing in the scanning electron microscope (SEM), apply it to the defined fundamental hexagonal crystal orientations, and detail the resultant deformation features and their relation to previous works.

2. Experimental

For this study, a Magnesium single crystal of high purity (better than 99.999%) was purchased from Goodfellow Cambridge Ltd. (Huntingdon, United Kingdom) to ascertain a homogenous chemistry and avoid any influences from chemical segregation or inhomogeneities on the results. To expose the crystalline faces and directions of interest for testing in this study, thin slices were cut from the pre-oriented single crystal using a slow speed diamond wire saw and shaped to resemble the geometry of half-moon samples for transmission electron microscopy (TEM). These samples were subsequently etched in a water-free solution consisting of 5 mL HNO_3 and 95 mL ethanol to avoid unintended oxidation of Mg and remove material potentially deformed due to the previous sawing process. Subsequently, miniaturised compression and tensile samples were fabricated along the chemically thinned wedge as shown exemplarily in Figure 1 and insets therein using a focussed ion beam (FIB, Leo 1540 XB, Carl Zeiss, Oberkochen, Germany). The acceleration voltage of the Ga ion beam was 30 keV, and the ion beam currents were gradually decreased from starting 5 nA for rough milling down to 10 pA for final polishing of the specimen surfaces. All samples were placed on somewhat larger pedestals seen in the overview in Figure 1 to provide a clear measure of potential sample sink-in [22] during in-situ testing. Furthermore, micro-compression samples have typical aspect ratios between 4:1 and 3:1 to avoid potential plastic buckling [22], while micro-tensile specimens have commonly aspect ratios of 5:1 [23]. For details on these sample fabrication strategies, please refer to [24,25]. By fabricating the samples along the wedge, many of them can be placed in close proximity of a few microns, while at the same time a free line of sight is ensured in order to facilitate in-situ testing or other means of analysis.

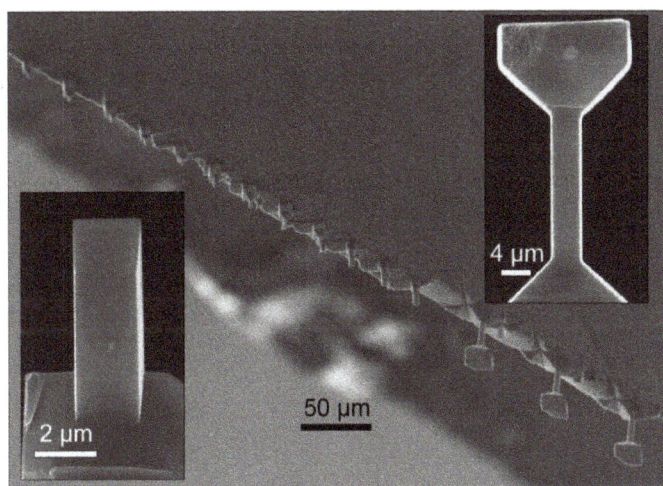

Figure 1. Low magnification SEM overview image of a series of differently sized compression and tensile samples before testing. The insets show a representative tensile and compression specimen, respectively.

Furthermore, it is essential to fabricate the samples with a well-defined orientation in a rigorous coordinate system and avoid all geometrical deficiencies. At the same time, the specimens should be transferrable between the FIB employed for specimen fabrication, the dedicated in-situ testing scanning electron microscope (SEM, DSM982, Carl Zeiss AG, Oberkochen, Germany), as well as the TEM (JEM 2100F, Jeol Ltd., Tokyo, Japan) for additional high resolution characterisation. Thus, we developed a staggered holder design as shown in Figure 2. The actual TEM-inspired sample has a flat along its circumference to ensure rotational alignment. Thus, even after repeated placement in a TEM, rotational alignment is kept [26]. This half-disk is mounted in a holder suited to fit the Bruker Hysitron Picoindenter PI-85 SEM as well as the PI-95 TEM nanoindenter with a clamping grip. As stated before, with the rotational degree of freedom fixed, the alignment during uniaxial testing is given. Finally, this whole SEM/TEM testing compatible holder can be placed in a larger mounting device that is on the one hand compatible to the Asmec UNAT 1 in-situ SEM microindenter, while on the other hand, this holder is used for the FIB processing, as it allows for controlled rotations of the mounted sample by ±45° and ±90°, as well as tilt along the long axis by 90°. Thus, by re-mounting the specimen at various tilts, perfect rectangular and orthogonal specimens with constant sample cross-sections can be fabricated. Moreover, utilising the ±45° tilts allows the specimens to be rotated by 45° with respect to the wedge, which provides a view on two faces rather than a single one during in-situ testing, which is a particular benefit during in-situ SEM testing [4,5].

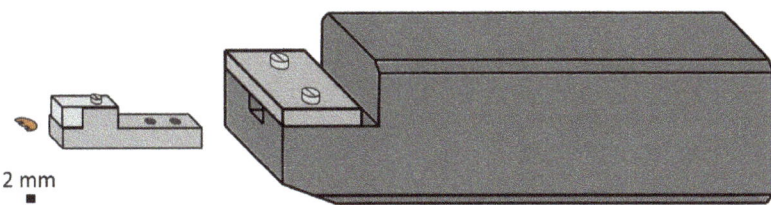

Figure 2. Specimen mounting and transfer concept. The actual sample follows a TEM specimen geometry, with adaptive holders to fit to different in situ TEM/SEM testing platforms as well as general focussed ion beam (FIB) fabrication and post testing SEM analysis.

The specimen sizes discussed here will be kept in the micrometre regime, spanning between 2 µm and 4 µm, respectively, taking into account that the emphasis of the present study is placed on bulk deformation mechanisms rather than effects specific to a nanoscale sample nature. While the strength values might thus be to some extent affected by the specimen size, the dimensions are above the typical dislocation spacing in well annealed crystals [27]. As such, we expect largely bulk deformation mechanisms to be examined [28].

The transferable sample mounting scheme proves a valuable asset also to examine the initial crystal orientation. Deviations from the planned crystallography could cause activation of unintended slip systems. So could the presence of unnoticed growth twins or occurring small angle boundaries [29]. While the crystallographic orientation of the specimen can be routinely accessed using electron backscatter diffraction (EBSD), with some effort even in a quasi in-situ manner [11], the orientation information from diffraction information in the TEM is much more sensitive, in particular with respect to low angle crystal defects that are challenging to resolve by standard EBSD. Furthermore, the effect of surface oxides or presence of initial dislocations can be easily assessed. Here, the specimens were routinely analysed using TEM, and an example for a 4-µm Mg tensile specimen is shown in Figure 3. Of course the sample is not fully electron transparent over the entire cross-section, but the whole length along both edges can be inspected for crystallographic defects, as exemplarily shown in Figure 3a,b for the bottom and head part of the sample, and the diffraction information in these regions is assessable, as shown for regions 3 and 4 in Figure 3c,d, to detail the crystallographic loading direction. In analogy, samples can be inspected post testing in the TEM.

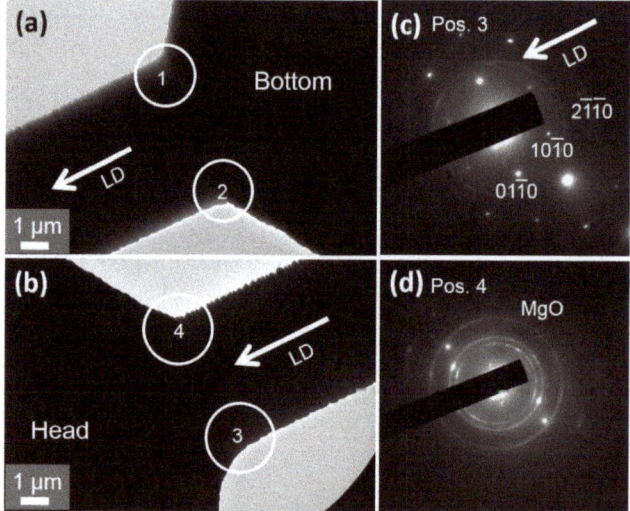

Figure 3. Pre-testing TEM analysis of a 4 µm thick Mg tensile sample, showing the gauge section towards the bottom (**a**) and head part (**b**), respectively. The diffraction patterns in (**c**) show the expected crystallography, while the image in (**d**) indicates redeposition of a nanocrystalline MgO layer. A white arrow indicates loading direction in real and reciprocal space.

To cover a number of possible deformation mechanisms [30], we examined loading directions along Mg$[2\bar{1}\bar{1}2]$, Mg$[0001]$, and Mg$[2\bar{1}\bar{1}0]$, respectively. An overview of the respective orientations, loading modes (compression/tension) and the Schmid factors corresponding on the one hand to dislocation plasticity on the basal, prismatic or pyramidal plane, and on the other hand to tension or contraction twinning, respectively, are given in Table 1. In addition, the deformation mode noted in the in-situ experiments as detailed later is noted. To relate this crystal orientation information to the testing geometry, in

Figure 4 the different crystal orientations are visualised in combination with the unit cell and the expected deformation mechanisms based on the actual Schmid [30].

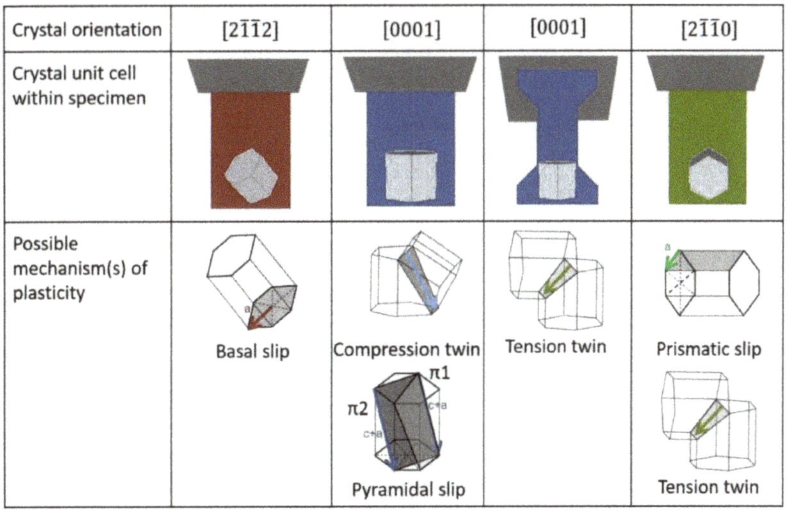

Crystal orientation	[2̄11̄2]	[0001]	[0001]	[2̄11̄0]
Crystal unit cell within specimen				
Possible mechanism(s) of plasticity	Basal slip	Compression twin π1 π2 Pyramidal slip	Tension twin	Prismatic slip Tension twin

Figure 4. Schematic overview of the studied crystal orientations, the related unit cell orientation with respect to the uniaxial testing volume, and the conceivable deformation mechanism(s). The introduced colour code is kept consistent throughout the manuscript.

To load the compression samples, a conical conductive diamond tip with a 16-μm diameter flat punch (Synton MDP, Nidau, Switzerland) was used, while for tensile testing, a gripping tool FIB fabricated from a heavily drawn W wire was employed [23]. Before running the actual deformation experiments, the stiffness of the respective lamella was determined for later compliance corrections. The miniaturised in-situ compression and tensile tests were conducted under displacement control at a nominally constant strain rate of $1 \cdot 10^{-3}$ s^{-1}, and a corresponding video of the experiment was recorded using a frame grabber at a repetition rate of 1 Hz. This allows for a trade-off between temporal resolution upon direct observation and sufficient signal-to-noise ratio for later digital image analysis, pillar sink-in correction, and analysis of local strain, respectively.

Table 1. Summary of Mg crystal orientations tested, loading mode, nominal Schmid factors, and experimentally observed deformation mode.

Loading Direction	Loading Mode	Schmid Factor for Feformation by						Observed Deformation Mode
		<a> Basal Slip	<a> Prismatic Slip	<c+a> Pyramidal (π1) Slip	<c+a> Pyramidal (π2) Slip	Tension Twinning	Contraction Twinning	
[-]	[-]							
Mg[2̄11̄2]	Compression	0.5	0.2	0.39	0.30	-	0.32	Basal slip
Mg[0001]	Compression	0	0	0.4	0.45	-	0.42	Pyramidal slip
Mg[0001]	Tension	0	0	0.4	0.45	0.5	-	Tensile twin
Mg[2̄11̄0]	Compression	0	0.43	0.4	0.45	0.37	-	Tensile twin and basal slip in the twin

3. Results and Discussion

Considering the TEM analysis in Figure 3a,b, it is possible to note that the specimen corner is not as sharp and smooth as one might expect. This results from the very high sputtering rates of Mg in conjunction with the broadening of the ion beam. Moreover, we note that the roughness is more pronounced along the upper sides, related to diffraction analysis spots one and four, respectively. From the indicated loading direction and the electron diffraction information in Figure 3c, it is evident that the sample is loaded along the intended $[2\bar{1}\bar{1}0]$ direction. While there is slight indication of additional diffraction rings in this image, they are much more pronounced in Figure 3d and can be indexed as MgO. Thus, we consider that the bottom side of the tensile sample was polished last during FIB preparation. As such, there is a rather smooth edge and only a minimal amount of nanocrystalline MgO on the surface. Contrarily, on the upper side that was finished before, redeposition of sputtered Mg could take place, causing a rougher surface that is more prone to oxidation, thereby explaining the rougher appearance and more distinct MgO diffraction signal. Thus, we conclude that we can assure that the crystal orientation for our samples is uniform and consistent with expectations, but there is a nanocrystalline MgO layer that could affect the observability of surface features during in-situ SEM testing and might in the worst case act as a confining hard shell. Mg is known to form an oxide easily, and we observed such an MgO layer previously, in particular upon heat treatments of nanoscale Mg pillars [31]. However, for the given situation of few micrometre sized pillars and few nm thick oxide layer, we do not expect pronounced influences in terms of a confining layer.

We start with the deformation mode easiest to activate in Mg, basal slip. As evident for the 4-µm sized Mg $[2\bar{1}\bar{1}2]$ micro-compression specimen shown in Figure 5, the flow curve in Figure 5a exhibits some minor and one major event during plastic deformation, which takes place at flow stresses in the range of 60–80 MPa. These events can be directly correlated to slip steps emerging on the sample surface during testing, as shown in the inset to Figure 5a. This is further supported by inclined post-mortem SEM images taken from the front and back, where one large and one smaller slip step are well discerned (Figure 5c,d). This localised slip behaviour was also reported earlier [12] and is in excellent agreement to common observations for dislocation slip based on a limited number of dislocation sources in other metals exhibiting a low Peierls barrier [23]. Given the sample crystallography and observed slip plane inclination, it is certain to ascribe this plastic deformation to basal slip [12]. A feature of note here is the oxide layer well displayed on the detail of an approximately 400 nm wide mayor slip step (Figure 5b). This MgO layer has a thickness of ~10 nm [31] and stretches conformal over the emerged slip step, which required roughly 1200 basal plane dislocations with $[11\bar{2}0]$ direction to exit in order to form the slip step. The important implication is that this MgO layer does not break easily and thus has the tendency to cover faint surface steps. Thus, it cannot be determined with certainty whether the other smaller features along the stress–strain curve correspond to reactivation of dislocation sources that already operated and produced a multitude of dislocations resulting in observable slip steps, or the activation of new sources with resultant slip steps that might be not well discernible on the specimen surface yet due to this oxide layer. However, considering that it is considerably harder to nucleate dislocations from an interface than to mobilise pre-existing ones, it seems likely that some residual defects allowed for generation of a dislocation source (classically Frank–Read type, double cross-slip type or single armed source type) in the interior of the specimens.

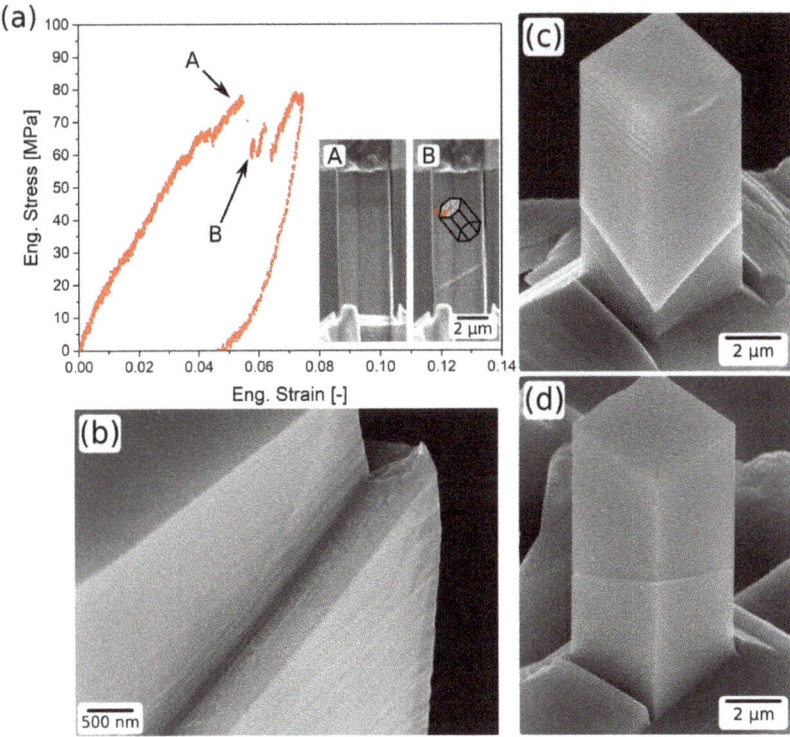

Figure 5. Micro-compression test on a 4-μm Mg$[2\bar{1}\bar{1}2]$ pillar. (**a**) Stress–strain behaviour showing some minor plastic events and one mayor stress drop. The inset depicts still images of the sample at points A and B of the experiment, as well as the crystal unit cell. (**b**) Detailed post mortem inspection of a large slip offset. (**c**,**d**) Front and backside view of the sample after testing, indicating a large and a smaller surface steps on the specimen surface.

Turning to the 4-μm Mg[0001]-oriented sample shown in Figure 6, we note that the stress–strain data in Figure 6a exhibit an initial linear elastic regime followed by a yield point at ~275 MPa, to be continued by a plastic deformation period exhibiting pronounced strain hardening. The loading portion of the curve exhibits a noticeable bend indicative of slight misalignment between sample and flat punch, while the deduced elastic modulus amounts to ~40 GPa, which is in reasonable accordance with calculated orientation dependent values [32] based on experimental elastic constants [33] and lattice spacing [34] as summarised in Table 2. Since local plastic deformation should correct for the local contact issues at the initial loading, the lower experimental value upon elastic unloading from a plastically deformed sample might result from a non-perfect accounting of parasitic compliance effects, which could be eliminated, for example, by measuring the deformation directly on the sample surface using digital image correlation [35]. Alternatively, this could originate from the reverse motion and straightening of bulged out dislocations upon unloading, similar to what is frequently referred to as a Bauschinger effect in bending geometries [36]. In that case in-situ TEM observations could help to clarify the origin, but this goes beyond the scope of this work. The images recorded during in-situ deformation as well as after loading (Figure 6b) display a distinct shortening of the sample, but only minimal traces of deformation on the surface, in fact the most pronounced notion is a certain sample barrelling. Loading along this crystal direction should suppress basal and prismatic slip, as well as $\{10\bar{1}2\}$ twinning. Still, macroscopic compression tests on Mg[0001] reported a certain amount of twinning to take place due to barrelling and friction effects [37]. Turning to the present miniaturised compression test, we note only a slight discontinuity at a stress of ~275 MPa, but the absence of a flow plateau that could be indicative of a limited propagation of a $\{10\bar{1}1\}$ compression twin. This is not too surprising, as during compression twinning the twinned part would have to rotate significantly out of the loading axis (Figure 4), which is constrained by the laterally stiff sample setup and the friction between specimen and diamond flat punch [38]. However, a significant amount of work hardening is displayed, which favourably corresponds to the activation of pyramidal slip. As post mortem EBSD investigation further confirmed the absence of twinning, this discrete burst is explained by the avalanche-like activation of pyramidal slip on various planes, leading to the increased dislocation density and resultant work hardening in these microscale specimens. In general, it should be recognised that the formation and growth of a compression twin during compressive loading (see Figure 4) would yield a favourable configuration for subsequent basal slip, which is not observed and therefore further validates the assumption that the majority of deformation is carried by dislocation plasticity on pyramidal ($\pi 1$ or $\pi 2$) slip planes.

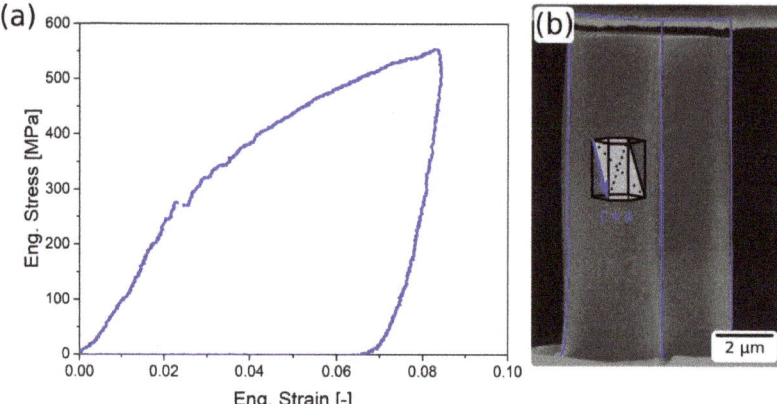

Figure 6. Micro-compression test on a 4-µm Mg[0001] specimen. (**a**) Stress-strain characteristic and (**b**) deformed specimen. Blue lines indicate original specimen geometry, while the inset indicates the unit cell and slip crystallography.

Table 2. Orientation dependent Young's moduli and Poisson's ratios for the cylindrically symmetrical Mg single crystal.

Crystal Orientation	Young's Modulus E (GPa)	Poisson's Ratio ν (-)
$<0001>$	48.3	0.214
$<10\bar{1}0>, <11\bar{2}0>$	46.2	0.343

Relating this behaviour to the previously shown $[2\bar{1}\bar{1}2]$ specimen, the strains at the first deviation from linear elasticity are similar, but the stress is about a factor of four higher in the [0001] specimen, which is expected in light of the different mechanisms (pyramidal slip versus basal slip). The fact that there are no pronounced slip bands is in accordance with the argument that deformation is mostly driven by pyramidal slip, as the higher number of available slip planes can lead to dislocation interaction and formation of kinks or jogs. Moreover, due to the large Burgers vector of <c+a> dislocations, they are dissociated into two partial dislocations. A recent study showed that those are intrinsically transformed to basal-dissociated immobile dislocation structures, which cannot contribute to plastic deformation, but serve as strong obstacles to the motion of all other dislocations inside the pillar [14], resulting in the strong strain hardening observed in the [0001] specimen. This is clearly different to the $[2\bar{1}\bar{1}2]$ specimen, where the predominant deformation occurs on the basal plane, having only two independent slip systems. Furthermore, the lack of "strain-burst" type behavior in conjunction with the high strain hardening suggest no or only a minor occurrence of twinning, as previously shown macroscopically by Syed et al. [37].

To underline the previous point regarding difficult to activate deformation mechanisms and deformation constraints, we consider another miniaturised compression test on a Mg[1000] specimen with 2 μm width in Figure 7. Similar to the previous example, the stress-strain data shown in Figure 7a does not exhibit a distinct yield event, but a gradual elastic–plastic transition. At a stress of ~185 MPa a flow plateau is observed, corresponding to a distortion of the micro-compression sample evident in Figure 7c. After the distinct load-drop at a strain of 0.07, we observe the emergence of a distinct slip step at the position of the previous specimen distortion (Figure 7d), which grows during deformation at a constant flow stress of about 160 MPa. While the initial unloading slope appears quite similar to that upon initial loading, the unloading curve develops a remarkable bend back, indicative of a significant amount of reverse plasticity. Alternatively, this could be ascribed to geometric changes occurring during the experiment. We attribute the gradual elastic–plastic transition again to pyramidal slip activity, and the constant flow plateau to the activation of basal slip in the locally modified crystallography of the deformed specimen. Initially, this basal slip is confined by the surface oxide layer. Once this breaks at the distinct stress drop, the slip step becomes well discernible on the sample surface and grows with continuing deformation.

Coming back to the previously mentioned aspect of confinement effects on twinning, we conducted related micro-tensile tests on an Mg[0001] crystal orientation using a 4-μm tensile sample in Figure 8, as this would permit the formation of a tension twin. As reported earlier, the longer aspect ratio tensile geometry is more compliant and allows for lateral adjustments [38]. In Figure 8a, we observe an elastic–plastic transition followed by a rather constant flow stress in the range of 80 to 90 MPa. This corresponds well to the nucleation and growth of a tensile twin seen in in Figure 8d,e. The pronounced load drop at a strain of 0.08 relates to the nucleation of a second tensile twin on the left hand side of the tensile specimen, as also indicated in the in-situ image shown as Figure 8f. Taking the specimen back into the TEM after unloading, in Figure 8b we observe the distinct crystal orientation change at the twin front in the area indicated by a dashed rectangle in Figure 8f. Thus, this setup allows for the unhindered nucleation and propagation of tensile twins.

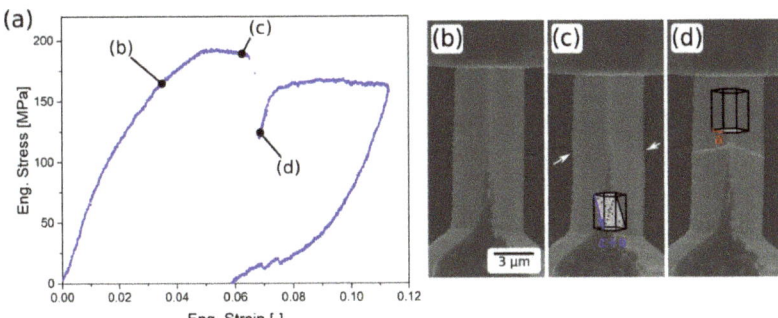

Figure 7. Micro-compression testing on a 2-μm Mg[0001] sample. (**a**) Stress-strain curve showing an extended elastic–plastic transition, followed by a distinct load drop and a flow plateau. Notably, the unloading portion exhibits a remarkable amount of reverse plasticity. (**b**–**d**) In-situ images in correlation to the flow curve, documenting the onset of plasticity by pyramidal slip, the nucleation of basal slip and the continuous growth of the related almost horizontal slip step.

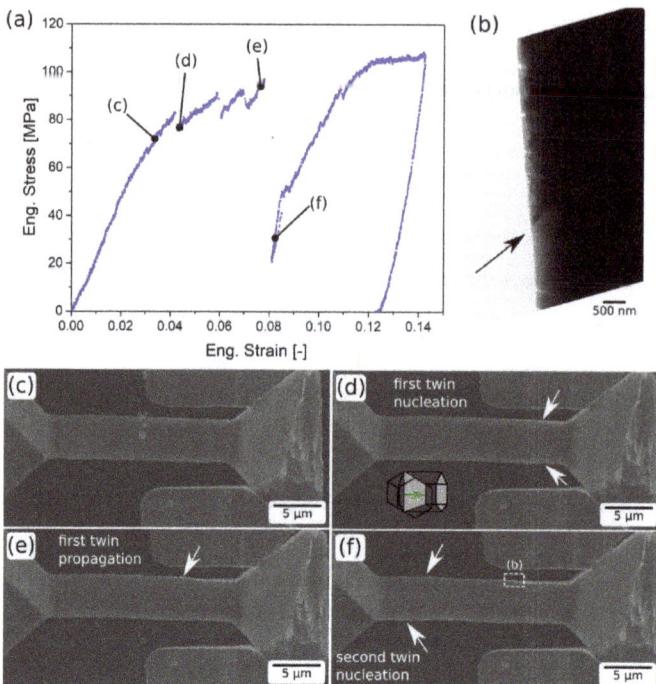

Figure 8. Micro-tensile test of a 4-µm Mg[0001] sample. (**a**) Stress–strain curve and (**b**) post mortem TEM image showing the formed twin boundary indicated by an arrow. (**c**–**f**) In-situ SEM images as indicated in (**a**), corresponding to the yield point, initiation of the first tensile twin, propagation of the twin, and nucleation of a second twin, respectively. The area inspected by TEM in (**b**) is indicated in (**f**).

It is well known that Mg exhibits two distinctly different twins when compressed ($\{10\bar{1}1\}$ contraction twin) or elongated ($\{10\bar{1}2\}$ tensile twin) along the [0001] axis. While the so called contraction twin is only rarely observed because of the higher activation threshold, tensile twinning is observed frequently and considered a major carrier of deformation in Mg. The slight twist in the tensile specimen might have worked as a catalyst for twin nucleation, but it is definitely the more favourable loading direction for tensile twinning. After first and second twinning events occur, the loading axis of the pillar is changed from [0001] to $\sim[10\bar{1}0]$ because the crystal is reoriented by 86° upon twinning. Under $[10\bar{1}0]$ tensile loading, the contraction twin would be crystallographically favoured, but the critical resolved shear stress for contraction twinning is much higher than 90 MPa. This makes activation of another twin within the twinned part very unlikely.

An interesting notion is that the threshold for twinning does not seem to be constant. Previously, we showed that the nucleation of deformation twins is not a purely stress driven process, but is triggered by dislocation bulges on the prismatic plane [39], which are dependent on the very local stress and strain state and can therefore not always be derived just from considering a global stress value.

To compare the tensile twinning observed in tension to the activation of the same mechanism in compression, the last example concerns a miniaturised compression test of a 2-µm wide specimen along the Mg$[2\bar{1}\bar{1}0]$ direction. For this crystallographic orientation, due to the specific crystal rotation created by the twin, lateral confinement effects should be less pronounced and thus also allow twinning to take place during compression. In fact, this specific orientation was previously studied by our group in a combined in-situ SEM and in-situ TEM fashion and we will therefore be rather brief and only mention the essential facts, while all details can be found elsewhere [39]. In Figure 9a, we again observe a clear elastic regime followed by a distinct burst and load drop at a stress of about 400 MPa, followed by an overshoot of the indenter due to a not sufficiently fast feedback control, thereby releasing the strain energy stored in the system. Distinct distortion of the specimen geometry, in fact a slight kink of the top part due to the twinning reorientation by 86°, is seen in point B in the inset upon elastic reloading. Thus, we can assign this again to the nucleation and propagation of a tensile twin. After elastic reloading, we observe a constant stress plateau at ~200 MPa, where slip steps emerge as seen in the inset corresponding to point C along the stress–strain data. This corresponds to basal slip in the twinned part of the specimen, as reported earlier [39]. In Figure 9b an inclined post mortem SEM image shows the slip steps in more detail.

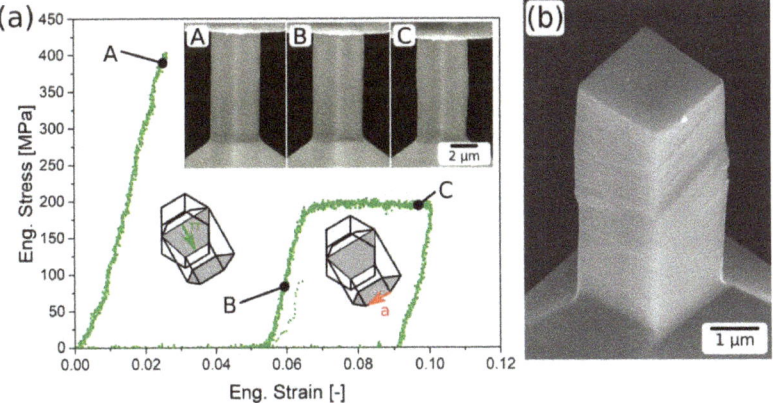

Figure 9. In-situ SEM micro-compression test of a 2-μm Mg $[2\bar{1}\bar{1}0]$ sample. (**a**) Stress–strain data showing elastic limit, a distinct plastic event followed by reloading, a plateau stress, and final unloading. Corresponding still images from the in-situ video relating to points A–C indicated in the stress–strain curve are provided as insets. (**b**) Inclined post mortem image of the deformed specimen, showing a slight tilt of the twinned part, as well as slip steps within this sample portion.

It is noteworthy that for the smaller 2-μm specimen examined here, the flow stress for basal slip is in the range of 200 MPa, while we measured only about 70 MPa for the 4-μm sample in Figure 5. This is mostly due to the sample size effect, with additional contributions from different Schmid factors to be accounted for.

To depict all discussed specimens in a comparative framework, the different critical resolved shear stresses (CRSS) were calculated, based on the Schmid factors summarised in Table 1, and are shown with respect to the different observed deformation mechanisms in Figure 10. There, the full symbols correspond to the first evident deviation from linear elastic loading as an indication for the beginning of the respective yield phenomena, while the open symbols correspond to the subsequent evident features of deformation: load drop corresponding to the large slip step for the $[2\bar{1}\bar{1}2]$ specimen, load drop corresponding to the second evident twin formation for the [0001] tension specimen, and basal slip plateau following twinning for the $[2\bar{1}\bar{1}0]$ specimen, respectively. The initial pyramidal slip CRSS occurring in the [0001] compression specimen was calculated for both, π1 and π2 slip, as the exact configuration is not known. However, based on the fact that π1 slip results in a lower CRSS, it can be assumed that it is easier to activate and therefore the majority contribution to plasticity in this specimen. This would also be in agreement with current observations by Xie et al. [40].

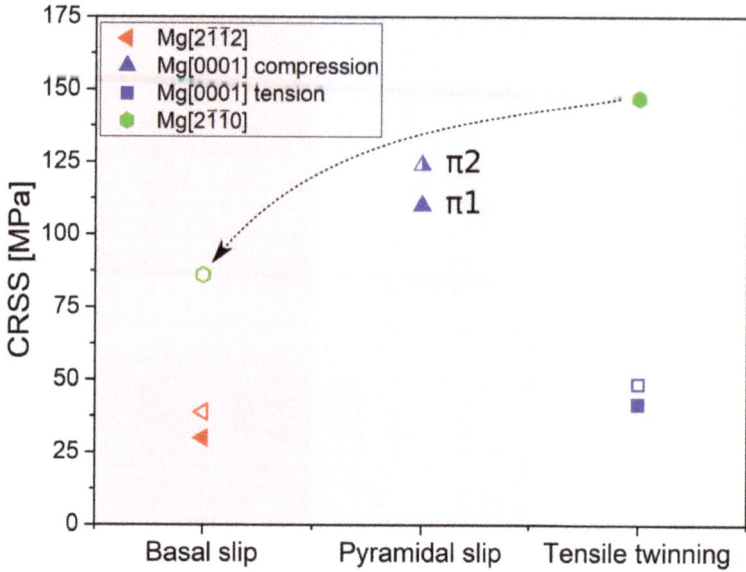

Figure 10. Critical resolved shear stresses (CRSS) of the previously detailed samples with regard to deformation mechanisms. The filled symbols represent initial deviation of the elastic load-displacement regime, while the open symbols depict evident deformation features, such as large load drops, which correspond to the respective mechanisms. For the Mg[0001] compression specimen, CRSS was calculated for both pyramidal planes (π1, π2) for comparison.

Comparing now the CRSS for tensile twinning in tension and compression as provided in Figure 10, it is evident that the necessary nucleation stress is considerably higher (~250%) for the 2-μm compression configuration than for the 4-μm tensile configuration. This could be partially attributed to the smaller geometric dimensions but seems unlikely to be a satisfying single cause explanation for this pronounced difference. Considering, however, not resolved stress but strain one could argue that the tensile strain along the [0001] direction is $\varepsilon_{tensile} = \frac{\sigma}{E_{<0001>}} = 0.00173$ in the tensile specimen, while it is $\varepsilon_{compression} = \frac{\sigma}{E_{<11\bar{2}0>}} \nu_{<11\bar{2}0>} = 0.00298$ for the compression specimen, which is only an increase of ~60% for a sample size reduction by a factor of 2. Furthermore, this train of thought suggests that tensile twinning in Mg is rather a strain driven than a stress driven process.

In comparison with literature values concerning CRSS as summarised by Zhang and Joshi [30], our values are a factor 5–10 higher for basal slip and tensile twinning, and about a factor 2–3 higher for pyramidal slip. This could, even though not desired in the first place, again be related to be resultant of the small size of the samples. However, specifically in basal slip configuration with typical CRSS values in the single digit MPa range, it could also be a result of the high lateral stiffness in compression testing which causes a constrained deformation field [41], or additional pile-up contributions from FIB damage [42,43] or the thin MgO layer [31].

As evident from the previously shown experiments, the remarkable difference between easy and hard deformation modes in anisotropic materials such as Mg renders it rather challenging to probe specific mechanisms individually. However, as shown in the examples given above, it is well feasible, in particular when complemented by in-situ observations or respective post characterisation. In contrast, for a bulk experiment on a polycrystalline sample such differentiations would not be possible, as local stresses would always cause the unwanted activation of easy deformation mechanisms such as basal slip, either long before the actual hard mode is activated, or right after the crystal lattice was changed to a favourable orientation by local lattice distortions or twinning.

In terms of detailed insights and concrete dislocation or twinning mechanisms, a combined in-situ SEM/TEM study might be preferred, as the larger SEM experiments deliver bulk quantities if conducted on large enough specimens [28,44], while the TEM observations identify specific dislocation character or twinning relationships [39]. It should, however, be kept in mind that the different mechanisms could exhibit various strength scaling characteristics [10,18]. As such, for small enough specimens a transition between plastic mechanisms could occur [44] and mislead the data interpretation. Even changes in dislocation dynamics due to the high stresses in such highly confined nanoscale volumes are conceivable [45]. Therefore, such an undertaking ideally encompasses a complete size-dependent assessment.

From the more general application perspective, in our present work we turned to extraction of differently oriented lamellas from a bulk piece of single crystalline material. For experimental polycrystalline materials or alloys, this can easily be further reduced with respect to required material, say down to a specific single grain. Therefore, the extraction of lift-out lamellas from the grain of interest under specific orientations with respect to the surface normal, or from different grains with suited orientations, in a pick and place operation using a micromanipulator would be a feasible way [46]. As this most certainly requires larger amounts of surrounding material to be removed, modern developments such as Xe-FIBs or femtosecond laser machining [47,48] might prove helpful. Alternatively, different sample holders could be 3D printed to tilt the grain to the different selected orientations [7]. The only challenge remaining in that case is to FIB machine the sample top surface perpendicular with respect to the loading axis of the specimen situated within the tilted sample surface, as this requires a 90° tilting of the mounted sample.

4. Conclusions

In the present work, on the example of the hexagonal metal Mg, we attempted a depictive overview to highlight that markedly different deformation mechanisms in anisotropic materials can be quantitatively examined. To do so, it is instructive to choose suited crystallographic orientations for the small scale testing from the bulk of interest, and to avoid mechanical constraints that could promote activation of unintended deformation mechanisms. Furthermore, the crystal reorientation due to twinning and related changes in the preferably activated deformation mechanisms should be kept in mind. Finally, while the mechanical signature and post mortem observations should suffice for analysing the respective deformation mechanisms, direct in-situ observation lends itself to simplify this process significantly and provide better insight into the dynamics and temporal sequence.

Author Contributions: Conceptualization, D.K. and S.H.O.; methodology, D.K., J.J., S.H.O.; formal analysis, M.A.; investigation, J.J., M.A., R.K.; writing—original draft preparation, D.K.; writing—review and editing, D.K., J.J., M.A., R.K., S.H.O.; visualization, J.J., M.A.; supervision, D.K., S.H.O. All authors have read and agreed to the published version of the manuscript.

Funding: The authors acknowledge financial support by the Austrian Science Fund FWF (project number I-1020) as well as the European Research Council (ERC) under the European Union's Horizon 2020 research and innovation programme (Grant No. 771146 TOUGHIT). S.H.O. was supported by the National Research Foundation of Korea (NRF) grant funded by the Ministry of Science and ICT (MSIT) (NRF-2020R1A2C2101735), the Creative Materials Discovery Program through (NRF-2019M3D1A1078296), Bio-inspired Innovation Technology Development Project (NRF-2018M3C1B7021994).

Conflicts of Interest: The authors declare no conflict of interest.

References

1. Uchic, M.D.; Dimiduk, D.M.; Florando, J.N.; Nix, W.D. Sample dimensions influence strength and crystal plasticity. *Science* **2004**, *305*, 986–989. [CrossRef] [PubMed]
2. Uchic, M.D.; Shade, P.A.; Dimiduk, D. Plasticity of micrometer-scale single crystals in compression. *Ann. Rev. Mater. Res.* **2009**, *39*, 361–386. [CrossRef]

3. Greer, J.R.; de Hosson, J.T.M. Plasticity in small-sized metallic systems: Intrinsic versus extrinsic size effect. *Prog. Mater. Sci.* **2011**, *56*, 654–724. [CrossRef]
4. Malyar, N.V.; Micha, J.S.; Dehm, G.; Kirchlechner, C. Size effect in bi-crystalline micropillars with a penetrable high angle grain boundary. *Acta Mater.* **2017**, *129*, 312–320. [CrossRef]
5. Fritz, R.; Maier-Kiener, V.; Lutz, D.; Kiener, D. Interplay between sample size and grain size: Single crystalline vs. ultrafine-grained chromium micropillars. *Mater. Sci. Eng. A* **2016**, *674*, 626–633. [CrossRef]
6. Kheradmand, N.; Vehoff, H. Orientation Gradients at Boundaries in Micron-Sized Bicrystals. *Adv. Eng. Mater.* **2012**, *14*, 153–161. [CrossRef]
7. Liebig, J.P.; Krauß, S.; Göken, M.; Merle, B. Influence of stacking fault energy and dislocation character on slip transfer at coherent twin boundaries studied by micropillar compression. *Acta Mater.* **2018**, *154*, 261–272. [CrossRef]
8. Byer, C.M.; Li, B.; Cao, B.; Ramesh, K.T. Microcompression of single-crystal magnesium. *Scr. Mater.* **2010**, *62*, 536–539. [CrossRef]
9. Lilleodden, E. Microcompression study of Mg (0 0 0 1) single crystal. *Scr. Mater.* **2010**, *62*, 532–535. [CrossRef]
10. Yu, Q.; Shan, Z.-W.; Li, J.; Huang, X.; Xiao, L.; Sun, J.; Ma, E. Strong crystal size effect on deformation twinning. *Nature* **2010**, *463*, 335–338. [CrossRef]
11. Guo, Y.; Schwiedrzik, J.; Michler, J.; Maeder, X. On the nucleation and growth of {$11\bar{2}2$} twin in commercial purity titanium: In situ investigation of the local stress field and dislocation density distribution. *Acta Mater.* **2016**, *120*, 292–301. [CrossRef]
12. Ye, J.; Mishra, R.K.; Sachdev, A.K.; Minor, A.M. In situ TEM compression testing of Mg and Mg—0.2 wt.% Ce single crystals. *Scr. Mater.* **2011**, *64*, 292–295. [CrossRef]
13. Wu, Z.; Ahmad, R.; Yin, B.; Sandlöbes, S.; Curtin, W.A. Mechanistic origin and prediction of enhanced ductility in magnesium alloys. *Science* **2018**, *359*, 447–452. [CrossRef] [PubMed]
14. Wu, Z.; Curtin, W.A. The origins of high hardening and low ductility in magnesium. *Nature* **2015**, *526*, 62–67. [CrossRef] [PubMed]
15. Liu, B.-Y.; Liu, F.; Yang, N.; Zhai, X.-B.; Zhang, L.; Yang, Y.; Li, B.; Li, J.; Ma, E.; Nie, J.-F.; et al. Large plasticity in magnesium mediated by pyramidal dislocations. *Science* **2019**, *365*, 73–75. [CrossRef]
16. Li, B.; Ma, E. Atomic shuffling dominated mechanism for deformation twinning in magnesium. *Phys. Rev. Lett.* **2009**, *103*, 035503. [CrossRef]
17. Ostapovets, A.; Molnár, P. On the relationship between the "shuffling-dominated" and "shear-dominated" mechanisms for twinning in magnesium. *Scr. Mater.* **2013**, *69*, 287–290. [CrossRef]
18. Liu, Y.; Li, N.; Kumar, M.A.; Pathak, S.; Wang, J.; McCabe, R.J.; Mara, N.A.; Tomé, C.N. Experimentally quantifying critical stresses associated with basal slip and twinning in magnesium using micropillars. *Acta Mater.* **2017**, *135*, 411–421. [CrossRef]
19. Prasad, K.E.; Rajesh, K.; Ramamurty, U. Micropillar and macropillar compression responses of magnesium single crystals oriented for single slip or extension twinning. *Acta Mater.* **2014**, *65*, 316–325. [CrossRef]
20. Wang, J.; Molina-Aldareguía, J.M.; Llorca, J. Effect of Al content on the critical resolved shear stress for twin nucleation and growth in Mg alloys. *Acta Mater.* **2020**, *188*, 215–227. [CrossRef]
21. Wang, J.; Ramajayam, M.; Charrault, E.; Stanford, N. Quantification of precipitate hardening of twin nucleation and growth in Mg and Mg-5Zn using micro-pillar compression. *Acta Mater.* **2019**, *163*, 68–77. [CrossRef]
22. Zhang, H.; Schuster, B.E.; Wei, Q.; Ramesh, K.T. The design of accurate micro-compression experiments. *Scr. Mater.* **2006**, *54*, 181–186. [CrossRef]
23. Kiener, D.; Grosinger, W.; Dehm, G.; Pippan, R. A further step towards an understanding of size-dependent crystal plasticity: In-situ tension experiments of miniaturized single crystal copper samples. *Acta Mater.* **2008**, *56*, 580–592. [CrossRef]
24. Moser, G.; Felber, H.; Rashkova, B.; Imrich, P.J.; Kirchlechner, C.; Grosinger, W.; Motz, C.; Dehm, G.; Kiener, D. Sample Preparation by Metallography and Focused Ion Beam for Nanomechanical Testing. *Pract. Metallogr.* **2012**, *49*, 343–355. [CrossRef]
25. Wurster, S.; Treml, R.; Fritz, R.; Kapp, M.W.; Langs, E.; Alfreider, M.; Ruhs, C.; Imrich, P.J.; Felber, G.; Kiener, D. Novel Methods for the Site Specific Preparation of Micromechanical Structures. *Pract. Metallogr.* **2015**, *52*, 131–146. [CrossRef]
26. Kiener, D.; Zhang, Z.; Sturm, S.; Cazottes, S.; Imrich, P.J.; Kirchlechner, C.; Dehm, G. Advanced nanomechanics in the TEM: Effects of thermal annealing on FIB prepared Cu samples. *Philos. Mag. A* **2012**, *92*, 3269–3289. [CrossRef]
27. Schneider, A.S.; Kiener, D.; Yakacki, C.M.; Maier, H.J.; Gruber, P.A.; Tamura, N.; Kunz, M.; Minor, A.M.; Frick, C.P. Influence of bulk pre-straining on the size effect in nickel compression pillars. *Mater. Sci. Eng. A* **2013**, *559*, 147–158. [CrossRef]
28. El-Awady, J.A. Unravelling the physics of size-dependent dislocation-mediated plasticity. *Nat. Commun.* **2015**, *6*, 5926. [CrossRef]
29. Maass, R.; van Petegem, S.; Grolimund, D.; van Swygenhoven, H.; Uchic, M.D. A strong micropillar containing a low angle grain boundary. *Appl. Phys. Lett.* **2007**, *91*, 131909. [CrossRef]
30. Zhang, J.; Joshi, S.P. Phenomenological crystal plasticity modeling and detailed micromechanical investigations of pure magnesium. *J. Mech. Phys. Solids* **2012**, *60*, 945–972. [CrossRef]
31. Jeong, J.; Lee, S.; Kim, Y.; Han, S.M.; Kiener, D.; Kang, Y.-B.; Oh, S.H. Microstructural evolution of a focused ion beam fabricated Mg nanopillar at high temperatures: Defect annihilation and sublimation. *Scr. Mater.* **2014**, *86*, 44–47. [CrossRef]
32. Zhang, J.-M.; Zhang, Y.; Xu, K.-W.; Ji, V. Anisotropic elasticity in hexagonal crystals. *Thin Solid Films* **2007**, *515*, 7020–7024. [CrossRef]
33. Slutsky, L.J.; Garland, C.W. Elastic Constants of Magnesium from 4.2 K to 300 K. *Phys. Rev.* **1957**, *107*, 972–976. [CrossRef]
34. Von Batchelder, F.W.; Raeuchle, R.F. Lattice Constants and Brillouin Zone Overlap in Dilute Magnesium Alloys. *Phys. Rev.* **1957**, *105*, 59–61. [CrossRef]

35. Moser, B.; Wasmer, K.; Barbieri, L.; Michler, J. Strength and fracture of Si micropillars: A new scanning electron microscopy-based micro-compression test. *J. Mater. Res.* **2007**, *22*, 1004–1011. [CrossRef]
36. Kapp, M.W.; Kirchlechner, C.; Pippan, R.; Dehm, G. Importance of dislocation pile-ups on the mechanical properties and the Bauschinger effect in microcantilevers. *J. Mater. Res.* **2015**, *30*, 791–797. [CrossRef]
37. Syed, B.; Geng, J.; Mishra, R.K.; Kumar, K.S. Compression response at room temperature of single-crystal magnesium. *Scr. Mater.* **2012**, *67*, 700–703. [CrossRef]
38. Kiener, D.; Grosinger, W.; Dehm, G. On the importance of sample compliance in uniaxial microtesting. *Scr. Mater.* **2009**, *60*, 148–151. [CrossRef]
39. Jeong, J.; Alfreider, M.; Konetschnik, R.; Kiener, D.; Oh, S.H. In-situ TEM observation of twin-dominated deformation of Mg pillars: Twinning mechanism, size effects and rate dependency. *Acta Mater.* **2018**, *158*, 407–421. [CrossRef]
40. Xie, K.Y.; Alam, Z.; Caffee, A.; Hemker, K.J. Pyramidal I slip in c-axis compressed Mg single crystals. *Scr. Mater.* **2016**, *112*, 75–78. [CrossRef]
41. Kiener, D.; Motz, C.; Dehm, G. Micro-compression testing: A critical discussion of experimental constraints. *Mater. Sci. Eng. A* **2009**, *505*, 79–87. [CrossRef]
42. Kiener, D.; Motz, C.; Rester, M.; Dehm, G. FIB damage of Cu and possible consequences for miniaturized mechanical tests. *Mater. Sci. Eng. A* **2007**, *459*, 262–272. [CrossRef]
43. Hütsch, J.; Lilleodden, E.T. The influence of focused-ion beam preparation technique on microcompression investigations: Lathe vs. annular milling. *Scr. Mater.* **2014**, *77*, 49–51. [CrossRef]
44. Kiener, D.; Hosemann, P.; Maloy, S.A.; Minor, A.M. In situ nanocompression testing of irradiated copper. *Nat. Mater.* **2011**, *10*, 608–613. [PubMed]
45. Wang, J.W.; Zeng, Z.; Weinberger, C.R.; Zhang, Z.; Zhu, T.; Mao, S.X. In situ atomic-scale observation of twinning-dominated deformation in nanoscale body-centred cubic tungsten. *Nat. Mater.* **2015**, *14*, 594–600. [PubMed]
46. Hirakata, H.; Takahashi, Y.; Truong, D.; Kitamura, T. Role of plasticity on interface crack initiation from a free edge and propagation in a nano-component. *Int. J. Fract.* **2007**, *145*, 261–271.
47. Echlin, M.P.; Mottura, A.; Torbet, C.J.; Pollock, T.M. A new TriBeam system for three-dimensional multimodal materials analysis. *Rev. Sci. Instrum.* **2017**, *83*, 023701. [CrossRef]
48. Pfeifenberger, M.J.; Mangang, M.; Wurster, S.; Reiser, J.; Hohenwarter, A.; Pfleging, W.; Kiener, D.; Pippan, R. The use of femtosecond laser ablation as a novel tool for rapid micro-mechanical sample preparation. *Mater. Des.* **2017**, *121*, 109–118. [CrossRef]

Article

Crystallographic Orientation Dependence of Mechanical Responses of FeCrAl Micropillars

Dongyue Xie, Binqiang Wei, Wenqian Wu and Jian Wang *

Department of Mechanical and Materials Engineering, University of Nebraska-Lincoln, Lincoln, NE 68588, USA; dxie@huskers.unl.edu (D.X.); bwei5@huskers.unl.edu (B.W.); wwu13@huskers.unl.edu (W.W.)
* Correspondence: jianwang@unl.edu

Received: 4 October 2020; Accepted: 14 October 2020; Published: 16 October 2020

Abstract: Iron-chromium-aluminum (FeCrAl) alloys are used in automobile exhaust gas purifying systems and nuclear reactors due to its superior high-temperature oxidation and excellent corrosion resistance. Single-phase FeCrAl alloys with a body centered cubic structure plastically deform through dislocation slips at room temperature. Here, we investigated the orientation dependence of mechanical responses of FeCrAl alloy through testing single-crystal and bi-crystal micropillars in a scanning electron microscopy at room temperature. Single-crystal micropillars were fabricated with specific orientations which favor the activity of single slip system or two slip systems or multiple slip systems. The strain hardening rate and flow strength increase with increasing the number of activated slip system in micropillars. Bi-crystal micropillars with respect to the continuity of slip systems across grain boundary were fabricated to study the effect of grain boundary on slip transmission. The high geometrical compatibility factor corresponds to a high flow strength and strain hardening rate. Experimental results provide insight into understanding mechanical response of FeCrAl alloy and developing the mechanisms-based constitutive laws for FeCrAl polycrystalline aggregates.

Keywords: FeCrAl; micropillar; dislocation; grain boundary; strain hardening

1. Introduction

Iron-chromium-aluminum (FeCrAl) alloys was initially developed by General Electric (GE) Corporation in the 1960s. This material exhibits superior high-temperature oxidation and excellent corrosion resistance, and thus were used in automobile exhaust gas purifying systems and nuclear reactors [1,2]. Single-phase FeCrAl alloys with a body centered cubic (BCC) structure plastically deform through dislocation slips. Predominate slip systems include {110}<111> and {112}<111> at room temperature [3,4]. The development of the constitutive laws for FeCrAl polycrystalline aggregates is in urgent demand for designing and predicting mechanical response of FeCrAl-made structural components. Numerous efforts were made to develop constitutive models based on the knowledge of crystal defects, such as dislocations, twins and grain boundaries [5–12]. Yielding strength of a material is determined by the glide of dislocations, in turn, the glide resistance of dislocations associated with different slip systems is the essential parameter. Strain hardening effect is mainly ascribed to dislocation interactions and/or dislocation-grain boundary interactions.

The glide resistance of dislocations, i.e., critical resolved shear stress (CRSS), can be estimated using different methods. First, CRSS can be estimated by analysis of orientation and slip trace on polycrystal sample after macro mechanical tests. In this method, the orientations of grains are analyzed by using transmission electron microscopy (TEM) [13] or electron backscatter diffraction (EBSD) [14]. The slip system of traces and Schmid factors are calculated with help of orientation information. Then the ratio of CRSS can be calculated by counting the frequency of slip traces. Coupled with in situ observation of deformation process, this method can also provide absolute estimation of CRSS [15].

However, the stress state in each grain may be different from macroscopic stress, which will lead to inaccurate measurement of CRSS. Second, CRSS can be calculated by mechanical tests and data fitting of crystal plasticity modeling. To use this method, the mechanical behavior of material is measured by using mechanical tests on polycrystalline specimens with known texture [16,17] or indentation tests on grains with measured orientation (also known as inverse indentation analysis) [18,19]. Then CRSS and other parameters in the model are fitted to the results of experiments. However, previous study shows that this method yields different CRSS for same material due to the difference in constitutive law [16,17]. The third method is based on in situ far-field high-energy X-ray diffraction microscopy (FF-HEDM) [20,21]. During mechanical tests, FF-HEDM can identify the activated slip system in each grain, measure the corresponding shear stress. This method requires expensive high energy synchrotron X-rays and is not suitable for the measurement of small irradiated region. Another method is directly applying load on single crystal with orientation favorable for a specific slip system. The orientation of grains is characterized by using EBSD. The Schmid factors in each grain can be calculated. In situ micromechanical tests can be performed on the pillars. The activated slip system can be identified by slip traces and the CRSS of slip systems can be directly obtained [22–24].

Dislocations interactions have been studied by different methods, such as dislocation dynamics simulations, molecular dynamics simulations, and experimental testing. For example, to estimate the interaction coefficients of slip systems in BCC iron, which are defined by Franciosi [25] and Devincre [26], the researchers built up the interaction matrix of slip systems by analyzing the crystal structure, and preformed dislocation dynamics simulations [27,28]. Meanwhile, molecular dynamics simulations also were used to study the interactions of dislocations [29,30]. The detailed process of dislocations interaction is closely analyzed in atomic scale. In addition to numerical simulations, dislocations interactions were also investigated by using micro and macro mechanical tests. Based on traditional macro mechanical tests, a lot of effort was made to measure the hardening effect of materials [31–34]. Most easy way is to measure the response of polycrystal materials and fit it with empirical equations, e.g., Ludwik equation [35], Hollomon equation [36], Swift equation [37], and Voce equation [38]. To get more detailed understand on the effect of slip on a specific system on other system, latent hardening tests was employed [25,39,40]. Single crystal sample was loaded in a specific direction to activate a specific primary slip system, and then loaded in another direction to activate another secondary slip system. The CRSS change of secondary slip system can be measured. In recent years, the development of micromechanical test enables us to study mechanical properties of small samples; for example, flow behavior and strain hardening rate by micro-pillar compression testing [41,42].

Micropillar compression/tension testing has been widely used to study mechanical properties of materials with sub-micron and micro-sized microstructural features [43–45], particularly for nuclear materials because limited ion penetration depths, a few microns for typical self-ions (2–10 MeV) and tens of microns for typical light ions (1–3 MeV), make micro-scale mechanical tests a necessity. In situ micromechanical testing is essential for correctly extracting experimental data, because shear instability often happens associated with local stress/strain concentration. In this study, we investigate the orientation dependence of mechanical responses of FeCrAl alloy using micromechanical testing of single-crystal and bi-crystal micropillars in a scanning electron microscopy at room temperature. Single-crystal micropillars were fabricated with designed orientation which favors the activity of single slip system or two slip systems or multiple slip systems in order to study the orientation dependence of mechanical responses of FeCrAl alloy. We evaluated the strain hardening rate by extracting the stress-strain data, i.e., $\Theta = d\sigma/d\epsilon$. By tailoring the orientation of single-crystal micro-pillars, we characterized the strain hardening behavior of micropillars associated with the activity of (a) one slip system, (b) two slip systems, and (c) multiple slip systems. Bi-crystal micropillars with respect to the continuity of slip systems across two high-angle GBs were fabricated to study the effect of grain boundary on slip transmission. These results provide insight into understanding mechanical response of FeCrAl alloy.

2. Experimental Methods

2.1. Materials

The FeCrAl alloy tested in the study has a nominal composition of Fe-13Cr-5Al-2Mo-0.2Si-0.05Y (wt.%). The ingot of alloy was prepared by arc-melting under an argon atmosphere using elemental ingredients with a purity above 99.9 at. %. The ingot was homogenized, hot-rolled at 1100 °C, warm-rolled at 800 °C in three steps from its initial thickness ~5 mm to a final thickness ~2 mm and annealed at 800 °C for 1 h. The grain size of the materials is around 50 µm. The sample was mechanically polished with silicon carbide grinding papers from grit 800# to 4000# and electropolished in a solution of 10 mL perchloric acid and 90 mL methanol at −5 °C for 15 s at 15 V. The following SEM observation and FIB fabrication were conducted on a FEI Helios NanoLab 660 dual beam system. The orientation of polished sample was characterized by using electron backscatter diffraction (EBSD). The orientation information was collected with an EBSD detector from EDAX in this dual-beam system. The OIM software was used to control the data collection process, during which a step size of 0.5 µm was adopted.

Based on the orientation mapping from EBSD, we choose the grain with specific orientation where SF of slip systems with designed interaction are maximized. On these grains, we prepared micropillars by using FIB. The acceleration voltage of Ga$^+$ beam was 30 kV. The currents for initial cut and final cut are 64 nA and 0.24 nA respectively. The height-to-diameter ratio of each pillar was 1.5–2.5. The taper angle is within 2 to 5 degrees. The in situ mechanical tests were performed by using a Hysitron PI85 PicoIndentor equipped with a 20-µm flat punch tip. The tests were displacement-controlled while the loading rate was set to reach a strain rate of 10^{-3} s^{-1}. During in situ testing, the indenter was controlled to minimize the misalignment between the tip and the top surface of the pillars, and a minimum of three tests were performed for each type of pillar to ensure the reproducibility of interested phenomena. The in situ compression testing stops when obvious slip bands can be observed on the surface of the samples.

Slip trace analysis based on self-developed Matlab program is used to identify the activated slip system in all deformed micropillars. The true stress is estimated using a full width half maximum (FWHM) approach [22,46], where the diameter in the middle of the pillar is estimated from known (1) top surface diameter, (2) taper angle, and (3) total length of the pillar. The true strain is obtained by correcting the engineering strain with the Young's modulus correction formula for a pillar [47,48].

2.2. Orientation Selection of Micro-Pillars

Three families of slip systems, {110}<111>, {112}<111>, and {123}<111>, are usually observed in BCC materials, as depicted in Figure 1a. For most BCC metals (e.g., Fe, V, and W), {110}<111> and {112}<111> slip systems are generally activated but {123}<111> slip system is rarely activated at room temperature. {123}<111> slip system is observed in Na and K at room temperature [49] and in Fe at high temperature [50]. For BCC FeCrAl alloy, the predominant slip systems at room temperature are {110}<111> and {112}<111> slip systems [51,52].

There will be 300 combinations amongst {110}<111> and {112}<111> slip systems, but only 17 configurations of interactions are unique as considering the symmetry of crystal [27]. As depicted in Figure 1b, these dislocations interactions are correspondingly classified into four types, self-, dipolar-, junction-, and collinear-interactions. Most interactions are belonging to junction interaction where both slip plane and slip direction of two slip systems are different. Dipolar interaction is the interaction between dislocations on the same slip plane with different slip vectors, and thus only happen in {110}<111> slip systems because two slip directions are available on a slip plane {110}. Collinear interaction is the interaction between dislocations on two slip planes with same slip direction. Self-interaction refers to the interaction between dislocations on same slip system.

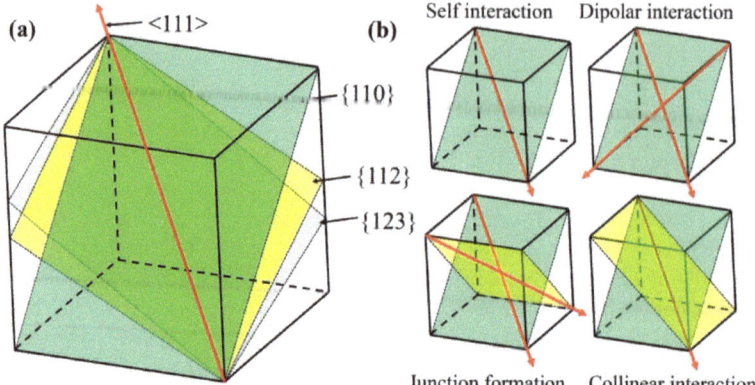

Figure 1. (a) Schematic diagram of three slip systems in a body centered cubic (BCC) unit cell; (b) Schematic diagrams show four categories of interactions between slip systems.

For single-crystal micropillars, strain gradients are generally lacking because of free surface [44,53]. When a single slip system is predominately activated in a micropillar, self-interaction is to a great extend reduced, and an obvious yielding and continuous shearing with weak hardening is generally observed. Thus, maximizing the Schmid factor (SF) of one specific slip system can be used to measure the glide resistance of dislocations associated with the slip system. Meanwhile, an apparent strain hardening behavior was expected to occur when the least two slip systems could be activated in the micropillar. Correspondingly, we choose the orientation of micropillars with the following criteria: (a) maximizing the SF of one slip systems in order to assess glide resistance of dislocations in a specific slip system, and (b) maximizing the SFs of two slip systems in order to study the effect of junction interaction on strain hardening. As a comparison, micropillars with multiple activated slip systems were fabricated and tested.

For bi-crystal micropillars, GBs generally act as pinning barriers for impeding dislocation propagation. Dislocation-GB interactions are heavily dependent on the transferability or continuity of slip systems in the two grains across a GB. Geometrical compatibility factor (m') is successfully used to estimate the deformation compatibility across GBs [54–58], where m'=cos(φ)×cos(k), φ is the angle between the slip plane normal directions and k is the angle between the slip directions. We chose bi-crystals with m' varying from 0.44 to 0.84.

3. Results and Discussion

3.1. Weak Hardening Associated with One Activated Slip System

The self-interaction of dislocations refers to the interaction between dislocations belonging to the same slip system. To study the mechanical response of micropillars with one dominant slip system, we selected grains with specific orientation in a polycrystalline sample according to the SF analysis on each slip system with Euler angle measured from EBSD mapping. As shown in Figure 2a, we chosen a grain with the orientation where [$\bar{5}\,\bar{1}\,5$] direction is parallel to the loading direction. Under uniaxial loading, the SF of (12-1)[-111] slip system reaches the maximum value of 0.5, the SF of (0-11)[-111] slip system is the second highest and equals 0.438, and SFs of other slip systems are lower than 0.428. Therefore, the (1-1)[-111] slip system would be primarily activated. Similarly, the grain with the orientation where [$\overline{20}\,\overline{9}\,2$] is parallel to the loading direction and shown in Figure 2b favors (10-1)[111] slip system, because this slip system has the maximum SF of 0.5, the second highest SF is 0.4683 associated with the slip system (101)[11-1], and SFs of other slip systems are lower than 0.435. The orientation of the two grains is demonstrated by the inverse pole figure in Figure 2c. Figure 2d,e show the pillars with initial diameters of 8.9 μm and 8.5 μm. Figure 2f,g show the compressed pillars where the apparent slip traces well match the slip traces on (12-1) plane and (10-1) plane, respectively.

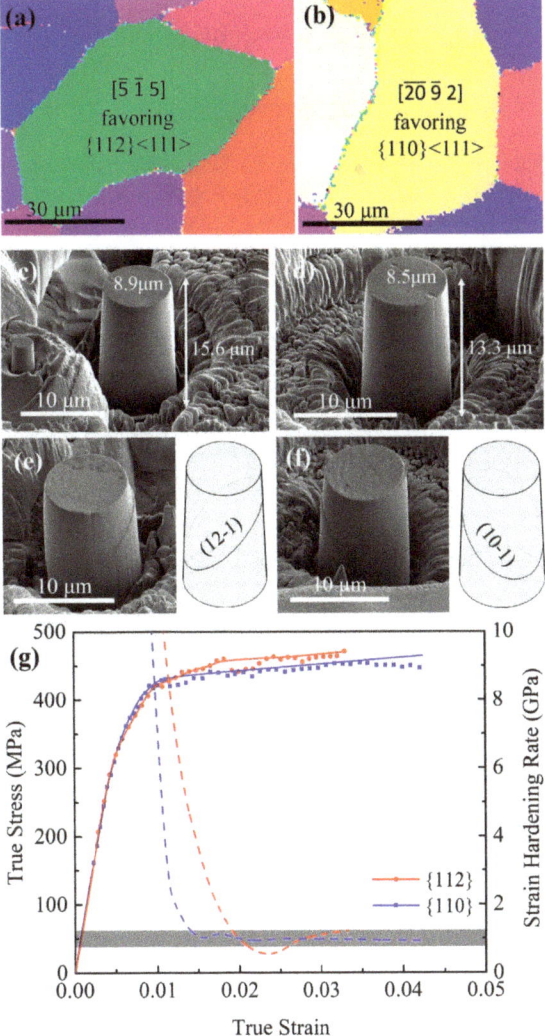

Figure 2. Electron backscatter diffraction (EBSD) IPF mappings showing the grains with the orientation of (**a**) [5̄15] favoring slip system {112}<111> and (**b**) [2̄0 9̄ 2] favoring slip system {110}<111>. (**c**,**d**) Micropillars FIBed in the two grains. (**e**,**f**) Compressed micropillars and schematic diagrams with traces. (**g**) Stress-strain curves and strain hardening rates.

We thus concluded that (12-1)[-111] slip system and (10-1)[111] slip system were obviously activated in each pillar, respectively. Figure 2h shows the typical stress-strain curves. It is noted there is no apparent strain hardening behavior once continuous plastic flow occurs. Several slight stress drops correspond to shear instability due to free surface. The weak hardening effect is ascribed to the lack of dislocation pileup because the size of the pillar is too small to form dislocation pileup. Videos 1 and 2 record the in situ compression testing of the two pillars. Corresponding to the 0.2% offset yield strengths of 460 MPa and 450 MPa, the glide resistance of dislocations are estimated to be 230 MPa and

220 MPa for the two slip systems {12-1}<-111> and {10-1}<111>, respectively. The similar resistance for the two slip systems is consistent with our previous study [52]. The strain hardening rate is around 1.0 GPa for the two slip systems {12-1}<-111> and {10-1}<111> as the strain exceeds 0.015.

3.2. Intermediate Hardening Associated with Two Activated Slip Systems

Micropillars with two orientations were prepared to study the interactions between two {1-21}<111> slip systems and between two {110}<111> slip systems, respectively. When the grain is orientated along the crystallographic direction [1 48 8], the (1-21)[111] and (-1-21)[-111] slip systems have the largest SFs of 0.494 and 0.487, and other slip systems have the SF smaller than 0.462. Figure 3a,d show two micro-pillars with diameters of 5.8 μm and 8.7 μm. Figure 3g shows the stress-strain curves, revealing apparent strain hardening with a hardening rate of ~1.9 GPa for the strain exceeding 0.02. Figure 3h schematically show two slip traces associated with the slip planes (1-21) and (-1-21). Apparent slip traces are identified to be associated with dislocation motion on the slip plane (1-21), as observed in the compressed pillars in Figure 3b,e. The slip trace associated with the slip plane (-1-21) was not obvious. Thus, we magnified the red square regions in Figure 3b,e′ which is the backside of the pillar. The high magnification images in Figure 3c,f reveal the slip traces associated with dislocation motion on the slip plane (-1-21). Thus, the slightly increased strain hardening rate is ascribed to the junction formation interaction between (1-21)[111] and (-1-21)[-111] slip systems. Video 3 records the in situ compression testing of the pillar with a diameter of 8.7 μm. Corresponding to the 0.2% offset yield strength of 430 MPa in Figure 3g, the critical resistance for dislocation glide on {121} slip planes is calculated to be 212 MPa, which is consistent with the result obtained from Section 3.1.

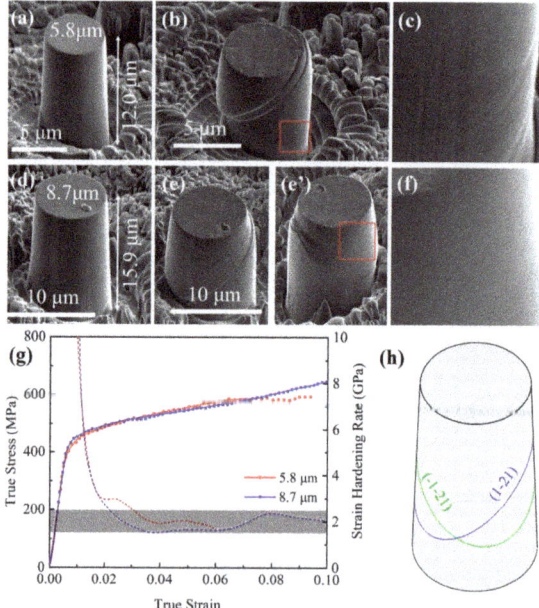

Figure 3. (**a,b**) Micropillars with a diameter of 5.8 μm before and after compression, showing apparent slip trace associated with slip plane (1-21); (**c**) the magnified image of the red square in (**b**), showing slip traces associated with slip plane (-1-21). (**d,e**) The micro-pillar with a diameter of 8.7 μm before and after compression; (**e′**) the back-side of the image (**e**); (**f**) The magnified image of the red square in (**e′**), showing slip traces associated with two slip planes (1-21) and (-1-21). (**g**) Stress-strain curves. (**h**) A schematic showing slip traces associated with slip planes (1-21) in blue and (-1-21) in green.

When the grain is orientated along the crystallographic direction [-6 9 -5], the (01-1)<1-1-1> and (-110)<-1-11> slip systems have the largest SFs of 0.399 and 0.344, and other slip systems have the SF smaller than 0.31. Figure 4a and d show two micro-pillars with diameters of 5.9 μm and 8.9 μm. Figure 4h schematically show two sets of slip traces associated with the slip planes (0 1 -1) and (-1 1 0). Apparent slip traces are associated with dislocation motion on the two slip planes (0 1 -1) and (-1 1 0), as observed in the compressed pillars in Figure 4b,e and the magnified images in Figure 4c,f. Especially in Figure 4c, the two slip systems seem equally activated. The stress-strain curves in Figure 4g show apparent strain hardening with a hardening rate around 4.0 GPa for the strain exceeding 0.02. Thus, the strain hardening rate is ascribed to the junction formation interaction between (01-1)[1-1-1] and (-110)[-1-11] slip systems, and is greater than the case associated with the interaction between two {110}<111> slip systems where one slip system is more activated. Video 4 records the in situ compression testing of the pillar with a diameter of 5.9 μm. Corresponding to the 0.2% offset yield strength of 580 MPa in Figure 4c, the critical resistance for dislocation glide is 231 MPa for (01-1)[111] slip family, which is consistent with the results in Section 3.1.

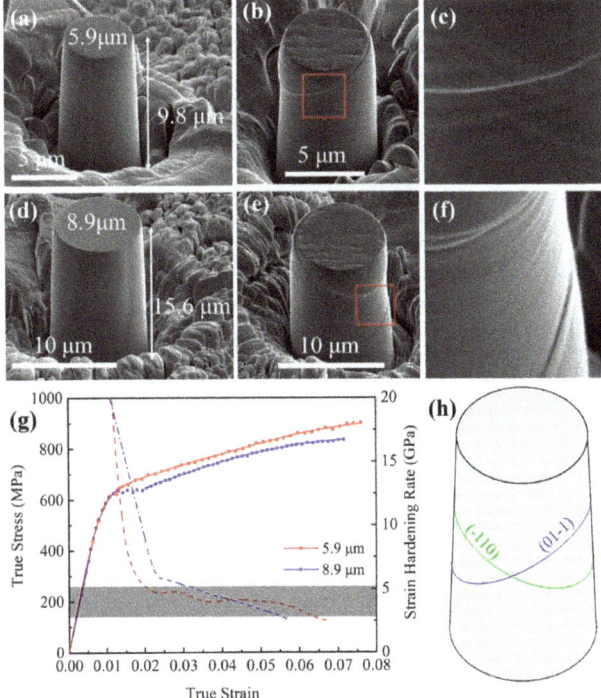

Figure 4. (**a,d**) Micro-pillars with a diameter of 5.9 μm and 8.9 μm. (**b,e**) The compressed micro-pillars. (**c,f**) The magnified images of the compressed micro-pillars, showing apparent slip traces. (**g**) Stress strain curves. (**h**) A schematic showing traces associated with slips on (-110) plane in green and (01-1) plane in blue.

3.3. Strong Hardening Behaviors Associated with Multiple Activated Slip Systems

To reveal hardening behaviors of micropillars associated with multiple activated slip systems, we fabricated two types of micropillars with the crystallographic orientations [100] and [111]. When the micropillars are oriented along the [100], four slip systems have comparable SFs around 0.47, including (-1-1-2)[11-1], (-1-12)[111], (-11-2)[1-1-1], and (-112)[1-11], as shown in Figure 5f. Figure 5a,c show the two pillars with a diameter of 5.8 μm and 8.5 μm, respectively. Figure 5b,d show the pillars at a compression strain of 0.1. Figure 5e shows the stress strain curves. Although the 0.2% offset yield strength is approximately same about 430 MPa for the two micropillars, the strain hardening behaviors are different. The micropillar with the bigger diameter of 8.5 µm shows a high strain hardening rate varying from 20 GPa to 2.0 GPa as the strain increases from 0.01 to 0.02. A relatively stable strain hardening rate is 2.0 GPa as the strain exceeds 0.025. A homogenous deformation was observed before the strain reaches 0.02, indicating multiple slip systems were activated. Video 5 records the in situ compression testing of the pillar with a diameter of 8.5 µm. The pillar with a diameter of 5.8 µm shows a low strain hardening rate varying from 5.0 GPa to 2.0 GPa for the strain from 0.01 to 0.02. Video 6 records the in situ compression testing of the pillar with a diameter of 5.8 µm, showing homogenous deformation which implies the activation of multiple slip systems. The lower strain hardening rate is thus ascribed to the smaller diameter.

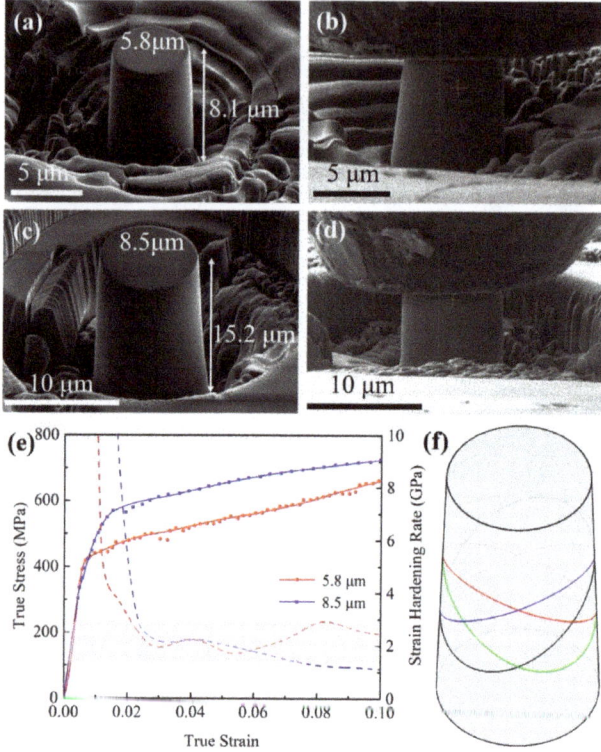

Figure 5. (a,b) The micropillar with a diameter of 5.8 µm before and after compression at a strain of 0.1. (c,d) The micropillar with a diameter of 8.5 µm before and after compression at a strain of 0.1. (e) Stress strain curves. (f) A schematic showing four slip systems, (-1-1-2)[11-1] in blue, (-1-12)[111] in green, (-11-2)[1-1-1] in red, and (-112)[1-11] in black.

A much stronger hardening behavior was observed in Figure 6a when the micropillar is oriented along the [111]. The strain hardening rate decreases from 35 GPa to 5 GPa as the strain increase from 0.01 to 0.05. Four slip systems, (1-1-2)[1-11], (1-2-1)[11-1], (10-1)[1-11] and (1-10)[11-1] as shown in Figure 6b, have comparable SFs in the range from 0.32 to 0.35, while other slip systems have SFs less than 0.29. Corresponding to the 0.2% offset yield strength of 600 MPa, these slip systems have the glide resistance of about 213 MPa which is consistent with the results from other tests. Figure 6c shows a dog-bone shaped micro-pillar with the cross-section of 3.3 µm × 3.0 µm. Figure 6d,e shows two snapshots at a strain of 0.075 and 0.085, respectively, indicating homogeneous deformation before cracking takes place. Video 7 records the in situ tension testing of the pillar, showing homogenous deformation during tension testing and implying the activation of multiple slip systems before the pillar failures at a strain of 0.08.

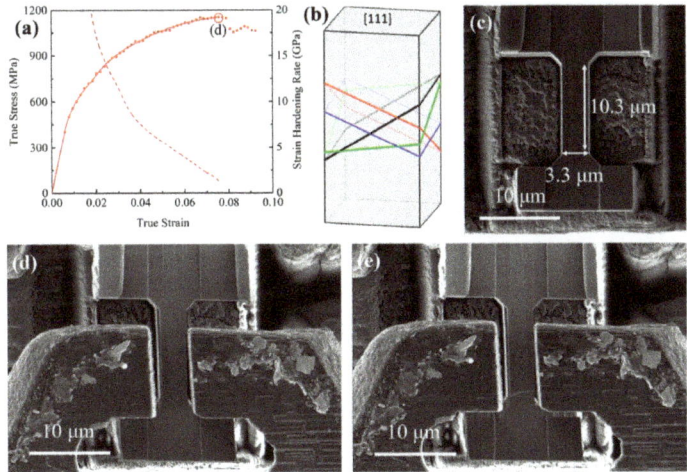

Figure 6. (**a**) Stress strain curve and strain hardening rate. (**b**) A schematic showing four slip systems, (1-1-2)[1-11] in blue, (1-2-1)[11-1] in green, (10-1)[1-11] in red, and (1-10)[11-1] in black. (**c**) A dog-bone micropillar. (**d**,**e**) Two snapshots at strain of 0.075 and 0.085, respectively.

3.4. Grain Boundary Effects on Mechanical Behavior

Grain boundary (GB) strengthening mechanism is based on the observation that GBs impede dislocation movement. Since the adjacent grains differ in orientation, it requires more energy for a dislocation to change slip direction and slip plane and transfer into the adjacent grain [59,60]. The transferability or continuity of slip systems in the two grains across a GB is described by a geometrical compatibility factor (GCF), m'=cos(φ)×cos(k) [54–58]. The bigger the GCF is, the easier slip transmission happens. φ is the angle between the slip plane normal directions and k is the angle between the slip directions. We examined this in two bi-crystal micropillars.

The bi-crystal 1 has the orientation relation: $[403]_A$//$[1]_B$, $(010)_A$//$(010)_B$ between Grains A and B. The compression direction is parallel to $[403]_A$ and $[1]_B$. The maximum SF in Grain A is 0.478 for $(121)_A$ and the second highest SF is 0.455 for $(1\bar{2}1)_A$. The maximum SF in Grain B is 0.483 for $(\bar{1}12)_B$ and the second highest SF is 0.479 for $(112)_B$. Video 8 records the in situ compression testing of the pillar, showing slip transmission across the GB. The morphology of the micropillar after compression is shown in Figure 7a. The slip traces on the surface of the deformed pillar enable us to identify the activated slip planes, $(1\bar{2}1)_A$ with a SF of 0.455 in grain A and $(0\bar{1}1)_B$ with a SF of 0.355 in grain B, as depicted in Figure 7b. We computed geometrical compatibility factors associated with any pair of slip systems $(1\bar{2}1)_A/\{011\}_B$ and $(1\bar{2}1)_A/\{112\}$ across the GB, the factor associated with the pair of slip systems $(1\bar{2}1)_A/(0\bar{1}1)_B$ is the largest, m' = 0.84, which accounts for the observed slip transmission across the GB in Figure 7a and no apparent strain hardening during the compression as shown in Figure 7e.

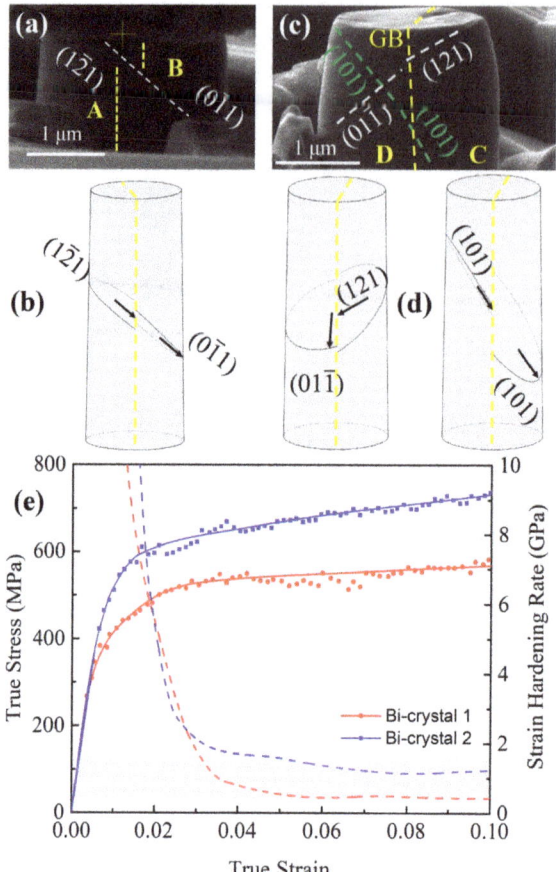

Figure 7. (**a**,**c**) Two bi-crystal micropillars after compression at a strain of 0.3. (**b**) A schematic of bi-crystal 1 showing two preferred slip systems. (**d**) Two schematics of bi-crystal 2 showing two pairs of two preferred slip systems. Yellow dashed lines indicate grain boundaries. The arrows show the slip vectors. (**e**) True stress-strain curves and strain hardening rates.

The bi-crystal 2 has the orientation relation: $[11\ 5\ 9]_C$ // $[1\bar{3}5]_D$ and $(\bar{9}\ 4\ 2)$//$(2\ 1\ \bar{10})$ between Grains C and D. The compression direction is parallel to $[11\ 5\ 9]_C$ and $[1\bar{3}5]_D$. The highest SF in Grain C is 0.466 for $(121)_C$ and the second highest SF is 0.428 for $(110)_C$. The highest SF in Grain D is 0.483 for $(2\bar{1}1)_D$ and the second highest SF is 0.482 for $(101)_D$. Video 9 records the in situ compression testing of the pillar, showing slip transmission across the GB. The morphology of the micropillar after compression in Figure 7c shows two sets of slip transmission across the GB of the bi-crystal. According to the slip traces on the surface of the deformed pillar, the activated slip systems are identified to be $(101)_D$ with a SF of 0.482 and $(101)_C$ with a SF of 0.25, as illustrated in Figure 7d. The geometrical compatibility factor associated with the two slip systems $(101)_D/(101)_C$ is 0.44. We computed the geometrical compatibility factors associated with all pairs of slip systems $(101)_D/\{110\}_C$ and $(101)_D/\{112\}_C$, the largest factor m′ = 0.74 is associated with the pair of slip systems $(101)_D/(1\bar{2}1)_C$. However, this pair of slips did not apparently activate. One possible reason is ascribed to the low SF of 0.15 associated with the slip system $(1\bar{2}1)_C$<111>. The another pair of slips are identified to be $(121)_C$ with the highest SF of 0.466 in grain C and $(01\bar{1})_D$ slip with a SF of 0.24 in grain D, as illustrated in Figure 7d. The geometrical compatibility factor associated with the two slip systems is 0.75. We computed the geometrical compatibility factors associated with all pairs of slip systems $(121)_C/\{110\}_D$ and $(121)_C/\{112\}_D$, the largest factor m' = 0.79 is associated with the pair of slip systems $(121)_C/(11\bar{2})_D$, but the slip system $(11\bar{2})_D<111>$ has a low SF of 0.20. Compared to the case of bi-crystal 1, there are two sets of slip transmissions associated with slip systems $(101)_D/(101)_C$ and $(121)_C/(01\bar{1})_D$, the pair of $(101)_D/(101)_C$ has small geometrical compatibility factor but large SFs and the pair of $(121)_C/(01\bar{1})_D$ has big geometrical compatibility factor but small SFs. Thus, the bi-crystal 2 exhibits a high flow strength, as shown in Figure 7e.

4. Conclusions

In this study, we investigated the mechanical response of single crystal and bi-crystal FeCrAl alloys using in situ micropillar compression and tension testing. According to the EBSD data, we selected grains with specific orientations which favor the type and number of activated slip systems. In order to explicitly envision the influence of the orientation and microstructures on the stress strain responses of micropillars, Figure 8 shows the mechanical responses of single crystal micropillars with one activated slip system, two activated slip systems, and multiple activated slip systems and bi-crystal micropillars. Firstly, it is expected that these micropillars will show different stress strain responses, corresponding to plastic anisotropy associated with crystal plasticity. However, we obtained a very close glide resistance of about 225 MPa for slips on {110} and {112} planes when the 0.2% offset yield strength and the largest Schmid factor were used to estimate the glide resistance. This conclusion is consistent with our systematic study in single crystal micropillars [52]. Secondly, the strain hardening rates derived from these tests clearly indicated that the weakest hardening is associated with pillars with one activated slip system, the largest hardening is observed in pillars with multiple activated slip systems. With increasing the number of activated slip systems, the strain hardening rate increases. Thirdly, the critical strain for reaching a stable strain hardening rate increases with the number of activated slip systems in micropillars and pillars' diameter. Lastly, for the bi-crystal pillars, the strain hardening rate is also related to the geometrical compatibility factor, the higher the factor is, the lower the strain hardening rate and the weaker the strengthening effect are. These results can be used to develop the mechanisms-based meso/micro/macro-scale constitutive laws for FeCrAl polycrystalline aggregates in order to accelerate designing and predicting mechanical response of structural components [61,62].

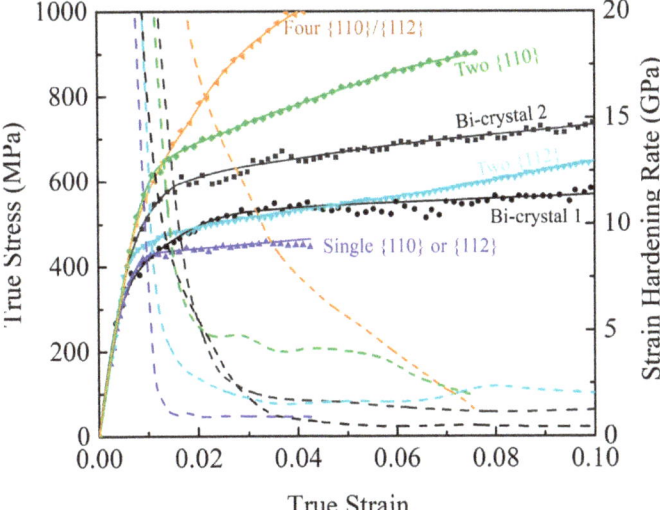

Figure 8. Comparison of stress-strain responses and corresponding strain hardening rates of micropillars with different orientations.

Author Contributions: Conceptualization and methodology, J.W.; software, D.X. and B.W.; formal analysis, D.X., B.W. and W.W.; investigation, D.X. and B.W.; resources, J.W.; data curation, D.X., B.W. and W.W.; writing—original draft preparation, D.X.; writing—review and editing, J.W.; supervision, J.W.; project administration, J.W.; funding acquisition, J.W. All authors have read and agreed to the published version of the manuscript.

Funding: This work was fully supported by the Department of Energy (DOE) Office of Nuclear Energy and Nuclear Energy University Program through Award No. DE-NEUP-18-15703.

Acknowledgments: The research was performed in part in the Nebraska Nanoscale Facility: National Nanotechnology Coordinated Infrastructure and the Nebraska Center for Materials and Nanoscience (and/or NERCF), which are supported by the National Science Foundation under Award ECCS: 2025298, and the Nebraska Research Initiative. Valuable discussions with Kaisheng Ming and Shun Xu is really appreciated.

Conflicts of Interest: The authors declare no conflict of interest.

References

1. He, Y.; Liu, J.; Han, Z.; Deng, Z.; Su, X.; Ji, Y. Phase transformation and precipitation during solidification of FeCrAl alloy for automobile exhaust gas purifying systems. *J. Alloy. Comp.* **2017**, *714*, 251–257. [CrossRef]
2. Field, K.G.; Briggs, S.A.; Edmondson, P.D.; Haley, J.C.; Howard, R.H.; Hu, X.; Littrell, K.C.; Parish, C.M.; Yamamoto, Y. *Database on Performance of Neutron Irradiated FeCrAl Alloys*; Oak Ridge National Laboratory (ORNL): Oak Ridge, TN, USA, 2016.
3. Du, C.; Maresca, F.; Geers, M.G.; Hoefnagels, J.P. Ferrite slip system activation investigated by uniaxial micro-tensile tests and simulations. *Acta Mater.* **2018**, *146*, 314–327. [CrossRef]
4. Sun, Z.; Yamamoto, Y.; Chen, X. Impact toughness of commercial and model FeCrAl alloys. *Mater. Sci. Eng. A* **2018**, *734*, 93–101. [CrossRef]
5. Peirce, D.; Asaro, R.J.; Needleman, A. An analysis of nonuniform and localized deformation in ductile single crystals. *Acta Metal.* **1982**, *30*, 1087–1119. [CrossRef]
6. Follansbee, P.S.; Kocks, U.F. A constitutive description of the deformation of copper based on the use of the mechanical threshold stress as an internal state variable. *Acta Metal.* **1988**, *36*, 81–93. [CrossRef]
7. Zerilli, F.J.; Armstrong, R.W. Dislocation-mechanics-based constitutive relations for material dynamics calculations. *J. Appl. Phys.* **1987**, *61*, 1816–1825. [CrossRef]
8. Lee, M.G.; Lim, H.; Adams, B.L.; Hirth, J.P.; Wagoner, R.H. A dislocation density-based single crystal constitutive equation. *Int. J. Plast.* **2010**, *26*, 925–938. [CrossRef]
9. Beyerlein, I.J.; Tomé, C.N. A dislocation-based constitutive law for pure Zr including temperature effects. *Int. J. Plast.* **2008**, *24*, 867–895. [CrossRef]
10. Bertin, N.; Capolungo, L.; Beyerlein, I.J. Hybrid dislocation dynamics based strain hardening constitutive model. *Int. J. Plast.* **2013**, *49*, 119–144. [CrossRef]
11. Lin, Y.C.; Wen, D.-X.; Huang, Y.-C.; Chen, X.-M.; Chen, X.-W. A unified physically based constitutive model for describing strain hardening effect and dynamic recovery behavior of a Ni-based superalloy. *J. Mater. Res.* **2015**, *30*, 3784–3794. [CrossRef]
12. Ma, A.; Roters, F.; Raabe, D. On the consideration of interactions between dislocations and grain boundaries in crystal plasticity finite element modeling—Theory, experiments, and simulations. *Acta Mater.* **2006**, *54*, 2181–2194. [CrossRef]
13. Zaefferer, S. A study of active deformation systems in titanium alloys: Dependence on alloy composition and correlation with deformation texture. *Mater. Sci. Eng. A* **2003**, *344*, 20–30. [CrossRef]
14. Li, H.; Mason, D.E.; Bieler, T.R.; Boehlert, C.J.; Crimp, M.A. Methodology for estimating the critical resolved shear stress ratios of α-phase Ti using EBSD-based trace analysis. *Acta Mater.* **2013**, *61*, 7555–7567. [CrossRef]
15. Barkia, B.; Doquet, V.; Couzinié, J.P.; Guillot, I.; Héripré, E. In Situ monitoring of the deformation mechanisms in titanium with different oxygen contents. *Mater. Sci. Eng. A* **2015**, *636*, 91–102. [CrossRef]
16. Wu, X.; Kalidindi, S.R.; Necker, C.; Salem, A.A. Prediction of crystallographic texture evolution and anisotropic stress-strain curves during large plastic strains in high purity α-titanium using a Taylor-type crystal plasticity model. *Acta Mater.* **2007**, *55*, 423–432. [CrossRef]
17. Knezevic, M.; Lebensohn, R.A.; Cazacu, O.; Revil-Baudard, B.; Proust, G.; Vogel, S.C.; Nixon, M.E. Modeling bending of α-titanium with embedded polycrystal plasticity in implicit finite elements. *Mater. Sci. Eng. A* **2013**, *564*, 116–126. [CrossRef]

18. Chakraborty, A.; Eisenlohr, P. Evaluation of an inverse methodology for estimating constitutive parameters in face-centered cubic materials from single crystal indentations. *Eur. J. Mech. A/Solids* **2017**, *66*, 114–124. [CrossRef]
19. Sánchez-Martín, R.; Pérez-Prado, M.T.; Segurado, J.; Bohlen, J.; Gutiérrez-Urrutia, I.; Llorca, J.; Molina-Aldareguia, J.M. Measuring the critical resolved shear stresses in Mg alloys by instrumented nanoindentation. *Acta Mater.* **2014**, *71*, 283–292. [CrossRef]
20. Wang, L.; Zheng, Z.; Phukan, H.; Kenesei, P.; Park, J.S.; Lind, J.; Suter, R.M.; Bieler, T.R. Direct measurement of critical resolved shear stress of prismatic and basal slip in polycrystalline Ti using high energy X-ray diffraction microscopy. *Acta Mater.* **2017**, *132*, 598–610. [CrossRef]
21. Pagan, D.C.; Shade, P.A.; Barton, N.R.; Park, J.-S.; Kenesei, P.; Menasche, D.B.; Bernier, J.V. Modeling slip system strength evolution in Ti-7Al informed by in-situ grain stress measurements. *Acta Mater.* **2017**, *128*, 406–417. [CrossRef]
22. Liu, Y.; Li, N.; Arul Kumar, M.; Pathak, S.; Wang, J.; McCabe, R.J.; Mara, N.A.; Tomé, C.N. Experimentally quantifying critical stresses associated with basal slip and twinning in magnesium using micropillars. *Acta Mater.* **2017**, *135*, 411–421. [CrossRef]
23. Gong, J.; Wilkinson, A.J. Anisotropy in the plastic flow properties of single-crystal α titanium determined from micro-cantilever beams. *Acta Mater.* **2009**, *57*, 5693–5705. [CrossRef]
24. Kim, J.-Y.; Greer, J.R. Tensile and compressive behavior of gold and molybdenum single crystals at the nano-scale. *Acta Mater.* **2009**, *57*, 5245–5253. [CrossRef]
25. Franciosi, P.; Berveiller, M.; Zaoui, A. Latent hardening in copper and aluminium single crystals. *Acta Metal.* **1980**, *28*, 273–283. [CrossRef]
26. Devincre, B.; Kubin, L.; Hoc, T. Physical analyses of crystal plasticity by DD simulations. *Scr. Mater.* **2006**, *54*, 741–746. [CrossRef]
27. Po, G.; Mohamed, M.S.; Crosby, T.; Erel, C.; El-Azab, A.; Ghoniem, N. Recent progress in discrete dislocation dynamics and its applications to micro plasticity. *JOM* **2014**, *66*, 2108–2120. [CrossRef]
28. Queyreau, S.; Monnet, G.; Devincre, B. Slip systems interactions in α-iron determined by dislocation dynamics simulations. *Int. J. Plast.* **2009**, *25*, 361–377. [CrossRef]
29. Nomoto, A.; Soneda, N.; Takahashi, A.; Ishino, S. Interaction analysis between edge dislocation and self interstitial type dislocation loop in BCC iron using molecular dynamics. *Mater. Trans.* **2005**, *46*, 463–468. [CrossRef]
30. Buehler, M.J.; Hartmaier, A.; Gao, H.; Duchaineau, M.A.; Abraham, F.F. The dynamical complexity of work-hardening: A large-scale molecular dynamics simulation. *Acta Mech. Sin.* **2005**, *21*, 103–111. [CrossRef]
31. Choudhary, B.K.; Samuel, E.I.; Bhanu Sankara Rao, K.; Mannan, S.L. Tensile stress–strain and work hardening behaviour of 316LN austenitic stainless steel. *Mater. Sci. Technol.* **2001**, *17*, 223–231. [CrossRef]
32. Sainath, G.; Choudhary, B.K.; Christopher, J.; Isaac Samuel, E.; Mathew, M.D. Applicability of Voce equation for tensile flow and work hardening behaviour of P92 ferritic steel. *Int. J. Press. Vessel. Pip.* **2015**, *132*, 1–9. [CrossRef]
33. Kocks, U.F. Laws for work-hardening and low-temperature creep. *J. Eng. Mater. Technol.* **1976**, *98*, 76–85. [CrossRef]
34. Sengupta, A.; Putatunda, S.K.; Bartosiewicz, L.; Hangas, J.; Nailos, P.J.; Peputapeck, M.; Alberts, F.E. Tensile behavior of a new single-crystal nickel-based superalloy (CMSX-4) at room and elevated temperatures. *J. Mater. Eng. Perfor.* **1994**, *3*, 73–81. [CrossRef]
35. Ludwik, P. Fließvorgänge bei einfachen Beanspruchungen. In *Elemente der Technologischen Mechanik*; Springer: Berlin/Heidelberg, Germany, 1909; pp. 11–35.
36. Hollomon, J.H. Tensile deformation. *Aime Trans.* **1945**, *12*, 1–22.
37. Swift, H.W. Plastic instability under plane stress. *J. Mech. Phys. Solid.* **1952**, *1*, 1–18. [CrossRef]
38. Voce, E. The relationship between stress and strain for homogeneous deformation. *J. Inst. Met.* **1948**, *74*, 537–562.
39. Wu, T.-Y.; Bassani, J.L.; Laird, C. Latent hardening in single crystals—I. Theory and experiments. *Proc. Roy. Soc. London. A Math. Phys. Sci.* **1991**, *435*, 1–19.
40. Wang, M.; Shi, L.; Li, C.; Wang, Z.G.; Xiao, J.M. A study of latent hardening behavior in aluminium single crystals. *Scr. Mater.* **1996**, *35*, 1183–1188.

41. Frick, C.P.; Clark, B.G.; Orso, S.; Schneider, A.S.; Arzt, E. Size effect on strength and strain hardening of small-scale [111] nickel compression pillars. *Mater. Sci. Eng. A* **2008**, *489*, 319–329. [CrossRef]
42. Kiener, D.; Guruprasad, P.J.; Keralavarma, S.M.; Dehm, G.; Benzerga, A.A. Work hardening in micropillar compression: In situ experiments and modeling. *Acta Mater.* **2011**, *59*, 3825–3840. [CrossRef]
43. Greer, J.R.; De Hosson, J.T.M. Plasticity in small-sized metallic systems: Intrinsic versus extrinsic size effect. *Prog. Mater. Sci.* **2011**, *56*, 654–724. [CrossRef]
44. Uchic, M.D.; Dimiduk, D.M.; Florando, J.N.; Nix, W.D. Sample dimensions influence strength and crystal plasticity. *Science* **2004**, *305*, 986–989. [CrossRef]
45. Uchic, M.D.; Shade, P.A.; Dimiduk, D.M. Plasticity of micrometer-scale single crystals in compression. *Ann. Rev. Mater. Res.* **2009**, *39*, 361–386. [CrossRef]
46. Budiman, A.S.; Narayanan, K.R.; Li, N.; Wang, J.; Tamura, N.; Kunz, M.; Misra, A. Plasticity evolution in nanoscale Cu/Nb single-crystal multilayers as revealed by synchrotron X-ray microdiffraction. *Mater. Sci. Eng. A* **2015**, *635*, 6–12. [CrossRef]
47. Fei, H.; Abraham, A.; Chawla, N.; Jiang, H. Evaluation of micro-pillar compression tests for accurate determination of elastic-plastic constitutive relations. *J. Appl. Mech.* **2012**, *79*, 061011. [CrossRef]
48. Ming, K.; Gu, C.; Su, Q.; Wang, Y.; Zare, A.; Lucca, D.A.; Nastasi, M.; Wang, J. Strength and plasticity of amorphous silicon oxycarbide. *J. Nucl. Mater.* **2019**, *516*, 289–296. [CrossRef]
49. Tsien, L.; Chow, Y. The glide of single crystals of molybdenum. *Proc. Roy. Soc. London. A Math. Phys. Sci.* **1937**, *163*, 19–28.
50. Barrett, C.S.; Ansel, G.; Mehl, R. Slip, twinning and cleavage in iron and silicon ferrite. *Trans. ASM* **1937**, *25*, 702.
51. Sun, Z.; Yamamoto, Y. Processability evaluation of a Mo-containing FeCrAl alloy for seamless thin-wall tube fabrication. *Mater. Sci. Eng. A* **2017**, *700*, 554–561. [CrossRef]
52. Xu, S.; Xie, D.; Liu, G.; Ming, K.; Wang, J. Quantifying the resistance to dislocation Gglide in single phase FeCrAl alloy. *Int. J. Plast.* **2020**, *132*, 102770. [CrossRef]
53. Greer, J.R.; Nix, W.D. Nanoscale gold pillars strengthened through dislocation starvation. *Phys. Rev. B* **2006**, *73*, 245410. [CrossRef]
54. Luster, J.; Morris, M. Compatibility of deformation in two-phase Ti-Al alloys: Dependence on microstructure and orientation relationships. *Metal. Mater. Trans. A* **1995**, *26*, 1745–1756. [CrossRef]
55. Kacher, J.; Sabisch, J.E.; Minor, A.M. Statistical analysis of twin/grain boundary interactions in pure rhenium. *Acta Mater.* **2019**, *173*, 44–51. [CrossRef]
56. Wang, L.; Eisenlohr, P.; Yang, Y.; Bieler, T.; Crimp, M. Nucleation of paired twins at grain boundaries in titanium. *Scr. Mater.* **2010**, *63*, 827–830. [CrossRef]
57. Xin, R.; Liu, Z.; Sun, Y.; Wang, H.; Guo, C.; Ren, W.; Liu, Q. Understanding common grain boundary twins in Mg alloys by a composite Schmid factor. *Int. J. Plast.* **2019**, *123*, 208–223. [CrossRef]
58. Beyerlein, I.J.; Wang, J.; Kang, K.; Zheng, S.; Mara, N. Twinnability of bimetal interfaces in nanostructured composites. *Mater. Res. Lett.* **2013**, *1*, 89–95. [CrossRef]
59. Wang, J. Atomistic simulations of dislocation pileup: Grain boundaries interaction. *JOM* **2015**, *67*, 1515–1525. [CrossRef]
60. Molodov, K.D.; Molodov, D.A. Grain boundary mediated plasticity: On the evaluation of grain boundary migration—Shear coupling. *Acta Mater.* **2018**, *153*, 336–353. [CrossRef]
61. Liu, G.; Xie, D.; Wang, S.; Misra, A.; Wang, J. Mesoscale crystal plasticity modeling of nanoscale Al–Al2Cu eutectic alloy. *Int. J. Plast.* **2019**, *121*, 134–152. [CrossRef]
62. Wang, H.; Wu, P.; Wang, J.; Tomé, C.N. A crystal plasticity model for hexagonal close packed (HCP) crystals including twinning and de-twinning mechanisms. *Int. J. Plast.* **2013**, *49*, 36–52. [CrossRef]

Publisher's Note: MDPI stays neutral with regard to jurisdictional claims in published maps and institutional affiliations.

© 2020 by the authors. Licensee MDPI, Basel, Switzerland. This article is an open access article distributed under the terms and conditions of the Creative Commons Attribution (CC BY) license (http://creativecommons.org/licenses/by/4.0/).

Article

Mechanical Behavior of Al–Al₂Cu–Si and Al–Al₂Cu Eutectic Alloys

Qian Lei [1,2], Jian Wang [3,*] and Amit Misra [1,*]

[1] Department of Materials Science and Engineering, College of Engineering, University of Michigan-Ann Arbor, Ann Arbor, MI 48109, USA; leiqian@csu.edu.cn
[2] State Key Laboratory for Powder Metallurgy, Central South University, Changsha 410083, China
[3] Department of Mechanical and Materials Engineering, University of Nebraska-Lincoln, Lincoln, NE 68588, USA
* Correspondence: jianwang@unl.edu (J.W.); amitmis@umich.edu (A.M.)

Abstract: In this study, laser rapid solidification technique was used to refine the microstructure of ternary Al–Cu–Si and binary Al–Cu eutectic alloys to nanoscales. Micropillar compression testing was performed to measure the stress–strain response of the samples with characteristic microstructure in the melt pool regions. The laser-remelted Al–Al₂Cu–Si ternary alloy was observed to reach the compressive strength of 1.59 GPa before failure at a strain of 28.5%, which is significantly better than the as-cast alloy with a maximum strength of 0.48 GPa at a failure strain of 4.8%. The laser-remelted Al–Cu binary alloy was observed to reach the compressive strength of 2.07 GPa before failure at a strain of 26.5%, which is significantly better than the as-cast alloy with maximum strength of 0.74 GPa at a failure strain of 3.3%. The enhanced compressive strength and improved compressive plasticity were interpreted in terms of microstructural refinement and hierarchical eutectic morphology.

Keywords: rapid solidification; compression; micropillar

1. Introduction

Metal-based eutectic composites exhibit periodic microstructures with lamellar, fibrous, degenerate, or other morphologies and are widely studied as model systems for understanding microstructure evolution under solidification conditions [1]. These microstructures typically contain intermetallic or other hard phases, which results in high hardness, suggesting potential for technological applications at ambient and elevated temperatures [2,3]. A limitation for all the as-cast metal-intermetallic eutectic systems has been the lack of plastic deformability [2,3], often even under compression loading. Recent work on a chill cast or laser surface-remelted eutectics has shown that high strength at room temperature can be achieved without loss of plastic deformability in a wide range of refined eutectic microstructures in Al–, Ti–, Ni–, Zn–, and Fe-based alloys [4–16].

In our earlier work, in situ micro-pillar compression testing in a scanning electron microscope (SEM) was used to characterize the stress–strain response of laser surface remelted Al–Al₂Cu eutectic alloys [15]. For micro-pillar compression from within a single eutectic colony, the plasticity mechanisms were observed to depend strongly on the orientation of the lamellae with respect to the loading direction. For compression axis 90° to the lamellae, the highest yield strength was observed with negligible global plasticity due to localized shear failure, whereas for compression axis close to 0° to the lamellae, buckling and kinking occurred. In the case of a loading axis near 45° to lamellae, sliding along Al–Al₂Cu lamellar interfaces was reported with the lowest yield strength [15]. Polycrystalline eutectics with the inter-lamellar spacing of ~20 nm exhibited a maximum flow strength of approximately 1.6 GPa and plasticity of approximately 17%. Bimodal morphologies in Al–Al₂Cu containing degenerate, wavy, morphology dispersed with nanoscale lamellar eutectic resulted in high plasticity that was uniformly distributed throughout the height of the compression sample at a strength level of 1.36 GPa [15]. The role of the mixed and bimodal morphology in promoting high strength and plasticity was not clearly elucidated, particularly for binary vs. ternary systems.

In this work, laser surface remelting was used to refine the eutectic morphology of Al–Al$_2$Cu–Si eutectic and compared with binary Al–Al$_2$Cu, processed and tested with the same approach. A focused ion beam (FIB) was employed to fabricate cylindrical micro-pillars from laser-remelted alloys. Nano-indenter was used to compress the micropillars. The compression behaviors of the micropillars were investigated systematically; the cracking and co-deformation mechanism of the binary and ternary eutectics are discussed below.

2. Experimental Procedure

The Al–Al$_2$Cu–Si (Al$_{81}$Cu$_{13}$Si$_6$, at%) and Al–Al$_2$Cu (Al$_{81}$Cu$_{19}$, at%) eutectic as-cast ingots were fabricated by an arc-melter at the Materials Preparation Center, Ames Laboratory, Iowa State University, under a protective argon gas environment. Plates with dimensions of $20 \times 10 \times 2$ mm^3 were cut from the as-cast ingot. The plates were mechanically ground and polished. Laser surface processing experiments were conducted on a solid-state disk laser (TRUMPF Laser HLD 4002 (Trumpf Laser, Plymouth, MI, USA) at a wavelength of 1.03 µm. Details of synthesis and laser re-melting have also been published elsewhere [17,18]. In an earlier study [18], the effects of the processing parameters on the microstructures of Al–Cu–Si ternary eutectic were reported. The laser power was varied between 500 and 1000 W and the laser spot diameter and scan speed ranged from 0.8 to 1.2 mm and 2.54 to 101.4 mm/s, respectively. In this study, samples processed with nominal parameters of 500 W laser power, 2.54 mm/s scan speed, and 1 mm spot size were used for micropillar compression testing. Argon shielding gas (flow rate of 9.4 L/min) was used during the laser melting process to prevent oxidation. The micropillar compression tests were performed on a TI-750 TriboIndenter (Bruker, Minneapolis, MN, USA) with a conical indenter in displacement-control mode, and the experimental compressive strain rate was $2.5 \times 10^{-2} \cdot s^{-1}$. An FEI Helios 650 NanoLab dual-beam scanning electron microscope (SEM) was used to capture the microstructures of the eutectic micropillars before and after compression tests. Thin foil samples fabricated by FIB were studied on a JEOL 3011 transmission electron microscope (TEM) with an operating voltage of 300 kV.

3. Experimental Results

Figure 1 shows the cross-section view of the laser-remelted molten and local regions in Al–Al$_2$Cu and Al–Al$_2$Cu–Si eutectics. As shown in Figure 1a, the as-cast Al–Al$_2$Cu eutectic structures can be divided into lamellae and degenerates, resulting from different heat transfer characteristics during solidification [9]. However, only lamellae Al–Al$_2$Cu eutectics can be observed in the laser-remelted sample (Figure 1b). In the backscattered electron (BSE) images (Figure 1a,b), the phases in the bright and dark contrasts are Al$_2$Cu and Al, respectively. The inter-lamellar spacing (λ) was statistically analyzed and was observed to decrease from 1400 ± 200 nm (as-cast) to 39 ± 4 nm (laser-treated region). The refined Al–Al$_2$Cu eutectics should be ascribed to the high heating and cooling rate of the laser surface processing technique [7,8,13]. In the Al–Al$_2$Cu–Si eutectics, Al (dark), Al$_2$Cu (bright), and Si (gray) were detected in the as-cast Al–Al$_2$Cu–Si samples (Figure 1c), while there were bimodal structures with dendrite structure of Al (dark) + Al$_2$Cu (bright) and lamellae structure (Al–Al$_2$Cu–Si) in the laser-remelted Al$_2$Cu–Si eutectics (Figure 1d).

Four Al–Al$_2$Cu–Si micropillars cut from the laser-treated (LT) region and as-cast base are named laser-remelted and as-cast lamellae, respectively. Three Al–Al$_2$Cu micropillars cut from the laser-treated (LT) region and as-cast base are labelled as-cast lamellae, as-cast degenerate, and laser-remelted lamellae. The applied compression force (F) and the immediate height (h$_I$) of the pillar were recorded. An average diameter (d) of the top, middle, and bottom of each pillar was employed to calculate the stress and the strain. The engineering stress (σ_E) was computed by dividing the applied force (F) by the average cross-section area ($\pi \times d \times d/4$), $\sigma_E = F/(\pi \times d \times d/4)$. The engineering strain (ε_E) was obtained by dividing the height variation of the pillar by the original height of the pillar (h_0), $\varepsilon_E = (h_I-h_0)/h_0$. The engineering stress (σ_E) versus engineering strain (ε_E) curves of the Al–Al$_2$Cu–Si and Al–Al$_2$Cu eutectic micro-pillars are depicted in Figure 2, in which the as-cast Al–Al$_2$Cu + Si eutectics exhibit ultimate engineering stress of 480 MPa and a total strain to fracture of 4.8%, while the laser-treated Al–Al$_2$Cu + Si eutectics with bimodal structures exhibit both the highest ultimate engineering stress of 1586 MPa and the highest fracture strain of 28.5%. The as-cast Al–Al$_2$Cu lamellae and degenerate eutectics exhibit ultimate engineering stress of 742 MPa and 1180 MPa and total strains to fracture of 3.3% and 5.6% respectively. The laser-treated Al–Al$_2$Cu lamellae exhibit ultimate engineering stress of 2075 MPa and a total strain to fracture of 26.5%, as summarized in Table 1. This suggests that laser surface processing can effectively improve both the strength and plastic deformability of Al–Al$_2$Cu–Si and Al–Al$_2$Cu eutectics. The high strength of laser-treated Al–Al$_2$Cu lamellae eutectics should be ascribed to the nanoscale lamellar spacing. Decreasing the inter-lamellar spacing results in more Al–Al$_2$Cu phase boundaries (PB) being introduced, facilitates the dislocation-PB interactions, and affords more room for dislocation storage, which sustains more pronounced strain hardening in the Al–Al$_2$Cu eutectics [16,19].

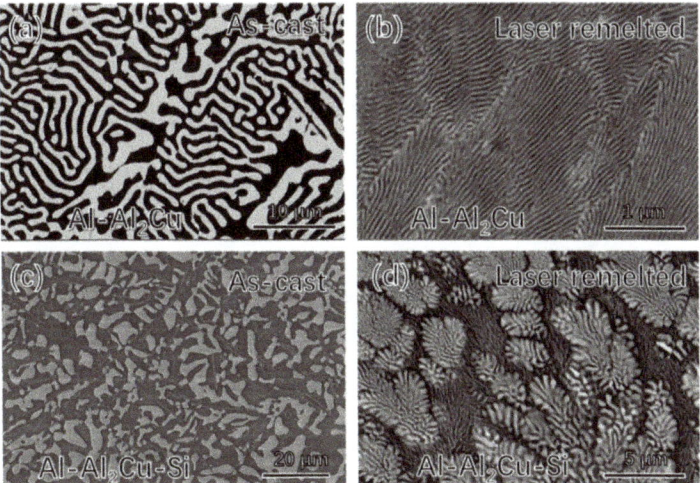

Figure 1. Backscattered electron (BSE) images of the microstructure of as-cast Al–Al$_2$Cu and Al–Al$_2$Cu–Si eutectics; (**a**) as-cast Al–Al$_2$Cu; (**b**) laser-remelted Al–Al$_2$Cu; (**c**) as-cast Al–Al$_2$Cu–Si; (**d**) laser-remelted Al–Al$_2$Cu–Si.

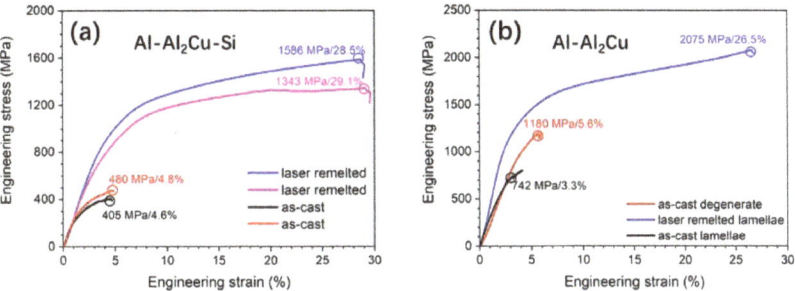

Figure 2. Compressive engineering stress—engineering strain curves of micropillars with different microstructures: (**a**) Al–Al$_2$Cu–Si eutectic; (**b**) Al–Al$_2$Cu eutectic. The strain to fracture and the maximum compress flow stress before fracture are presented.

Table 1. Microstructure and compressive properties of the studied eutectic alloys.

Eutectics	Processing	Morphology	Strain to Fracture (%)	Maximum Strength (MPa)
Al–Al$_2$Cu	As-cast	Lamellae	3.3	742
Al–Al$_2$Cu	As-cast	Degenerate	5.6	1180
Al–Al$_2$Cu	Laser-remelted	Lamellae	26.5	2075
Al–Al$_2$Cu–Si	As-cast	Mixed	4.8	480
Al–Al$_2$Cu–Si	As-cast	Mixed	4.6	405
Al–Al$_2$Cu–Si	Laser-remelted	Bimodal	29.1	1343
Al–Al$_2$Cu–Si	Laser-remelted	Bimodal	28.5	1586

Figure 3 shows the macro- and micro-structures of the Al–Al$_2$Cu–Si pillars before and after compression tests. Microstructures of the as-cast Al–Al$_2$Cu–Si eutectic micropillars before and after the compression test are shown in Figure 3a1,a2. Microstructures of the laser-treated Al–Al$_2$Cu–Si eutectic micropillars before and after the compression test are shown in Figure 3b,c. Micro-cracks were detected in the compressed as-cast Al–Al$_2$Cu–Si eutectic micropillars (Figure 3a2–a4). The detected micro-cracks on the Si and Al$_2$Cu phases indicated that the stress concentration occurred on the boundaries between Si and the other as-cast eutectics, while there were no micro-cracks detected in the laser-treated Al–Al$_2$Cu+Si eutectic micropillars (Figure 3b2–c4). The laser-treated Al–Al$_2$Cu–Si eutectics presented good deformation capability.

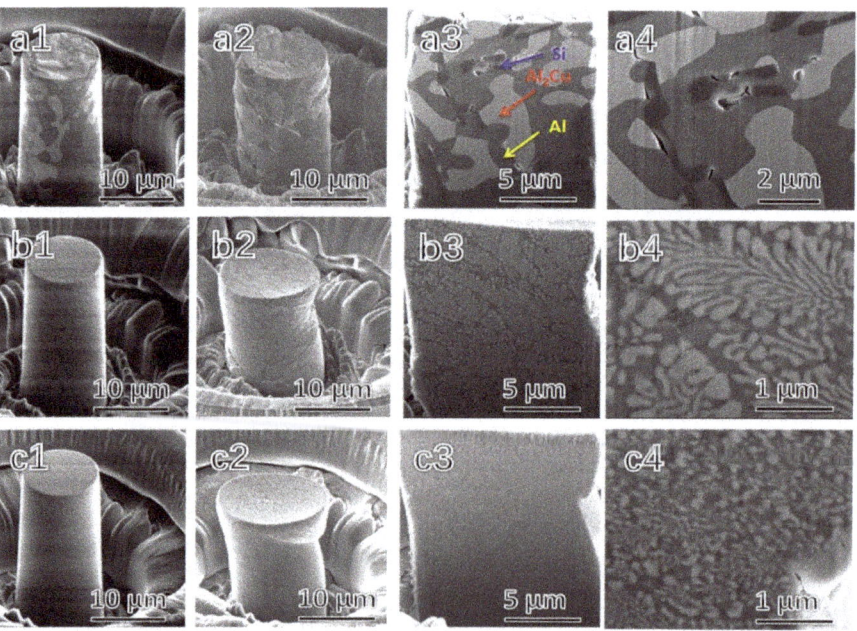

Figure 3. Marco- and micro-structures of the Al–Al$_2$Cu–Si pillars before and after compression test; (**a1–c1**) the as-cast, laser treated, refined laser remelted pillars before compression; (**a2–c2**) the pillars after compression; (**a3–c3**) the cross-section of the pillars after compression; (**a4–c4**) the high magnification images of the images from a3–c3. All the images were captured in the SEM with a tilt angle of 52°.

Figure 4 shows the macro- and micro-structures of the Al–Al$_2$Cu pillars before and after compression tests. The phases in the bright and dark contrasts are Al$_2$Cu and Al in the BSE images (Figure 4) and TEM image (Figure 4c3). Before compressive tests, the size and shape of these micro-pillars are nearly identical. Both lamellae and degenerate as-cast eutectics show the distinct interface of α–Al and θ–Al$_2$Cu phases (Figure 4a1,b1); however, the interface cannot be distinguished clearly for the laser-treated lamellae eutectics because of their nanoscale inter-lamellar spacing (Figure 4c1). The microstructures of the as-cast Al–Al$_2$Cu lamellar eutectics micropillars before and after the compression tests are shown in Figure 4a1,a2, and Figure 4a3 is the section view of the compressed pillar. Micro-cracks were detected on the Al$_2$Cu lamellae inside the compressed micropillar (Figure 4a2,a3), and two voids were formed inside the Al$_2$Cu (Figure 4a3). The microstructures of the laser-treated Al–Al$_2$Cu+Si degenerate eutectics micropillars before and after the compression tests are shown in Figure 4b1,b2. Micro-cracks were also detected on the Al$_2$Cu lamellae. The microstructures of the laser-treated Al–Al$_2$Cu+Si eutectics micropillars before and after the compression test are shown in Figure 4c1,c2. There were no micro-cracks detected in the laser-treated Al–Al$_2$Cu eutectics micropillars (Figure 4c2). Further TEM characterization in Figure 4c3 revealed that cracking in θ–Al$_2$Cu phase was suppressed in the laser-treated nanoscale Al–Al$_2$Cu eutectics; α–Al and θ–Al$_2$Cu phases were observed to co-deform together with some distinct dislocation transmission across Al-Al$_2$Cu interfaces, as pointed out by red arrows. After compressive tests, apparent cracks could be observed in θ–Al$_2$Cu layers for as-cast lamellae eutectics with 3.22% reduction and as-cast degenerate eutectics with 4.69% reduction. A similar phenomenon was observed in our previous works, in which after indention tests, Al–Al$_2$Cu eutectics with microscale inter-lamellar spacing exhibited cracking in the θ–Al$_2$Cu phase, owing to their brittle nature at room temperature [2,3]. The angles of the crack propagation direction were detected to be 48.5° and 47.7° tilted along the loading direction for as-cast lamellae and as-cast degenerated, respectively (Figure 4a2,b2). However, no crack was detected in the laser-treated lamellae eutectics with a 10.36% reduction. The microstructure of the compressed micropillars indicated that the laser-treated Al–Al$_2$Cu eutectics present much higher deformation capability than the as-cast Al–Al$_2$Cu eutectics. The high strength of laser-treated lamellae should be ascribed to the nanoscale lamellar spacing. Decreasing the inter-lamellar spacing results in more Al–Al$_2$Cu phase boundaries (PB) being introduced, facilitates the dislocation-PB interactions, and affords more room for dislocation storage, which sustains more pronounced strain hardening in the Al–Al$_2$Cu eutectics [19].

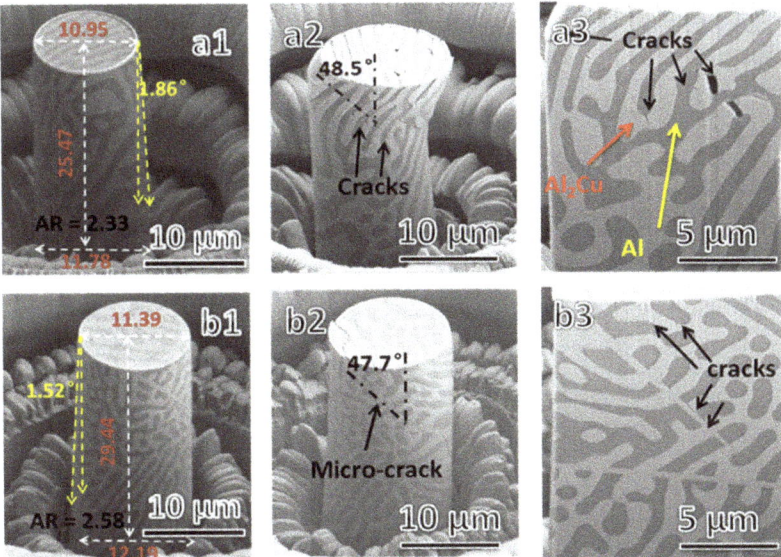

Figure 4. *Cont.*

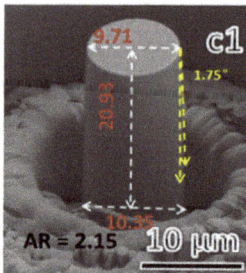
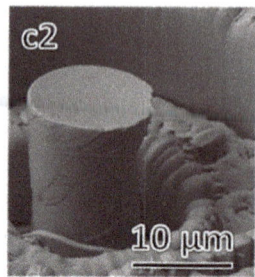
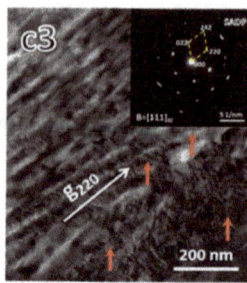

Figure 4. Macro- and micro-structures of micropillars before and after compressive tests from the Al–Al$_2$Cu eutectics with different structures of (**a**) as-cast lamellae, (**b**) as-cast degenerate, and (**c**) laser-remelted lamellae; (**a1–c1**) the micropillars before compression; (**a2–c2**) the micropillars after compression; (**a3,b3**) the cross-section views; (**c3**) the TEM image of the cross-section view. All images except c3 were captured in the SEM with a tilt angle of 52°.

4. Discussion

4.1. Strengthening Mechanisms

To further understand the strengthening mechanism of the laser-treated Al–Al$_2$Cu–Si and Al–Al$_2$Cu eutectics, it is essential to consider the roles of interfaces (soft Al phase and hard Al$_2$Cu or Si phase) when the interspacing is refined to nanometer size. In this case, the empirical rule of the mixture is invalid. In this work, the volume fractions of each phase in the studied eutectics are the same even though the length scale has been changed.

A classical constitutive model of the composite structure containing the soft phase and the hard phase could be written as follows:

$$\sigma_m = E_m \cdot \varepsilon_m, \quad \sigma_m < \sigma_m^{yield} \tag{1}$$

$$\sigma_m = E_m \cdot \varepsilon_m + K_m \cdot \varepsilon_m, \quad \sigma_m > \sigma_m^{yield} \tag{2}$$

where the subscript (m) infers to either soft phase (s) or hard phase (h). Here, the uniform stress at the ends of the hard phase, and the soft phase is unaltered with the displacement. It is important to convert the stress and strain into a more fundamental quantity (e.g., energy density). For individual phase, the total work density can be expressed by integrating the area under the stress–strain curve as follows:

$$W_m = \frac{1}{2} \frac{\left(\sigma_m^{yield}\right)^2}{E_m} + \sigma_m^{yield} \cdot \left(\frac{\sigma_m^{yield}}{E_m} - \varepsilon_{max}\right) + \int_{\frac{\sigma_m^{yield}}{E_m}}^{\varepsilon_{max}} [K_m(\varepsilon_m)^{n_m}] - \sigma_i^{yield}] \cdot d\varepsilon - \frac{1}{2} \frac{P_{max}}{A}(\varepsilon_{max} - \varepsilon_{pi}) \tag{3}$$

where the subscript (m) infers to either the soft phase (s) or the hard phases (h). ε_p, ε_{max}, and P_{max} are the plastic parts of the true-strain and maximum true-strain induced in the experiment and the corresponding maximum indentation load, respectively. We hypothesize that the energies of interaction at the lamellar interfaces (soft Al and hard Al$_2$Cu or Si) play significant roles in the tensile test's deformation process. The total work density of the composites ($\overline{W}$) can be divided into a plastic part and an elastic part, with respective interfacial energy terms (denoted by superscript "interface"), as follows:

$$\overline{W} = \left[W_e^h + W_e^s + W_e^{interface}\right] + \left[W_p^h + W_p^s + W_p^{interface}\right] \tag{4}$$

The elastic part of the energy of the effective composite equates to the total elastic energy, since there is no other phase combination, therefore a corresponding equation can be written as follows:

$$\frac{1}{2} \cdot \overline{E} \cdot \varepsilon^2 (h_s + h_h) = \frac{1}{2} E_m \varepsilon^2 h_s + \frac{1}{2} E_f \varepsilon^2 h_h + W_e^{interface}, \quad \varepsilon \ll \frac{\sigma_i^{yield}}{E_i} \tag{5}$$

where h_s is the thickness of the soft phase (Al) and h_h is the thickness of the hard phase (Al$_2$Cu or Si), and $W_e^{interface}$ is the elastic energy contribution of the interface. The inter-lamellar spacing, $\lambda=(h_1+h_2)/2$, becomes the function of the lamellar geometry. The plastic part of the energy of the composite can be equated with total plastic energy due to the pure phase combination, therefore a corresponding equation can be written as follows:

$$\left[\overline{\sigma}^{yield}\varepsilon_e + \int_{\varepsilon_e}^{\varepsilon}\overline{K\varepsilon^n}d\varepsilon\right](h_1+h_2) \cong \left[\overline{\sigma}_m^{yield}\varepsilon_e^m + \int_{\varepsilon_e}^{\varepsilon}\overline{K_m(\varepsilon_m)^{n^m}}d\varepsilon\right]h_1 + \left[\overline{\sigma}_f^{yield}\varepsilon_e^f + \int_{\varepsilon_e}^{\varepsilon}\overline{K_f(\varepsilon_f)^{n^f}}d\varepsilon\right]h_2 + W_p^{interface} \quad (6)$$

Combining the above two energy equations as described in Equation (4), a total energy balance is obtained. This allows us to evaluate the effect of the length scale of microstructure as expressed through k on the mechanical properties. Therefore, with the knowledge of inter-lamellar spacing k and the properties of the composite and pure phase, one can estimate the contribution of the interface to the overall elastic energy balance.

In order to verify the above relationships, the Al–Al$_2$Cu+Si eutectic was selected as an example. This eutectic comprises alternate plates of soft phase (Al) and hard phase (Al$_2$Cu and Si) intermetallic phases due to the variation in the lamellar spacing with the cooling rate. The mechanical properties of localized regions were further evaluated from a load-displacement curve obtained by the use of micropillar compression technique. The measured true-stress versus true-strain curves are shown in Figure 2.

4.2. Fracture Mechanisms

Eutectic alloys often exhibit poor fracture strength, which hinders their structural applications [9]. Recent studies have revealed significant improvements in fracture strength due to refinements of the microstructure [10–12]. Laser surface processing is an effective way to refine the microstructure, and it leads to a reduction in interlamellar spacing; it is evidence that microstructure refinement improves the plastic or fracture strain. These findings can be explained by two kinds of theoretical methods: one is a dislocation pile-up-based approach, and the other is a fracture-based approach. According to the dislocation-based approach, the deformation of eutectics is controlled by dislocation pile-ups at the interphase boundaries [20–22]. The interface between the soft phase (Al) and hard phase (Al$_2$Cu or Si) provides an obstacle for dislocation pile-ed up, which controlled the plasticity in this model. The interface hardening controls the yield strength. The stress arises irrespective of the detailed mechanism of the slip, since the plasticity is highly localized in such cases; here, the stress at the tip of the pile-up is given by [2].

$$\tau_P = \frac{K \cdot D \cdot \varepsilon_p}{b \cdot \sqrt{2\lambda}} \quad (7)$$

where K is the constant, D is the distance between the two dislocations, b is the Burger vector ($b = 0.285$ nm), λ is the lamellar spacing, and ε_p is the strain. Equation (6) indicates that the shear stress increases at a given strain with the decrease of inter-lamellar spacing, resulting in improvements in the strength of the eutectic alloy. This agrees well with the experimental data in this work.

Fracture is thought to be controlled by dislocation and interface behavior [23]. Griffith relation was applied in the other model based on crack propagation [3]. The critical crack size for cleavage crack propagation can be calculated as follows:

$$\sigma_F = \left[\frac{4 \cdot E \cdot \gamma}{\pi \cdot a \cdot (1-v^2)}\right]^{0.5} \quad (8)$$

where σ_F is the fracture stress, E is the Young's modulus, γ is the surface energy (219.5 mJ/m^2 [24], a is the critical crack length, and v is the Poisson's ratio ($v_{Al} = 0.35$ and $v_{Al2Cu} = 0.34$). Here, the Al–Al$_2$Cu eutectic is taken as an example for the analysis of the fracture mechanism. The Al$_2$Cu has a higher modulus (103 GPa) compared to Al (69 GPa). Substituting the above value, the fracture stress could be calculated: for Al lamellar, $\sigma_F^{Al} = 101.34a^{-1/2}$, while for Al$_2$Cu, $\sigma_F^{Al2Cu} = 180.415a^{-1/2}$. Thus, the fracture stress of Al$_2$Cu lamellar is larger than Al lamellar. As the eutectic micropillars were compressed, both Al lamellar and Al$_2$Cu lamellar were compressed, leading to plastic deformation. Since plastic deformation in Al lamella is easier than that in Al$_2$Cu lamella, the load transfers to Al$_2$Cu lamella; correspondingly, cracks were observed in Al$_2$Cu lamella (Figure 4a2,a3). For bulk materials, the critical crack length is mostly around a few millimeters to tens of millimeters, while the eutectic microstructure had two kinds of different ductility phases, whose critical crack length was limited by the inter-lamellar spacing. Supporting that the critical crack length is equal to the inter-lamellar spacing, the fracture stress increased dramatically as the inter-lamellar spacing was reduced from microscale to nanoscale. Thus, the laser-treated Al–Al$_2$Cu–Si and Al–Al$_2$Cu eutectics exhibit higher fracture strength and higher total strain to failure.

In summary, laser surface processing was conducted on Al–Al$_2$Cu–Si and Al–Al$_2$Cu eutectics to refine and manipulate the microstructures. Both laser-remelted ternary Al–Al$_2$Cu–Si and binary Al–Al$_2$Cu eutectics showed high strengths with maximum compressive strength of 1586 MPa and 2075 MPa and improved compressive plasticity with a failure strain of around 26%. By comparison, the as-cast ternary and binary eutectics with coarse microstructures exhibited low strength (lower than 740 MPa) and poor compressive plasticity (failure strain less than 5%). The enhanced compressive strength and improved compressive plasticity were interpreted in terms of microstructural refinement and hierarchical eutectic morphology. Laser-processed nanoscale eutectics show a bright promise of achieving ultra-high strength without loss of plastic deformability.

Author Contributions: Writing—original draft, Q.L.; Writing—review & editing, J.W. and A.M. All authors have read and agreed to the published version of the manuscript.

Funding: This research was funded by U.S. Department of Energy: DE-SC0016808.

Institutional Review Board Statement: Not applicable.

Informed Consent Statement: Not applicable.

Data Availability Statement: Not applicable.

Acknowledgments: This research was sponsored by DOE, Office of Science, Office of Basic Energy Sciences, grant DE-SC0016808. The authors acknowledge the assistance of J. Mazumder, B.P. Prashanth, and Y.C. Wang at the University of Michigan in material synthesis and laser surface treatment. Nanomechanical testing and electron microscopy were performed at the Michigan Center for Materials Characterization at the University of Michigan.

Conflicts of Interest: The authors declare no conflict of interest.

References

1. Chanda, B.; Potnis, G.; Jana, P.P.; Das, J. A review on nano-/ultrafine advanced eutectic alloys. *J. Alloy. Compd.* **2020**, *827*, 154226. [CrossRef]
2. Cantor, B.; Chadwick, G.A. The tensile deformation of unidirectionally solidified Al–Al$_3$Ni and Al–Al$_2$Cu eutectics. *J. Mater. Sci.* **1975**, *10*, 578–588. [CrossRef]
3. Cantor, B.; May, G.J.; Chadwick, G.A. The tensile fracture behavior of the aligned Al–Al$_3$Ni and Al–CuAl$_2$ eutectics at various temperatures. *J. Mater. Sci.* **1973**, *8*, 830–838. [CrossRef]
4. Park, J.M.; Mattern, N.; Kühn, U.; Eckert, J.; Kim, K.B.; Kim, W.T.; Chattopadhyay, K.; Kim, D.H. High-strength bulk Al-based bimodal ultrafine eutectic composite with enhanced plasticity. *J. Mater. Res.* **2009**, *24*, 2605–2609. [CrossRef]
5. Han, J.H.; Kim, K.B.; Yi, S.; Park, J.M.; Sohn, S.W.; Kim, T.E.; Kim, D.H.; Das, J.; Eckert, J. Formation of a bimodal eutectic structure in Ti–Fe–Sn alloys with enhanced plasticity. *Appl. Phys. Lett.* **2008**, *93*, 141901. [CrossRef]
6. Park, J.M.; Kim, T.E.; Sohn, S.W.; Kim, D.H.; Kim, K.B.; Kim, W.T.; Eckert, J. High strength Ni–Zr binary ultrafine eutectic-dendrite composite with large plastic deformability. *Appl. Phys. Lett.* **2008**, *93*, 031913. [CrossRef]
7. Wang, Z.; Lin, X.; Cao, Y.; Huang, W. Microstructure evolution in laser surface remelting of Ni–33 wt.%Sn alloy. *J. Alloy. Compd.* **2013**, *577*, 309–314. [CrossRef]
8. Wang, Z.; Zhang, Q.; Guo, P.; Gao, X.; Yang, L.; Song, Z. Effects of laser surface remelting on microstructure and properties of biodegradable Zn-Zr alloy. *Mater. Lett.* **2018**, *226*, 52–54. [CrossRef]

9. Park, J.; Kim, K.; Kim, D.; Mattern, N.; Li, R.; Liu, G.; Eckert, J. Multi-phase Al-based ultrafine composite with multi-scale microstructure. *Intermetallics* **2010**, *18*, 1829–1833. [CrossRef]
10. Park, J.M.; Kim, D.H.; Kim, K.B.; Eckert, J.M. Improving the plasticity of a high strength Fe–Si–Ti ultrafine composite by introduction of an immiscible element. *Appl. Phys. Lett.* **2010**, *97*, 251915. [CrossRef]
11. Tiwari, C.S.; Roy Mahapatra, D.; Chattopadhyay, K. *Appl. Phys. Lett.* **2012**, *101*, 171901. [CrossRef]
12. Zhang, L.; Lu, H.-B.; Mickel, C.; Eckert, J. Ductile ultrafine-grained Ti-based alloys with high yield strength. *Appl. Phys. Lett.* **2007**, *91*, 51906. [CrossRef]
13. Gouveia, G.L.; Kakitani, R.; Gomes, L.F.; Afonso, C.R.M.; Cheung, N.; Spinelli, J.E. Slow and rapid cooling of Al–Cu–Si ultrafine eutectic composites: Interplay of cooling rate and microstructure in mechanical properties. *J. Mater. Res.* **2019**, *34*, 1381–1384. [CrossRef]
14. Kim, J.T.; Lee, S.W.; Hong, S.H.; Park, H.J.; Park, J.Y.; Lee, N.; Seo, Y.; Wang, W.M.; Park, J.M.; Kim, K.B. Understanding the relationship between microstructure and mechanical properties of Al–Cu–Si ultrafine eutectic composites. *Mater. Des.* **2016**, *92*, 1038–1045. [CrossRef]
15. Wang, S.; Xie, D.; Wang, J.; Misra, A. Deformation behavior of nanoscale Al–Al_2Cu eutectics studied by in situ micropillar compression. *Mater. Sci. Eng. A* **2021**, *800*, 140311. [CrossRef]
16. Lien, H.-H.; Mazumder, J.; Wang, J.; Misra, A. Ultrahigh strength and plasticity in laser rapid solidified Al–Si nanoscale eutectics. *Mater. Res. Lett.* **2020**, *8*, 291–298. [CrossRef]
17. Lei, Q.; Ramakrishnan, B.P.; Wang, S.; Wang, Y.; Mazumder, J.; Misra, A. Structural refinement and nanomechanical response of laser remelted Al–Al_2Cu lamellar eutectic. *Mater. Sci. Eng. A* **2017**, *706*, 115–125. [CrossRef]
18. Ramakrishnan, B.P.; Lei, Q.; Misra, A.; Mazumder, J. Effect of laser surface remelting on the microstructure and properties of Al–Al_2Cu–Si ternary eutectic alloy. *Sci. Rep.* **2017**, *7*, 13468. [CrossRef]
19. Wang, S.; Liu, G.; Xie, D.; Lei, Q.; Ramakrishnan, B.; Mazumder, J.; Wang, J.; Misra, A. Plasticity of laser-processed nanoscale Al–Al_2Cu eutectic alloy. *Acta Mater.* **2018**, *156*, 52–63. [CrossRef]
20. Liu, G.; Xie, D.; Wang, S.; Misra, A.; Wang, J. Mesoscale crystal plasticity modeling of nanoscale Al–Al_2Cu eutectic alloy. *Int. J. Plast.* **2019**, *121*, 134–152. [CrossRef]
21. Wang, J.; Misra, A. Strain hardening in nanolayered thin films. *Curr. Opin. Solid State Mater. Sci.* **2014**, *18*, 19–28. [CrossRef]
22. Armstrong, R.W. Size effects on material yield strength/deformation/fracturing properties. *J. Mater. Res.* **2019**, *34*, 2161–2176. [CrossRef]
23. Porter, D.; Easterling, K.; Smith, G. Dynamic studies of the tensile deformation and fracture of pearlite. *Acta Met.* **1978**, *26*, 1405–1422. [CrossRef]
24. Lewandowski, J.; Thompson, A. Micromechanisms of cleavage fracture in fully pearlitic microstructures. *Acta Metall.* **1987**, *35*, 1453–1462. [CrossRef]

Article

Plasticity through De-Twinning in Twinned BCC Nanowires

G. Sainath [1,*], Sunil Goyal [1,2] and A. Nagesha [1,2]

[1] Materials Development and Technology Division, Indira Gandhi Centre for Atomic Research, Kalpakkam, Tamilnadu 603102, India
[2] Homi Bhabha National Institute, Indira Gandhi Centre for Atomic Research, Kalpakkam, Tamilnadu 603102, India
* Correspondence: sg@igcar.gov.in or mohansainathan@gmail.com

Received: 18 March 2020; Accepted: 18 April 2020; Published: 1 May 2020

Abstract: The deformation behaviour of twinned FCC nanowires has been extensively investigated in recent years. However, the same is not true for their BCC counterparts. Very few studies exist concerning the deformation behaviour of twinned BCC nanowires. In view of this, molecular dynamics (MD) simulations have been performed to understand the deformation mechanisms in twinned BCC Fe nanowires. The twin boundaries (TBs) were oriented parallel to the loading direction [110] and the number of TBs is varied from one to three. MD simulation results indicate that deformation under the compressive loading of twinned BCC Fe nanowires is dominated by a unique de-twinning mechanism involving the migration of a special twin–twin junction. This de-twinning mechanism results in the complete annihilation of pre-existing TBs along with reorientation of the nanowire. Further, it has been observed that the annihilation of pre-existing TBs has occurred through two different mechanisms, one without any resolved shear stress and other with finite and small resolved shear stress. The present study enhances our understanding of de-twinning in BCC nanowires.

Keywords: molecular dynamics simulations; BCC Fe nanowires; twin boundaries; de-twinning

1. Introduction

In recent years, twinned nanowires have attracted considerable attention for research due to their superior physical properties as compared to their perfect counterparts. Moreover, since twin boundaries (TBs) possess high symmetry and lowest interface energy, materials containing TBs exhibit better properties compared to those containing any other boundaries [1]. TBs can act both as barrier and carrier for dislocation motion, thus resulting in a simultaneous increase of strength and ductility [2]. TBs can also exhibit high thermal and mechanical stability [3], improve corrosion resistance, fracture toughness and strain rate sensitivity [4,5]. In view of this, the TBs are desirable planar defects in nanowires/nanocrystalline materials and many efforts have been made to synthesize and investigate the twinned nanowires and other nanocrystalline materials [6].

In order to understand the deformation behaviour of twinned FCC nanowires/nanopillars, many experimental and atomistic simulation studies have been carried out in the literature [2,6–9]. The results have shown that depending on the orientation of the TBs with respect to the loading direction, deformation mechanisms vary in nanowires and also many novel deformation mechanisms have been reported. For example, in nanopillars with TBs perpendicularly to the loading direction (transverse TBs), the dislocation–twin interactions like dislocation transmission, absorption, multiplication and stair-rod formations have been observed [2,6]. On the other hand, an extensive de-twinning is reported in nanopillars with slanted TBs [6]. The de-twinning is due to the easy glide of twinning partials along the TBs under finite shear stress. In addition to de-twinning, a novel pseudo-elasticity and shape memory effects have been discovered in nanowires with slanted TBs [10], which are distinct from those exhibited by the perfect nanowires [11]. The de-twinning mechanism has also been observed in nanowires with TBs parallel to the loading axis [12]. Unlike the nanowires with slanted TBs, the de-twinning in nanopillar containing axial TBs is surprising due to zero resolved shear stress on the pre-existing twin boundary. However, using experiments and atomistic simulations, Cheng et al. [12] have reported a de-twinning mechanism in nanopillars with axial twin boundary placed closed to the surface. This de-twinning resulted in a complete annihilation of the existing twin boundary, giving rise to defect-free nanopillar. The observation of unique de-twinning mechanism has been attributed to the migration of a junction consisting of twin boundary and other high angle grain boundary [12]. Similar annihilation of twin boundary due to de-twinning mechanisms has been observed during the bending of Ni nanowires with twin thickness less than 1 nm [13]. These observations clearly indicate that the de-twinning is quite common in twinned FCC nanowires with the exception of traverse twin boundaries.

However, all the above studies were focused on twinned FCC nanowires/nanopillars and little attention has been paid towards twinned BCC nanowires. There are only a couple of investigations in the literature on BCC nanowires containing twin boundaries [14,15]. These studies on BCC nanowires with transverse TBs have shown a strong tension-compression asymmetry in deformation mechanisms. With respect to axial TBs, Li et al. [16] have investigated the role of six-fold twin on the torsional deformation of BCC Fe nanowires. Except this, no study exists on the role of axial TBs in BCC nanopillars. Further, it is interesting to examine the possibility of the occurrence of de-twinning mechanism in twinned BCC nanowires. In view of this, the present study aimed at understanding the deformation mechanisms in axially twinned BCC Fe nanowires using molecular dynamics (MD) simulations. The results show an intersecting de-twinning mechanism involving the migration of twin–twin junction, which results in a complete annihilation of pre-existing TBs and also leads to the reorientation of the nanowire.

2. MD Simulation Details

The deformation behaviour of [110] BCC Fe nanowires with twin boundaries oriented parallel to the loading direction were considered in this study (Figure 1). The nanopillar is enclosed by [1-11] and [1-1-2] type side surfaces. The number of TBs is varied from one to three (Figure 1a–c). All the nanowires had a square cross section width (d) = 8.5 nm and length (l) was twice the cross section width, i.e., l = 17 nm. In order to mimic an infinitely long nanowire, periodic boundary conditions were chosen only along the wire axis, while the other directions were kept free. On these twinned nanowires, compression loading has been simulated using molecular dynamics (MD) simulations. MD simulations have been carried out in LAMMPS package [17] employing an embedded atom method (EAM) potential for BCC Fe given by Mendelev et al. [18]. The Mendelev EAM potential has been chosen mainly because several predictions obtained using this potential are in good agreement with experimental observations. For example, this potential correctly predicted deformation by twinning and dislocation slip based on nanopillar orientation in BCC Fe [19,20], which is quite close to the experimental findings in ultra-thin BCC W nanopillars [21]. Before applying the compressive load, energy minimization has been performed by conjugate gradient method followed by an equilibration to a required temperature of 10 K in NVT ensemble with a Nose-Hoover thermostat. Velocity verlet algorithm has been used to integrate the equations of motion with a time step of 2 femto seconds. The compressive deformation was carried out at a constant strain rate of 1×10^8 s^{-1} with respect to initial box length. The visualization of TBs and dislocations is accomplished in AtomEye [22] and OVITO [23] packages using centro-symmetry parameter (CSP) and common neighbour analysis (CNA).

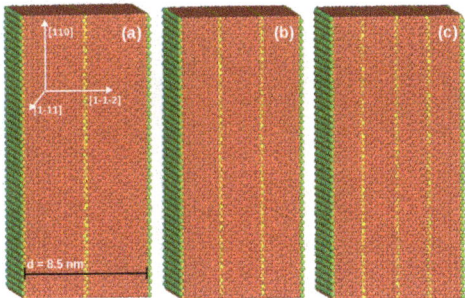

Figure 1. The initial configuration of BCC Fe nanowires containing (**a**) one, (**b**) two, and (**c**) three twin boundaries. The corresponding twin boundary spacings is of 4.25, 2.83 and 2.12 nm, respectively. The nanowire orientation and cross-section width (*d*) are indicated in (**a**). The atoms are coloured according to their centro-symmetry parameter (CSP).

3. Results And Discussion

Figure 2a–h shows a typical deformation behaviour in axially twinned [110] BCC Fe nanowires containing two pre-existing twin boundaries. Similar behaviour has been observed in nanowires containing one and three TBs and therefore not shown here. It can be seen that the plastic deformation initiates by the nucleation of a twin from the nanowire corners (Figure 2a). With increasing deformation, the leading front of the twin approaches the existing twin boundary and penetrates into neighbouring grain (Figure 2b). Here, it is interesting to see that the twin penetration happens in two different ways; (1) it can penetrate directly to the next grain without any deviation in twinning plane and (2) it can pass onto a plane symmetrical to the original twinning plane (Figure 2b). Following the penetration, the leading twin front again passes through the second pre-existing twin boundary and results in the formation of twinned rhombohedron, which is enclosed by twin boundaries (Figure 2c). Also, the initially nucleated twin (Figure 2a) has grown significantly cutting across the pre-existing TBs (Figure 2c). It can be seen that the twin penetration and its growth has led to the annihilation and also the migration of pre-existing TBs (Figure 2c,d). With increasing strain, the twin growth dominates the deformation leading to the complete annihilation of pre-existing TBs (Figure 2e,f). Following the annihilation, the growing twin completely sweeps the nanowire length and leads to the reorientation of the nanowire from initial <110>/{111}{112} orientation to <001>/{010}{310} orientation (Figure 2g,h). Thus, in axially twinned BCC Fe nanowires there is an annihilation of pre-existing TBs. In other words, the twinned nanowires become twin free single crystalline nanowires with different orientation.

The deformation by twinning and reorientation in [110] twinned BCC Fe nanowires is similar to that observed in perfect [110] nanowires [19]. However, the annihilation of pre-existing TBs and the observation of reorientation in twinned nanowires is interesting. Similar to the present study, the annihilation of twin boundary due to de-twinning has also been observed recently in FCC nanowires [12]. However, this annihilation in FCC system has been observed only when the nanowire contains a single axial twin boundary located close to surface [12]. In the present study, the annihilation of TBs has been observed in nanowires containing one, two and three TBs irrespective of their location with respect to the surface. Further, there is an experimental evidence for the annihilation of twin boundaries due to de-twinning in FCC nanowires [12]. However, due to limited experimental studies on BCC nanowires compared to FCC, the annihilation of twin boundaries as a result of de-twinning has not been reported experimentally in BCC nanowires. Interestingly, a reversible twinning, which is also called de-twinning has been reported experimentally in BCC W nanowires [21]. Using in situ high-resolution transmission electron microscopy, it has been demonstrated that during loading, the W nanowire undergo deformation by twinning leading to twin growth. However, upon unloading, deformation proceeds by de-twinning, which is same as twinning, but in the reverse direction. As a result, the twin thickness is gradually reduced [21].

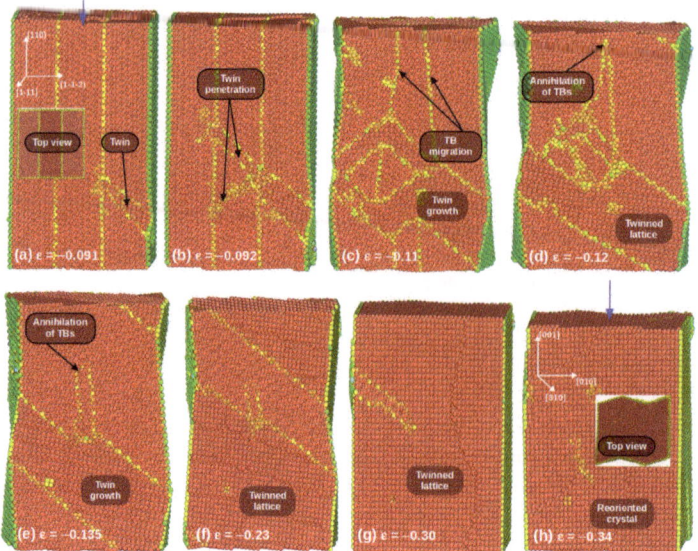

Figure 2. The atomic snapshots displaying the typical deformation behaviour in twinned BCC Fe nanowires at different strains; (**a**) twin nucleation, (**b**,**c**) twin penetration through existing TBs, (**c**,**d**) twin boundary migration and annihilation, (**d**–**g**) twin growth, and (**h**) reoriented nanowire. The atoms are coloured according to their centro-symmetry parameter (CSP). The insets in (**a**,**h**) shows the top view of the nanowire.

The annihilation of pre-existing TBs in Figure 2a–h has occurred through two different mechanisms, one without any resolved shear stress (Figure 3) and other with finite and small resolved shear stress (Figure 4). Figure 3 shows a mechanism where a pre-existing twin boundary gets annihilated step by step due to the glide of 1/6<111> partial dislocations on newly formed TBs. The 1/6<111> partial dislocations nucleate from a twin–twin junction (Figure 3b,f,j) and glide on TBs (Figure 3c–d,g–h,k–l) whose migration results in twin growth. As a result of twin boundary migration and twin growth, the twin–twin junction moves step by step and leads to the annihilation of pre-existing twin boundary (Figure 3a,e,f). During this process, there is no apparent resolved shear stress on the pre-existing twin boundary as it is completely parallel to the loading axis (Figure 3a,e). It remains parallel till the partial dislocation activity is equal on both left and right TBs (Figure 3a–h). However, with increasing strain, due to the presence of some barriers, the partial dislocation activity on one twin boundary (left in this case) dominates over the other (Figure 3k,l). As a result, the pre-existing twin boundary slightly bends with respect to the loading axis (Figure 3i) which introduces a small resolved shear stress. Due to this shear stress, the partial dislocation activity also commences on the pre-existing twin boundary (Figure 4a,e). This partial dislocation activity further bends the twin boundary towards the other pre-existing twin boundary (Figure 4b,c,f,g). As a result, they meet each other and annihilate over time (Figure 4d,h). This annihilation occurs under small resolved shear stress. Thus, two different mechanisms combinedly contribute to de-twinning in twinned BCC Fe nanowires (Figure 2). In contrast, the de-twinning in FCC nanowires occurs through a single mechanism of twin embryo nucleation followed by its expansion due to the migration of a special junction consisting of two TBs and one high angle grain boundary [12]. During this complete process, the existing twin boundary is always parallel to the loading axis, thereby getting annihilated under zero resolved shear stress [12].

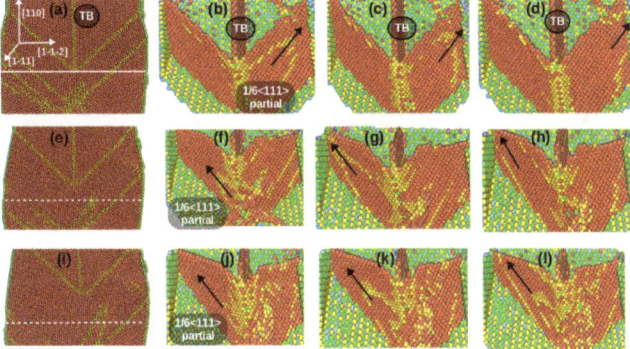

Figure 3. The annihilation of pre-existing twin boundary due to the step-by-step migration of the twin–twin junction (**a**,**e**,**i**). The movement of twin-twin junction is aided by the continuous nucleation (from the twin–twin junction) (**b**,**f**,**j**) and glide of 1/6<111> partial dislocations on newly formed twin boundaries(**c**,**d**,**g**–**h**,**k**,**l**). The atoms are coloured according to their centro-symmetry parameter (CSP). For clarity, the atoms in perfect BCC structure were removed in (**b**–**d**), (**f**–**h**) and (**j**–**l**).

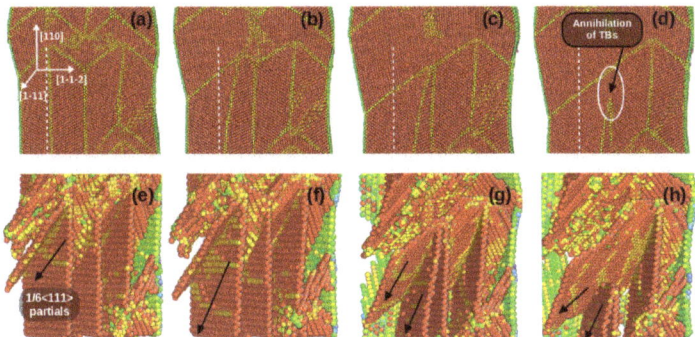

Figure 4. The annihilation of pre-existing twin boundary due to the bending and subsequent migration of pre-existing twin boundaries (**a**–**d**). This process is aided the by the continuous glide of 1/6<111> partial dislocations on the existing TBs. The atoms are coloured according to their centro-symmetry parameter (CSP). For clarity, the atoms in perfect BCC structure were removed in (**e**–**h**).

An investigation on the deformation behaviour of twinned nanowires also offers valuable insights into twin–twin interactions. Figure 5 shows two different types of twin–twin junctions observed in the present study. In one junction, three TBs meet at an angle of 120° with respect each other as shown in Figure 5a (Y-junction), while in other junction, they meet at an angle of 60 and 240° forming an arrow (↓) like junction (Figure 5b). Further, the annihilation of twin boundary with zero resolved shear stress has occurred mainly near arrow like twin–twin junctions (Figure 3), which participates more actively in de-twinning mechanisms. In the past, a junction containing six TBs (six-fold twin) has been inserted and studied for its evolution under torsion in α-Fe [16]. However, no Y-junction or arrow like junctions have been observed. It is well known that in BCC systems, the TBs lie on {112} planes and also three of the {112} planes have the same <111> zone axis as depicted in Figure 5c. As a result of three {112} planes having the same zone axis, the twin–twin junctions such as those shown in Figure 5a,b and also the six fold twins in Ref. [16] were feasible in BCC systems. In contrast, the six-fold twins were not compatible in FCC systems, where only up to five-fold twins were observed [24]. The arrangement of three {112} planes as shown in Figure 5c is also responsible for the direct as well as symmetrical transmission of twin across the existing twin boundary as seen in Figure 2b. In a previous study [14], it has been shown that, like twins, dislocations can also either directly transmit through the twin boundary without any deviation in slip plane or they can transmit to symmetrical plane.

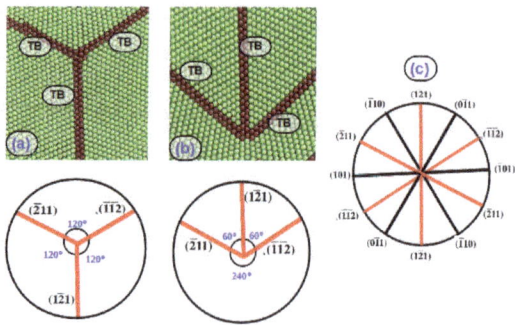

Figure 5. Two different twin–twin junctions observed in the present study; (**a**) Y-junction, where three TBs meet at an angle of 120° with respect each other, (**b**) arrow (↓) like junction, where three TBs meet at an angle of 60 and 240°, (**c**) the arrangement of {112} slip planes, when viewed along <111> direction. The atoms are coloured according to their common neighbour analysis (CNA).

4. Conclusions

Molecular dynamics simulations performed on the compressive loading of axially twinned [110] BCC Fe nanowire indicate that the deformation is dominated by a de-twinning mechanism involving the migration of a twin–twin junction. In the present study, two types of twin–twin junctions were observed. One is Y-junction where three TBs meet at an angle of 120° with respect to each other and the other is an arrow (↓)-like junction, where TBs meet at an angle of 60 and 240°. The results indicate that compared to the Y-junction, the arrow like junction participates more actively in de-twinning mechanisms. In all the nanowires, the de-twinning mechanism results in a complete annihilation of pre-existing TBs along with a change in nanowire orientation from initial <110> to <001>. Further, it was observed that the annihilation of pre-existing TBs occurs through two different mechanisms, one without any resolved shear stress and other with a finite and small resolved shear stress. The annihilation of pre-existing TBs without any resolved shear stress has occurred through the step-by-step movement of the twin–twin junction due to the glide of 1/6<111> partial dislocations on newly formed twin boundaries. The annihilation of pre-existing TBs with finite shear stress occurred due to the slight bending of twin boundaries at higher strains. The present study significantly improves our understanding of de-twinning and also the stability of twin boundaries in BCC nanowires.

Author Contributions: G.S.: Conceptualization, Methodology, Data curation, Formal analysis and Writing-original draft; S.G.: Resources and Writing - review & editing; A.N.: Resources and Writing-review & editing. All authors have read and agreed to the published version of the manuscript.

Funding: This research received no external funding.

Conflicts of Interest: The authors declare no conflict of interest.

References

1. Lu, K.; Lu, L.; Suresh, S. Strengthening materials by engineering coherent internal boundaries at the nanoscale. *Science* **2019**, *324*, 349–352. [CrossRef]
2. Cao, A.J.; Wei, Y.G.; Mao, S.X. Deformation mechanisms of face-centered-cubic metal nanowires with twin boundaries. *Appl. Phys. Lett.* **2007**, *90*, 151909. [CrossRef]

3. Wang, J.; Li, N.; Misra, A. Structure and stability of Σ3 grain boundaries in face centered cubic metals. *Philos. Mag.* **2012**, *93*, 315–327. [CrossRef]
4. Singh, A.; Tang, L.; Dao, M.; Lu, L.; Suresh, S. Fracture toughness and fatigue crack growth characteristics of nanotwinned copper. *Acta Mater.* **2011**, *59*, 2437–2446. [CrossRef]
5. Deng, C.; Sansoz, F. Effects of twin and surface facet on strain-rate sensitivity of gold nanowires at different temperatures. *Phys. Rev. B* **2010**, *81*, 155430. [CrossRef]
6. Jang, D.; Li, X.; Gao, H.; Greer, J.R. Deformation mechanisms in nanotwinned metal nanopillars. *Nat. Nanotechnol.* **2012**, *7*, 594–601. [CrossRef]
7. Sainath, G.; Choudhary, B.K. Molecular dynamics simulation of twin boundary effect on deformation of Cu nanopillars. *Phys. Lett. A* **2015**, *379*, 1902–1905. [CrossRef]
8. Sun, J.; Li, C.; Han, J.; Shao, X.; Yang, X.; Liu, H.; Song, D.; Ma, A. Size effect and deformation mechanism in twinned Cu nanowires. *Metals* **2017**, *7*, 438. [CrossRef]
9. Rohith, P.; Sainath, G.; Sunil, G.; Nagesha, A.; Srinivasan, V.S. Role of axial twin boundaries on deformation dechanisms in Cu nanopillars. *Philos. Mag.* **2020**, *100*, 529–550. [CrossRef]
10. Deng, C.; Sansoz, F. A new form of pseudo-elasticity in small-scale nanotwinned gold. *Extrem. Mech. Lett.* **2016**, *8*, 201–207. [CrossRef]
11. Liang, W.; Zhou, M. Atomistic simulations reveal shape memory of fcc metal nanowires. *Phys. Rev. B* **2006**, *73*, 115409. [CrossRef]
12. Cheng, G.; Yin, S.; Chang, T.-H.; Richter, G.; Gao, H.; Zhu, Y. Anomalous tensile detwinning in twinned nanowires. *Phys. Rev. Lett.* **2017**, *119*, 256101. [CrossRef]
13. Wang, L.; Lu, Y.; Kong, D.; Xiao, L.; Sha, X.; Sun, J.; Zhang, Z.; Han, X. Dynamic and atomic-scale understanding of the twin thickness effect on dislocation nucleation and propagation activities by in situ bending of Ni nanowires. *Acta Mater.* **2015**, *90*, 194–203. [CrossRef]
14. Sainath, G.; Choudhary, B.K. Deformation behaviour of body centered cubic iron nanopillars containing coherent twin boundaries. *Philos. Mag.* **2016**, *96*, 3502–3523. [CrossRef]
15. Xu, S.; Startt, J.K.; Payne, T.G.; Deo, C.S.; McDowell, D.L. Size-dependent plastic deformation of twinned nanopillars in body-centered cubic tungsten. *J. Appl. Phys.* **2017**, *121*, 175101. [CrossRef]
16. Li, S.; Salje, E.K.H.; Jun, S.; Ding, X. Large recovery of six-fold twinned nanowires of α-Fe. *Acta Mater.* **2017**, *125*, 296–302. [CrossRef]
17. Plimpton, S. Fast parallel algorithms for short-range molecular dynamics. *J. Comp. Phy.* **1995**, *117*, 1–19. [CrossRef]
18. Mendelev, M.I.; Han, S.; Srolovitz, D.J.; Ackland, G.J.; Sun, D.Y.; Asta, M. Development of new interatomic potentials appropriate for crystalline and liquid iron. *Philos. Mag.* **2003**, *83*, 3977–3994. [CrossRef]
19. Sainath, G.; Choudhary, B.K. Orientation dependent deformation behaviour of BCC iron nanowires. *Comput. Mater. Sci.* **2016**, *111*, 406–415. [CrossRef]
20. Dutta, A. Compressive deformation of Fe nanopillar at high strain rate: Modalities of dislocation dynamics. *Acta Mater.* **2017**, *125*, 219–230. [CrossRef]
21. Wang, J.; Zeng, Z.; Weinberger, C.R.; Zhang, Z.; Zhu, T.; Mao, S.X. In situ atomic-scale observation of twinning-dominated deformation in nanoscale body-centred cubic tungsten. *Nat. Mater.* **2015**, *14*, 594–600. [CrossRef] [PubMed]
22. Li, J. AtomEye: An efficient atomistic configuration viewer. *Model. Simul. Mater. Sci. Eng.* **2003**, *11*, 173. [CrossRef]
23. Stukowski, A. Visualization and analysis of atomistic simulation data with OVITO - the Open Visualization Tool. *Modell. Simul. Mater. Sci. Eng.* **2009**, *18*, 015012. [CrossRef]
24. Zhang, Z.; Huang, S.; Chen, L.; Zhu, Z.; Guo, D. Formation mechanism of fivefold deformation twins in a face centered cubic alloy. *Sci. Rep.* **2017**, *7*, 45405. [CrossRef] [PubMed]

© 2020 by the authors. Licensee MDPI, Basel, Switzerland. This article is an open access article distributed under the terms and conditions of the Creative Commons Attribution (CC BY) license (http://creativecommons.org/licenses/by/4.0/).

Article

Atomic Force Microscopy Study of Discrete Dislocation Pile-ups at Grain Boundaries in Bi-Crystalline Micro-Pillars

Xiaolei Chen [1,2], Thiebaud Richeton [1,*], Christian Motz [2] and and Stéphane Berbenni [1]

1. Université de Lorraine, CNRS, Arts et Métiers ParisTech, LEM3, F-57000 Metz, France; xiaolei.chen@univ-lorraine.fr (X.C.); stephane.berbenni@univ-lorraine.fr (S.B.)
2. Department of Materials Science and Engineering, Saarland University, 66123 Saarbrücken, Germany; motz@matsci.uni-sb.de
* Correspondence: thiebaud.richeton@univ-lorraine.fr; Tel.: +33-372747802

Received: 20 April 2020; Accepted: 18 May 2020; Published: 20 May 2020

Abstract: Compression tests at low strains were performed to theoretically analyze the effects of anisotropic elasticity, misorientation, grain boundary (GB) stiffness, interfacial dislocations, free surfaces, and critical force on dislocation pile-ups in micro-sized Face-Centered Cubic (FCC) Nickel (Ni) and α-Brass bi-crystals. The spatial variations of slip heights due to localized slip bands terminating at GB were measured by Atomic Force Microscopy (AFM) to determine the Burgers vector distributions in the dislocation pile-ups. These distributions were then simulated by discrete pile-up micromechanical calculations in anisotropic bi-crystals consistent with the experimentally measured material parameters. The computations were based on the image decomposition method considering the effects of interphase GB and free surfaces in multilayered materials. For Ni and α-Brass, it was found that the best predicted step height spatial profiles were obtained considering anisotropic elasticity, free surface effects, a homogeneous external stress and a certain critical force in the material to equilibrate the dislocation pile-ups.

Keywords: micromechanical testing; micro-pillar; bi-crystal; discrete dislocation pile-up; grain boundary; free surface; anisotropic elasticity; crystallographic slip

1. Introduction

The plasticity of crystalline materials results primarily from motion and multiplication of dislocations. Both theoretical and experimental investigations show that the mechanical properties of metals depend on the density, the distribution, the nucleation, and the mobility of dislocations. In particular, the mechanical properties of polycrystals depend on the presence of grain boundaries (GBs), such as the elastic limit and strain hardening. More specifically, these properties are greatly dependent on the interaction mechanisms between dislocations and GBs (dislocation transmission or absorption at GB, formation of a dislocation pile-up, etc.). Indeed, GBs generally present themselves as obstacles to dislocation motion.

In the case of grain size reduction, which means the increase of GB fraction in the material, the material strength is increased following the Hall–Petch's relationship [1–3]. Experimental observations of collective dislocation distributions in the neighborhood of GBs have also made it possible to characterize slip lines using electron microscopy. The interaction between dislocations and GB can be directly observed through in-situ mechanical tests in a Transmission Electron Microscope (TEM). From TEM, it is also possible to obtain the position of each dislocation in a discrete dislocation pile-up, which can be used to calculate stress concentrations at GB [4]. Electron Back Scattered Diffraction (EBSD) measurements in Scanning Electron Microscopes (SEM) are able to measure the local crystallographic orientations during deformation and therefore the evolution of intra-granular lattice rotations and geometrically dislocation densities [5,6]. Moreover, the coupling of Electron Channelling Contrast Imaging (ECCI) and EBSD permits characterizing single dislocations, dislocation densities, and dislocation substructures in deformed bulk materials [7]. Recently, Atomic Force Microscopy (AFM) makes it possible to analyze the topography and surface roughness of a material following the emergence of dislocations (slip step heights) during mechanical loading of single and poly-crystals [8–10]. Moreover, based on experimental observations, multi-scale simulations involving Crystal Plasticity Finite Element Method (CPFEM), Discrete Dislocation Dynamics (DDD), and Molecular Dynamics (MD) have been developed during these last decades. These methods aim at deeply understanding slip mechanisms and the interactions between dislocations and GBs, and to predict mechanical properties of crystalline materials. However, CPFEM can only consider averaged slip in continuum mechanics, which cannot be used to investigate the discrete interactions between dislocations and GBs. DDD can be performed to study the motion of each individual dislocation, but the complex properties of GBs are missing in this method and elastic anisotropy calculations remain tedious. MD can show the precise structure of dislocation cores and GBs at atomic scale, but it can be only used for simulations at the nanoscale and within a small period of time due to intensive required computing resources.

As the simplest system containing GB, the bi-crystal configuration has been extensively studied. Dehm et al. [11] have reported a comprehensive review on micro-/nanomechanical testing at small scale, especially about GB effect in plasticity and the interactions between dislocations and GBs. Based on many experimental results, there are mainly four mechanisms of dislocation-GB interactions which have been studied: dislocation transmission across GB [12], GB as a dislocation source for lattice dislocation [13], formation of a dislocation pile-up [4] and dislocation absorption at GB [14]. The dislocation transmission is largely investigated through the geometrical transmission factor as reviewed by Bayerschen et al. [15] and the critical stress for transmission as reviewed by Hunter et al. [16]. The geometrical transmission factor has been firstly proposed by Livingston and Chalmers [17] considering only slip system orientations. Then, Shen et al. [4] have proposed another transmission factor considering both slip systems and GB orientations, which has been successfully applied to the prediction of slip activity due to a dislocation pile-up in a 304 stainless steel. Afterwards, Beyerlein et al. [18] have proposed another transmission factor considering threshold angles for the slip systems and GB. It is conjectured that the slip transfer is not possible if at least one of these two angles is bigger than a critical angle [18,19]. The critical transmission stress is generally studied based on image force on dislocations due to the moduli mismatch [20–24]. Furthermore, the interaction between dislocations and special boundaries like twin boundaries has been investigated by many researchers [25–28]. As discussed above, it is believed that a critical stress is needed for the dislocation to get from one grain to the other grain across the GB. The GB will dominate the overall material's behavior when this critical stress is higher than the size-dependent single crystalline flow stress. However, Imrich et al. [25] have showed that, at a critical small sample size, the coherent Σ3 twin GB may not have an impact on the overall mechanical property of micro-beam since the critical stress for transmission is quite low in this case, as shown also by Malyar et al. [28]. In general, however, a large angle random GB does have an evident influence on slip transmission mechanism [29].

In this study, the objective is to develop a new experimental/theoretical methodology suitable for investigating the effects of elastic and plastic anisotropies on the interactions between dislocations and GBs in bi-crystalline micro-pillars. We will consider the effects of GB stiffness, free surfaces, and critical force (lattice friction force and other forces due to obstacles to dislocation motion) on discrete dislocation pile-ups at grain boundaries. In order to reach this objective, both an experimental study based on compression tests on bi-crystalline micro-pillars and a theoretical investigation of discrete dislocation pile-ups in heterogeneous and anisotropic media are performed. Among the different possible mechanisms involving collective dislocation behavior, dislocation pile-ups at GBs and slip transfer are essentially studied. The experimental procedure of in-situ micro-compression tests on Nickel (Ni) and α-Brass bi-crystals is first described in Section 2. Then, slip line height profiles analyzed by SEM and AFM are presented in Section 3. Moreover, the simulation of the slip step height profiles using an anisotropic and heterogeneous discrete dislocation pile-up model in a bi-crystalline configuration are given in Section 4. All the simulation results and the important material parameters for both Ni and α-Brass samples are discussed. Section 5 provides the conclusions.

2. Experiments

2.1. Materials

The materials used for bi-crystalline micro-pillars are Ni with very high purity (99.999%) and α-Brass (70%Cu-30%Zn, wt%) with impurities (Fe, Pb, P and As, etc.) less than 0.001%. The elastic stiffness constants of Ni are $C_{11} = 246.5$ GPa, $C_{12} = 147.3$ GPa, $C_{44} = 124.7$ GPa [30] and those of α-Brass are $C_{11} = 139.21$ GPa, $C_{12} = 104.51$ GPa, $C_{44} = 71.3$ GPa [31,32]. Thus, these two materials have an elastic anisotropy Zener ratio $A = 2C_{44}/(C_{11} - C_{12})$ of the order of 2.51 and 4.11, respectively. Furthermore, the stacking fault energies for α-Brass and Ni are about 14 mJ/m^2 [33] and 90 mJ/m^2 [34], respectively. The stacking fault energy of α-Brass is low which promotes planar slip and thus facilitates the observation of localized slip. Although the stacking fault energy of Ni is higher, planar slip was still well observed in Ni bi-crystals during compression tests [35].

2.2. Sample Preparation and Characterization

Macrosamples were prepared in the form of cubes of 20 × 20 × 5 mm^3. These macrosamples were ground with grit paper and then homogenized by heat treatment. For Ni, the produced samples were homogenized at 1100 °C for three days in vacuum. However, for α-Brass, in order to prevent dezincified phenomenon in vacuum [36], the heat treatment was realized in argon at 980 °C for 1 min with a special sample holder. Then, the samples were ground again and polished using a diamond suspension of particle sizes until 1 micron on zeta cloth. After that, the sample surfaces were finally polished by OP-U-NonDry solution with water on chem-cloth for 5 mins. At the end, the samples were electropolished at a voltage of 24 V for 20 s with electrolyte A2 (65~86% Ethanol, 10~15% 2-Butoxyethanol and 5~15% water) for Ni and with electrolyte D2 (15~35% Phosphoric acid, 15~25% Ethanol, <10% Propane-1-OL, <1% urea and 50~70% water) for α-Brass in order to eliminate hardening and damage (including scratches) occurred during the mechanical polishing.

Each grain orientation was measured by EBSD (Oxford, UK) in a Carl Zeiss SIGMA series SEM (Oberkochen, Germany). Orientation maps were acquired with a 20 kV acceleration voltage and a spatial step size of 20 μm. Data processing was performed by Flamenco Channel 5 software (HKL Technology, Oxford Instruments, Abingdon, UK) with the indexing rate always greater than 99%. In order to investigate dislocation pile-ups behavior and their interaction with GBs by AFM measurements and slip lines characterization on the upper surface, special GBs were chosen from the hereinafter described conditions following the surface notations presented in Figure 1b,d. The first condition is that GB is roughly perpendicular to the upper surface. The second one is that with a mechanical loading parallel to GB plane, for the target grain, the Burgers vector of the likely activated slip system (i.e., the one with the maximum Schmid factor) must not be perpendicular to GB. Meanwhile, the slip plane of this slip system must not be parallel to the upper surface. These conditions ensure that the Burgers vector has a component perpendicular to the upper surface where AFM measurements will be performed. Based on all the above conditions, one GB of each material was used to analyze the dislocation distributions and the slip transmission phenomenon.

2.3. Micro-Pillar Production

The chosen GBs were cut out at the sample edge in order to facilitate the FIB cutting. During the FIB process, the top surface should indeed be polished from two sides in order to ensure that it is perpendicular to the upper surface. Only in that case is it a pure compression test. If not, there are some bending stresses. In the present study, it was crucial to have a pure compression in order to get accurate loading stresses, which were used in the simulations. Finally, the pre-prepared sample was cut into a micro-pillar by FIB with FEI Versa 3D Dual Beam system (Hillsboro, OR, USA) using ion beam currents of 15 nA for rough cutting and 1 nA at 30 kV for fine polishing as shown in Figure 1b,d for Ni and α-Brass, respectively. In the present paper, the grain at the edge side of the macrosample is named crystal I and the other one is named crystal II. For the Ni sample, the average length of the micro-pillar was about 15.46 µm and the average width was about 8.03 µm with 4.30 µm for crystal I and 3.73 µm for crystal II. While for the α-Brass sample, the average length of the micro-pillar was about 15.37 µm and the average width was about 9.02 µm with 3.97 µm for crystal I and 5.05 µm for crystal II. Then, the orientations of the bi-crystal were acquired again by EBSD with a spatial step resolution of 0.1 µm. Data processing was performed by AZtec software (2.0, Oxford Instruments, Abingdon, UK) with an indexing rate greater than 99%. The microstructures of the micro-pillars are presented by IPF in Figure 1a for Ni and in Figure 1c for α-Brass. Meanwhile, the slip analysis configuration is presented in Figure 1e. The inclination angle of GB are $\varepsilon = 0°$ for the Ni micro-pillar and $\varepsilon = 6.4°$ for the α-Brass micro-pillar (see Figure 1e).

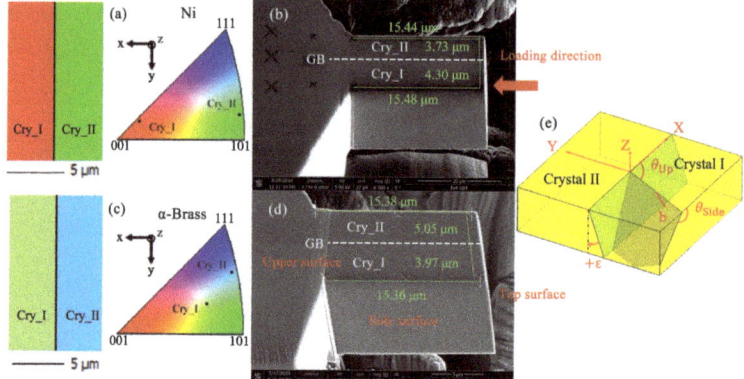

Figure 1. The crystallographic orientations of both crystals, Cry_I and Cry_II, are given in the standard IPF in the direction of the y-axis which is parallel to GB and is also the loading direction of the compression test for (**a**) Ni micro-pillar and (**c**) α-Brass micro-pillar. SEM micrograph of a micro-pillar containing GB cut by FIB for (**b**) Ni sample and (**d**) α-Brass sample; (**e**) schematic figure of slip line analyses: The sample coordinates are set as y perpendicular to GB line on the upper surface pointing towards crystal II, z perpendicular to the upper surface and $x = y \times z$ parallel to the GB line. The angle between the slip line and the GB line in a positive x-direction on the upper surface is noted as θ_{Up}. The angle between the slip line and the upper edge of sample in a positive x-direction on the side surface is noted as θ_{Side}. The inclination angle of GB is noted as ε.

2.4. Micro-Compression Test in SEM

The in-situ compression tests were carried out in high vacuum at room temperature in a Carl Zeiss ΣIGMA series SEM, by an in-situ nanoindenter (UNAT-SEM II) which had a load noise floor level at about 0.002 mN. The micro-pillars were compressed by a diamond flat punch. Furthermore, in order to control the orientation of the micro-pillar well for a pure compression test, such as the loading direction parallel to GB and the top surface of sample perpendicular to flat punch, an additional rotation stage for the sample holder performing rotations in two dimensions was developed. Because the original sample holder could only rotate in one direction, it was necessary to combine it with a complementary rotation stage with two more degrees of freedom. All the compression tests were carried out in displacement controlled mode with several cycles of loading and unloading with small increment as shown in Figure 2a for Ni and in Figure 2b for α-Brass. The idea of loading and unloading cycles is to have enough time to make an SEM picture of high quality in order to well observe slip lines. Once slip lines were observed or yield stress was reached, the compression test was stopped during the unloading step in order to prevent from the extra force on the pillar coming from the tip vibration when the test was manually interrupted. The loading force and corresponding displacement were recorded during the mechanical tests, so that the stress–strain curves can be calculated after the tests. Even though the cyclic loadings were performed with small increments, the tests were stopped when the first slip event was activated. The first slip event was identified when slip lines were observed in SEM or when yield stress of the bi-crystal was reached. Thus, only the last cycle (marked as red line in Figures 2a,b) contributed to plastic deformation, and all the previous cycles were in the elastic state.

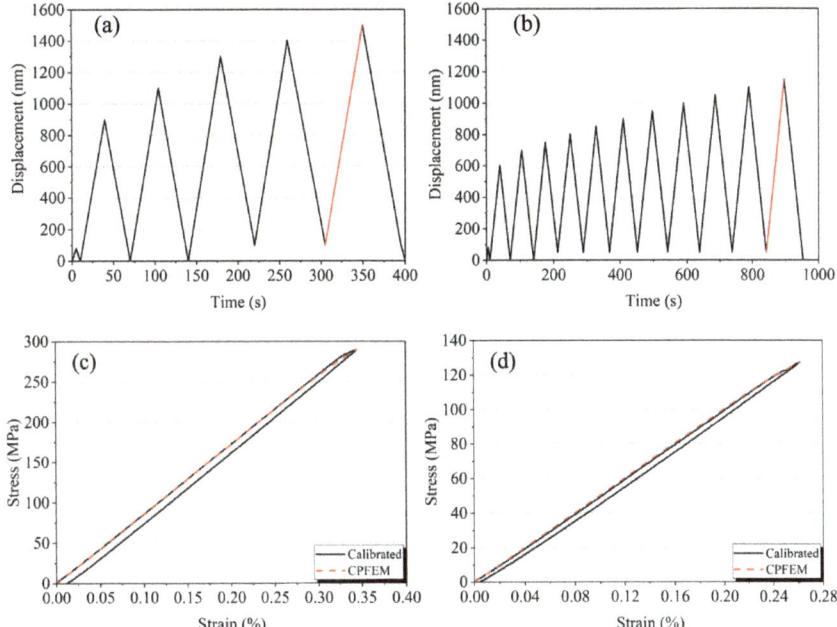

Figure 2. Loading curves obtained with displacement control mode for the compression tests of (**a**) Ni and (**b**) α-Brass. Only red lines contribute to plastic deformation, and all the previous cycles are in the elastic state. Stress–strain curves were corrected by CPFEM simulations for (**c**) the Ni micro-pillar and (**d**) the α-Brass micro-pillar. The applied stress at the end of the compression test is about 289.4 MPa for the Ni sample and 127.4 MPa for the α-Brass sample.

2.5. Stress-Strain Responses and Overall Stresses at Plastic Yielding

Figure 2c,d show the stress–strain responses of the compression tests for Ni and α-Brass, respectively. As designed, most of the deformation was in the elastic state, and only the last small part corresponded to plasticity. However, the yield point occurred at about 8.5% strain for Ni and 6.3% for α-Brass from the originally experimental measurement. These strains are of course too large for characterizing the yield point of such metals. This error came from the drift of the tip during the compression test. Hence, the displacement measurement was not accurate, but the force measurement was accurate. Therefore, CPFEM simulations of the compression tests with the same configuration were performed for both Ni and α-Brass samples. Then, the experimentally measured stress–strain curves were corrected by using CPFEM simulations such that both elastic parts match together. After the correction, the yield point occurred at the stain about 0.33% for Ni and 0.24% for α-Brass, which are reasonable values for such metals. Thus, the effective average loading strain rate was about $7.7 \times 10^{-5} \text{s}^{-1}$ for Ni and $5.0 \times 10^{-5} \text{s}^{-1}$ for α-Brass. The applied stress at the end of the compression test was about 289.4 MPa for Ni and 127.4 MPa for α-Brass. These values will be used in Section 4.

3. Experimental Results

3.1. Preliminary Slip Analysis Using Schmid Factors and Incompatibility Stresses

Based on the local crystalline orientation, the mechanical and geometrical information of each slip system in the studied grains were firstly analyzed. The slip analysis results for Ni sample and α-Brass sample are reported in Tables 1 and 2, respectively. The results include the Schmid factors, the resolved shear stresses normalized by the applied stress considering elastic incompatibility stresses as detailed in [37,38], the angles between the slip line and the GB line on the upper surface θ_{Up}, the angles between the slip line and the upper edge of sample on the side surface θ_{Side} (see Figure 1e) and the maximum transmission factor with corresponding slip system in the adjacent grain. Here, slip systems in FCC crystals are indicated by Schmid and Boas's notation [39]. The analysis of Schmid factors and resolved shear stresses can be used to predict the first active slip systems (with the maximum Schmid factor or the maximum resolved shear stress) during compressive loading. Schmid factor are computed directly from the projection of the homogeneous external applied stress. However, for the analysis of the resolved shear stresses considering incompatibility stresses, the stresses in the two crystals are the sum of the external applied stress and incompatibility stresses due to heterogeneous elasticity coming from misorientation and anisotropic elasticity [37,38,40]. In some cases, the first active slip system with maximum Schmid factor can be different from the one obtained when considering incompatibility stresses [38,40]. After the mechanical tests, the activated slip lines could be observed by SEM (see Figure 3a for Ni and Figure 4a for α-Brass) and by AFM (see Figure 5a for Ni and Figure 6a for α-Brass). Then, both angles θ_{Up} and θ_{Side} were experimentally determined as shown in Figure 3a (right) for Ni and in Figure 4a (right) for α-Brass. By comparing these angles to the theoretical analyses as described in Tables 1 and 2 (see also Figure 3b for Ni and Figure 4b for α-Brass), the active slip planes were identified. For the present experiments, the onset of plasticity occurred in crystal II for the Ni sample and in crystal I for the α-Brass sample. In both grains, the first active slip system is always the same, either considering a simple Schmid analysis or elastic incompatibility stresses [37,38,40], as marked in red in Tables 1 and 2. However, for the α-Brass sample, combining the analyses provided by Schmid factor and incompatibility stresses, a second slip system could be active as well and is marked in blue in Table 2.

Table 1. Slip system analyses for Ni micro-pillar. The first active slip system is marked in red. The compressive direction is along X, the GB normal along Y and the upper face normal along Z as presented in Figure 1e.

Slip Systesm	Ni Crystal I					Ni Crystal II				
	Schmid Factor	RSS/IS*1	θ_{Up} (°)	θ_{Side} (°)	MTF/CSS*2	Schmid Factor	RSS/IS*1	θ_{Up} (°)	θ_{Side} (°)	MTF/CSS*2
A2	0.0910	0.0246			0.8129/C5	0.0536	0.1087			0.7108/A2
A3	0.3500	0.2960	132.16	40.98	0.8919/C3	0.0373	0.0051	6.11	10.27	0.8250/A3
A6	0.4410	0.3205			0.7393/C1	0.0909	0.1138			0.7049/A6
B2	0.0929	0.1054			0.8635/B5	0.4263	0.5068			0.8984/D4
B4	0.4461	0.3624	58.87	142.12	0.8892/B2	0.4377	0.5270	95.09	133.01	0.8435/B5
B5	0.3532	0.2570			0.8435/B4	0.0113	0.0202			0.8635/B2
C1	0.0051	0.0419			0.9179/C1	0.3468	0.4131			0.9179/C1
C3	0.4355	0.3354	49.30	61.63	0.9594/C3	0.3606	0.4269	88.56	62.62	0.9594/C3
C5	0.4304	0.2935			0.7803/C5	0.0138	0.0137			0.8129/A2
D1	0.0032	0.0390			0.7801/B4	0.0259	0.0151			0.6515/A6
D4	0.3394	0.2689	150.26	137.01	0.8984/B2	0.0398	0.0950	175.07	5.97	0.6161/B5
D6	0.3394	0.2689			0.8984/B2	0.0658	0.0799			0.6715/A6

*1 RSS: Resolved shear stress normalized by the applied stress when considering elastic incompatibility stresses (IS) following the formula reported by Richeton and Berbenni [37,38,40]. *2 MTF: Maximum transmission factor based on the formula given by Shen et al. [4] with corresponding slip system (CSS) [39] in the adjacent grain.

Table 2. Slip system analyses for α-Brass micro-pillar. The first active slip system is marked in red and the second one in blue. The compressive direction is along X, the GB normal along Y and the upper face normal along Z as presented in Figure 1e.

Slip System	α-Brass Crystal I					α-Brass Crystal II				
	Schmid Factor	RSS/IS	θ_{Up} (°)	θ_{Side} (°)	MTF/CSS	Schmid Factor	RSS/IS	θ_{Up} (°)	θ_{Side} (°)	MTF/CSS
A2	0.2339	0.2075			0.7850/B5	0.1578	0.3233			0.9326/D6
A3	0.2010	0.1073	115.22	109.00	0.7433/C1	0.0460	0.0431	20.11	15.87	0.9214/D1
A6	0.4349	0.3148			0.7750/B2	0.2038	0.2802			0.9351/D4
B2	0.0305	0.0522			0.7253/B5	0.3944	0.3717			0.7750/A6
B4	0.1086	0.0218	16.08	7.04	0.9287/A6	0.3992	0.3585	64.19	145.40	0.6673/A3
B5	0.0781	0.0740			0.7785/D4	0.0048	0.0132			0.7850/A2
C1	0.3262	0.3366			0.9197/C3	0.1954	0.0783			0.8949/C3
C3	0.1650	0.0084	45.20	91.97	0.8949/C1	0.2040	0.1278	94.97	77.13	0.9197/C1
C5	0.4913	0.3450			0.7182/C5	0.0085	0.0495			0.7833/A2
D1	0.1228	0.1813			0.9214/A3	0.0412	0.0299			0.7006/D1
D4	0.2573	0.1376	159.65	22.64	0.9417/D6	0.1492	0.2738	167.81	39.88	0.7785/B5
D6	0.1345	0.0437			0.9326/A2	0.1904	0.2439			0.9417/D4

3.2. Slip Analysis by SEM

After the in-situ compression test, the slip lines were firstly analyzed by SEM. Figures 3a and 4a show the observed slip lines in SEM for Ni and α-Brass, respectively. For Ni, the slip lines were very weak on the upper surface and there were only two obvious parallel slip lines for each crystal. Based on the direction of slip lines on the upper surface, the active slip planes were identified as B planes for each crystal as shown in Figure 3. Moreover, the active incoming slip system in the crystal II was identified to be B4 as shown in Figure 3 since it corresponds to the slip system with the highest Schmid factor or the highest RSS given by incompatibility stress formula [37] (see Table 1). Slip system B4 was then considered in the discrete dislocation pile-up simulations (see Section 4). However, for α-Brass, there were two obvious slip lines intersecting the upper surface for crystal I and two obvious slip lines intersecting the side surface for crystal II. Therefore, multiple active slip systems were present in both grains for the α-Brass sample. Based on the analysis of the slip line directions, the active slip planes were determined as A and C for crystal I and B and D for crystal II as shown in Figure 4. Moreover, from the analysis of Schmid factor or incompatibility stresses as described in Table 2, the active incoming slip systems were identified as A6 and C5 for crystal I. Slip system A6 was then considered in the discrete dislocation pile-up simulations due the specific direction of the studied pile-up (see Figure 6).

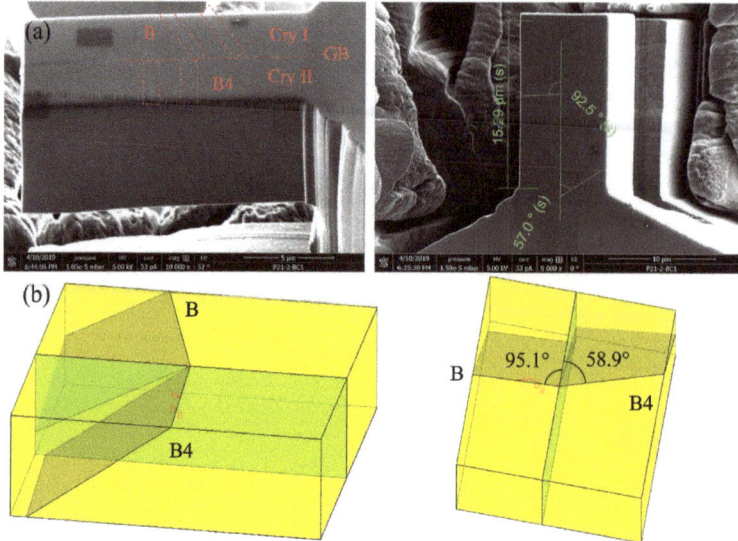

Figure 3. (**a**) crystallographic analysis of slip lines from SEM picture for the Ni sample; (**b**) theoretical analysis of slip systems based on microstructural data (see Table 1).

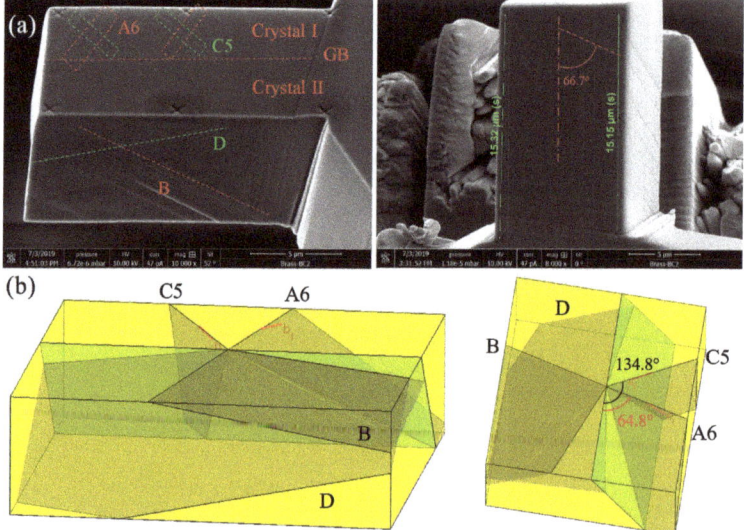

Figure 4. (**a**) crystallographic analysis of slip lines from SEM picture for the α-Brass sample; (**b**) theoretical analysis of slip systems based on microstructural data (see Table 2).

3.3. Slip Analysis by Ex-Situ AFM

AFM measurements were performed on the upper surface by Dimension FastScan with ScanAsyst (Bruker, (Billerica, Mass., USA)) with tapping mode, so that the 3D topography of slip lines can be obtained and used to analyze dislocation distributions. It should be pointed out that these AFM measurements were carried out post-mortem after unloading (ex-situ measurements). The step size of scan was 30 nm with a scan rate of 1 Hz. The results were analyzed by NanoScope Analysis (v180r2sr3, Bruker, Billerica, Mass., USA). They were first of all flattened by polynomial fit of the second order, and then they were treated by a median filter with 9 × 9 matrix operation. Figures 5a and 6a show the 3D topography of the upper surface of Ni and α-Brass samples, respectively. The slip steps correspond to the slip lines presented in the SEM picture (Figure 3a for Ni and 4a for α-Brass). For the Ni sample, a slip step with classic features of dislocation pile-up was selected for analysis. The heights of the top line and the bottom line of this step were measured along the slip direction from GB which are marked as red and black lines, respectively, in Figure 5a. The measured data are presented in Figure 5c with the corresponding name "Top line" and "Bottom line". Then, they were fitted by the polynomial method which are marked as "Fit h^{Top}" and "Fit h^{Bottom}" in Figure 5c. The relative height of this slip step Δh was calculated as the difference of the fitted heights between these two lines as shown in Figure 5c with the blue line. It was found that the height difference between the two lines at GB, $\Delta h^{GB} \approx 0.86$ nm, was not zero, which is due to a weak slip transmission and/or dislocation absorption as observed in Figure 5a. The slip step height increased from GB along the slip direction and reached its maximum value $\Delta h \approx 9.06$ nm at about $d \approx 3.28$ μm.

A similar analysis was performed in the crystal I of the α-Brass sample as shown in Figure 6. The observed pile-up had a typical dislocation configuration corresponding to slip on system A6. The slip transmission and/or dislocation absorption were even more intense compared to the Ni sample as $\Delta h^{GB} \approx 4.29$ nm, but the propagation of dislocations in the adjoining grain had a shorter distance. The maximum of slip step height was $\Delta h \approx 10.03$ nm at about $d \approx 2.44$ μm. There was an obvious peak valley in the middle part of the curve of slip step height, which might be caused by the intersection with another non coplanar slip line.

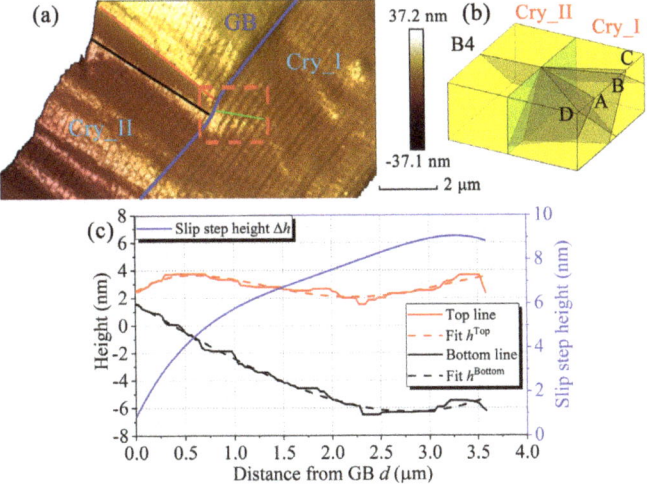

Figure 5. (**a**) ex-situ AFM topographic measurement for the Ni sample. The transmission phenomenon and/or dislocation absorption at GB are surrounded by the frame with red dashed lines. The green line indicates the direction of slip transmission. By comparison of this direction with the schematic presentation of the four slip planes in the crystal I (**b**) and by analysis of the transmission factors in Table 1 (B5 has the largest transmission factor), it can be assumed that the transmission phenomenon occurs in the B slip plane; (**c**) the result of slip step height measurement: red and black solid lines indicate the measured height of the top line and the bottom line of the step, respectively. The dash lines with corresponding color indicate the fitted results by the polynomial method with the correlation coefficient $R^2 = 0.8176$ for the top line and $R^2 = 0.9898$ for the bottom line. The blue line indicates the relative height of the slip step calculated as the difference of the fitted heights.

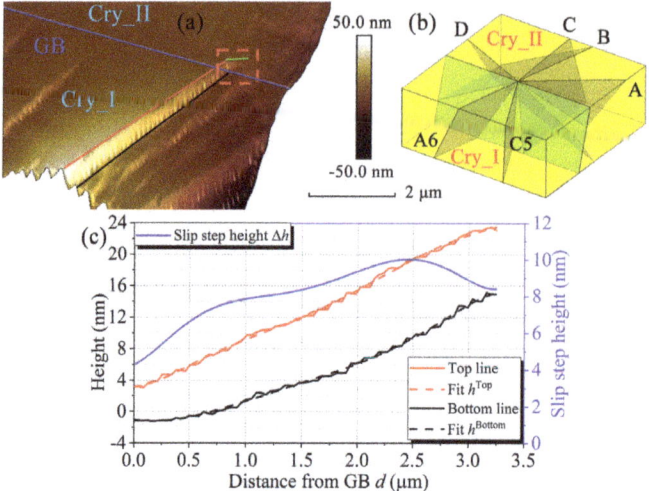

Figure 6. (a) ex-situ AFM topographic measurement for the α-Brass sample. The transmission phenomenon and/or dislocation absorption at GB are surrounded by the frame with red dashed lines. The green line indicates the direction of slip transmission. By comparison of this direction with the schematic presentation of the four slip planes in the crystal II (b), it can be assumed that the transmission phenomenon occurs in the A slip plane; (c) the result of slip step height measurement: red and black solid lines indicate the measured height of the top line and the bottom line of the step, respectively. The dash lines with corresponding color indicate the fitted results by the polynomial method with the correlation coefficient $R^2 = 0.9982$ for the top line and $R^2 = 0.9973$ for the bottom line. The blue line indicates the relative height of the slip step calculated as the difference of the fitted heights.

4. Discrete Dislocation Pile-Up Simulations and Discussion

4.1. Discrete Dislocation Pile-Up Configuration in Heterogeneous and Anisotropic Elasticity

First of all, let us consider the single slip configuration of a single dislocation-ended discrete pile-up in a bi-crystal model accounting for heterogeneous anisotropic elasticity. There are N infinite straight dislocations in the pile-up, which are all parallel to the x_3-axis, having the same Burgers vector b and lying in the same slip plane as shown in Figure 7a.

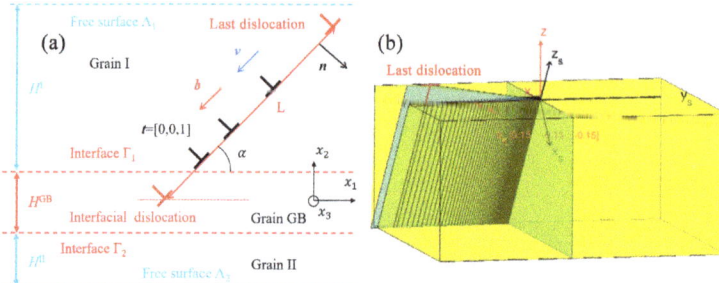

Figure 7. (a) α-inclined single-ended discrete dislocation pile-up in a slip plane of unit normal n for N equilibrated edge dislocations with Burgers vector b and line vector t in a heterogeneous anisotropic bi-crystalline micro-pillar. GB is regarded as an interphase of thickness H^{GB} bounded by two planar interfaces Γ_1 and Γ_2. If the free surfaces Λ_1 and Λ_2 are considered, Grain I and Grain II are finite with the corresponding thickness H^I and H^{II}. Conversely, when free surfaces are not considered, Grain I and Grain II are infinite, thus there are only two interfaces Γ_1 and Γ_2 and the free surfaces Λ_1 and Λ_2 do not exist; (b) schematic figure of the simulation configuration adopted for the experiment. (x_S, y_S, z_S) indicate the simulation coordinates system, while (X, Y, Z) indicates the global sample coordinates system.

The equilibrium positions of the N dislocations can be found out by minimizing the component of the Peach–Koehler (P–K) force along the slip direction for each dislocation to a material's critical force F_c as follows:

$$F^{(\gamma)} = \text{abs}(\{(\sigma_{int}(X_1(\gamma), X_2(\gamma)) + \sigma_{ext}) \cdot b \times t\} \cdot v) \xrightarrow{\text{Minimize}} F_c \qquad (1)$$

where $(X_1(\gamma), X_2(\gamma))$ denotes the position of the γth dislocation. v is a unit vector that belongs to the slip plane and is normal to the dislocation line. It indicates the glide direction of all the dislocations. It is considered to be directed towards the GB so that a pile-up can form. σ_{ext} is a homogeneous applied stress tensor to the bi-crystal and $\sigma_{int} = \sigma_{im} + \sigma_{dis}$ is the internal stress tensor produced by all the other dislocations, σ_{dis}, and the image stress tensor σ_{im} on this particular dislocation coming from all elastic heterogeneities, see the detailed calculations in [41] using the Leknitskii–Eshelby–Stroh (L–E–S) formalism for two-dimensional anisotropic elasticity. Here, the elastic heterogeneities include the GB seen as an interphase [41], both grains related by a misorientation and two free surfaces (denoted as Λ_1 and Λ_2 in Figure 7a). The interphase is supposed to be the region between two planar interfaces Γ_1 and Γ_2 with a thickness H^{GB} as shown in Figure 7a. The image stress produced by these heterogeneities can be calculated using the image decomposition method for anisotropic multilayers [42]. In this method, all the interfaces and the free surfaces are regarded as a distribution of image dislocation densities which can be resolved through boundary conditions [42]. The interphase model allows considering a non-zero thickness in the nanometer range and a specific elastic stiffness tensor for the GB region. While the configuration with two free surfaces can be used to study size effects [43], more discussions about the interphase model and the effect of free surfaces can be found in [44]. At the end, the critical force F_c may include the lattice friction force (primary) and other forces due to obstacles to dislocation motion (solute atoms, precipitates, etc.). Meanwhile, this critical force can be converted into a shear stress on a dislocation by dividing by the corresponding Burgers vector magnitude as $\tau_c = F_c / |b|$. Most past studies assume $F_c = 0$ N/m in Equation (1) because of the low value of the lattice friction stress in pure FCC crystals, which is around $1 \sim 2$ MPa [45]. However, in the present study, it is found that a non-zero critical force has a crucial effect on the discrete distribution of dislocations in the presence of GB and free surfaces. The reason is that there are already lattice defects in the material after sample preparation, such as the Ga^+ ions from FIB [46]. Furthermore, for α-Brass, the theoretical value of the lattice friction force should be higher than in pure FCC crystals, like pure Ni. Thus, the critical force cannot be ignored in Equation (1) for realistic discrete pile-up calculations. The calculations of the discrete dislocation pile-up equilibrium positions are thus obtained by following an iterative relaxation scheme that minimizes all the $F^{(\gamma)}$ after an initial configuration is specified [41,47].

4.2. Simulation Configuration of Experiments

Combined with the discussion of image stress effects [41,44], the slip step heights due to a dislocation pile-up measured in experiments (see Figure 5c for Ni and Figure 6c for α-Brass) can be simulated in a bi-crystal configuration containing a GB and two free surfaces using a discrete dislocation pile-up modelling in heterogeneous anisotropic elasticity. The hypotheses and configurations are set as follows and presented in Figure 7:

1. The dislocation lines are supposed to be infinite straight lines. They are parallel to each other and also parallel to the GB plane.
2. For the simulation coordinates system, the x_3 direction is set to be the direction of dislocation line z_S, and the x_2 direction is set to be the direction of GB normal y_S as presented in Figure 7b. Then, x_1 is determined by the vector product $x_1 = x_2 \times x_3$ which is presented as x_S in Figure 7b. Furthermore, all the used vectors and tensors, such as the Burgers vector of active slip system, the slip direction vector, and the elastic stiffness tensor, etc. are transformed into the simulation coordinates system.

3. The position of the maximum slip step height measured in the experiment is considered to be the end of the discrete dislocation pile-up. However, in the model, the last dislocation is fixed at the position of the observed maximum slip step height. Hence, the length of the dislocation pile-up as shown in Figure 7 is the same in all the simulations. The P–K force on this last dislocation is not zero. It is assumed that this dislocation is locked on some material defect not considered in the simulation. Furthermore, it is supposed that there is no more dislocation between the maximum slip step height and the side free surface as if the dislocations in this area have escaped through the side free surface.

4. All the presented simulations are performed by the image decomposition method for anisotropic multilayers problem which only considers linear anisotropic elasticity [42,44]. In the present study, a three layers configuration is used. GB is always regarded as an elastic interphase with a thickness $H^{GB} = 0.9$ nm (the second layer in the model, marked as Grain GB in Figure 7a). This value has been obtained from Molecular Statics (MS) simulations on the Ni sample. The stiffness tensor of this interphase is here simply modeled as a first approximation by $C^{GB}_{ijkl} = \left(C^{I}_{ijkl} + C^{II}_{ijkl} \right)/2$, which is the average of the two grains' elastic stiffness tensors.

5. As the theory is two-dimensional (invariance along the dislocation line), dislocations with different line directions cannot be considered at the same time. Thus, the transmitted and/or absorbed dislocations are modeled as an interfacial super-dislocation fixed in the middle of the GB interphase. Therefore, $b^{Tran} = N^{Tran} \times b^{GB}$, where b^{Tran} is the Burgers vector of the super-dislocation, N^{Tran} the number of transmitted and/or absorbed dislocations, and b^{GB} is the Burgers vector of dislocations stored at GB. b^{GB} can be equal to the Burgers vector of the incoming dislocations, or equal to the residual Burgers vector between incoming slip system (b^{In}) and one of the 12 outgoing slip systems (b^{Out}) in the adjacent grain defined as $b^{GB} = b^{In} - b^{Out}$.

6. For the present results, the applied stress tensor σ_{ext} is always considered as homogeneous in both grains without considering incompatibility stresses [37,38,40,41]. Thus, σ_{ext} has only one non-zero component σ_X in the global sample coordinate system (see Figure 7b), which is considered as a uniaxial compression test along the GB direction. The value of σ_X for each sample is the maximum applied stress in the compression test, which can be obtained from experimental stress–strain curves, thus $\sigma_X = 289.4$ MPa for the Ni sample and $\sigma_X = 127.4$ MPa for the α-Brass sample (see Figure 2c,d).

7. If not specifically stated in the text, the default value of the critical force is equal to zero, as $F_c = 0$ N/m. This default value will only correspond to Simulations 1 and 2 in Section 4.3.

8. The equilibrium positions of the dislocations in the pile-up are determined by Equation (1), then the slip step height at a given position along the slip direction can be calculated by:

$$\Delta h(d) = N(d) \times b(Z) \qquad (2)$$

where d is the distance from GB along slip direction, N the number of dislocations in the dislocation pile-up from GB to the position d, and $b(Z)$ is the out-of-surface component of the Burgers vector (along Z direction, which is perpendicular to the upper surface as shown in Figure 7b).

9. In the following parts, the measured slip step height is calibrated to be zero at GB, which means $\Delta h = h^{Top} - h^{Bottom} - \Delta h^{GB}$. Similarly, the simulated slip step height profile is considered to be zero at GB as $\Delta h(0) = 0$.

4.3. Simulation Results and Discussions

4.3.1. Results for Ni Sample

Based on the hypotheses described in Section 4.2 and considering the B4 slip system in the crystal II of the studied Ni sample, the out-of-surface component of the Burgers vector is about 0.15 nm. The maximum slip step height is about 9.1 nm at $d \approx 3.28$ µm and the slip step height at GB due to slip transmission and/or dislocation absorption is about 0.86 nm. Thus, based on Equation (2), the number of dislocations in the pile-up is 61 and the number of transmitted and/or absorbed dislocations is 6. Therefore, the number of dislocations in the pile-up is equal to 55. The applied stress is 289.4 MPa as measured in experiment, and it corresponds to a resolved shear stress $\tau = 129.0$ MPa. The thicknesses of Grains I and II are $H^I = 4.30$ µm and $H^{II} = 3.73$ µm as shown in Figure 1b. With all these parameters, the slip step height distribution due to dislocations in the pile-up was simulated for different critical forces while considering or not the effect of free surfaces as shown in Figure 8a.

Comparing the result of Simulation 1 without free surfaces to Simulation 2 with two free surfaces, it is found that the free surfaces have an influence on dislocations behavior. From the theory, it is known that free surfaces have always an attractive effect on dislocations. When the dislocations are closer to the first free surface Λ_1 than the second one Λ_2 (see Figure 7a), the total force without considering the stress field of the last dislocation is always towards Λ_1, and so dislocations move towards the last dislocation to reach equilibrium. However, the dislocations around GB are nearly in the middle of both free surfaces, thus the effects of the two free surfaces are balanced out by each other. Therefore, the free surfaces have much more effects on the dislocations which are near the free surface rather than the ones which are located around GB. These two simulation results are close to the experimental measurement, but there are still some discrepancies. Compared to the experimental measurement, the dislocations are closer to GB without free surfaces, while they are closer to the first free surface when considering the effect of free surfaces.

As discussed in Section 4.1, it is actually necessary to consider the effect of critical force F_c (see Equation (1)). After the analysis of the results with different value of F_c, it is found that F_c moves the dislocations towards the GB. When considering the effect of free surfaces with a material's critical force $F_c = 0.003$ N/m in Equation (1) (equivalent to a resolved shear stress $\tau_c = 12$ MPa), the simulation result is closer to the experimental measurement as shown in Figure 8a with Simulation 3. Here, the value is higher than the theoretical value of lattice friction for pure FCC crystals which is around $1 \sim 2$ MPa [45]. The reason might be due to the sample preparation, such as defects coming from FIB polishing. Furthermore, as discussed in Section 4.2 (3), the P–K force is not zero on the last fixed dislocation. The total resolved shear stresses on this fixed dislocation are 6.6 MPa for Simulation 1, 175.0 MPa for Simulation 2 and 71.2 MPa for Simulation 3. The value found in Simulation 2 is quite high for a FCC crystal and appears unrealistic.

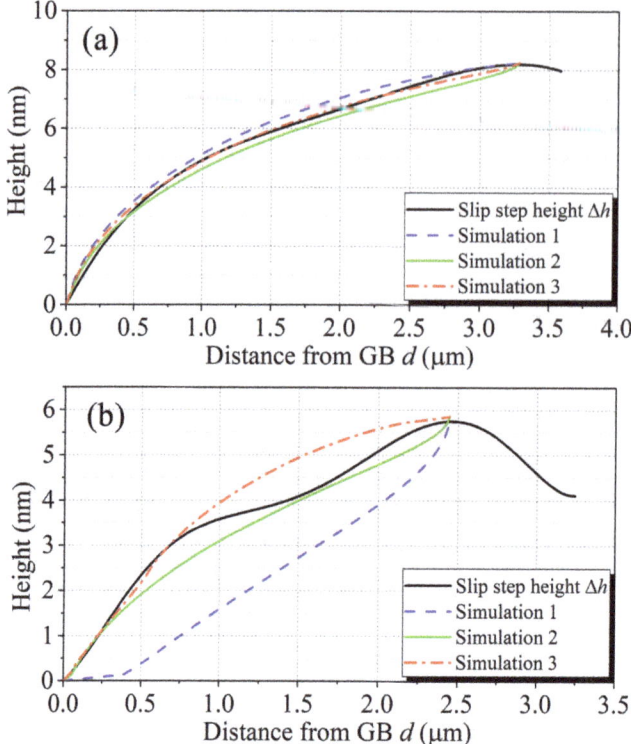

Figure 8. (a) simulation of slip step height profile for Ni with different discrete dislocation pile-up simulation conditions (b^{GB} is always the same as b^{In}). Simulation 1: without free surfaces. Simulation 2: with two free surfaces. Simulation 3: with two free surfaces and a non zero material dependent critical force $F_c = 0.003$ N/m; (b) simulation of slip step height profile for α-Brass with different discrete dislocation pile-up simulation conditions (always with two free surfaces). Simulation 1: $b^{GB} = b^{In}$. Simulation 2: $b^{GB} = b^{In} - b^{Out}$ (A6). Simulation 3: $b^{GB} = b^{In} - b^{Out}$ (A6) with a non zero material dependent critical force $F_c = 0.011$ N/m.

4.3.2. Results for the α-Brass Sample

Similarly to the Ni sample, the same simulations have been performed for the A6 slip system in the crystal I of the α-Brass sample. In this case, the out-of-surface component of the Burgers vector is about 0.14 nm. The maximum slip step height is about 10.03 nm at $d \approx 2.44$ µm and the slip step height at GB due to slip transmission and/or dislocation absorption is about 4.29 nm. Based on Equation (2), the number of dislocations in the pile-up is 72 and the number of transmitted and/or absorbed dislocations is 31. Therefore, the number of dislocations in the pile-up is equal to 41. The applied stress is 127.4 MPa following experimental results, and it corresponds to a resolved shear stress $\tau = 55.4$ MPa. The thicknesses of Grains I and II are $H^I = 3.97$ µm and $H^{II} = 5.05$ µm as shown in Figure 1d.

The effects of free surfaces and critical force for the α-Brass sample are the same as for the Ni sample. However, the number of transmitted and/or absorbed dislocations is much larger in the α-Brass sample than in the Ni one, i.e., 31 instead of 6. Thus, the repulsive force from these interfacial dislocations is more important if their Burgers vector is considered as the same as the incoming dislocations (see the Simulation 1 in Figure 8b). All dislocations are moved far away from GB into the direction of free surface due to this repulsive force. The transmission mechanism observed in Figure 6a leads us to consider a different residual Burgers vector for interfacial dislocations. After verifying all the possible residual Burgers vectors with the 12 slip systems in the adjacent grain, it is found that the best solution is with A6 slip system in the adjacent grain while considering free surfaces and a reasonable critical force $F_c = 0.011$ N/m ($\tau_c = 43$ MPa) as shown in Figure 8b with Simulation 3. Furthermore, the theoretical analysis of the slip trace of A6 slip system on the upper surface of the adjacent grain agrees well with the experimental observation as shown in Figure 6a compared to other slip systems, such as B2 slip system with the maximum transmission factor [4] or B4 slip system with the maximum Schmid factor in the adjacent grain (see Table 2). As discussed in Section 3.3 for the α-Brass sample, the peak valley in the middle part of the measured curve is certainly caused by the crossing with a second non coplanar active slip system. In the present L–E–S two-dimensional anisotropic elastic theory, it is not possible to consider the effect of two dislocations whose dislocation lines are not parallel. Thus, this effect of peak valley part cannot be simulated within the present theory. However, it is important to well identify the part close to GB with $d < 0.75$ µm. Here, the suitable critical stress is $\tau_c = 43$ MPa. It is larger than the critical stress for the Ni sample, which may be due to alloying element hardening. Furthermore, the total resolved shear stresses for the last fixed dislocation are 997.3 MPa for Simulation 1, 258.9 MPa for Simulation 2 and 28.6 MPa for Simulation 3. Thus, Simulation 3 appears to be the most realistic one.

4.3.3. Discussion

Except the free surfaces effects, it is found that the Burgers vector of interfacial dislocations has an obvious influence on the dislocation distribution in the pile-up. It should be pointed out that, for the Ni sample, b^{GB} has been tried with all the 12 residual Burgers vectors, but the results are never better than in the case with $b^{GB} = b^{In}$. Thus, the Burgers vector of interfacial dislocations can be regarded as the same as incoming dislocations in case of a low number (6) of transmitted and/or absorbed dislocations in the Ni sample. However, in case of a high number (31) of transmitted and/or absorbed dislocations in the α-Brass sample, one needs to consider the residual Burgers vector of interfacial dislocations.

In addition to the effects of interfacial dislocations, the critical force is also another important parameter to find the correct equilibrated dislocation distribution. Comparing the simulation 2 ($F_c = 0$ N/m) with the Simulation 3 ($F_c \neq 0$ N/m) presented in Figure 8 for both Ni and α-Brass samples, it is found that the critical force does not have an obvious effect on dislocations that are located at a distance less than 0.3 µm from GB. From Equation (1), it is found that this critical force does not have a uniform effect on all the dislocations within the pile-up. For example, when the total force on a dislocation is towards free surface for positions close to the free surface, the critical force will point towards the GB. However, the stress state is more complicated in regions close to the GB. The total force on a dislocation here can point towards the GB due to repulsive forces from other dislocations, attractive misorientation [41], compliant GB [41], attractive force from interfacial dislocations, and applied stress. On the contrary, it can point towards the opposite direction due to repulsive misorientation [41], rigid GB [41], repulsive force from interfacial dislocations, other dislocations, and free surfaces. Hence, the total force depends on many physical parameters. Thus, the critical force could either move the dislocations towards GB or to the opposite direction.

As mentioned in Section 3.3, the ex-situ AFM measurements were carried out after compression and elastic unloading; thus, for the present static dislocation pile-up simulation, the applied stress should be set to zero. However, in the present study, the dislocations in the pile-up are supposed to be unchanged during unloading, which means that the equilibrium positions of dislocations obtained at the onset of plastic loading are consistent before and after unloading process due to a large critical force in the crystal. The found critical forces $F_c = 0.003$ N/m for Ni and $F_c = 0.011$ N/m for α-Brass are supposed to be the original critical force before slip activation with a low dislocation density in the crystal.

Furthermore, the effects of incompatibility stresses [37], elastic anisotropy compared to elastic isotropy, number of interfacial dislocations, applied stress, and misorientation can be also investigated with the developed model [44]. As a summary, for the present samples, the main effects come from the external stress, heterogeneous and anisotropic elasticity, and the number of interfacial dislocations which should be equal to the one of the transmitted and/or absorbed dislocations.

In Figure 8b, it was also found that the measured slip step height decreases significantly after its maximum value along the slip direction. This tendency might be the signature of a double-ended dislocation pile-up due to a less penetrable layer than the free surface coming from surface defects produced by FIB polishing. The effect of these surface defects seems to be stronger for the α-Brass sample compared to the Ni sample. Hence, the present model which was designed to account for single slip within a single-ended dislocation pile-up configuration is more suitable for the study of the Ni sample than the α-brass sample. In general, the present model can be applied to any crystalline structure where dislocation pile-ups can be characterized experimentally. However, it is more suitable for FCC materials where slip is relatively planar.

5. Conclusions

In this work, experimental studies of Burgers vector distribution in discrete dislocation pile-ups at GB were performed in Ni and α-Brass bi-crystalline micro-pillars. Ex-situ AFM measurements were carried out after a compression test with a flat punch nanoindenter. The measured slip step heights were simulated using a three anisotropic layers model combined with image stress calculation [42], which was more suitable for a single-ended dislocation pile-up configuration with single slip compared to multi-slip configurations. In this model, GB was regarded as an elastic interphase with a thickness $H^{GB} = 0.9$ nm, which was obtained from MS simulations. Meanwhile, the effects of two free surfaces around micro-pillars were considered. It was found that, for the Ni sample, the best solution of this simulation was obtained considering anisotropic elasticity, the effect of free surfaces, a critical shear stress $\tau_c = 12$ MPa, a homogeneous external stress, and the same Burgers vector of incoming dislocations and interfacial dislocations. However, for the α-Brass sample, still considering anisotropic elasticity, the best solution was obtained with free surfaces, a homogeneous external stress field, and a larger critical shear stress of $\tau_c = 43$ MPa. In addition, the Burgers vector of interfacial dislocations was modeled as a residual Burgers vector because there was a huge number of transmitted and/or absorbed dislocations.

Author Contributions: Conceptualization, X.C., T.R., C.M., and S.B.; methodology, X.C., T.R., C.M., and S.B.; software, X.C.; validation, T.R., C.M., and S.B.; formal analysis, X.C., T.R., C.M., and S.B.; investigation, X.C.; resources, C.M. and S.B.; data curation, X.C.; writing—original draft preparation, X.C.; writing—review and editing, T.R., C.M., and S.B.; visualization, X.C.; supervision and project administration, S.B.; funding acquisition, C.M. and S.B. All authors have read and agreed to the published version of the manuscript.

Funding: The authors are grateful to the French Ministry of Higher Education and Scientific Research and the French-German University (UFA-DFH) for financial supports. The authors are also grateful to Experimental Methodology in Materials Science (MWW) in Saarland University and the French State (ANR) through the program "Investment in the future" (LabEx "DAMAS" referenced as ANR-11-LABX-0008-01) for additional financial supports. We also would like to thank the Deutsche Forschungsgemeinschaft DFG for the financing of the AFM microscope, Grant No. INST 256/455-1 FUGG.

Acknowledgments: The authors would like to thank J. Rafael Velayarce for his help with the experiments.

Conflicts of Interest: The authors declare no conflict of interest.

Abbreviations

The following abbreviations are used in this manuscript:

GB(s)	Grain Boundary(ies)
FCC	Face-Centered Cubic
Ni	Nickel
AFM	Atomic Force Microscopy
TEM	Transmission Electron Microscope
EBSD	Electron Back Scattered Diffraction
SEM	Scanning Electron Microscope

ECCI	Electron Channelling Contrast Imaging
CPFEM	Crystal Plasticity Finite Element Method
DDD	Discrete Dislocation Dynamics
MD	Molecular Dynamics
MS	Molecular Statics
RSS	Resolved Shear Stress
IS	Incompatibility Stresses
CSS	Corresponding Slip System
P–K	Peach–Koehler
L-E-S	Leknitskii–Eshelby–Stroh

References

1. Hall, E.O. The deformation and ageing of mild steel: III discussion of results. *Proc. Phys. Soc. Sect. B* **1951**, *64*, 747–753. [CrossRef]
2. Petch, N.J. The cleavage strength of polycrystals. *J. Iron Steel Inst.* **1953**, *174*, 25–30.
3. Armstrong, R.W. The influence of polycrystal grain size on several mechanical properties of materials. *Metall. Mater. Trans. B* **1970**, *1*, 1169–1174. [CrossRef]
4. Shen, Z.; Wagoner, R.H.; Clark, W.A.T. Dislocation and grain boundary interactions in metals. *Acta Metall.* **1988**, *36*, 3231–3242. [CrossRef]
5. Kamaya, M.; Wilkinson, A.J.; Titchmarsh, J.M. Measurement of plastic strain of polycrystalline material by electron backscatter diffraction. *Nucl. Eng. Des.* **2005**, *235*, 713–725. [CrossRef]
6. Pantleon, W. Resolving the geometrically necessary dislocation content by conventional electron backscattering diffraction. *Scr. Mater.* **2008**, *58*, 994–997. [CrossRef]
7. Gutierrez-Urrutia, I.; Zaefferer, S.; Raabe, D. Coupling of Electron Channeling with EBSD: Toward the Quantitative Characterization of Deformation Structures in the SEM. *JOM* **2013**, *65*, 1229–1236. [CrossRef]
8. Perrin, C.; Berbenni, S.; Vehoff, H.; Berveiller, M. Role of discrete intra-granular slip on lattice rotations in polycrystalline Ni: Experimental and micromechanical studies. *Acta Mater.* **2010**, *58*, 4639–4649. [CrossRef]
9. Kahloun, C.; Badji, R.; Bacroix, B.; Bouabdallah, M. Contribution to crystallographic slip assessment by means of topographic measurements achieved with Atomic Force Microscopy. *Mater. Charact.* **2010**, *61*, 835–844. [CrossRef]
10. Kahloun, C.; Monnet, G.; Queyreau, S.; Le, L.T.; Franciosi, P. A comparison of collective dislocation motion from single slip quantitative topographic analysis during in-situ AFM room temperature tensile tests on Cu and Feα crystals. *Int. J. Plast.* **2016**, *84*, 277–298. [CrossRef]
11. Dehm, G.; Jaya, B.N.; Raghavan, R.; Kirchlechner, C. Overview on micro-and nanomechanical testing: New insights in interface plasticity and fracture at small length scales. *Acta Mater.* **2018**, *142*, 248–282. [CrossRef]
12. Clark, W.A.T.; Wise, C.E.; Shen, Z.; Wagoner, R.H. The use of the transmission electron-microscope in analyzing slip propagation across interfaces. *Ultramicroscopy* **1989**, *30*, 76–89. [CrossRef]
13. Gleiter, H.; Hornbogen, E.; Ro, G. The mechanism of grain boundary glide. *Acta Metall.* **1968**, *16*, 1053–1067. [CrossRef]
14. Pumphrey, P.H.; Gleiter, H. Annealing of dislocations in high-angle grain boundaries. *Philos. Mag. J. Theor. Exp. Appl. Phys.* **1974**, *30*, 593–602. [CrossRef]
15. Bayerschen, E.; McBride, A.T.; Reddy, B.D.; Böhlke, T. Review on slip transmission criteria in experiments and crystal plasticity models. *J. Mater. Sci.* **2016**, *51*, 2243–2258. [CrossRef]
16. Hunter, A.; Leu, B.; Beyerlein, I.J. A review of slip transfer: applications of mesoscale techniques. *J. Mater. Sci.* **2018**, *53*, 5584–5603. [CrossRef]
17. Livingston, J.D.; Chalmers, B. Difference fatigue cracking behaviors of Cu bicrystals with the same component grains but different twin boundaries. *Acta Metall.* **1957**, *5*, 322–327. [CrossRef]
18. Beyerlein, I.; Mara, N.; Wang, J.; Carpenter, J.; Zheng, S.; Han, W.; Zhang, R.; Kang, K.; Nizolek, T.; Pollock, T. Structure-property-functionality of bimetal interfaces. *JOM* **2012**, *64*, 1192–1207. [CrossRef]
19. Werner, E.; Prantl, W. Slip transfer across grain and phase boundaries. *Acta Metall. Mater.* **1990**, *38*, 533–537. [CrossRef]

20. Head, A.K. X. The Interaction of Dislocations and Boundaries. *Lond. Edinb. Dublin Philos. Mag. J. Sci.* **1953**, *44*, 92–94. [CrossRef]
21. Pacheco, E.S.; Mura, T. Interaction between a screw dislocation and a bimetallic interface. *J. Mech. Phys. Solids* **1969**, *17*, 163–170. [CrossRef]
22. Koehler, J.S. Attempt to Design a Strong Solid. *Phys. Rev. B* **1970**, *2*, 547–551. [CrossRef]
23. Krzanowski, J.E. The effect of composition profile shape on the strength of metallic multilayer structures. *Scr. Metall. Mater.* **1991**, *25*, 1465–1470. [CrossRef]
24. Anderson, P.M.; Xin, X.J. The critical shear stress to transmit a peierls screw dislocation across a non-slipping interface. In *Multiscale Deformation and Fracture in Materials and Structures. Solid Mechanics and Its Applications*; Chuang T.J., Rudnicki J.W., Eds.; Springer Science & Business Media: Berlin, Germany, 2000, Volume 84, pp. 87–105.
25. Imrich, P.J.; Kirchlechner, C.; Motz, C.; Dehm, G. Differences in deformation behavior of bicrystalline Cu micropillars containing a twin boundary or a large-angle grain boundary. *Acta Mater.* **2014**, *73*, 240–250. [CrossRef]
26. Imrich, P.J.; Kirchlechner, C.; Dehm, G. Influence of inclined twin boundaries on the deformation behavior of Cu micropillars. *Mater. Sci. Eng. A* **2015**, *642*, 65–70. [CrossRef]
27. Knorr, A.F.; Marx, M.; Schaefer, F. Crack initiation at twin boundaries due to slip system mismatch. *Scr. Mater.* **2015**, *94*, 48–51. [CrossRef]
28. Malyar, N.V.; Micha, J.-S.; Dehm, G.; Kirchlechner, C. Dislocation-twin boundary interaction in small scale Cu bi-crystals loaded in different crystallographic directions. *Acta Mater.* **2017**, *129*, 91–97. [CrossRef]
29. Kheradmand, N.; Vehoff, H.; Barnoush, A. An insight into the role of the grain boundary in plastic deformation by means of a bicrystalline pillar compression test and atomistic simulation. *Acta Mater.* **2013**, *61*, 7454–7465. [CrossRef]
30. Anderson, P.M.; Hirth, J.P.; Lothe J. *Theory of Dislocations*, 3rd Ed.; Cambridge University Press: Cambridge, UK, 2017.
31. Rayne, J.A. Elastic Constants of α-Brasses: Room-Temperature Variation with Solute Concentration. *Phys. Rev.* **1958**, *112*, 1125–1130. [CrossRef]
32. Rayne, J.A. Elastic Constants of α-Brasses: Variation with Solute Concentration from 4.2–300 °K. *Phys. Rev.* **1959**, *115*, 63–66. [CrossRef]
33. Zhao, Y.H.; Liao, X.Z.; Zhu, Y.T.; Horita, Z.; Langdon, T.G. Influence of stacking fault energy on nanostructure formation under high pressure torsion. *Mater. Sci. Eng. A* **2005**, *410–411*, 188–193. [CrossRef]
34. Rémy, L. Maclage et transformation martensitique CFC → HC induite par déformation plastique dans les alliages austénitiques à basse énergie de défaut d'empilement des sytèmes CO-NI-CR-MO et FE-MN-CR-C. Ph.D Thesis, Université de Paris-Sud, Orsay, France, 1975.
35. Tiba, I. Effets des Interfaces Cristallines sur les Champs méCaniques en Plasticité Cristalline et conséQuences sur le Glissement Dans des Micro-Pilliers Bi-Cristallins. Ph.D. Thesis, Université de Lorraine et Université de Saarbrücken, Metz, France, 2015.
36. Itoh, I.; Hikage, T. Dezincification Mechanism of Brass in Vacuum at High Temperature. *Trans. Jpn. Inst. Met.* **1976**, *17*, 165–169. [CrossRef]
37. Richeton, T.; Berbenni, S. Effects of heterogeneous elasticity coupled to plasticity on stresses and lattice rotations in bicrystals: A Field Dislocation Mechanics viewpoint. *Eur. J. Mech. A/Solids* **2013**, *37*, 231–247. [CrossRef]
38. Tiba, I.; Richeton, T.; Motz, C.; Vehoff, H.; Berbenni, S. Incompatibility stresses at grain boundaries in Ni bicrystalline micropillars analyzed by an anisotropic model and slip activity. *Acta Mater.* **2015**, *83*, 227–238. [CrossRef]
39. Schmid, E.; Boas, W. *Kristallplastizität mit Besonderer Berücksichtigung der Metalle*; Springer: New York, NY, USA, 1935.
40. Richeton, T.; Tiba, I.; Berbenni, S.; Bouaziz, O. Analytical expressions of incompatibility stresses at $\Sigma 3 \langle 111 \rangle$ twin boundaries and consequences on single-slip promotion parallel to twin plane. *Philos. Mag.* **2015**, *95*, 12–31. [CrossRef]
41. Chen, X.; Richeton, T.; Motz, C.; Berbenni, S. Elastic fields due to dislocations in anisotropic bi- and tri-materials: Applications to discrete dislocation pile-ups at grain boundaries. *Int. J. Solids Struct.* **2019**, *164*, 141–156. [CrossRef]

42. Wang, H.Y.; Wu, M.S.; Fan, H. Image decomposition method for the analysis of a mixed dislocation in a general multilayer. *Int. J. Solids Struct.* **2007**, *44*, 1563–1581. [CrossRef]
43. Rafael Velayarce, J.; Motz, C. Effect of Sample Size and Crystal Orientation on the Fatigue Behaviour of Single Crystalline Microbeams. *Materials* **2020**, *13*, 741. [CrossRef]
44. Chen, X. Experimental and Theoretical studies of Incompatibility and Dislocation Pile-up Stresses at Grain Boundaries Accounting for Elastic and Plastic Anisotropies. Ph.D. Thesis, Université de Lorraine et Université de Saarbrücken, Metz, France, 2020.
45. Suzuki, T.; Takeuchi, S. Correlation of Peierls-Nabarro Stress with Crystal Structure. *Rev. Phys. Appl.* **1988**, *23*, 685–685. [CrossRef]
46. Kheradmand, N. Grain Boundary-Dislocation Interaction: A local Investigation via Micron-Sized Bicrystals. Ph.D. Thesis, Universität des Saarlandes, Saarbrücken, Germany, 2012.
47. Wagoner, R.H. Calculating dislocation spacings in pile-ups at grain boundaries. *Metall. Trans. A* **1981**, *12*, 2015–2023. [CrossRef]

© 2020 by the authors. Licensee MDPI, Basel, Switzerland. This article is an open access article distributed under the terms and conditions of the Creative Commons Attribution (CC BY) license (http://creativecommons.org/licenses/by/4.0/).

Review

Emission of Dislocations from Grain Boundaries and Its Role in Nanomaterials

James C. M. Li [1,*], C. R. Feng [2] and Bhakta B. Rath [2]

[1] Department of Mechanical Engineering, University of Rochester, Rochester, NY 14627, USA
[2] U.S. Naval Research Laboratory, 4555 Overlook Ave SW, Washington, DC 20375, USA; jerry.feng@nrl.navy.mil (C.R.F.); drrath34@gmail.com (B.B.R.)
* Correspondence: jli@ur.rochester.edu

Abstract: The Frank-Read model, as a way of generating dislocations in metals and alloys, is widely accepted. In the early 1960s, Li proposed an alternate mechanism. Namely, grain boundary sources for dislocations, with the aim of providing a different model for the Hall-Petch relation without the need of dislocation pile-ups at grain boundaries, or Frank-Read sources inside the grain. This article provides a review of his model, and supporting evidence for grain boundaries or interfacial sources of dislocations, including direct observations using transmission electron microscopy. The Li model has acquired new interest with the recent development of nanomaterial and multilayers. It is now known that nanocrystalline metals/alloys show a behavior different from conventional polycrystalline materials. The role of grain boundary sources in nanomaterials is reviewed briefly.

Keywords: dislocation emission; grain boundaries; nanomaterials; Hall-Petch relation; metals and alloys

1. Introduction

To explain the properties of crystalline aggregates, such as crystal plasticity, Taylor [1,2] provided a theoretical construct of line defects in the atomic scale of the crystal lattice. With the use of the electron microscope, the sample presence of dislocations validated Taylor's theory. However, the questions remained on the source of dislocations and the conditions required for their generation and migration in the crystal. While Frank and Reed [3] presented a mechanism to show the generation of many dislocations when the crystal is subjected to an applied stress, electron microscopic observations provided the support for the proposed mechanism. The rare observation of the Frank-Read (FR) source could not be justified for the large density of dislocations seen in crystals.

Li [4] was the first to recognize this issue and put forth the concept of grain boundary ledges as sources for dislocations. Li proposed the emission of dislocations from grain boundary ledges of the type shown in Figure 1. He assumed that the density of ledges is about the same in the grain boundary, which implies that fine-grained materials have a greater dislocation density when they yield. Since then, there have been numerous confirmations of grain boundary dislocation sources under an applied stress or fluctuation of temperature. Direct evidence of these ledges has been found using transmission electron microscopy (TEM).

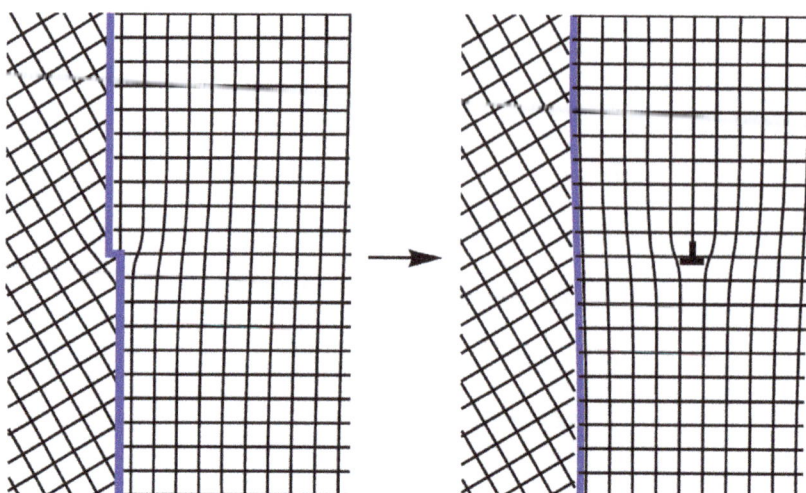

Figure 1. Grain boundary ledge model. (Schematic diagram) From Li [4].

Hall-Petch (HP) relation and the pile-up model were originally developed for iron and used the concept of pile-up of dislocations, originating from the Frank-Read source. Pile-up model has been frequently quoted to explain the HP relationship in other materials. However, it is somewhat ironic that the Frank-Read source and dislocation pile-ups were rarely, if ever found in iron. Li proposed the grain boundary source model. He has considered that dislocations are generated at grain boundary (GB) ledges and all of these dislocations migrate to dislocation forests in grain interiors. Li assumed that the density of ledges is about the same in the grain boundary, so the dislocation density will scale inversely with grain size. Dislocation density also scales as square of yield stress. The two relations, when combined, give the well-known Hall-Petch relation. In fact, Li compared the magnitude of stress in his model with that of the pile-up model with a reasonable agreement.

Ashby [5] described polycrystals as plastically non-homogeneous materials in which each grain deforms by different amounts, depending on its orientation and the constraints imposed by neighboring grains. If each grain were forced to undergo the same uniform strain, voids would form at grain boundaries. To avoid the creation of voids, and ensure strain compatibility, dislocations are introduced. Voids can be corrected by the same process by dislocations of opposite signs. Ashby [5] showed that the number of geometrically necessary dislocations (GNDs) generated at the boundaries and pumped into the individual grains is roughly De/4b, where D is the grain diameter, e is the average strain, and b the magnitude of the dislocation Burgers vector. In both two- and three-dimensional arrays of grains, this leads to a density of geometrically necessary dislocations. These geometrically necessary dislocations can be the origin of the grain size dependence of work-hardening in polycrystals. If it is assumed that hardening depends only on average dislocation density, then this approach yields a $\sigma \sim D^{-1/2}$ law for grain boundary hardening, where σ is the yield stress. The concept of generating geometrically necessary dislocations to ensure strain compatibility in polycrystalline materials eliminates the necessity to consider dislocation pile-ups to rationalize the flow of the stress-grain size relationship. It suggests that grain interiors behave essentially like single crystals of the same orientation deforming by single slip while the regions on either side of the grain boundaries undergo lattice rotation and secondary slip. This, in fact, has been observed experimentally by Essmann, et al. [6]. Emission of geometrically necessary dislocations occurs from the grain boundary, which must act as a major source of dislocations.

A comprehensive review of Li's model, its experimental verification, and its comparison with Frank-Read (FR) source models, especially pile-up models, have been given by Murr [7]. A more recent review of Li's model and its implications, vis-a-vis Hall-Petch relation, is also given by Naik and Walley [8]. They reviewed some of the factors that control the hardness of polycrystalline materials with grain sizes less than 1 μm, especially the fundamental physical mechanisms that govern the hardness of nanocrystalline materials. For grains less than 30 nm in size, there is evidence for a transition from dislocation-based plasticity to grain boundary sliding, rotation, or diffusion as the main mechanism responsible for hardness. However, we disagree with their conclusion that the evidence surrounding the inverse Hall-Petch phenomenon is inconclusive, and can be due to processing artefacts, grain growth effects, and errors associated with the conversion of hardness to yield strength in nanocrystalline (NC) materials. The inverse Hall-Petch phenomena is now well established and subject of numerous theoretical and experimental studies. For a review, see reference [8,9]. There were some doubts about its validity in early years when the specimens tested were not fully dense or had other defects. The role of dislocations in fine grain and nanomaterials has renewed interest in the concept of GB as source of complete or partial dislocations. It was found that for most metals with grain sizes in the nanometer regime, experiments have suggested a deviation away from the HP relation relating yield stress to average grain size (It is not clear if such deviations are a result of intrinsically different material properties of nanocrystalline (NC) systems or due simply to inherent difficulties in the preparation of fully dense NC samples and in their microstructural characterization. For a detailed discussion on classical HP and Inverse HP models see the review by Pande and Cooper [9] Therefore, we will not go into the details of Li's model, except for the case of fine-grained materials, where GB sources may play a prominent role.

The use of large-scale molecular dynamics (MD) in the study of the mechanical properties of NC materials provides a detailed picture of the atomic-scale processes during plastic deformation at room temperature. Using an MD model performed at room temperature, it is suggested that the GB accommodation mechanism can be identified with GB sliding, triggered by atomic movement and to some extent stress assisted diffusion [10].

We will briefly consider these issues in subsequent sections.

2. Modeling and Simulation of Dislocations from Grain Boundaries

Computer modeling of dislocations of dislocation emission from grain boundaries has been extensively studied in recent years. Excellent work can be attributed due to Yamakov et al. [10]. Using MD simulation, they examined the nucleation of extended dislocations from the GBs in NC aluminum. They showed the length of the stacking fault connecting the two Shockley partials that form the extended dislocation, i.e., the dislocation splitting distance, r_{split}, depended not only on the stacking-fault energy but also on the resolved nucleation stress. Their simulations were only for columnar grain microstructures with a grain diameter, d, of up to 70 nm. The magnitude of r_{split} relative to d was found to represent critical length scale controlling the low-temperature mechanical behavior of NC materials. They, thus, confirmed that the mechanical properties of NC materials, such as the yield stress, depend critically on the grain size.

It was shown that under high stresses, needed to nucleate extended dislocations from GBs, the magnitude of r_{split} can be significant, particularly in a NC fcc metal. In fact, it becomes comparable to the grain size, d. Thus, the splitting distance can be considered as a new length scale, in addition to d. A complete extended dislocation cannot be nucleated unless $d > r_{split}(\sigma)$. Therefore, it appears to be a necessary condition for the dislocation–glide mechanism to operate in a NC fcc metal. The authors have beautifully demonstrated the interplay between these two length scales and their effect on the mechanical properties of NC materials.

Recent simulations have confirmed the important role dislocations are expected to play in the deformation of NC fcc metals [11–13]. However, it should be noted that for the rather small grain diameters (of up to 12 nm) considered in these simulations, single partials produce stacking faults as they glide through the grain and become incorporated into the GBs. These stacking faults remain in the grains as planar defects which are not recovered. This process, most likely, will only operate during the initial stages of deformation, and is quite different from the usual slip mechanism in materials with large grain sizes. The largest grain sizes considered in these simulations was of $d = 12$ nm) [13], and there, the dislocations could account for only about 30% of the total deformation, and for even smaller grain sizes, practically no dislocations were observed. No dislocations were observed in grains smaller than 8 nm. What was the dominant deformation mechanism observed? Most MD simulations involved GB-mediated processes, such as GB sliding [11–13] or GB diffusion [14–16]. Although most of the studies of inverse HP has been for fcc metals there is no reason to believe that it will not apply to other structures. The role of dislocations coming from the Frank-Read (FR) source diminish as grain size decreases, discussed in detail in reference 9 and 45. For grain sizes below 12 nm, the only dislocation source active would be at grain boundaries.

In recent years, the interaction of dislocations with GBs has been subject of both experimental and theoretical studies. The theoretical work of Li led to the conclusion that the nucleation of a dislocation on a GB requires GB ledges [4,13]. On the other hand, experimental studies of the plastic deformation of polycrystals at low temperatures suggested that GBs become the dominant dislocation sources for relatively small grain sizes. The nucleation process may be expected to be independent of grain size [13,16]. Therefore, it is expected that the GBs should provide the necessary dislocation sources at least for the low-temperature deformation of NC materials. Hence, at the relatively high stresses required for the nucleation process, the dislocation-glide mechanism should, in principle, take place even in NC materials. For details of their simulation procedure in NC metals with columnar microstructure, see reference [17].

The process briefly is: (1) at a high enough stress, only a single 1/6 [112] partial is nucleated from a GB or a triple junction. (2) This partial dislocation, while still attached to GB, glides through the interior of the grain, connected to the GB by the stacking fault left behind by the glide. (3) For large grains, dislocation is annihilated at the opposite GB before a second partial can be formed, and a stacking fault transecting the entire grain remains. (4) If the grain is large enough, the energy of the propagating stacking fault eventually becomes too high and a second partial is nucleated to terminate the stacking fault; together with the first partial, an extended 1/2 [110] dislocation is thus formed.

It should be stated that only if the grain size is large enough [16] for the slip mechanism by dislocation glide to be fully developed, conventional HP hardening should occur due to the retarding effect of pile-ups. For the formation of these pile-ups in NC materials, and their effect on the HP mechanism, see reference [18–20]. It is shown that interaction between two extended dislocations should lead to another length scale—the critical distance for a dislocation pile-up—above which one can expect to observe the HP effect. Thus, it is clearly demonstrated that at such small grain sizes even a single complete extended dislocation is not expected. All of this leads us to conclude that when grain sizes are in nanoscale, the dominant sources of the dislocations are grain boundaries and FR mechanism is not present.

An important contribution to this field is the work of Borovokov [21]. In this MD study, they demonstrate that one of the mechanisms that can play an important role in the strength and plasticity of metallic polycrystalline materials is the heterogeneous nucleation and emission of dislocations from GB. This was done by considering the dislocation nucleation from copper bi-crystal with a number of $\langle 110 \rangle$ tilt GBs. These GBs covered a wide range of misorientation angles. It is conclusively shown that the mechanic behavior of GBs and the energy barrier of dislocation nucleation from GBs are closely related to the lattice crystallographic orientation, GB energy, and the intrinsic GB structures. Thus, one can conclude that atomistic analysis of the nucleation mechanisms can provide the exact mechanism of this kind of nucleation and emission process, and that can help us to better understand the role of GBs as a dislocation source. For details, see also reference [22].

We should also mention, in this context, a theoretical model that has been developed by Li et al. [23]. It deals with the effect of shear-coupled migration of grain boundaries on dislocation emission in NC materials. They showed that the dislocation emission can be enhanced considerably as shear-coupled migration. They also discuss factors that can lead to an optimal dislocation emission. Their conclusions have been confirmed quantitatively by the existing MD simulations based on the results of Shiue and Lee [24].

3. Experimental Verification of Dislocation Emission

We briefly mentioned before the use of TEM in thin films for detecting evidence of dislocation emission from GBs. Additionally, a brief description is given of the experiments, which were conducted to observe in-situ dynamics of dislocation emission from GBs for bulk specimens at elevated temperatures. An uncommonly known technique of photo-emission was used. The instrument used for this purpose, is known as the photo emission electron microscope (PEEM). A brief description of the instrument and an electron micrograph are presented below. Details of the experiment is reported by Li et al. [25]. A schematic diagram of PEEM is shown in Figure 2.

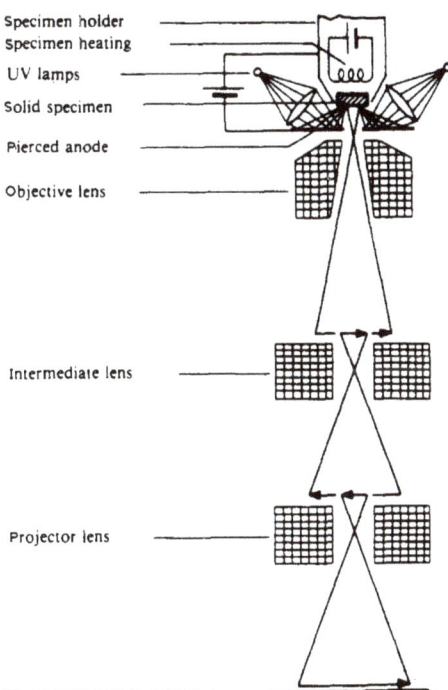

Figure 2. Schematic diagram of photoelectron emission microscopy.

A cylindrical sample of a titanium alloy was used as the target in the electron optical system. The sample was heated to a predetermined temperature. It served as a cathode in the electron accelerating field. A three-stage lens system of the microscope, as shown in Figure 2, was used to image the emitting photoelectrons on to a fluorescent screen, and recorded by a video camera, along with a television monitor. Details of the instrument and its applications are described elsewhere [26–35]. This system allows continuous monitoring of changes in the microstructure, in in bulk samples, up to a temperature of 1500 °C. Prior to the photo-electron emission experiment, the polished surface of the sample was cleaned by bombarding argon at a glancing angle, conducted in an adjoining chamber. The intensity of emitted electrons is a function of the crystallographic orientations, chemical composition of the sample, and surface topography, during heating or cooling, at a temperature above the transformation of the titanium alloy. A bulge emerging from the boundary appeared, as shown in Figure 3.

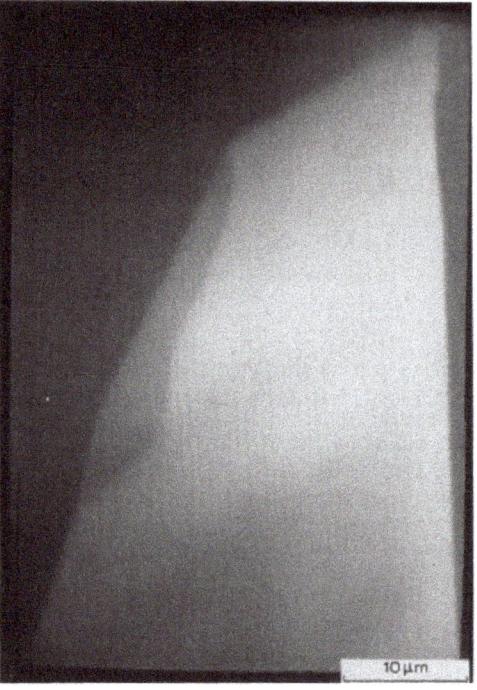

Figure 3. A bulge formed at a grain boundary of β-Ti6Al-2V-6Zr alloy during rapid heating (photoemission electron microscope PEEM).

The contrast seen in Figure 3 is interpreted as emission of defects from grain boundary ledges. Details of these observations are presented in [25].

Autruffe et al. [36], using TEM, with samples subjected to uniaxial tensile straining showed that dislocation profiles extending from the GBs, and associated with ledges on the boundary plane, increased in frequency with increasing strain. They conclude that GBs are the principal sources for dislocations. In these observations, dislocation profiles resembling dislocation pile-ups were directly observed to form at GB ledges.

Similar and very extensive direct experimental evidence of dislocation emission from GBs have been provided by Murr and his coworkers [7], showing dislocation emission from GB ledges in 304 stainless steel. (see Figure 4).

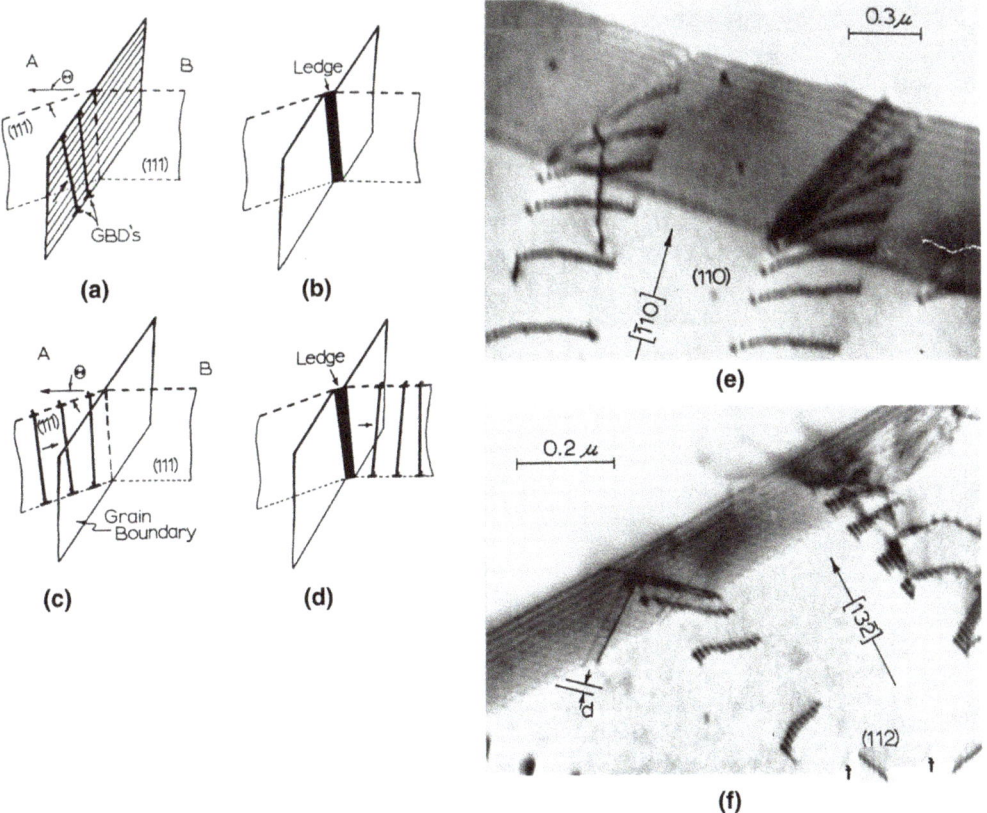

Figure 4. (a–d) schematics of grain boundary ledge formation, and (e,f) showing dislocation emission from grain boundary ledges in 304 stainless steel. From Murr [7].

4. Effects of Grain Boundary Disorder and Impurities on Dislocation Emission

Very small amounts of immiscible solutes can dramatically decrease the ability of dislocations to nucleate from GBs. It was recently reported [37] that segregation of Zr to GBs in nanocrystalline (NC) Cu can lead to the development of disorder in the intergranular structure. In this study, the authors employ atomistic computer simulations to investigate how this disorder affects dislocation nucleation from the GBs under applied stress. It was found that a fully disordered GB structure suppresses dislocation emission and significantly increases the yield stress.

The results of the MD simulations [38], which show that, at larger concentrations, the solute effect becomes non-trivial: there are concentration ranges where even a small addition of solutes considerably suppresses the dislocation nucleation from grain boundaries, and there are concentration ranges where the addition of new solutes almost does not change the dislocation nucleation. The authors also provide an atomic mechanism of these effects [38].

Solute additions are commonly used to stabilize NC materials against grain growth and can simultaneously enhance the strength of the material by impeding dislocation emission from the GBs. In this study [38], the authors demonstrate using molecular dynamics (MD) simulations that the effect of solutes on dislocation nucleation depends on the distribution of solutes at the GB. Solutes with a smaller positive size mismatch to the host can be more effective in suppressing dislocation emission from GBs, in comparison to others that have larger mismatch. These findings are relevant to the search for optimal solute additions, which can strengthen NC material by suppressing the nucleation of dislocation slip from GBs, while stabilizing it against grain growth.

Another study [39] reports using atomic simulation on GB dislocation sources in NC copper. The authors provide an insight into dislocation sources in NC copper. Atomistic studies of this type provide details of the emission sequence that enhance understanding of dislocation sources in high angle boundaries.

A three-dimensional model for the generation of split dislocations by GBs in NC Al is proposed in reference [40]. In terms of this model, rectangular glide split-dislocation half-loops nucleate at glide lattice dislocation loops pressed to GBs by an applied stress. The level of the applied stress and the grain size at which the emission of such dislocation half-loops becomes energetically favorable are determined.

A theoretical model is suggested that describes the emission of partial Shockley dislocations from triple junctions of GBs in deformed NC materials [41]. In the framework of the model, triple junctions accumulate dislocations due to GB sliding along adjacent GBs. The dislocation accumulation at triple junctions causes partial Shockley dislocations to be emitted from the dislocated triple junctions and, thus, accommodates GB sliding. Ranges of parameters (applied stress, grain size, etc.) are calculated, in which the emission events are energetically favorable in NC Al, Cu and Ni. The model accounts for the corresponding experimental data reported in the literature.

Finally, it is worth mentioning the work of Swygenhoven et al. [42], which deals with the atomic mechanism responsible for the emission of partial dislocations from GB in NC metals. It is shown that, in a 12 nm grain-size sample, GBs containing grain-boundary dislocations (GBDs) can emit a partial dislocation during deformation by local shuffling and stress assisted free-volume migration.

5. Dislocation Emission from Grain Boundaries in Nanocrystalline Materials

Finally, utilizing the brief discussion provided in previous sections we consider the Dislocation Emission from GBs in NC materials in some detail. Rapid development of NC materials has raised the question of the role that dislocation pile-ups and other modes of dislocation production play in determining the strength of the materials, vis-a-vis the HP effect. For this purpose, we divide the region in two subsections. First, we consider deformation mechanisms, ultrafine grains, typically between 20 to 100 nm, and later grains finer than that.

5.1. Deformation Mechanisms Ultrafine Grains

In this section, we consider grain sizes in the intermediate range, i.e., between 20 and 100 nm, where a departure from classical HP is first noticed. Even in nanoscale some features of pile-up characteristics are applicable, and can have a significant effect. These include the effect of a small number of dislocations, a continuum versus discrete crystal dislocation description and the influence of elastic anisotropy. For a detailed history of the first 60 years of HP relation, see the excellent review by Armstrong [43]. It is seen that, usually the pile-up of dislocations at GBs is envisioned as the key mechanism, which causes strength increase by grain refinement. However, several models predict the breakdown of the pile-up mechanism in NC metals. With even further grain refinement, some experiments evidence softening of the material (for a review see [9,44,45]). The HP effect is normally explained by appealing to dislocation pile-ups near the grain boundaries. Once the grain size drops below the equilibrium distance between dislocations in a pile-up, pile-ups are no longer possible, and the HP relation should cease to be valid. For example, Pande, Masumura and Armstrong 1993 [46] following the dislocation pile-up model proposed by Pande et al. (see reference [9]), estimated that about ten dislocations can be stored in the approximately 100 nm Au or Cu grains. Such short dislocation pile-ups may not easily induce enough shear stress concentration to break through the grain, although it can accommodate a certain extent of plastic strain.

Using the dislocation pile-up model for the HP relation, Armstrong [47] examined ultrafine grain sizes when only a few dislocations were involved in the pile-up, which were formed at applied stress levels close to the theoretical limit. Each dislocation added to the pile-up contributes a step reduction in the HP stress. Consequently, differences in dislocation configurations and types of pile-ups are easily recognized in the limit of small dislocation numbers.

Even if the micro grains only have a few dislocations, the source of the dislocations are not established with certainty. FR sources in such small spaces seems unlikely. Could they come from the GB? We know that the GBs certainly play a role in determining properties of metallic materials. For example, dislocation-GB interactions, GB sliding grain rotation, GB diffusion, and GB migration all influence mechanical properties of polycrystalline materials. In this region they may also provide the small number of whole dislocations needed in the models discussed above.

Typically, the HP plot is divided into three regions: (1) where classical HP plot is fully valid, typically over 30 nm, (2) where a departure from HP is first noticed, typically below 30nm and (3) where HP plot starts to fall drastically, typically below 15 nm (for details see reference [9]). As the average grain size in nanomaterial is lowered say below 10nm, the dislocations inside most likely do not exist, and there are probably no FR or GB sources operating. Although GB can still be a source of partial dislocations. We will consider this in next section.

5.2. Dislocation Emission from Grain Size Less than 20nm

By now, extensive experiments and simulations exist to show the important role that stress-assisted nanograin rotation play in GB-mediated deformation mode in NC materials. In fact, it is clear that the rotation of nanograins has a close correlation with the enhanced dislocation activity and plasticity in NC metals and alloys.

Li et al. [48] propose a theoretical model to investigate the dislocation emission behavior in NC face-centered cubic crystals. Noting that the nanograin rotation dominates the deformation in these materials, they investigate energy characteristics and the critical shear stress that is required to trigger the dislocation emission from GBs. Their important results show that the "nanograin rotation process can make the originally energetically unfavorable dislocation emission process favorable". They show that the critical stress can also be extraordinarily reduced as compared with the rotation-free case. For example, the proposed dislocation emission process can reduce the required external stress to almost zero if the rotation magnitude can reach 8.5 and 2.9 degrees for Al and Pt, respectively. The findings thus establish that nanograin rotation is a very effective dislocation-emission mechanism in NC materials. Thus, it possibly explains the experimentally observed rotation-dislocation correlation in NC materials. For example, Wang et al. [49] experimentally investigated the atomic-scale deformation dynamic behaviors of Pt nanocrystals with size of ∼18 nm in situ. They used a homemade device in a high-resolution TEM for this purpose. They found that the plastic deformation of the nanosized single crystalline Pt commenced first with dislocation "appearance" then followed by a dislocation "saturation" phenomenon. It was also clear that the magnitude of strain plays a key role on dislocation behaviors in these materials. More recently, these results were confirmed by extensive experiments and molecular dynamics simulations by Liu et al. [50]. They also find that stress-driven grain growth can greatly contribute to enhanced dislocation emission and tensile ductility in NC metals. What is the underlying mechanism behind this correlation? Liu et al. [50] propose a theoretical model based on the cooperative nanograin rotation and GB migration for grain growth to explain the enhanced dislocation emission from GBs.

One should also note the important results obtained by Swygenhoven and Derlet. They investigated [51], GB sliding using MD computer simulations. They used a model of Ni NC sample with a mean grain size of 12 nm under uniaxial tension. They found atomic shuffling and stress-assisted free-volume migration. The activated accommodation processes under high-stress and room temperature conditions are shown to be GB and triple-junction migration, and dislocation activity.

From these studies, it becomes abundantly clear that GB dislocations in GB and their emission plays a very significant role sliding or rotation of the GD. The sliding or rotation are necessary to explain the deformation behavior and dislocation emission of grains about 15 nm or less in NC materials.

6. Summary

We have showed the importance of the mechanism of emission of dislocations from GBs, especially in case of NC materials where FR dislocation sources are difficult to operate. We also briefly mentioned the direct observation of GB dislocation mechanism by experimental methods. Of course, in even finer grains, dislocations have no role to play, except at the GBs. Therefore, although it appears that the dislocation pile-up model is still the most frequently favored model to explain the HP relationship, the other models have their own merit and advantages. It is especially clear that the Li mechanism plays an important role in the deformation studies of many types of materials, especially in nanomaterials. Furthermore, one should note that the HP relationships applies not only to yield stress, but also to flow and fracture stresses and even to fatigue and creep. Thus, it appears that the HP type size dependence of strength is a rather general result of natural dislocation behavior. Revealing various aspects of physics behind the HP law, especially in nanomaterials, remains an attractive and challenging type of research.

Funding: This research did not receive any external funding.

Conflicts of Interest: The authors declare that there is no conflict of Interest.

References

1. Taylor, G.I. The Mechanism of Plastic Deformation of Crystals. Part I.—Theoretical. *Proc. R. Soc.* **1934**, *145*, 362–387.
2. Taylor, G.I. The Mechanism of Plastic Deformation of Crystals. Part II.—Comparison with observations. *Proc. R. Soc.* **1934**, *145*, 388–404.
3. Frank, F.C.; Read, J.W.T. Multiplication Processes for Slow Moving Dislocations. *Phys. Rev.* **2002**, *79*, 722–723. [CrossRef]
4. Li, J.C.M. Petch Relation and Grain Boundary Sources. *Trans. Metall. Soc. AIME* **1963**, *227*, 239–247.
5. Ashby, M.F.; Bullough, R.; Hartley, C.S.; Hirth, J.P. Dislocation Modelling of Physical Systems. In Proceedings of the International Conference, Gainesville, FL, USA, 22–27 June 1980.
6. Essmann, U.; Rapp, M.; Wilkins, M. The dislocation arrangement in cold-worked polycrystalline copper rods. *Acta MET* **1968**, *16*, 1275–1287. [CrossRef]
7. Murr, L. Dislocation Ledge Sources: Dispelling the Myth of Frank–Read Source Importance. *Met. Mater. Trans. A* **2016**, *47*, 5811–5826. [CrossRef]
8. Naik, S.N.; Walley, S.M. The Hall–Petch and inverse Hall–Petch relations and the hardness of nanocrystalline metals. *J. Mater. Sci.* **2019**, *55*, 2661–2681. [CrossRef]
9. Pande, C.; Cooper, K. Nanomechanics of Hall–Petch relationship in nanocrystalline materials. *Prog. Mater. Sci.* **2009**, *54*, 689–706. [CrossRef]
10. Yamakov, V. Deformation twinning in nanocrystalline Al by molecular-dynamics simulation. *Acta Mater.* **2002**, *50*, 5005–5020. [CrossRef]
11. Schiøtz, J.; Di Tolla, F.D.; Jacobsen, K.W. Softening of nanocrystalline metals at very small grain sizes. *Nat. Cell Biol.* **1998**, *391*, 561–563. [CrossRef]
12. Schiøtz, J.; Vegge, T.; Di Tolla, F.; Jacobsen, K.W. Atomic-scale simulations of the mechanical deformation of nanocrystalline metals. *Phys. Rev. B* **1999**, *60*, 11971–11983. [CrossRef]
13. Van Swygenhoven, H.; Spaczer, M.; Caro, A. Microscopic description of plasticity in computer generated metallic nanophase samples: A comparison between Cu and Ni. *Acta Mater.* **1999**, *47*, 3117–3126. [CrossRef]
14. Keblinski, P.; Phillpot, S.R.; Wolf, D.; Gleiter, H. On the Thermodynamic Stability of Amorphous Intergranular Films in Covalent Materials. *J. Am. Ceram. Soc.* **1997**, *80*, 717–732. [CrossRef]
15. Keblinski, P.; Wolf, D.; Phillpot, S.R.; Gleiter, H. Self-diffusion in high-angle fcc metal grain boundaries by molecular dynamics simulation. *Philos. Mag. A* **1999**, *79*, 2735–2761. [CrossRef]
16. Yamakov, V.; Phillpot, S.R.; Wolf, D.; Gleiter, H. Molecular Dynamics Simulation of Nanocrystalline Pd under Stress. In *Computer Simulation Studies in Condensed-Matter Physics XIII*; Landau, D.P., Lewis, S.P., Schüttler, H.B., Eds.; Springer Proceedings in Physics; Springer: Berlin/Heidelberg, Germany, 2001; Volume 86. [CrossRef]

17. Haslam, A.; Phillpot, S.R.; Wolf, D.; Moldovan, D.; Gleiter, H. Mechanisms of grain growth in nanocrystalline fcc metals by molecular-dynamics simulation. *Mater. Sci. Eng. A* **2001**, *318*, 293–312. [CrossRef]
18. Nieh, T.; Wadsworth, J. Hall-petch relation in nanocrystalline solids. *Scr. Met. Mater.* **1991**, *25*, 955–958. [CrossRef]
19. Scattergood, R.; Koch, C. A modified model for hall-petch behavior in nanocrystalline materials. *Scr. Met. Mater.* **1992**, *27*, 1195–1200. [CrossRef]
20. Wang, N.; Wang, Z.; Aust, K.; Erb, U. Effect of grain size on mechanical properties of nanocrystalline materials. *Acta Met. Mater.* **1995**, *43*, 519–528. [CrossRef]
21. Borovikov, V.; Mendelev, M.I.; King, A.H. Effects of solutes on dislocation nucleation from grain boundaries. *Int. J. Plast.* **2017**, *90*, 146–155. [CrossRef]
22. Zhang, L.; Lu, C.; Tieu, K.; Pei, L.; Zhao, X.; Cheng, K. Molecular dynamics study on the grain boundary dislocation source in nanocrystalline copper under tensile loading. *Mater. Res. Express* **2015**, *2*. [CrossRef]
23. Li, J.; Soh, A.K.; Wu, X. Enhancing dislocation emission in nanocrystalline materials through shear-coupled migration of grain boundaries. *Mater. Sci. Eng. A* **2014**, *601*, 153–158. [CrossRef]
24. Shiue, S.; Lee, S. The effect of grain size on fracture: Dislocation-free zone in the front of the finite crack tip. *J. Appl. Phys.* **1991**, *70*, 2947–2953. [CrossRef]
25. Li, J.C.M.; Imam, M.A.; Rath, B.B. Dislocation emission from grain boundaries during rapid heating or cooling. *J. Mater. Sci. Lett.* **1992**, *11*, 906–908. [CrossRef]
26. Wegmann, L. The photo-emission electron microscope: Its technique and applications. *J. Microsc.* **1972**, *96*, 1–23. [CrossRef]
27. Middleton, C.J.; Form, G.W. Direct Observation of an Austenite Memory Effect in Low-Alloy Steels. *Met. Sci.* **1975**, *9*, 521–528. [CrossRef]
28. Middleton, C.; Edmonds, D. The application of photoemission electron microscopy to the study of diffusional phase transformations in steels. *Metallography* **1977**, *10*, 55–87. [CrossRef]
29. Edmonds, V.; Honeycombe, R.W.K. Photoemission electron microscopy of growth of grain boundary ferrite allotriomorphs in chromium steel. *Met. Sci.* **1978**, *12*, 399–405. [CrossRef]
30. Pfefferkorn, G.; Schur, K. Bibliography on emission electron microscopy (EEM), especially on Photo EEM. In *Beiträge zur Elektronenmikroskopischen Direktabbildung von Oberflächen, Proceedings of the First International Conference on Emission Electron Microscopy*; R.A. Remy-Verlag: Münster, Germany, 1979; pp. 205–233.
31. Schwarzer, R. Evidence for vectorial photoelectric effect from PEEM study of biological thin sections. In *Beiträge zur Elektronenmikroskopischen Direktabbildung von Oberflächen, Proceedings of the First International Conference on Emission Electron Microscopy*; R.A. Remy-Verlag: Münster, Germany, 1979; pp. 165–170.
32. Crosbie, A. Controlled Rolling Simulation. Master's thesis, University of Leeds, Leeds, UK, 1979.
33. Hammond, C.; Ashraf Imam, M. Photoemission electron microscopy in metallurgical research: Applications using the balzers KE3 metioscope. *Ultramicroscopy.* **1991**, *36*, 173–185. [CrossRef]
34. Imam, M.A.; Rath, B.B.; Hammond, C. Arora: 6th World Conference on Titanium; Les Editions de Physique: Cannes, France, 1988; pp. 1313–1318.
35. Hillert, M.; Purdy, G.R. Chemically induced grain boundary migration. *Acta Met.* **1978**, *26*, 333–340. [CrossRef]
36. Autruffe, A.; Hagen, V.S.; Arnberg, L.; Di Sabatino, M. Dislocation generation at near-coincidence site lattice grain boundaries during silicon directional solidification. *J. Cryst. Growth* **2015**, *411*, 12–18. [CrossRef]
37. Borovikov, V.; Mendelev, M.I.; King, A.H. Effects of grain boundary disorder on dislocation emission. *Mater. Lett.* **2019**, *237*, 303–305. [CrossRef]
38. Borovikov, V.; Mendelev, M.I. Modification of dislocation emission sources at symmetric tilt grain boundaries in Ag by Cu solute atoms. *Mater. Lett.* **2018**, *223*, 243–245. [CrossRef]
39. Tschopp, M.; McDowell, D.L. Grain boundary dislocation sources in nanocrystalline copper. *Scr. Mater.* **2008**, *58*, 299–302. [CrossRef]
40. Bobylev, S.V.; Gutkin, M.Y.; Ovid'Ko, I.A. Generation of glide split-dislocation half-loops by grain boundaries in nanocrystalline Al. *Phys. Solid State* **2006**, *48*, 1495–1505. [CrossRef]
41. Gutkin, M.Y.; Ovid'Ko, I.A.; Skiba, N.V. Emission of partial dislocations from triple junctions of grain boundaries in nanocrystalline materials. *J. Phys. D Appl. Phys.* **2005**, *38*, 3921–3925. [CrossRef]
42. Derlet, P.M.; Van Swygenhoven, H.; Hasnaoui, A. Atomistic simulation of dislocation emission in nanosized grain boundaries. *Philos. Mag.* **2003**, *83*, 3569–3575. [CrossRef]
43. Armstrong, R.W. 60 Years of Hall-Petch: Past to Present Nano-Scale Connections. *Mater. Trans.* **2014**, *55*, 2–12. [CrossRef]
44. Gleiter, H. Nanocrystalline materials. *Prog. Mater. Sci.* **1989**, *33*, 223–315. [CrossRef]
45. Meyers, M.A.; Mishra, A.; Benson, D.J. Mechanical properties of nanocrystalline materials. *Prog. Mater. Sci.* **2006**, *51*, 427–556. [CrossRef]
46. Pande, C.; Masumura, R.; Armstrong, R. Pile-up based hall-petch relation for nanoscale materials. *Nanostruct. Mater.* **1993**, *2*, 323–331. [CrossRef]
47. Armstrong, R.W. Dislocation Pile-Ups, Material Strength Levels, and Thermal Activation. *Met. Mater. Trans. A* **2015**, *47*, 5801–5810. [CrossRef]

48. Li, J.; Chen, S.; Weng, G.J.; Liu, C. Stress-assisted grain-rotation-induced dislocation emission from grain boundaries in nanocrystalline face-centered-cubic metals. *Philos. Mag. Lett.* **2019**, *99*, 466–478. [CrossRef]
49. Wang, L.; Teng, J.; Sha, X.; Zou, J.; Zhang, Z.; Han, X. Plastic Deformation through Dislocation Saturation in Ultrasmall Pt Nanocrystals and Its in Situ Atomistic Mechanisms. *Nano Lett.* **2017**, *17*, 4733–4739. [CrossRef]
50. Liu, C.; Lu, W.; Weng, G.J.; Li, J. A cooperative nano-grain rotation and grain-boundary migration mechanism for enhanced dislocation emission and tensile ductility in nanocrystalline materials. *Mater. Sci. Eng. A* **2019**, *756*, 284–290. [CrossRef]
51. Van Swygenhoven, H.; Derlet, P.M. Grain-boundary sliding in nanocrystalline fcc metals. *Phys. Rev. B* **2001**, *64*, 224105. [CrossRef]

Article

Extraordinary Response of H-Charged and H-Free Coherent Grain Boundaries in Nickel to Multiaxial Loading

Petr Šesták [1,*], Miroslav Černý [1,2], Zhiliang Zhang [3] and and Jaroslav Pokluda [1,2,4]

1. Central European Institute of Technology, CEITEC BUT, Brno University of Technology, Purkyňova 123, CZ–612 00 Brno, Czech Republic; cerny.m@fme.vutbr.cz (M.Č.); pokluda@fme.vutbr.cz (J.P.)
2. Faculty of Mechanical Engineering, Brno University of Technology, Technická 2896/2, CZ-616 69 Brno, Czech Republic
3. Faculty of Engineering, Norwegian University of Science and Technology (NTNU), Rich. Birkelandsvei 1A, 7491 Trondheim, Norway; zhiliang.zhang@ntnu.no
4. Faculty of Special Technology, Alexander Dubcek University of Trencin, Pri parku 19, 911 06 Trenčín, Slovakia
* Correspondence: petr.sestak@ceitec.vutbr.cz

Received: 29 May 2020; Accepted: 6 July 2020; Published: 8 July 2020

Abstract: The cohesive strength of $\Sigma 3$, $\Sigma 5$, and $\Sigma 11$ grain boundaries (GBs) in clean and hydrogen-segregated fcc nickel was systematically studied as a function of the superimposed transverse biaxial stresses using ab initio methods. The obtained results for H-free GBs revealed a quite different response of the coherent twinning boundary $\Sigma 3$ to the applied transverse stresses in comparison to the other GB types. While the cohesive strength of $\Sigma 5$ and $\Sigma 11$ GBs increased with increasing level of tensile transverse stresses, the strength of $\Sigma 3$ GB remained constant for any applied levels of transverse stresses. In the case of GBs with segregated hydrogen, the cohesive strength of $\Sigma 3$ was distinctly reduced for all levels of transverse stresses, while the strength reduction of $\Sigma 5$ and $\Sigma 11$ GBs was significant only for a nearly isotropic (hydrostatic) triaxial loading. This extraordinary response explains a high susceptibility of $\Sigma 3$ GBs to crack initiation, as recently reported in an experimental study. Moreover, a highly triaxial stress at the fronts of microcracks initiated at $\Sigma 3$ boundaries caused a strength reduction of adjacent high-energy grain boundaries which thus became preferential sites for further crack propagation.

Keywords: ab initio calculations; hydrogen embrittlement; grain boundary; cohesive strength; multiaxial loading

1. Introduction

Hydrogen may cause a significant reduction of ductility of metallic materials which leads to a premature fracture of engineering components and structures. This so-called hydrogen embrittlement was already extensively studied in the last century with the help of experimental [1,2] as well as theoretical methods [3–13]. The most accepted theoretical concepts explaining the embrittlement at the atomistic level are the Hydrogen-Enhanced Decohesion (HEDE) [14] and the Hydrogen-Enhanced Localized Plasticity (HELP) [15]. The HEDE concept deals with a hydrogen-induced reduction of the cohesive strength of grain boundaries (GBs), while the HELP model is based on the hydrogen-enhanced dislocation mobility. At high tensile triaxialities, the HELP mechanism accelerates the necking failure of microvoids and, at low triaxialities, it induces their shearing coalescence and failure (see e.g., [16,17]). The relevance of HEDE and HELP damage mechanisms should be identified for each particular case of the hydrogen-assisted fracture.

Decohesion processes in polycrystalline metallic materials are mostly restricted to a vicinity of planar defects like grain boundaries and, as a rule, they are affected by a presence of hydrogen and impurity atoms segregating at these defects [18–20]. This was also the case of the H-charged nickel-based superalloy that exhibited quasi-brittle fracture surfaces of a mixed intergranular and transgranular morphology in Ni matrix as reported in the experimental study [1]. Surprisingly, the most frequent crack initiation sites reported by Seita et al. [1] were Σ3 coherent twin boundaries, in spite of their lowest energy and hydrogen solubility. Other low-Σ GBs ($\Sigma = 1 - 29$) were less susceptible to crack initiation but less resistant to crack propagation. General GBs ($\Sigma > 29$) exhibited the highest resistance to crack initiation but the lowest resistance to crack propagation. Since the plane of Σ3 GB is also the plane of dislocation glide and many of fractured Σ3 GBs were inclined by 60° to the tensile axis, the authors assumed that a shear deformation by dislocation glide in the GB plane assisted the crack initiation process, along with dislocations trapped to this plane from other (111) slip planes. Such a damage process resembles the shearing failure induced by the HELP mechanism at low triaxialities [16]. Further crack propagation followed along general grain boundaries due to their highest energy (lowest separation energy and cohesive strength) and the highest hydrogen concentration.

However, such an interpretation of the peculiar fracture behavior is certainly not exhaustive without exploring the effect of hydrogen segregation on the cohesive strength of GBs in H-charged nickel specimens—i.e., without also taking the HEDE mechanism into account. This is the main objective of our ab-initio study. In the first principles calculations, the mechanical properties of GBs are usually characterized by strengthening/embrittlening energies [18], cleavage energies (work of separation) [20,21], and/or by the cohesive strength related to uniaxial loading (or deformation). However, GBs in metallic engineering components are rather subjected to an external multiaxial loading/deformation. Moreover, a superposition of local stresses of various kind often leads to a multiaxial stress state, even in the case of uniaxial external loading. Tensile triaxial stresses ahead of crack fronts or tensile/compressive internal stresses induced by thermomechanical and surface treatments can serve as good examples of such local stresses. For the purpose of practical applications of grain boundary engineering, therefore, it is also useful to understand the effect of multiaxial loading on the cohesive strength of GBs. Let us note that cracks not only induce triaxial stress state but also act as stress concentrators that can significantly raise the local stress level.

In this paper, we present values of cohesive strength of clean and hydrogen segregated GBs in fcc nickel. We also studied the aforementioned effect of triaxiality of the stress state on the strength response of GBs in our ab initio predictions. Namely, we studied the Σ5, Σ3, and Σ11 coherent GBs. The Σ5 served as an example of GBs with an excess volume (void space) at GBs where impurities tend to segregate. One can therefore expect locally increased hydrogen concentration at the GB affecting its cohesion. The other two considered low-angle GBs have rather negligible excess volume. In such a case, one could expect that these GBs do not affect H distribution in the crystal. However, as was shown in the work of Stefano et al. [7], Σ3 GB can serve as a two-dimensional barrier for H migration. Moreover, the GBs included in our study have been studied theoretically and there is enough data for comparison in the literature.

2. Computational Details

The calculations were performed using the program VASP [22] (Vienna ab initio simulation program) developed at the Fakultät für Physik, Universität Wien. In our study, the electron interactions were described by the projector-augmented waves (PAW) potentials [23] supplied with the VASP code, and the exchange correlation energy was evaluated by means of the generalized gradient approximation (GGA) with parametrization of Perdew–Burke–Ernzerhof [24]. The Methfessel–Paxton method of the first order was adopted with a smearing width of 0.1 eV.

Sampling of the Brillouin zone was done using a Monkhorst-Pack [25] scheme with $6 \times 1 \times 10$, $3 \times 1 \times 7$, and $3 \times 1 \times 10$ k–point grids for Σ3, Σ5, and Σ11 GBs, respectively. The solution was considered self-consistent when the difference between energies of two subsequent steps was below 10^{-7} eV and the plane-wave basis set was expanded with the cutoff energy of 350 eV. Optimization of atomic positions in computational supercells was performed using the internal VASP procedure until all forces between atoms were lower than 10 meV/Å. For the optimization of the cell shape, we employed our own external program that cooperated with VASP via reading its output and writing new input files. This program allowed us to relax the stress tensor components to their targeted values within a tolerance of ± 0.1 GPa. In all presented calculations, ferromagnetic ordering of Ni was included via spin polarization.

Before introducing the studied GBs, let us define two quantities commonly used for their characterization [7,11,26]. The first one is the excess volume v_{exc}, which can be computed as

$$v_{exc} = \frac{V_{GB} - N_{GB} V_{Ni}}{2A}, \quad (1)$$

where V_{GB} represents the volume, N_{GB} is the number of atoms in a fully optimized supercell containing GB, V_{Ni} is the volume per atom in a perfect bulk crystal of Ni, and A is the GB area. In general, larger values of the excess volume v_{exc} represent more void space at GB. Another important characteristic quantity is the GB energy γ_{GB} determined as

$$\gamma_{GB} = \frac{E_{GB} - N_{GB} E_{Ni}}{2A}, \quad (2)$$

where the E_{GB} is the total energy of the optimized supercell with GB and E_{Ni} is the total energy per atom of the bulk fcc nickel crystal. Thus, γ_{GB} represents the energy necessary to create such a planar defect of a unit area in a perfect crystal structure. Since all simulation cells contain two identical GBs (as described in the next paragraph), right-hand sides of both equations must be divided by the factor of 2.

In our systematic study, we considered three types of symmetrical GBs in fcc nickel, namely, the $\Sigma 3(111)$ GB, $\Sigma 5(210)$ GB, and the $\Sigma 11(311)$ GB. Corresponding computational supercells that were constructed for the present study are illustrated in Figure 1. These supercells have orthorhombic symmetry and, in order to keep periodic boundary conditions also in the direction perpendicular to the GB plane, they contain two identical GBs. One is located in the center of the supercell and the other one at its edge (displayed as the dashed vertical lines in Figure 1).

All the computational cells were subjected to several types of tensile loading or deformation. Figure 2 illustrates the geometry of our tensile tests. The loading axis is parallel with the x-axis, which was set perpendicular to the GB plane in all our simulations (i.e., $x \parallel [111]$, $x \parallel [210]$, and $x \parallel [311]$ for the $\Sigma 3$, $\Sigma 5$, and $\Sigma 11$ GBs, respectively). Stresses σ_2 and σ_3 are thus the transverse stresses that were controlled by our computational procedure at each strain increment (of 0.01), optimizing both the cell shape and the ionic positions. Since the general loading with $\sigma_2 \neq \sigma_3$ would lead to an enormous number of triaxial stress states (and corresponding values of cohesive strength), we considered these stresses mutually dependent, keeping their ratio $k = \sigma_2/\sigma_3$ constant as discussed hereafter. For $\sigma_2 = \sigma_3 = 0$, the tensile test corresponds to the so-called uniaxial loading. Another special type of loading is the isotropic (or hydrostatic) one with $\sigma_2 = \sigma_1 = \sigma_3$. In this only case, we also controlled the axial stress σ_1. In all the other cases, we computed σ_1 as a function of the axial strain ϵ_1. For comparative purposes, we also simulated uniaxial deformation with $\epsilon_2 = \epsilon_3 = 0$ (which implies $\sigma_2 \neq \sigma_3$ due to the crystal anisotropy) and isotropic deformation with $\epsilon_2 = \epsilon_1 = \epsilon_3$. In all applied loading cases, the cohesive strength value was identified with the maximum of σ_1, hereafter denoted σ_{max}. A brief overview of all loading types is given in Table 1.

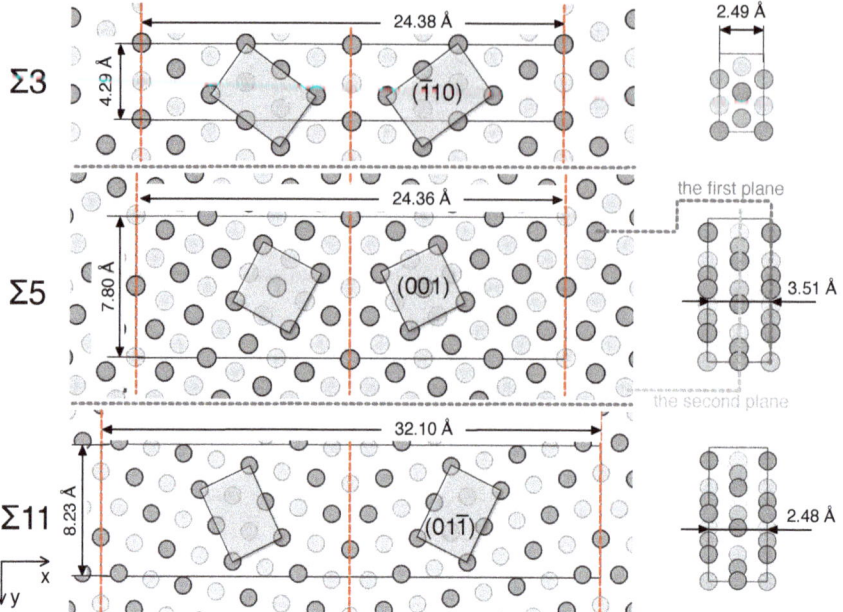

Figure 1. The supercells containing Σ3(111), Σ5(210), and Σ11(311) grain boundaries (GBs) used in the present ab initio calculations. The planes perpendicular to the rotation axis related to GBs are highlighted. Orientation of the supercells were $x \parallel [111]$, $y \parallel [11\bar{2}]$, and $z \parallel [\bar{1}10]$ for the Σ3 GB; $x \parallel [210]$, $y \parallel [\bar{1}20]$, and $z \parallel [001]$ for the Σ5 GB; and $x \parallel [311]$, $y \parallel [\bar{2}33]$, and $z \parallel [0\bar{1}1]$ for the Σ11 GB.

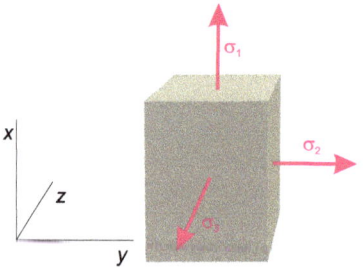

Figure 2. Illustration of the computational supercell under triaxial tensile loading. Stress tensor components are denoted using the Voigt notation.

Table 1. Types of the applied loading.

uniaxial loading	$\sigma_2 = \sigma_3 = 0$
uniaxial deformation	$\epsilon_2 = \epsilon_3 = 0$
triaxial loading	$\sigma_2 = k\sigma_3 \neq 0$
isotropic deformation	$\epsilon_2 = \epsilon_1 = \epsilon_3 \neq 0$
isotropic loading	$\sigma_2 = \sigma_1 = \sigma_3 \neq 0$

3. Results and Discussion

3.1. Structure Parameters of $\Sigma 3$, $\Sigma 5$, and $\Sigma 11$ GBs

3.1.1. Clean Grain Boundaries

All the constructed supercells were fully optimized in order to get their ground-state structures with equilibrium ionic positions and relaxed cell shape and dimensions. The structural parameters received from the optimization process as well as the excess volume v_{exc} and the grain boundary energy γ_{GB} are summarized in Table 2 and compared with available literature data.

Table 2. The excess volume v_{exc}, GB energy γ_{GB}, and the supercell parameters a_0, b_0, c_0 of clean $\Sigma 3$, $\Sigma 5$, and $\Sigma 11$ GBs in fcc nickel along with the k-points grid and number N in the simulation cell.

GB		Plane	Rotation Axis	Rotation Angle	v_{exc} (Å)	γ_{GB} (J/m²)	a_0 (Å)	b_0 (Å)	c_0 (Å)	k-Points Grid	N
$\Sigma 3$	present	{111}	⟨110⟩	109.5°	0.02	0.06	24.4	4.29	2.49	1×6×10	24
	Reference [7]				0.05	0.18	24.5	4.32	2.50	1×6×10	24
	Reference [26]				0.01	0.04	-	-	-	1×3×3	-
	Reference [11]				−0.11	0.09	24.3	4.73	4.94	1×4×4	48
$\Sigma 5$	present	{210}	⟨001⟩	36.9°	0.26	1.26	24.4	7.81	3.49	1×3×7	60
	Reference [7]				0.38	1.29	16.4	7.93	3.55	2×4×8	40
	Reference [18]				-	1.23	23.6	7.87	3.52	2×6×18	60
	Reference [26]				0.35	1.23	-	-	-	1×3×3	-
	Reference [11]				0.45	1.30	22.9	7.82	6.80	1×3×3	-
$\Sigma 11$	present	{311}	⟨100⟩	50.5°	0.08	0.43	32.1	8.23	2.48	1×3×10	60
	Reference [11]				0.06	0.47	21.7	8.22	4.93	1×4×4	80

All the GB parameters agree well with most of the ab initio results formerly published in literature [7,18,26]. Since the $\Sigma 3$ GB is a coherent GB, its volume excess v_{exc} and also the GB energy γ_{GB} are very small (almost zero) and, therefore, these values might be strongly influenced by computational parameters (choice of pseudopotentials, size of computational supercell, etc.) as well as convergence settings during the relaxation process. Note, for example, that v_{exc} computed by Chen et al. [11] is negative but their value of γ_{GB} is in a good agreement with other ab initio results and also with the result $\gamma_{GB} = 0.05$ J/m² obtained by Shiga et al. [27] using the embedded atom method (EAM). Values of γ_{GB} computed for the $\Sigma 5$ GB are significantly greater and in a very good mutual agreement. They also reasonably agree with the EAM result [27] of 1.34 eV. The excess volume of the $\Sigma 5$ GB is an order of magnitude greater than that of the $\Sigma 3$. The $\Sigma 11$ GB is also very compact, with very small v_{exc} and relatively low γ_{GB}. Former ab initio [11] as well as EAM calculations [27], giving the γ_{GB} value of 0.40 eV, are in agreement with the present result.

3.1.2. H-Charged Grain Boundaries

The most frequently used strategy to identify the preferred lattice sites for segregation of the hydrogen atoms is usually based on considering several energetically favorable positions (selected with the help of intuition or former experience) at each GB and finding the most favorable one. In the present study, we tested another possible strategy—taking the advantage of the first principles molecular dynamics simulations (FP-MD)—that allowed us to simulate elevated temperatures and to observe a migration of hydrogen atom across the GB. The main advantage of this strategy is the possibility of also finding energetically favorable positions that are not dictated by symmetry and might thus be overlooked in static approaches. Although this approach is more convenient for less-symmetrical GBs than those considered in our study, we decided to test its predictive potential. Moreover, this approach can also indicate positions that can be stabilized by entropy terms at elevated temperatures.

For the present FP-MD simulations, we used a special version of the VASP code compiled for the gamma point only. We enlarged the simulation supercells, repeating those described in Table 2 in the z-direction (in the case of Σ3 GB, also in the y-direction) to avoid artificial atomic interactions due to the limited supercell size. Thus, the numbers of Ni atoms in the enlarged supercells were increased to 144, 180, and 180 for Σ3, Σ5, and Σ11 GBs, respectively. To make the simulations feasible, the convergence criterion was reduced to 10^{-5} eV, the cutoff energy was set to 200 eV, and symmetrization of the charge density was switched off. The time step was set to 2 fs. Let us note that lattice parameters of the enlarged supercells were multiplied by the factor of 1.0065 (the atomic positions were set in fractional coordinates), which corresponded to the thermal expansion of a pure nickel from 0 K to 500 K. Then, we introduced a hydrogen atom to the GB and started the FP-MD calculations by a gradual increase in temperature from 0 K to 500 K within the 10,000 time steps, and proceeded with another 10,000 time steps at constant temperature of 500 K.

Hydrogen positions recorded during the constant-temperature range were subjected to a statistical analysis to identify the positions most frequently occupied by the H atom. Results for the Σ5 GB (with the highest energy and void space) can be seen from the histogram in Figure 3 displaying the frequency of occurrence of hydrogen atom in positions described by their coordinates (fractional coordinates with respect to the supercell dimensions). The interval on the horizontal axis for a construction of the histogram was set to 1×10^{-3}. The histogram contains only the data for y and z coordinates since the data for the x-coordinate exhibited only one sharp peak at 0.5. To label the preferred segregation sites, we use the same nomenclature (S_x) as Di Stefano et al. [7]. These positions are also marked in Figure 4, displaying the atomic configurations of all GBs. The peaks in Figure 3 reveal the preferred segregation sites labeled S_0 and S_2.

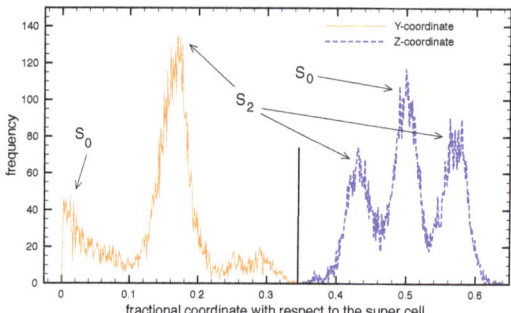

Figure 3. The histogram for the Σ5 GB. The solid and dashed lines correspond to frequencies of occurrence of the H atom along the y- and z-coordinates, respectively. The x-coordinate is not shown because there is only one strong peak at 0.5 which corresponds to the GB located in the middle of the supercell. The coordinates are in fractional units of the first principles molecular dynamics simulation (FP-MD) supercell (corresponding to that in Figure 2 repeated three times along z), and the positions S_0 and S_2 are defined in Figure 4.

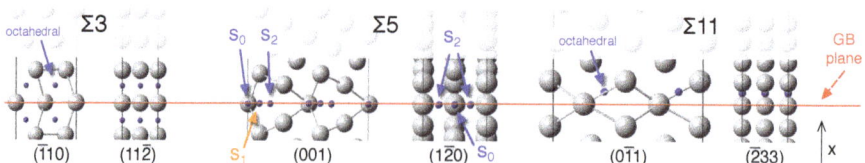

Figure 4. Details of the grain boundary configuration with indicated positions of the preferred segregation sites for hydrogen atoms. S_0 and S_2 are the preferred segregation positions found in the FP-MD simulations, and S_1 is another position considered in Reference [7].

Results for other GBs (though not included in Figure 3) were obtained the same way. Such established hydrogen positions and corresponding segregation energies were compared with those already published by Di Stefano et al. [7]. Since the present nomenclature is consistent, the S_2 positions in both studies are identical. However, instead of the position S_1 reported in [7], we found another position marked S_0. The segregation energy of -247 meV (-228 meV) obtained for S_0 (S_2) in the present work is close to the value of -230 meV calculated for S_1 in Reference [7]. Alvaro et al. [10] studied preferable hydrogen positions in the $\Sigma 5$ GB and suggested that hydrogen atoms prefer the octahedral-like positions similar to those in the Ni bulk. However, according to our results—as well as the work of Di Stefano et al. [7] (where these positions were marked as S_6 and S_7)—their segregation energies are higher than the energies of S_1 and S_2.

The hydrogen segregation for the remaining GBs of a smaller energy and void space is significantly reduced when compared to $\Sigma 5$. According to the FP-MD results obtained for $\Sigma 3$ and $\Sigma 11$ GBs, the hydrogen atoms tend to segregate only at octahedral sites near the GB planes. These positions are depicted in Figure 4 and marked by small (blue) spheres representing the segregated hydrogen atoms. This figure shows the hydrogen occupation sites for all three studied GB supercells that were used for the determination of cohesive strengths.

3.2. Cohesive Strength of Hydrogen-Free GBs

The computational tensile tests were first performed for a perfect crystal of Ni loaded in crystallographic directions corresponding to the orientations of the loading axes in GB models (perpendicular to the GB planes), i.e., $\langle 111 \rangle$, $\langle 210 \rangle$, and $\langle 311 \rangle$ for $\Sigma 3$, $\Sigma 5$, and $\Sigma 11$, respectively. We started with the uniaxial deformation keeping the transverse lattice parameters constant (see Table 1) and computed all the normal stresses as functions of the axial strain ϵ_1. The results are displayed in Figure 5, where the transverse stresses σ_2, σ_3 are plotted as functions of the axial stress σ_1 up to σ_{max}.

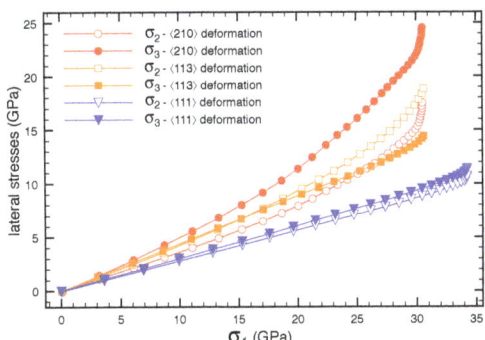

Figure 5. The relationship between the transverse stresses σ_2, σ_3 and the axial stress σ_1 computed for the uniaxial deformation of bulk fcc Ni in the $\langle 111 \rangle$, $\langle 210 \rangle$, and $\langle 311 \rangle$ crystallographic directions.

One can note that omitting the Poisson contraction during the uniaxial deformation induces tensile transverse stresses that are superimposed to σ_1, thus making the stress state triaxial. Values of σ_2 and σ_3 generally differ. They depend not only on the particular orientation of the x-axis but also on anisotropy of the perpendicular ({111}, {210}, and {311}) atomic planes. Therefore, we respected this anisotropy by keeping the same ratios σ_3/σ_2 also in the cases of triaxial loading (see Table 1) applied to supercells simulating perfect crystals as well as the crystals with GBs. The greatest ratio $\sigma_3/\sigma_2 = 1.4$ was obtained for the [210] direction, i.e., the same ratio was used in the triaxial tensile tests of supercells with the $\Sigma 5$ GB. The ratio obtained for the [111] deformation is approximately equal to 1, thus, this ratio was also used for the triaxial tensile tests of $\Sigma 3$ GB. The response of Ni crystal elongated in the {311} direction is somewhat complicated. For smaller strains (and the σ_1 values slightly above 15 GPa), the ratio is close to 1.0, but for greater strains (and the stresses close to the σ_{max} value) it decreases to $\sigma_3/\sigma_2 = 0.8$. The latter value was selected for the triaxial loading of Σ11 GB since the ratio close to the σ_{max} value is of a higher relevance.

The tensile tests for all the loading conditions listed in Table 1 were then applied to the optimized GBs and the computed data are summarized in Figure 6. The left panel displays results of the triaxial loading in terms of the cohesive strength value σ_{max} as a function of the applied transverse stress σ_2 for each GB model and the crystallographic direction. The right panel of Figure 6 shows the results received for special loading cases: uniaxial deformation, isotropic loading, and isotropic deformation.

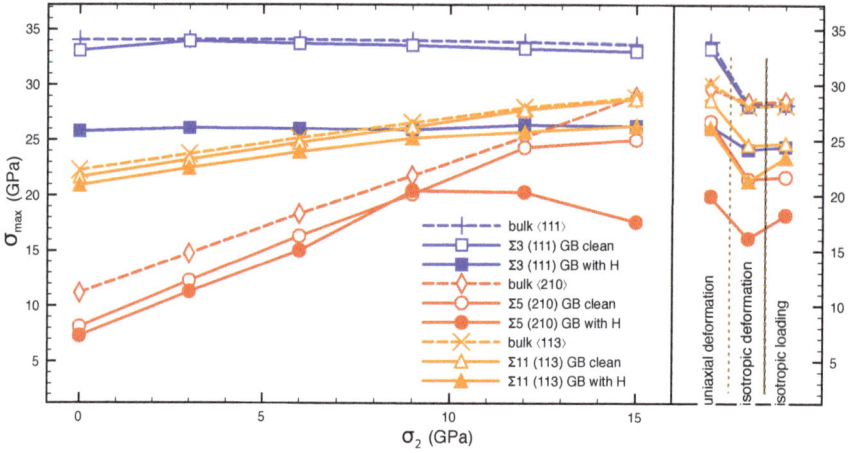

Figure 6. Computed values of the cohesive strength σ_{max} for the bulk crystal (symbols connected by dashed lines), clean GB (open symbols, solid lines), and H-charged GB (solid symbols, solid lines). The left panel shows σ_{max} as a function of one of the transverse stresses (σ_2) during triaxial loading. The right panel displays the σ_{max} values computed for other specific loading types. Note that the data for $\sigma_2 = 0$ correspond to results of uniaxial loading.

The results shown in the left panel of Figure 6 reveal that σ_{max} for Σ5 and Σ11 GBs linearly increases with increasing transverse stresses (as already reported for a majority of perfect cubic crystals and loading directions [28]), while it remains constant for the Σ3 GB and the bulk Ni crystal loaded in the [111] direction. Such an insensitivity to the superimposed transverse stresses is also indicated by the value of σ_{max} achieved via uniaxial deformation that is practically equal to σ_{max} from uniaxial loading (i.e., the value in the left panel of Figure 6 for $\sigma_2 = 0$).

One can also see that σ_{max} values of the bulk crystal loaded in the crystallographic directions corresponding to orientations of grains in individual GBs (plotted in Figure 6 by symbols connected by dashed lines) mostly follow the values computed for the related GBs. In the cases of Σ3(111) and Σ11(311) GBs, values for the bulk and clean GB are almost equal (naturally, with slightly higher σ_{max} values for the bulk crystal). This means that the presence of Σ3 and Σ11 GBs practically does not reduce the crystal strength. More remarkable reduction of σ_{max} is caused by the presence of Σ5(210) GB.

The right panel in Figure 6 shows that the isotropic (hydrostatic) loading as well as the isotropic deformation yield lower σ_{max} values than the uniaxial deformation. Let us note that, in the case of bulk crystals, both the isotropic loading and the isotropic deformation must lead to the same σ_{max} value, regardless of the orientation of the loading axis. The corresponding data points (each obtained using a differently oriented supercell or deformation model) displayed in Figure 6 confirm reliability of our computational procedures by their negligible differences. Interestingly, almost the same values were also obtained for the supercell with the clean Σ3 GB, thus indicating that its effect on crystal strength can be considered negligible. On the other hand, presence of the other two GBs (with higher energy γ_{GB}) significantly reduce σ_{max} under isotropic tensile loading. Therefore, one can presume that the cohesive strength of general GBs exhibiting greater GB energy will be even more reduced, particularly for highly triaxial stress states.

3.3. Cohesive Strength of Hydrogen-Charged GBs

The influence of hydrogen segregation on the cohesive strength was then studied using supercells charged with the hydrogen. Its amount introduced to the supercell was 4, 6, and 2 atoms for $\Sigma 3$, $\Sigma 5$, and $\Sigma 11$ GBs, respectively. These numbers correspond to an optimum coverage (all available interstitial positions are filled with hydrogen) but also to a locally-enhanced hydrogen concentration (much greater than can be expected for the rest of the crystal.

The supercells were subjected to tensile tests with all the considered loading conditions and the results were added to Figure 6. The H-charged $\Sigma 3$ GB again exhibits a quite different strength response than the other two GB types. Similarly to the results for hydrogen-free GBs, the cohesive strengths of $\Sigma 5$ and $\Sigma 11$ GBs increased with increasing transverse stresses while the value of the cohesive strength of $\Sigma 3$ GB remained constant in the whole range of the transverse stresses. More interestingly however, the cohesive strength of $\Sigma 3$ was distinctly reduced for all levels of the superimposed transverse stresses. The strength reduction for $\Sigma 5$ and $\Sigma 11$ GBs was significant only in the case of high biaxial stresses, i.e., for nearly isotropic (hydrostatic) loading cases.

To understand the effect of the compact $\Sigma 3$ GB (with a negligible excess volume) in the H-charged crystal, we also computed the tensile strength of H-charged bulk Ni (perfect crystal) subjected to uniaxial deformation along the [111] direction. As can be seen from Figure 6, the strength values for the clean bulk and the $\Sigma 3$ GB are very similar (33.9 GPa and 33.2 GPa, respectively). After introducing H to the $\Sigma 3$ GB, the strength decreased to 26.3 GPa, but the same amount of H introduced to the perfect lattice (using comparable supercell) reduced the strength only to 28.6 GPa. This suggests that the higher strength reduction of $\Sigma 3$ GB than that of the perfect lattice is caused by an interaction of H atoms that get closer to each other in the $\Sigma 3$ GB than in the perfect lattice. The values of cohesive strength are seemingly too high in comparison with typical levels of stress applied to the specimen in the instant of first occurrence of cracks. However, one must bear in mind that, due to the presence of stress concentrators, local stress levels are much higher and can reach the computed strength values.

Thus, our study reveals strong additional HEDE-based reasons for the highest susceptibility of special $\Sigma 3$ GBs to crack initiation, as well as for the lowest resistance of general GBs to crack propagation during uniaxial loading. Indeed, the presence of hydrogen reduces the cohesive strength of $\Sigma 3$ GBs to become closer to strength levels of special GBs which remain hydrogen-unaffected. Therefore, along with the movement and interaction of dislocations in the $\Sigma 3$ GB plane, such a decrease in the cohesive strength makes the $\Sigma 3$ GB plane a comprehensible preferential site for nucleation of microcracks. Once these cracks appear in the $\Sigma 3$ GB planes, the cohesive strength of other types of GBs adjacent to $\Sigma 3$ GBs also becomes reduced due to a highly triaxial tensile loading induced in the vicinity of the microcrack networks. Naturally, a further crack propagation preferentially occurs along general grain boundaries which exhibit the lowest cohesive strength and the highest hydrogen concentrations. The most probable fracture scenario of nickel polycrystal in the hydrogen environment is, therefore, a result of both HELP and HEDE damage mechanisms.

3.4. Work of Separation

Work of separation W_{oS} (also called cleavage energy or cohesive energy) is another quantity which can be (and often is) used to characterize the GB cohesion. It is calculated as an energy necessary to break the crystal along a specified cleavage plane, as expressed by the following formula

$$W_{oS} = \frac{E_{FS(+H)} - E_{GB(+H)}}{A}, \qquad (3)$$

where $E_{GB(+H)}$ is the total energy of the fully optimized supercell containing the GB (with or without hydrogen atoms), E_{FS} is the total energy of a supercell with two free surfaces (i.e., the fractured supercell), and A is the GB cross-section area. As has been formerly discussed [19,29], selection of the weakest cleavage (or fracture) plane is of key importance. There is no doubt about its position in the case of a clean GB but, in the case of H-decorated GB, one must consider possible redistribution of hydrogen on the created surfaces since it affects the $E_{FS(+H)}$ value. Our choice was based on the result of the tensile tests, namely, the uniaxial deformation which keeps the transverse dimensions constant and, therefore, is consistent with typical W_{oS} calculations. Values computed for GB configurations in Figure 4 are listed in Table 3. To illustrate the effect of surface relaxation, we list the values for both the unrelaxed (as created) and the relaxed surfaces (relaxation reduces the value of $E_{FS(+H)}$ in Equation (3)).

Table 3. Work of separation (in J/m^2) calculated for $\Sigma 3$, $\Sigma 5$, and $\Sigma 11$ GBs with and without hydrogen.

GB	Unrelaxed		Relaxed	
	Clean	With H	Clean	With H
$\Sigma 3(111)$	3.78	2.77	3.76	2.34
$\Sigma 5(210)$	4.00	3.41	3.59	1.82
$\Sigma 11(311)$	4.35	3.76	4.15	3.37

Despite the fact that the $\Sigma 3$ GB exhibits the greatest strength values (see Figure 6), its W_{oS} values for unrelaxed surfaces are lower than values computed for the other GBs. The greatest W_{oS} values were obtained for the $\Sigma 11$ GB. These values fall within the range of results of fracture energies computed by Tehranchi and Curtin [12] for seven other GBs.

Let us note that values of the work of separation for bulk differ from the W_{oS} values in Table 3 only by the GB energy (γ_{GB} in Table 2), and one can therefore easily calculate the surface energy $\gamma_{FS} = (W_{oS} + \gamma_{GB})/2$. Values of 2.43 J/m^2 for (210) and 1.91 J/m^2 for (111) surfaces obtained this way (using the relaxed W_{oS} values) agree well with the values of 2.40 J/m^2 and 1.92 J/m^2 reported by Tran et al. [30].

Alvaro et al. [10] and Chen et al. [11] calculated the work of separation for clean and H-charged GBs in Ni using a relationship differing from Equation (3) by a factor of 2 (the definition corresponded rather to the surface energy), therefore their values correspond to one half of our W_{oS} in Table 3. Values of 1.88 J/m^2 [10] and 1.86 J/m^2 [11] determined for the clean $\Sigma 3$ GB therefore agree very well with our results of relaxed calculations. In addition, the values of 1.75 J/m^2 [11] and 1.8 J/m^2 [10] for the clean $\Sigma 5$ GB and 2.04 J/m^2 [11] for the clean $\Sigma 11$ GB are in agreement with data in Table 3.

For H-charged GBs in fcc Ni, Chen et al. [11] predicted a complete decohesion of the $\Sigma 3$ GB fully covered with hydrogen atoms (forming a monolayer), since their relevant W_{oS} value was almost zero. However, our tensile tests predict only a reduction of the tensile strength (for loading of any kind). Our W_{oS} values in Table 3 also show that the presence of one hydrogen monolayer at the $\Sigma 3$ GB reduces its W_{oS} by one third. Although such a relative reduction of W_{oS} is greater than that of σ_{max}, it does not predict the catastrophic crystal decohesion reported in Reference [11]. Instead, it agrees much better with the value of 1.1 J/m^2 published by Alvaro et al. [10].

4. Conclusions

This article presents an ab initio study of the cohesive strength of selected types of special grain boundaries in hydrogen-free and hydrogen-charged nickel crystals under uniaxial and triaxial loading. The main motivation was to find out if not only the dislocation HELP mechanism, but also the HEDE might have been responsible for experimentally observed high susceptibility of $\Sigma 3$ coherent twin boundaries to crack initiation. The results indeed revealed that the presence of hydrogen reduces the cohesive strength of $\Sigma 3$ boundaries to become closer to strength levels of higher-energy GBs which, in contrary, remain hydrogen-unaffected. Thus, this HEDE (decohesion) mechanism makes, along with the previously reported dislocation (HELP) mechanism, the Σ3 GB plane a comprehensible preferential site for nucleation of microcracks. The results of this study also brought an additional HEDE-based explanation of a small resistance of higher-energy GBs to crack propagation. The highly-triaxial stress state at the tips of microcracks (initiated at Σ3 boundaries) caused an extreme reduction of cohesive strength of adjacent high-energy grain boundaries, especially those of a general kind.

Author Contributions: Conceptualization, J.P. and Z.Z.; methodology, P.Š. and M.Č.; ab initio calculations, P.Š.; formal analysis, M.Č. and J.P.; writing—original draft preparation, P.Š.; writing—review and editing, J.P., M.Č., and Z.Z.; visualization, P.Š. and M.Č. ; supervision, J.P.; funding acquisition, J.P. and M.Č. All authors have read and agreed to the published version of the manuscript.

Funding: This research was supported by the Czech Science Foundation (projects No. 17-18566S and No. 20-08130S) and by the Ministry of Education, Youths and Sports of the Czech Republic within special support paid from the National Programme for Sustainability II funds within CEITEC 2020 (project No. LQ1601). Z.Z. would like to thank the Research Council of Norway for the support via the M-HEAT (294689) and HyLine project. Computational resources were provided by the Ministry of Education, Youths and Sports of the Czech Republic under the Project IT4Innovations National Supercomputer Center (Project No. LM2015070) within the program Projects of Large Research, Development and Innovations Infrastructures.

Conflicts of Interest: The authors declare no conflict of interest.

References

1. Seita, M.; Hanson, J.P.; Gradečak, S.; Demkowicz, M.J. The dual role of coherent twin boundaries in hydrogen embrittlement. *Nat. Commun.* **2015**, *6*, 6164. [CrossRef] [PubMed]
2. Koyama, M.; Tasan, C.C.; Akiyama, E.; Tsuzaki, K.; Raabe, D. Hydrogen-assisted decohesion and localized plasticity in dual-phase steel. *Acta Mater.* **2014**, *70*, 174–187. [CrossRef]
3. Geng, W.T.; Freeman, A.J.; Wu, R.; Geller, C.B.; Raynolds, J.E. Embrittling and strengthening effects of hydrogen, boron, and phosphorus on a Σ5 nickel grain boundary. *Phys. Rev. B* **1999**, *60*, 7149–7155. [CrossRef]
4. Xu, X.; Wen, M.; Hu, Z.; Fukuyama, S.; Yokogawa, K. Atomistic process on hydrogen embrittlement of a single crystal of nickel by the embedded atom method. *Comput. Mater. Sci.* **2002**, *23*, 131–138. [CrossRef]
5. Wen, M.; Xu, X.J.; Omura, Y.; Fukuyama, S.; Yokogawa, K. Modeling of hydrogen embrittlement in single crystal Ni. *Comput. Mater. Sci.* **2004**, *30*, 202–211. [CrossRef]
6. Tahir, A.; Janisch, R.; Hartmaier, A. Hydrogen embrittlement of a carbon segregated symmetrical tilt grain boundary in α-Fe. *Mater. Sci. Eng. A* **2014**, *612*, 462–467. [CrossRef]
7. Di Stefano, D.; Mrovec, M.; Elsässer, C. First-principles investigation of hydrogen trapping and diffusion at grain boundaries in nickel. *Acta Mater.* **2015**, *98*, 306–312. [CrossRef]
8. Yamaguchi, M.; Shiga, M.; Kaburaki, H. First-Principles Study on Segregation Energy and Embrittling Potency of Hydrogen in NiΣ5(012) Tilt Grain Boundary. *J. Phys. Soc. Jpn.* **2004**, *73*, 441–449. [CrossRef]
9. Yu, H.; Olsen, J.S.; Alvaro, A.; Olden, V.; He, J.; Zhang, Z. A uniform hydrogen degradation law for high strength steels. *Eng. Fract. Mech.* **2016**, *157*, 56–71. [CrossRef]
10. Alvaro, A.; Jensen, I.T.; Kheradmand, N.; Løvvik, O.; Olden, V. Hydrogen embrittlement in nickel, visited by first principles modeling, cohesive zone simulation and nanomechanical testing. *Int. J. Hydrog. Energy* **2015**, *40*, 16892–16900. [CrossRef]
11. Chen, J.; Dongare, A.M. Role of grain boundary character on oxygen and hydrogen segregation-induced embrittlement in polycrystalline Ni. *J. Mater. Sci.* **2017**, *52*, 30–45. [CrossRef]
12. Tehranchi, A.; Curtin, W.A. Atomistic study of hydrogen embrittlement of grain boundaries in nickel: II. Decohesion. *Modell. Simul. Mater. Sci. Eng.* **2017**, *25*, 075013. [CrossRef]
13. He, S.; Ecker, W.; Pippan, R.; Razumovskiy, V.I. Hydrogen-enhanced decohesion mechanism of the special Sigma5(012)[100] grain boundary in Ni with Mo and C solutes. *Comput. Mater. Sci.* **2019**, *167*, 100–110. [CrossRef]
14. Oriani, R.A. A mechanistic theory of hydrogen embrittlement of steels. *Berichte Bunsenges. Phys. Chem.* **1972**, *76*, 848–857. [CrossRef]
15. Birnbaum, H.; Sofronis, P. Hydrogen-enhanced localized plasticity—A mechanism for hydrogen-related fracture. *Mater. Sci. Eng. A* **1994**, *176*, 191–202. [CrossRef]

16. Yu, H.; Olsen, J.S.; He, J.; Zhang, Z. Hydrogen-microvoid interactions at continuum scale. *Int. J. Hydrog. Energy* **2018**, *43*, 10104–10128. [CrossRef]
17. Zhao, K.; He, J.; Mayer, A.; Zhang, Z. Effect of hydrogen on the collective behavior of dislocations in the case of nanoindentation. *Acta Mater.* **2018**, *148*, 18–27. [CrossRef]
18. Všianská, M.; Šob, M. The effect of segregated *sp*-impurities on grain-boundary and surface structure, magnetism and embrittlement in nickel. *Prog. Mater. Sci.* **2011**, *56*, 817–840. [CrossRef]
19. Černý, M.; Šesták, P.; Řehák, P.; Všianská, M.; Šob, M. Ab initio tensile tests of grain boundaries in the fcc crystals of Ni and Co with segregated *sp*-impurities. *Mater. Sci. Eng. A* **2016**, *669*, 218–225. [CrossRef]
20. Tahir, A.M.; Janisch, R.; Hartmaier, A. Ab initio calculation of traction separation laws for a grain boundary in molybdenum with segregated C impurites. *Modell. Simul. Mater. Sci. Eng.* **2013**, *21*, 075005. [CrossRef]
21. Janisch, R.; Ahmed, N.; Hartmaier, A. *Ab initio* tensile tests of Al bulk crystals and grain boundaries: Universality of mechanical behavior. *Phys. Rev. B* **2010**, *81*, 184108. [CrossRef]
22. Kresse, G.; Furthmüller, J. Efficient iterative schemes for ab initio total-energy calculations using a plane-wave basis set. *Phys. Rev. B* **1996**, *54*, 11169–11186. [CrossRef]
23. Kresse, G.; Joubert, D. From ultrasoft pseudopotentials to the projector augmented-wave method. *Phys. Rev. B* **1999**, *59*, 1758–1775. [CrossRef]
24. Perdew, J.P.; Burke, K.; Ernzerhof, M. Generalized Gradient Approximation Made Simple. *Phys. Rev. Lett.* **1996**, *77*, 3865–3868. [CrossRef]
25. Monkhorst, H.J.; Pack, J.D. Special points for Brillouin-zone integrations. *Phys. Rev. B* **1976**, *13*, 5188–5192. [CrossRef]
26. Bean, J.J.; McKenna, K.P. Origin of differences in the excess volume of copper and nickel grain boundaries. *Acta Mater.* **2016**, *110*, 246–257. [CrossRef]
27. Shiga, M.; Yamaguchi, M.; Kaburaki, H. Structure and energetics of clean and hydrogenated Ni surfaces and symmetrical tilt grain boundaries using the embedded-atom method. *Phys. Rev. B* **2003**, *68*, 245402. [CrossRef]
28. Černý, M.; Pokluda, J. Ideal tensile strength of cubic crystals under superimposed transverse biaxial stresses from first principles. *Phys. Rev. B* **2010**, *82*, 174106. [CrossRef]
29. Černý, M.; Šesták, P.; Řehák, P.; Všianská, M.; Šob, M. Atomistic approaches to cleavage of interfaces. *Modell. Simul. Mater. Sci. Eng.* **2019**, *27*, 035007. [CrossRef]
30. Tran, R.; Xu, Z.; Radhakrishnan, B.; Winston, D.; Sun, W.; Persson, K.A.; Ong, S.P. Surface energies of elemental crystals. *Sci. Data* **2016**, *3*, 160080. [CrossRef]

© 2020 by the authors. Licensee MDPI, Basel, Switzerland. This article is an open access article distributed under the terms and conditions of the Creative Commons Attribution (CC BY) license (http://creativecommons.org/licenses/by/4.0/).

Article

Interdependent Linear Complexion Structure and Dislocation Mechanics in Fe-Ni

Vladyslav Turlo [1,2] and Timothy J. Rupert [2,3,*]

1. Laboratory for Advanced Materials Processing, Empa, 3603 Thun, Switzerland; Vladyslav.Turlo@empa.ch
2. Department of Mechanical and Aerospace Engineering, University of California, Irvine, CA 92697, USA
3. Department of Materials Science and Engineering, University of California, Irvine, CA 92697, USA
* Correspondence: trupert@uci.edu

Received: 13 November 2020; Accepted: 8 December 2020; Published: 11 December 2020

Abstract: Using large-scale atomistic simulations, dislocation mechanics in the presence of linear complexions are investigated in an Fe-Ni alloy, where the complexions appear as nanoparticle arrays along edge dislocation lines. When mechanical shear stress is applied to drive dislocation motion, a strong pinning effect is observed where the defects are restricted by their own linear complexion structures. This pinning effect becomes weaker after the first dislocation break-away event, leading to a stress-strain curve with a profound initial yield point, similar to the static strain aging behavior observed experimentally for Fe-Mn alloys with the same type of linear complexions. The existence of such a response can be explained by local diffusion-less and lattice distortive transformations corresponding to $L1_0$-to-B2 phase transitions within the linear complexion nanoparticles. As such, an interdependence between a linear complexion structure and dislocation mechanics is found.

Keywords: linear complexions; dislocations; strength; lattice distortive transformations

1. Introduction

Linear complexions are thermodynamically-stable nanoscale phases recently discovered at dislocations in the Fe-9 at.% Mn alloy [1]. Similar to interfacial complexions confined to grain boundary regions [2–4], linear complexions are defined by a structure and chemistry that are different from the matrix, but can only exist in the presence of crystalline defects, with dislocations serving that role for linear complexions. Using atomistic simulations, the authors of this work recently predicted a wide variety of linear complexions in body centered cubic (BCC) and face centered cubic (FCC) metals [5–7]. One interesting feature of linear complexions in BCC Fe-based alloys is the presence of a metastable phase in the dislocation segregation zone, which maintains coherent interfaces with the matrix phase. These metastable phases have been reported for the Fe-Ni system with simulations [5] and for the Fe-Mn system with experiments [8]. Other interesting features have been predicted for FCC alloys, such as the formation of 2D platelet phases that can form platelet arrays along partial dislocations or replace the dislocation stacking fault [7]. While some of these complexion types still require experimental validation, it is clear that linear complexions at dislocations represent a new exciting materials research area for crystalline solids, since this topic has the potential to enable new materials with unique properties.

While the effects of grain boundary complexions on various material properties have been studied extensively [9–12], similar research on the influence of linear complexions is limited. One example of existing work in this area is the study of Kwiatkovski de Silva et al. [13], who demonstrated a static strain aging effect in single crystal Fe-Mn samples containing linear complexions. Specifically, the atomic-scale details of dislocation-linear complexion interactions and the associated mechanical behavior are not known. Atomistic simulations are proven to be a powerful tool for investigating the nanoscale mechanics involving dislocation interactions with alloying elements [14], grain boundaries and grain boundary complexions [15,16], nanoscale precipitates [17,18], ceramic nanoparticles [19], Guinier-Preston zones [20,21], and vacancy clusters [22]. Atomistic simulations act in these situations as a digital microscope, providing a great level of detail on structural and chemical transitions as well as deformation mechanisms at the nanoscale. For example, multi-principal element alloys have intriguing mechanical properties. However, their deformation physics are complicated by the compositional complexity of the lattice. Jian et al. [23] reported on the roles of lattice distortion and chemical short range order on dislocation behavior, finding that these factors can result in enhanced glide resistance. Xu et al. [24] explored the local slip resistances in a BCC multi-principal element alloy on a variety of slip planes and with a variety of Burgers vectors, observing that these alloys could deform by a multiplicity of slip modes. The work of Wang et al. [14] provided experimental validation of such a plasticity mechanism and connected this behavior to the observation of a strength plateau at intermediate temperatures rather than the rapidly decreasing strength of traditional BCC alloys with an increasing testing temperature.

Due to the recent discovery of linear complexions, a comprehensive investigation of their effect on dislocation propagation and pinning is missing in the literature. In this paper, we provide the first mechanistic insight into the effect of nanoparticle array linear complexions in a BCC Fe-Ni alloy on mechanical behavior, with a solid solution of the same composition providing a point for comparison. The atomistic mechanisms associated with dislocation pinning and unpinning events during the shear deformation are investigated in detail, and connected to the shape of the stress-strain curve. Finally, the structure of the linear complexion is found to change as the dislocation and its local stress field moves away, resulting in an interdependence of dislocation behavior and complexion structure. The results shown here highlight that linear complexions are defect states that both alter and react to the dislocation environment, providing a pathway for the direct manipulation of mechanical behavior.

2. Materials and Methods

Atomistic simulations, including molecular statics, molecular dynamics (MD), and hybrid Monte Carlo (MC)/MD, were performed using the Large-scale Atomic/Molecular Massively Parallel Simulator (LAMMPS) software (Sandia National Laboratories, Albuquerque, NM, USA) [25] with an embedded-atom method (EAM) potential parametrized to reproduce the binary Fe-Ni system used to model atomic interactions [26]. All MD simulations used a 1 fs integration timestep. Atomic snapshots were analyzed and visualized using the OVITO software (OVITO GmbH, Darmstadt, Germany) [27]. The crystalline structure and chemical ordering were analyzed using the Polyhedral Template Matching method [28], while the positions of dislocation lines were identified using the Dislocation Extraction Algorithm [29].

The Fe-Ni system and the given potential were chosen for several reasons. First, the Fe-Ni phase diagram is similar to the phase diagram of the Fe-Mn system in which linear complexions were first discovered experimentally. This is important because only two Fe-Mn potentials currently can be found in the literature, but neither is appropriate for the goals of this study. Bonny et al. [30] developed an Fe-Mn EAM potential while Kim et al. [31] developed an Fe-Mn modified embedded atom method (MEAM) potential but neither was rigorously fitted to reproduce the bulk phase diagram. Moreover, MEAM potentials are not currently compatible with the hybrid MC/MD code used here. Since it is critical to reproduce the relevant phases for the alloy system, the Fe-Ni potential from Bonny et al. [26] was chosen for this work since it was fitted based on the experimental phase diagram for Fe-Ni and was found to reproduce the stable intermetallic phases ($L1_0$-FeNi and $L1_2$-FeNi$_3$). The main weakness of the potential is that the solubility limit of Ni in BCC Fe is overestimated, meaning that exact composition values for complexion transitions should not be compared with experiments. In addition, Domain and Becquart [32] performed density functional theory (DFT) calculations of segregation to self-interstitial atom clusters, finding that Ni may be able to segregate to sites under both local compressive and tensile stresses, while the EAM potential only predicted segregation to the tensile regions. This suggests that some segregation of Ni dopants to both sides of the dislocation may occur during the initial segregation stages. However, the final linear complexion structure is not expected to be affected by this, and, in fact, the simulated linear complexions resemble those observed experimentally in Fe-Mn [5,6]. Strong variations of Ni composition along the dislocation line were observed at the compression side in the dislocation segregation zone with the composition near precipitates approaching ~50 at.% Ni while, in between precipitates, the composition was near the global composition in the system. Based on these observations, we conclude that the final form of the complexion and solute distribution around the dislocation core is controlled by the second-phase precipitation and growth, and not by the initial solute segregation. Next, we provide the simulation details for a representative computational cell containing linear complexions, prepared for mechanical testing.

First, an initial simulation cell with two edge dislocations was prepared, as shown in Figure 1a. The dislocations were inserted by removing one-half of the atomic plane in the middle of the sample and relaxing the atomic structure using the conjugate gradient descent method implemented in LAMMPS. Next, 2 at.% Ni atoms were randomly distributed within the sample by replacing Fe atoms, which was followed by MD equilibration with an NPT ensemble (constant number of atoms, constant pressure, and constant temperature) at zero pressure and 300 K temperature for 20 ps. Three thermodynamically-equivalent initial solid solution configurations with different random distributions of solutes were prepared with this procedure, providing a baseline for comparison against samples containing linear complexions. To induce linear complexion formation, the alloy samples were equilibrated at 500 K using the hybrid MC/MD method, which allows for both chemical segregation as well as local structural relaxations. The MC steps were performed after every 0.1 ps of MD relaxation time, using a variance-constrained semi-grand canonical ensemble that can stabilize alloy systems with coexisting phases [33,34]. This method has been used in a number of recent modeling studies to capture complexion transitions in metallic alloys [35–37]. The MC/MD procedure led to Ni segregation to the compressive side of the dislocations and then formation of linear complexions in the form of nanoscale precipitate arrays composed of metastable B2-FeNi and stable $L1_0$-FeNi phases, as shown in Figure 1. The presence of nanoscale precipitates reduces the compressive stresses on the side of the dislocation with the extra half-plane of atoms (see the zoomed view of the bottom dislocation in Figure 2). More details about the MC/MD procedure for linear complexions in the Fe-Ni system can be found in our previous studies of equilibrium complexion states [5,6]. The MC/MD simulations were determined to be in equilibrium once the rate of evolution of total simulation cell energy fell below 1 eV/ns, even though the procedure was continued for additional time to sample three distinct yet thermodynamically equivalent samples. These samples were then cooled to 300 K over 20 ps with MD for subsequent mechanical testing.

The solid solution specimens and the samples containing linear complexions were then deformed to promote dislocation slip by applying the XY shear deformation to the simulation cell. The deformation of the cell was performed with the non-equilibrium MD method [38,39], using an engineering strain rate of 10^8 s^{-1} at 300 K. To test the effect of deformation rate, an additional simulation was performed at an engineering strain rate of 10^7 s^{-1}. The deformation of the cell was stopped after 10% applied shear strain, and the stress-strain curves were extracted and analyzed in the context of atomic-scale deformation mechanisms and the local structure of the linear complexions.

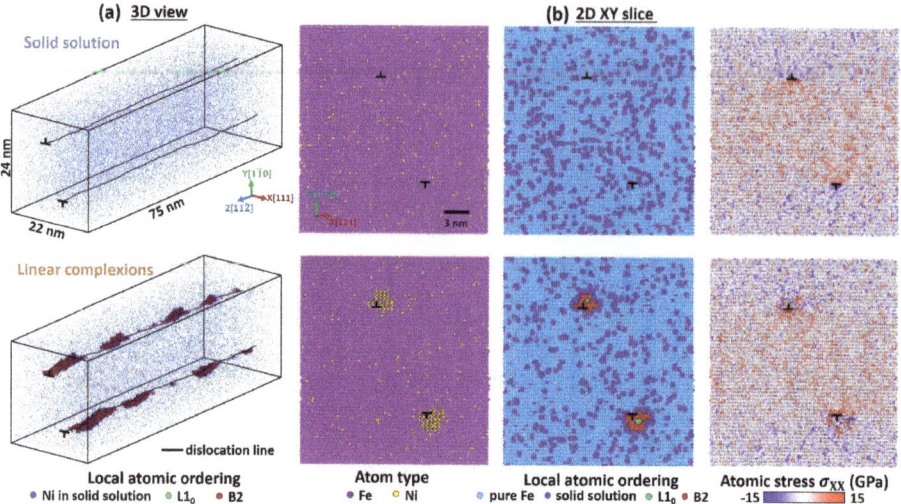

Figure 1. (a) 3D perspective view of the simulation cells with the random distribution of solutes, and with linear complexions comprised of nanoscale precipitates containing B2 (red) and L1$_0$ (green) phases. Atoms are colored by their local atomic order and the matrix Fe atoms are deleted for visualization purposes. The dislocations are shown as black lines. (b) XY slices of the system with Z-position of 28 nm and a thickness of 0.26 nm. The atoms are colored as follows: magenta–Fe, yellow–Ni, light blue–BCC pure Fe, dark blue–BCC Fe-Ni solid solution, red–B2-FeNi, and green–L1$_0$-FeNi. The XX component of the atomic stresses are shown in a red-white-blue color scheme with red indicating tensile stresses and blue indicating compressive stresses.

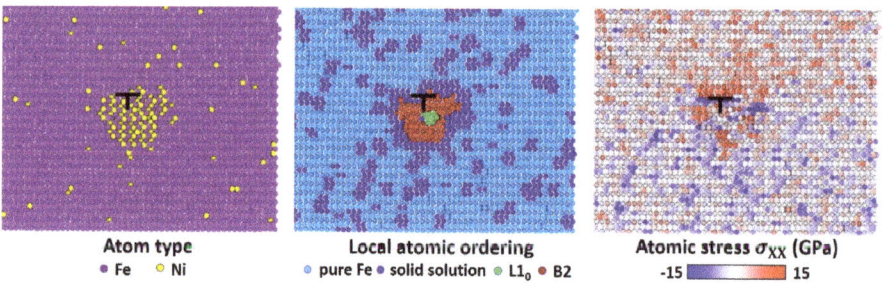

Figure 2. Zoomed view of the bottom dislocation in the system with linear complexions that was previously shown in Figure 1b.

3. Results and Discussion

Figure 3a shows the obtained stress-strain curves for the samples with linear complexions. We observe profound initial stress peaks for all the samples, indicated by black circles. The initial plastic events are followed by smaller peaks, indicated by black triangles, that represent the flow stress of the dislocations passing by the linear complexions. Similar stress-strain curves with a profound first peak have been previously reported in experimental work on linear complexions in an Fe-9 at. % Mn alloy [13]. The substantial strain aging is observed in both experimental Fe-Mn and modeled Fe-Ni systems with linear complexions. The average values for both the initial break-away stress and the flow stress for the simulated Fe-Ni samples with linear complexions are presented in Figure 3b. The mean initial break-away stress of 586 MPa is 45% higher than the mean flow stress of 404 MPa.

We note that these values should not be directly compared to experimental measurements, as other aspects of alloy strengthening from defects, such as grain boundaries, are not present in the simulation cell, which isolates the dislocation pair.

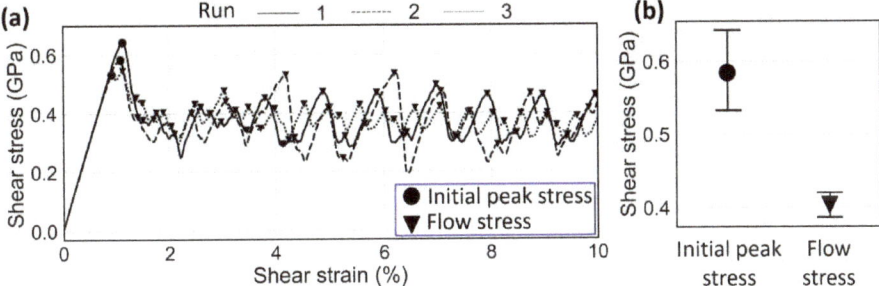

Figure 3. (a) The stress-strain curves for three samples with linear complexions. The initial peak stresses and flow stresses are denoted by circle and triangle symbols, respectively. (b) Mean values and corresponding standard deviations for the initial peak stress and flow stress.

To understand how deformation differs between the solid solution and linear complexion states, Figure 4 presents the initial flow events for representative examples of each of the two sample types. Figure 4a shows the elastic loading and initial flow event, where it is clearly observed that the dislocation can move much more easily in the solid solution sample. Figure 4b shows the dislocation pair at three different simulation times (or, equivalently, applied strains since the strain rate is controlled). The two dislocations in the solid solution remain relatively straight during the motion, with the top dislocation shifting to the right and the bottom dislocation moving to the left. The solutes in the solution act as obstacles that must be locally overcome, leading to very little change in the dislocation shape. The bottom dislocation is shown from a top view in the lower half of Figure 4b, demonstrating the progressive migration that leads to a temporary stress drop as a set of local, solute obstacles are overcome. The local obstacles are easily overcome, which is why the initial peak stress for the solid solution specimens is relatively low. In general, the dislocation in the solid solution sample behaves in a "textbook" fashion, with few features of interest.

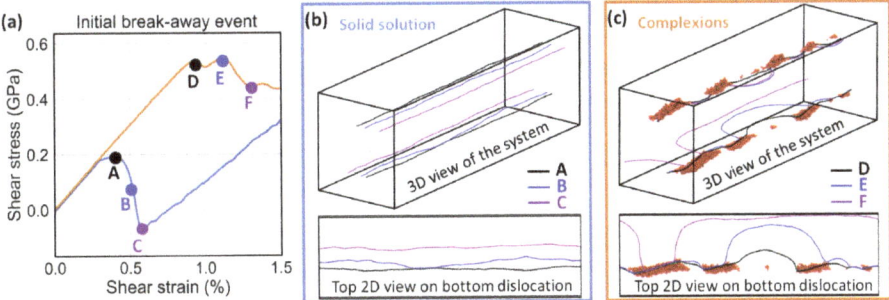

Figure 4. (a) Representative stress-strain curves showing the initial break-away event for a sample with solid solution (blue) and a sample with linear complexions (orange). 3D perspective views of the entire computational cell and 2D views of the bottom dislocation are shown for (b) the solid solution and (c) the sample with linear complexions. The position of the dislocations are shown as solid curves of different colors corresponding to different times, as shown in section (a).

The mechanism of the dislocation motion is drastically different for the specimen with linear complexions with dislocation bowing and unpinning from the nanoscale precipitates one by one, starting in the region with the largest distance between particles, controlling the yield event shown in Figure 4c. It is important to note that only one dislocation, which is the defect at the bottom of the cell, is moving in the linear complexion sample. The other dislocation at the top of the cell bows under increasing stress but remains pinned by the nanoparticle array linear complexion. One dislocation is favored to move over the other because the nanoparticle arrays, while similar, are not exactly identical. A bowing mechanism is easiest in the location where there is the largest spacing between obstacles, which occurs near the middle of the lower dislocation. While dislocation bowing is a common mechanism for overcoming precipitates in conventional alloys, an important distinction is found for the linear complexion state. The obstacle is not in the dislocation's slip plane. Traditional Orowan bowing occurs when an impenetrable obstacle (e.g., a precipitate with a different crystal structure than the matrix) impedes the dislocation's slip path, requiring bowing to move past the obstacle in a way that leaves dislocation loops around the obstacle. However, in the case of linear complexions, the nanoparticles are primarily above or below the dislocation slip plane (depending on whether one is looking at the top or bottom dislocation). Even the small portion of the B2-FeNi intermetallic phase that crosses the slip plane in Figure 1 has the same BCC crystal structure as the matrix phase and a lattice parameter that is similar. The linear complexion is not an impenetrable obstacle and, in fact, no new dislocation loops or segments are formed as the dislocation pulls away. The strong initial pinning can be explained from the same energetic perspective that is used to describe the complexion nucleation. Segregation of Ni occurs to the compressed region near the edge dislocation until the local composition is enriched enough that a complexion transition can occur [5,6]. Although the Gibbs free energy of the $L1_0$ phase is lower, which is why this structure appears on the bulk phase diagram, the restriction to a nanoscale region and the large energy cost for an incoherent BCC-$L1_0$ interface results in the formation of a "shell" of a metastable B2 phase surrounding the $L1_0$ core [6]. Fundamentally, the local stress field near the dislocation is relaxed by this transformation and the driving force for motion under shear stress (i.e., the Peach-Koehler force [40,41]) is, therefore, reduced. We do note that segregation of Ni to the dislocation is needed to create the linear complexion states, lowering the solute composition in the matrix solid solution and removing some weak obstacles. However, the net effect is still a notable strengthening increment, as the strong linear complexion obstacles more than make up for losing some amount of solid solution strengthening.

Since periodic boundary conditions are used here, the dislocation exits the cell after it breaks away from the linear complexion and then re-enters on the other side, where it eventually interacts with the complexion again. Therefore, subsequent dislocation-complexions can be observed by continuing the simulation and investigating the flow events at larger strains. A detailed look at multiple flow events is shown in Figure 5 for one of the linear complexion samples. In addition to the stress-strain curve shown in black, measurements of the dislocation length at any given time are also extracted and presented as the blue curve. Three separate events are labeled A-B, C-D, and E-F and isolated into separate parts of the figure. Similar to the initial yield event in Figure 4, only one dislocation (bottom) is moving while the other (top) remains pinned by the nanoscale precipitates. Figure 5 shows that there are cyclic undulations in the shear stress, which correlate with the observation of repeated bursts in the dislocation density associated with dislocation bowing. Furthermore, the shapes of the dislocation length bursts are extremely similar for each cycle, suggesting that the physical events are similar as well. Snapshots A-B show the final pinned segment of the dislocation, breaking away from a large particle in the complexion array, and then entering from the other side and actually being attracted to the particles, as shown by the fact that the dislocation is pulled closer to the precipitates first in frame B. Snapshots C-D show the first bowing event in a new cycle, which occurs at the location with the largest spacing between particles and is reminiscent of the events presented in Figure 4c. We do remind the reader that this bowing does occur at lower stresses than the initial yield event, suggesting that something about the complexion has evolved. Finally, snapshots E-F show an intermediate bowing event, neither the first nor the last in a cycle.

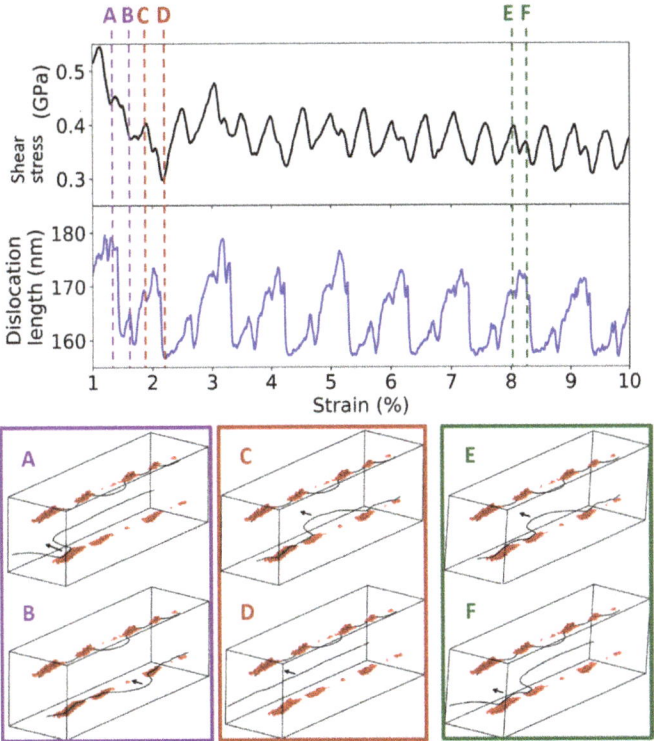

Figure 5. The evolution of the simulation cell shear stress (top, black) and dislocation length (bottom, blue) for the sample with linear complexions, with (**A–F**) different atomic configurations sampled. The atomic configurations are grouped by different passes of the dislocation that happen due to the periodic boundary conditions of the simulation cell. Black arrows denote the direction of motion for mobile dislocation at the bottom of the cell.

To investigate the complexion structure during the deformation simulation, the number of B2 and $L1_0$ atoms in the simulation cell were tracked and are shown in Figure 6a. A cyclic pattern is again observed, suggesting a connection to the repetitive dislocation motion through the cell. Most notable is that reductions in $L1_0$ atoms are generally aligned with increases in the number of B2 atoms (and vice versa). Figure 6b–e show atomistic snapshots of important dislocation interactions with a linear complexion particle in an effort to understand this cyclic behavior. In Figure 6b, the dislocation is still pinned next to the $L1_0$ precipitate, so the complexion contains the expected $L1_0$ core and B2 shell. The dislocation started to pull away from the particle in Figure 6c, and close inspection of the complexion particle shows a reduction in the size of the green $L1_0$ region. However, the transition to the B2 structure does not happen immediately, as some time is apparently needed, although short, since it is captured on MD time scales. Figure 6d shows a later time when the dislocation has moved fully away from the particle, and the linear complexion is almost entirely composed of the B2 structure in this frame (a very small number of green $L1_0$ atoms remain, but the number is dramatically reduced). This observation provides further support for the concept that the two-phase complexion structure is caused by the dislocation's hydrostatic stress field. When that stress field is no longer present, a diffusion-less and lattice distortive transformation from an FCC-like structure ($L1_0$) to a BCC-like structure (B2) occurs. Figure 6e shows that the $L1_0$ region of the precipitate starts to be recovered as the dislocation, and its stress field, arrives at the other side. We note that, while a cyclic behavior is observed in Figure 6a, the number of B2 atoms trend downward as the process continues. This could be a sign that the linear complexion particles are becoming smaller in subsequent cycles (or at least during the transitions from the first few cycles to the steady-state cycling), which would provide an explanation for why the initial yield event is more difficult than later dislocation flow. Figure 6 generally highlights the close connection between the structure of the linear complexion and its interactions with the dislocation. The complexions restrict the dislocation and make it harder to move, while the structure of the complexion relies on the local stress field from the dislocation to find a local equilibrium state. There is, hence, an interdependence of these two defect structures, which truly sets linear complexions apart from traditional obstacles to dislocation motion.

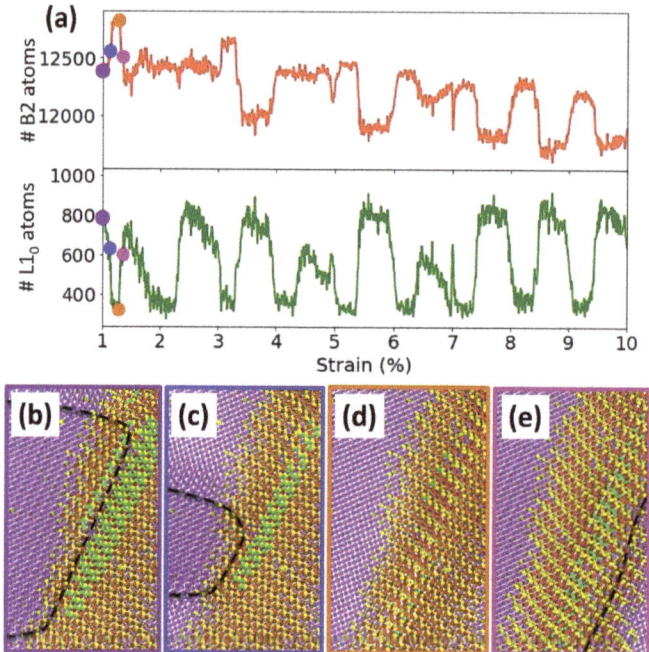

Figure 6. (a) The number of B2 (red) and $L1_0$ (green) atoms in the computational cell containing linear complexions, as the deformation simulation progresses. (b–e) A region perpendicular to the Y-direction from the dislocation slip plane to the middle of the precipitate. Yellow spheres represent Ni atoms, while violet, red, and green spheres represent Fe atoms in the matrix, B2, and L10 phases, respectively. The dashed line shows the position of the dislocation line. A diffusion-less and lattice distortive transformation is observed with the complexion structure being dependent on the dislocation position.

Finally, to show that this local transition within the complexion is not dependent on the deformation rate used here, an additional deformation simulation was run at one order of magnitude slower strain rate (10^7 s^{-1}). The results of this simulation are presented in Figure 7, capturing all of the same important features described above. Loading is initially elastic until a high stress is reached, when the dislocation is able to bow out at the region with the largest spacing between linear complexion particles. The dislocation re-enters the simulation cell and becomes pinned, and subsequent dislocation unpinning and motion follows the same mechanism. A reduction of the number of $L1_0$ atoms in the system and a corresponding increase in the number of B2 atoms is observed each time the dislocation is able to break away. These trends agree with the $L1_0$-to-B2 lattice distortive transformation shown in Figure 6.

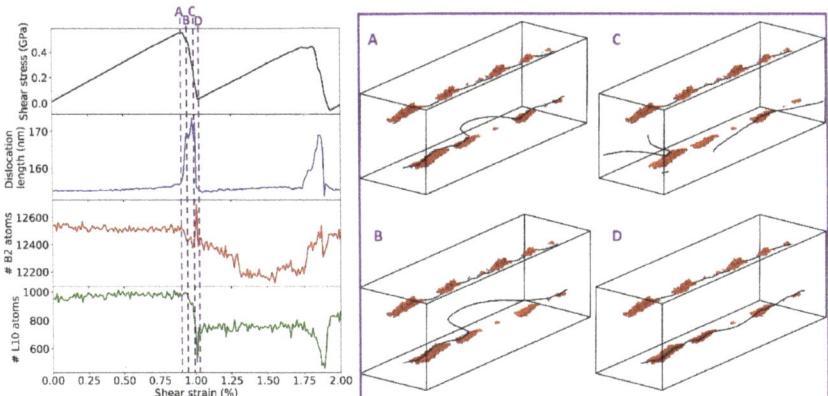

Figure 7. The evolution of the simulation cell shear stress (black), dislocation length (blue), the number of B2 (red), and L10 (green) atoms for the sample with linear complexions deformed at a strain rate of 10^7 s^{-1}, with (**A**–**D**) different atomic configurations sampled on the right side of the figure for the initial break-away event of the dislocation.

4. Conclusions

This paper presents the first atomistic study of the effect of linear complexions on the mechanical behavior of metallic alloys, with Fe-Ni used as a model system. A strong pinning effect of linear complexions on their host dislocations is observed, which is connected to the alteration of the dislocation's stress field in the crystal. This pinning effect leads to a substantial increase in the initial breakaway stress with a pronounced initial peak stress, aligning with experimental observations from alloys with similar complexion structures. Dislocation motion away from the nanoparticle arrays leads to an $L1_0$-to-B2 lattice distortive transformation as the stress field, which stabilized the original complexion structure, is removed. These findings provide additional understanding and context for nanoscale phase transformations induced by dislocation stress fields and their effect on the mechanical properties of materials. Additional studies to obtain a complete understanding of linear complexion thermodynamics and the deformation physics associated with dislocation complexion-interactions are needed to enable "defects-by-design," which can be used to tailor mechanical response.

Author Contributions: Conceptualization, Methodology, Data curation, Formal analysis, and Writing-original draft, V.T. Conceptualization, Methodology, Resources and Writing—review & editing, T.J.R. All authors have read and agreed to the published version of the manuscript.

Funding: This research was funded by the U.S. Army Research Office, grant number W911NF-16-1-0369.

Acknowledgments: The authors express appreciation to Daniel S. Gianola (University of California, Santa Barbara) for fruitful discussions on this topic.

Conflicts of Interest: The authors declare no conflict of interest.

References

1. Kuzmina, M.; Herbig, M.; Ponge, D.; Sandlöbes, S.; Raabe, D. Linear complexions: Confined chemical and structural states at dislocations. *Science (80-)* **2015**, *349*, 1080–1083. [CrossRef] [PubMed]
2. Straumal, B.B.; Mazilkin, A.A.; Baretzky, B. Grain boundary complexions and pseudopartial wetting. *Curr. Opin. Solid State Mater. Sci.* **2016**, *20*, 247–256. [CrossRef]
3. Cantwell, P.R.; Tang, M.; Dillon, S.J.; Luo, J.; Rohrer, G.S.; Harmer, M.P. Grain boundary complexions. *Acta Mater.* **2014**, *62*, 1–48. [CrossRef]
4. Dillon, S.J.; Tang, M.; Carter, W.C.; Harmer, M.P. Complexion: A new concept for kinetic engineering in materials science. *Acta Mater.* **2007**, *55*, 6208–6218. [CrossRef]
5. Turlo, V.; Rupert, T.J. Dislocation-assisted linear complexion formation driven by segregation. *Scr. Mater.* **2018**, *154*. [CrossRef]
6. Turlo, V.; Rupert, T.J. Linear Complexions: Metastable Phase Formation and Coexistence at Dislocations. *Phys. Rev. Lett.* **2019**, *122*. [CrossRef]
7. Turlo, V.; Rupert, T.J. Prediction of a wide variety of linear complexions in face centered cubic alloys. *Acta Mater.* **2020**, *185*, 129–141. [CrossRef]
8. Kwiatkowski Da Silva, A.; Ponge, D.; Peng, Z.; Inden, G.; Lu, Y.; Breen, A.; Gault, B.; Raabe, D. Phase nucleation through confined spinodal fluctuations at crystal defects evidenced in Fe-Mn alloys. *Nat. Commun.* **2018**, *9*, 1–11. [CrossRef]
9. Dillon, S.J.; Tai, K.; Chen, S. The importance of grain boundary complexions in affecting physical properties of polycrystals. *Curr. Opin. Solid State Mater. Sci.* **2016**, *20*, 324–335. [CrossRef]
10. Rupert, T.J. The role of complexions in metallic nano-grain stability and deformation. *Curr. Opin. Solid State Mater. Sci.* **2016**, *20*, 257–267. [CrossRef]
11. Khalajhedayati, A.; Pan, Z.; Rupert, T.J. Manipulating the interfacial structure of nanomaterials to achieve a unique combination of strength and ductility. *Nat. Commun.* **2016**, *7*, 10802. [CrossRef] [PubMed]
12. Frolov, T.; Divinski, S.V.; Asta, M.; Mishin, Y. Effect of interface phase transformations on diffusion and segregation in high-angle grain boundaries. *Phys. Rev. Lett.* **2013**, *110*, 1–5. [CrossRef] [PubMed]
13. Kwiatkowski da Silva, A.; Leyson, G.; Kuzmina, M.; Ponge, D.; Herbig, M.; Sandlöbes, S.; Gault, B.; Neugebauer, J.; Raabe, D. Confined chemical and structural states at dislocations in Fe-9wt%Mn steels: A correlative TEM-atom probe study combined with multiscale modelling. *Acta Mater.* **2017**, *124*, 305–315. [CrossRef]
14. Wang, F.; Balbus, G.H.; Xu, S.; Su, Y.; Shin, J.; Rottmann, P.F.; Knipling, K.E.; Stinville, J.C.; Mills, L.H.; Senkov, O.N.; et al. Multiplicity of dislocation pathways in a refractory multiprincipal element alloy. *Science (80-)* **2020**, *370*, 95–101. [CrossRef]
15. Van Swygenhoven, H.; Derlet, P.M.; Frøseth, A.G. Nucleation and propagation of dislocations in nanocrystalline fcc metals. *Acta Mater.* **2006**, *54*, 1975–1983. [CrossRef]
16. Turlo, V.; Rupert, T.J. Grain boundary complexions and the strength of nanocrystalline metals: Dislocation emission and propagation. *Acta Mater.* **2018**. [CrossRef]
17. Vaid, A.; Guénolé, J.; Prakash, A.; Korte Kerzel, S.; Bitzek, E. Atomistic simulations of basal dislocations in Mg interacting with Mg17Al12 precipitates. *Materialia* **2019**, *7*, 100355. [CrossRef]
18. Kirchmayer, A.; Lyu, H.; Pröbstle, M.; Houllé, F.; Förner, A.; Huenert, D.; Göken, M.; Felfer, P.J.; Bitzek, E.; Neumeier, S. Combining Experiments and Atom Probe Tomography-Informed Simulations on γ′ Precipitation Strengthening in the Polycrystalline Ni-Base Superalloy A718Plus. *Adv. Eng. Mater.* **2020**, *22*. [CrossRef]
19. Li, J.; Liu, B.; Fang, Q.H.; Huang, Z.W.; Liu, Y.W. Atomic-scale strengthening mechanism of dislocation-obstacle interaction in silicon carbide particle-reinforced copper matrix nanocomposites. *Ceram. Int.* **2017**, *43*, 3839–3846. [CrossRef]
20. Singh, C.V.; Mateos, A.J.; Warner, D.H. Atomistic simulations of dislocation-precipitate interactions emphasize importance of cross-slip. *Scr. Mater.* **2011**, *64*, 398–401. [CrossRef]
21. Singh, C.V.; Warner, D.H. Mechanisms of Guinier-Preston zone hardening in the athermal limit. *Acta Mater.* **2010**, *58*, 5797–5805. [CrossRef]
22. Landeiro Dos Reis, M.; Proville, L.; Marinica, M.C.; Sauzay, M. Atomic scale simulations for the diffusion-assisted crossing of dislocation anchored by vacancy clusters. *Phys. Rev. Mater.* **2020**, *4*, 103603. [CrossRef]

23. Jian, W.R.; Xie, Z.; Xu, S.; Su, Y.; Yao, X.; Beyerlein, I.J. Effects of lattice distortion and chemical short-range order on the mechanisms of deformation in medium entropy alloy CoCrNi. *Acta Mater.* **2020**, *199*, 352–369. [CrossRef]
24. Xu, S.; Su, Y.; Jian, W.-R.; Beyerlein, I.J. Local slip resistances in equal-molar MoNbTi multi-principal element alloy. *Acta Mater.* **2020**, *202*, 68–79. [CrossRef]
25. Plimpton, S. Fast Parallel Algorithms for Short-Range Molecular Dynamics. *J. Comput. Phys.* **1995**, *117*, 1–19. [CrossRef]
26. Bonny, G.; Pasianot, R.C.; Malerba, L. Fe-Ni many-body potential for metallurgical applications. *Model. Simul. Mater. Sci. Eng.* **2009**, *17*. [CrossRef]
27. Stukowski, A. Visualization and analysis of atomistic simulation data with OVITO–the Open Visualization Tool. *Model. Simul. Mater. Sci. Eng.* **2010**, *18*, 015012. [CrossRef]
28. Larsen, P.M.; Schmidt, S.; SchiØtz, J. Robust structural identification via polyhedral template matching. *Model. Simul. Mater. Sci. Eng.* **2016**, *24*, 055007. [CrossRef]
29. Stukowski, A.; Albe, K. Extracting dislocations and non-dislocation crystal defects from atomistic simulation data. *Model. Simul. Mater. Sci. Eng.* **2010**, *18*, 015012. [CrossRef]
30. Bonny, G.; Terentyev, D.; Bakaev, A.; Zhurkin, E.E.; Hou, M.; Van Neck, D.; Malerba, L. On the thermal stability of late blooming phases in reactor pressure vessel steels: An atomistic study. *J. Nucl. Mater.* **2013**, *442*, 282–291. [CrossRef]
31. Kim, Y.M.; Shin, Y.H.; Lee, B.J. Modified embedded-atom method interatomic potentials for pure Mn and the Fe-Mn system. *Acta Mater.* **2009**, *57*, 474–482. [CrossRef]
32. Domain, C.; Becquart, C.S. Solute–<111> interstitial loop interaction in α-Fe: A DFT study. *J. Nucl. Mater.* **2018**, *499*, 582–594. [CrossRef]
33. Sadigh, B.; Erhart, P.; Stukowski, A.; Caro, A.; Martinez, E.; Zepeda-Ruiz, L. Scalable parallel Monte Carlo algorithm for atomistic simulations of precipitation in alloys. *Phys. Rev. B-Condens. Matter Mater. Phys.* **2012**, *85*, 1–11. [CrossRef]
34. Sadigh, B.; Erhart, P. Calculation of excess free energies of precipitates via direct thermodynamic integration across phase boundaries. *Phys. Rev. B-Condens. Matter Mater. Phys.* **2012**, *86*, 1–8. [CrossRef]
35. Pan, Z.; Rupert, T.J. Effect of grain boundary character on segregation-induced structural transitions. *Phys. Rev. B-Condens. Matter Mater. Phys.* **2016**, *93*, 1–15. [CrossRef]
36. Frolov, T.; Asta, M.; Mishin, Y. Segregation-induced phase transformations in grain boundaries. *Phys. Rev. B* **2015**, *92*, 020103. [CrossRef]
37. Hu, Y.; Schuler, J.D.; Rupert, T.J. Identifying interatomic potentials for the accurate modeling of interfacial segregation and structural transitions. *Comput. Mater. Sci.* **2018**, *148*, 10–20. [CrossRef]
38. Evans, D.J.; Morriss, G.P. Nonlinear-response theory for steady planar Couette flow. *Phys. Rev. A* **1984**, *30*, 1528–1530. [CrossRef]
39. Daivis, P.J.; Todd, B.D. A simple, direct derivation and proof of the validity of the SLLOD equations of motion for generalized homogeneous flows. *J. Chem. Phys.* **2006**, *124*. [CrossRef]
40. Weertman, J. The Peach–Koehler equation for the force on a dislocation modified for hydrostatic pressure. *Philos. Mag.* **1965**, *11*, 1217–1223. [CrossRef]
41. Lubarda, V.A. Dislocation Burgers vector and the Peach-Koehler force: A review. *J. Mater. Res. Technol.* **2019**, *8*, 1550–1565. [CrossRef]

Publisher's Note: MDPI stays neutral with regard to jurisdictional claims in published maps and institutional affiliations.

© 2020 by the authors. Licensee MDPI, Basel, Switzerland. This article is an open access article distributed under the terms and conditions of the Creative Commons Attribution (CC BY) license (http://creativecommons.org/licenses/by/4.0/).

Review

Review of γ′ Rafting Behavior in Nickel-Based Superalloys: Crystal Plasticity and Phase-Field Simulation

Zhiyuan Yu [1], Xinmei Wang [1], Fuqian Yang [2,*], Zhufeng Yue [1] and James C. M. Li [3]

[1] School of Mechanics, Civil Engineering and Architecture, Northwestern Polytechnical University, Xi'an 710072, China; yuzhiyuan93@mail.nwpu.edu.cn (Z.Y.); wangxinmei@nwpu.edu.cn (X.W.); zfyue@nwpu.edu.cn (Z.Y.)
[2] Materials Program, Department of Chemical and Materials Engineering, University of Kentucky, Lexington, KY 40506, USA
[3] Department of Mechanical Engineering, University of Rochester, Rochester, NY 14627, USA; jli@ur.rochester.edu
* Correspondence: fuqian.yang@uky.edu

Received: 7 November 2020; Accepted: 27 November 2020; Published: 29 November 2020

Abstract: Rafting is an important phenomenon of the microstructure evolution in nickel-based single crystal superalloys at elevated temperature. Understanding the rafting mechanism and its effect on the microstructure evolution is of great importance in determining the structural stability and applications of the single crystal superalloys. Phase-field method, which is an excellent tool to analyze the microstructure evolution at mesoscale, has been gradually used to investigate the rafting behavior. In this work, we review the crystal plasticity theory and phase-field method and discuss the application of the crystal plasticity theory and phase-field method in the analysis of the creep deformation and microstructure evolution of the single crystal superalloys.

Keywords: rafting behavior; phase-field simulation; crystal plasticity theory; mechanical property

---

1. Introduction

Turbine blades are critical structural components in modern aircraft engines and function under extreme service conditions, such as high temperature, high pressure, and high stress level. The performance of turbine blades plays an important role in determining the stability and service life of aircrafts. There is a great need to develop high-performing materials for turbine blades.

Nickel-based single crystal superalloys are currently used in turbine blades due to their excellent mechanical properties at elevated temperature, including high mechanical strength and creep resistance. The excellent mechanical properties of the superalloys are attributed to the two-phase microstructure, i.e., γ′ precipitate phase and γ matrix phase, with $L1_2$ ordered γ′ phase being coherently embedded in γ phase of face-centered cubic structure (Figure 1a). The ordered γ′ phase with high volume fraction (about 60–70% in most cases) serves as the strengthening phase in the superalloys and hinders the motion of dislocations during creep deformation [1–3].

Under mechanical loading at high temperature, the morphology of the γ′ precipitate evolves from cubic shape to plate-like shape, as shown in Figure 1, which is referred to as directional coarsening or rafting. The rafting is a complex process, which can contribute to the hardening of the materials, since the raft structure inhibits dislocation climb around the γ′ phase and restricts the motion of dislocations in γ channels [2–4]. However, there are reports that the rafting behavior can also lead to the softening of materials [5,6]. The onset of rafting widens the width of some γ channels which reduces the blocking effect of Orowan stress and causes the bowing of dislocations in the γ channels (Figure 2). This behavior results in the increase of the plastic strain in the corresponding γ channels and the nearby γ/γ′ interfaces. Note that the plastic deformation around the γ/γ′ interfaces plays a key role in the rafting process. Therefore, the rafting can lead to the strengthening and softening of nickel-based superalloys during creep, which necessitates the understanding of the evolution of the raft structure under mechanical loading at elevated temperature.

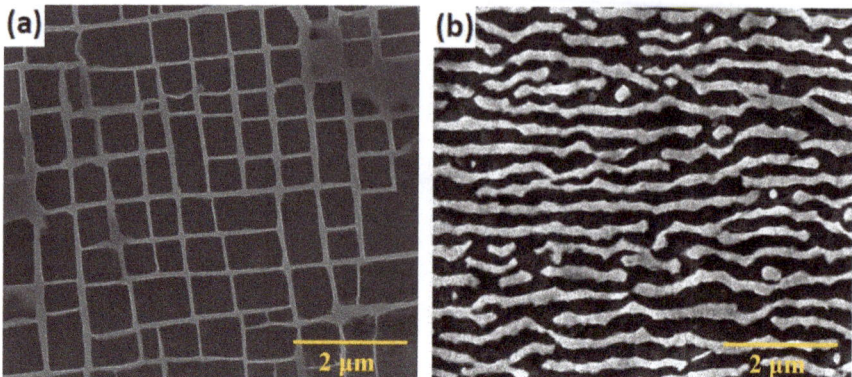

Figure 1. SEM images of the typical microstructure of nickel-based single crystal superalloy (DD5): (a) initial microstructure without deformation, and (b) raft structure after the creep deformation at 980 °C.

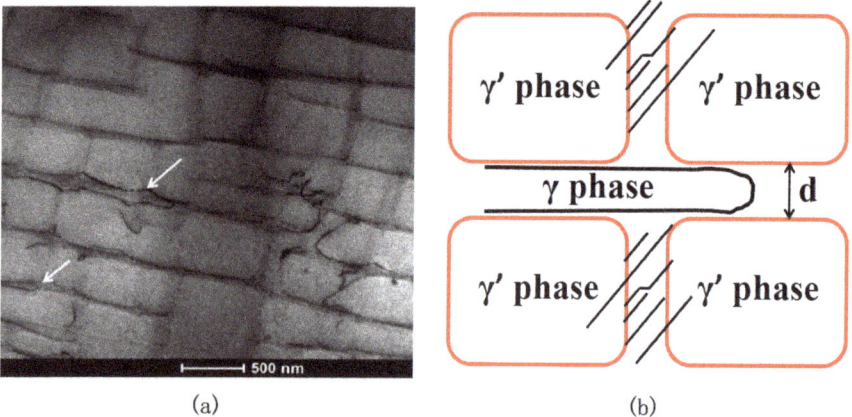

Figure 2. Dislocations bowing through γ channels after long-term aging at 1050 °C for 100 h: (a) TEM image, and (b) schematic of dislocation bowing. In the field of view, some dislocations are present in the γ' phase, which is attributed to the dislocation climb [7] (Reproduced with permission from [8] © 2020 Elsevier).

There are a variety of methods available to investigate the rafting behavior, which can be divided into two categories: experimental methods [1,9] and numerical simulation [10–13]. In general, it is very costly and impractical to use conventional experimental methods to study the rafting behavior due to the need to "record" the changes of the microstructures during creep deformation. Numerical methods, such as finite element method [10], Monte Carlo method [11], and phase-field method [12], have become a major approach to understand the microstructure evolution associated with the rafting process. Among the numerical methods, the phase-field method has become an important technique to analyze the microstructure evolution, which involves phase transition at mesoscale for a long-time period without explicit interface tracking [14–20], and has been used efficiently to simulate the rafting process. Realizing the presence of plastic deformation during the rafting, the theory of crystal plasticity is usually incorporated in the phase-field method to investigate the plastic deformation concurrently presented during the rafting.

In this review, we briefly introduce the rafting behavior in nickel-based single crystal superalloys, together with the crystal plasticity theory and the phase-field method. The models, which are based on the crystal plasticity theory and the phase-field method, respectively, for the analyses of plastic deformation and microstructure evolution are summarized. The applications of the crystal plasticity theory and the phase-field method in studying the rafting behavior are then discussed. Finally, a brief summary is given.

2. γ' Rafting Behavior in Superalloys

2.1. Types of Rafting

Rafting is an important phenomenon of the microstructure evolution in $L1_2$ hardened superalloys, such as nickel- and cobalt-based superalloys [21–23]. There are extensive studies on the formation mechanism in single crystal superalloys, which are also applicable to individual grains in polycrystalline superalloys [24]. Here, we are mainly focused on the rafting behavior in single crystals.

Under mechanical loading in <001> direction, raft structures can form in single crystals in the direction either parallel (P-type) or normal (N-type) to the direction of external loading. The factors controlling the rafting orientation mainly include the sign of applied stress, σ^A (positive for tensile loading and negative for compressive loading), the γ/γ' lattice misfit, δ, and the difference in the mechanical behavior between γ and γ' phase, ΔG [25,26]. Here, $\delta = 2(a_{\gamma'} - a_\gamma)/(a_{\gamma'} + a_\gamma)$, with $a_{\gamma'}$ and a_γ being the lattice constants of γ' phase and γ phase, respectively. ΔG is usually used to represent the hardness ratio of the γ' precipitate to the γ matrix with $\Delta G > 1$ for hard precipitates and $\Delta G < 1$ for soft precipitates. The P-type raft structures form under $\sigma^A \delta (1 - \Delta G) < 0$, and the N-type raft structures form under $\sigma^A \delta (1 - \Delta G) > 0$, as shown in Figure 3. Schmidt and Gross [25] summarized several cases for the raft structures presented in regions (1–7) in Figure 3, in which regions (5–7) represent extreme cases. Note that the raft structures presented in region (5) are only observed in very soft precipitates under compressive loading, and the raft structures presented in regions (6) and (7) are observed in two-phase materials with different Poisson's ratio and/or Zener anisotropy parameter [26].

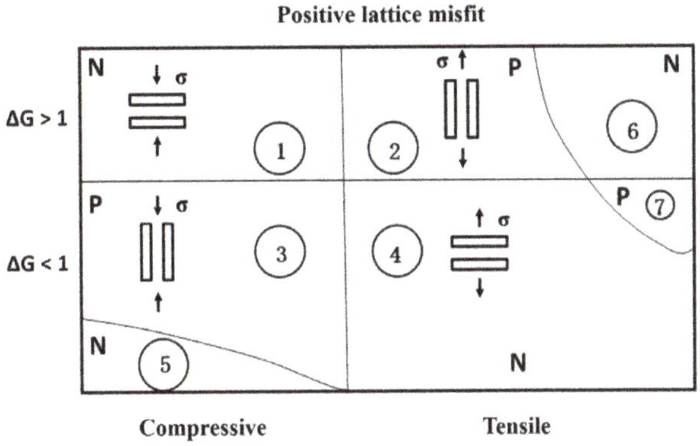

Figure 3. Types of raft structures under mechanical loading in <001> orientation of single crystals and positive lattice misfit. ΔG represents the hardness ratio of the γ' precipitate to the γ matrix. Seven different cases are summarized and marked with corresponding numbers in the figure. 1—N-type structure under compression for $\Delta G > 1$; 2—P-type structure under tension for $\Delta G > 1$; 3—P-type structure under compression for $\Delta G < 1$; 4—N-type structure under tension for $\Delta G < 1$; 5—N-type structure under compression for $\Delta G < 1$; 6—N-type structure under tension for $\Delta G > 1$; 7—P-type structure under tension for $\Delta G < 1$. Adapted from [25].

The rafting direction plays an important role in determining the mechanical behavior of superalloys. In most cases, the N-type raft structures can reduce the low-cycle fatigue strength of materials and the P-type raft structures generally possess high creep resistance and low-cycle fatigue strength [27–29]. Note that the N-type raft structures can also increase the creep resistance by effectively hindering the motions of dislocations at high temperature under the action of small stress [28] in comparison to the corresponding alloys with cubic γ' precipitates. Mughrabi and Tetzlaff [28] reported that the prerafting to form the P-type raft structures in superalloys with negative γ/γ' lattice misfit also resulted in the improvement in fatigue strength at elevated temperature. Compromise needs to be made in controlling the rafting behavior of superalloys with negative γ/γ' lattice misfit.

Commercial nickel-based superalloys possess negative lattice misfit at room temperature, whose magnitude decreases with the increase of environmental temperature [30,31]. To better describe the effect of lattice misfit on the rafting behavior, cobalt-based superalloys are taken as an example for comparison. Cobalt-based superalloys possess positive lattice misfit at room temperature, whose magnitude also decreases with the increase of environmental temperature, but the lattice misfit remains positive even at temperature of 1000 °C [22,31]. Negative lattice misfit in nickel-based superalloys and positive lattice misfit in cobalt-based superalloys at such a high temperature result in the formation of N-type and P-type raft structures under tension, respectively. The opposite behavior is observed under compression.

It is known that a superalloy of single crystal exhibits different mechanical behavior from the corresponding superalloy of polycrystal, and its creep and fatigue properties are extremely sensitive to crystal orientation [32,33]. The complex geometry of single-crystal turbine blades and the multiaxial stress state under service necessitate careful characterization and analyses of the mechanical behavior and γ/γ' microstructure evolution of single crystal superalloys in typical crystal orientations, including <011> and <111>. Under a tensile loading in <011> direction, directional coarsening of γ' phases was found parallel to (010) plane, while equiaxed coarsening behavior was observed in (100) plane, which exhibits a 45-degree rafting [34,35]. Under a tensile loading in <011> direction, no rafting behavior was present [34,36]. However, there always exists a slight misorientation between actual loading direction and <011> or <111> crystal orientation, which can lead to the formation of raft structures different from theoretical results. If the loading direction slightly deviates from <011> direction, see Figure 4 as an example, directional coarsening of γ' phases is still found parallel to (010) plane, but the coarsening behavior in (100) plane is not equiaxed, tending to align along [010] or [001] direction. In the case of loading direction slightly deviating from <111> direction, rafting behavior along one main direction is also observed (Figure 5). These unexpected phenomena were also reported in some works [37–39], which is believed to be related to asymmetric stress distribution in γ/γ' microstructures, caused by a slight misorientation from the loading direction and microdefects randomly distributed in the material. Therefore, the rafting in a superalloy of single crystal is sensitive to the effective loading direction.

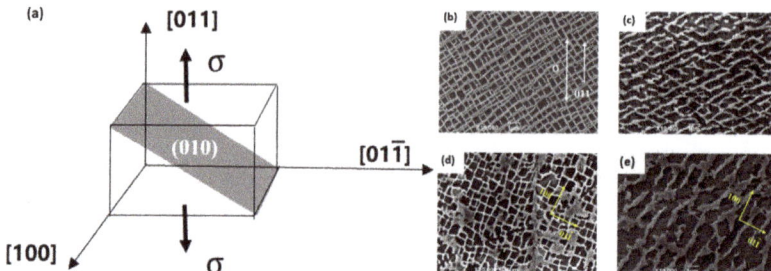

Figure 4. Microstructures of a nickel-based single crystal superalloy under a tensile loading of 250 MPa at 1253 K, with a 5° misorientation from [011] direction. (**a**) Loading diagram of a cubical cell, (**b**) initial microstructure in (100) plane, (**c**) micrograph in (100) plane for 150 h, (**d**) micrograph in (011) plane for 20 h, and (**e**) micrograph in (011) plane for 100 h.

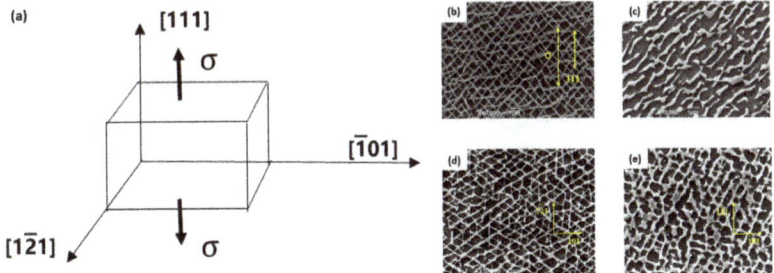

Figure 5. Microstructures of a nickel-based single crystal superalloy under a tensile loading of 250 MPa at 1253 K with a 5° misorientation from [111] direction. (**a**) Loading diagram of a cubical cell, (**b**) micrograph of initial microstructure in ($1\bar{2}1$) plane, (**c**) micrograph in ($1\bar{2}1$) plane for 300 h, (**d**) micrograph in (111) plane for 20 h, and (**e**) micrograph in (111) plane for 200 h.

2.2. Kinetics of Rafting

In the heart of rafting is a stress-limited diffusion process, depending on applied stress, σ^A, lattice misfit, δ, and the difference between the mechanical property of γ and γ' phases, ΔG [3,40]. Both applied stress and inherent misfit stress play important roles in the distribution of internal stress in the grains/crystals consisting of γ and γ' phases due to the dependence of chemical potential on local stress state [40]. Under tensile loading, the difference of the chemical potentials between the γ' phase and the γ phase (γ channels) drives the forming atoms of the γ' phase (Al, Ti, Ta, etc.) in horizontal γ channels into vertical γ channels and drives nearly insoluble atoms (Co, Cr, Mo, Re, W, etc.) away from vertical γ channels. This results in the formation of raft structures perpendicular to the tensile loading, as shown in Figure 6 [41,42].

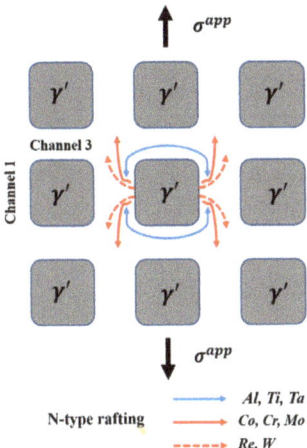

Figure 6. Schematic for the migration of solute atoms under tension in <001> direction.

Assuming that the stress-limited diffusion process is the dominant mechanism controlling the rafting, Fan et al. [34] proposed a von-Mises stress-based criterion in determining the rafting direction. Under concurrent action of external stress and misfit stress, the von-Mises stress in different γ channels can be calculated according to the formulations in Table 1. Dislocations can easily accumulate in γ channels under a large stress, leading to the relief of the misfit stress through the migration of the γ/γ' interface and the directional coarsening in the γ channels with maximum von-Mises stress [34,43].

Table 1. Calculation of the von-Mises stress in γ channels under uniaxial loading. Here, σ_0 is applied stress in the global coordinate system, σ_i is the magnitude of the misfit stress, and α is a reduction factor. Adapted from [34].

	Tension along <001>	Tension along <110>	Tension along <111>
Loading Diagram	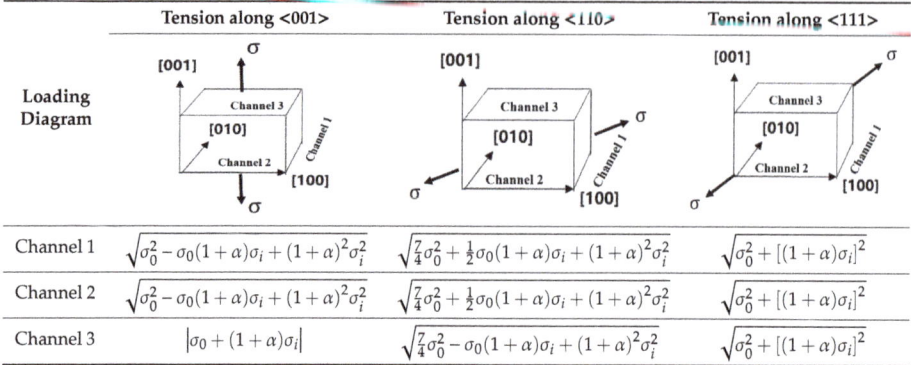		
Channel 1	$\sqrt{\sigma_0^2 - \sigma_0(1+\alpha)\sigma_i + (1+\alpha)^2\sigma_i^2}$	$\sqrt{\tfrac{7}{4}\sigma_0^2 + \tfrac{1}{2}\sigma_0(1+\alpha)\sigma_i + (1+\alpha)^2\sigma_i^2}$	$\sqrt{\sigma_0^2 + [(1+\alpha)\sigma_i]^2}$
Channel 2	$\sqrt{\sigma_0^2 - \sigma_0(1+\alpha)\sigma_i + (1+\alpha)^2\sigma_i^2}$	$\sqrt{\tfrac{7}{4}\sigma_0^2 + \tfrac{1}{2}\sigma_0(1+\alpha)\sigma_i + (1+\alpha)^2\sigma_i^2}$	$\sqrt{\sigma_0^2 + [(1+\alpha)\sigma_i]^2}$
Channel 3	$|\sigma_0 + (1+\alpha)\sigma_i|$	$\sqrt{\tfrac{7}{4}\sigma_0^2 - \sigma_0(1+\alpha)\sigma_i + (1+\alpha)^2\sigma_i^2}$	$\sqrt{\sigma_0^2 + [(1+\alpha)\sigma_i]^2}$

Geometrical characteristics of the microstructures have been used to determine the degree of rafting, such as the width of γ channels [34,44], the dimension of γ′ phase [45], and the shape of γ′ phase [46]. The images of the microstructures can be captured through scanning electron microscopy (SEM) and transmission electron microscopy (TEM), and the geometrical characteristics of the microstructures can be then analyzed via image-processing algorithms. There are a few algorithms commonly used to analyze the SEM and TEM images, including Connectivity number method [46], AutoCorrelation Method [47,48], Rotational Intercept Method [48], Fourier analysis [9], and Moment invariants method [49].

Fedelich et al. [44] introduced a dimensionless variable ξ for the analysis of rafting as

$$\xi = \frac{w(t) - w_{cube}}{w_{raft} - w_{cube}} \tag{1}$$

where $w(t)$ is the width of the γ channel at time t, w_{cube} and w_{raft} are the channel widths of the cubic structure and the raft structure, respectively. Their numerical values are correlated to the volume fraction of γ′ phase, $f_{\gamma'}$, and the microstructure periodicity, λ. The dimensionless variable, ξ, varies in a range of 0 (initial cubic morphology) to 1 (complete raft morphology).

Tinga et al. [45] proposed an evolution law for the dimension of γ′ phase under the action of a multiaxial stress as

$$\dot{L}_i = \frac{-3}{2} L_i \left[\frac{\sigma'_{ii}}{\sigma_M + \sigma_\delta} \right] \left| \frac{A^*}{L_0} \right| \exp\left[\frac{-Q - \sigma_M U_T}{RT} \right], \quad (i = 1, 2, 3) \tag{2}$$

where L_i are the dimensions of γ′ phase along three orthogonal directions, σ'_{ii} are the diagonal components of deviatoric stress tensor, σ_M is von-Mises stress, and U_T is thermal shear-activation volume, Q, R, and T are activation energy, gas constant, and absolute temperature, respectively, A^* and σ_δ are constants, and L_0 is the dimension of γ′ precipitate in cubic shape. Desmorat et al. [50] used the width of γ channel and the dimensions of γ′ phase as internal variables in the framework of thermodynamics with elasto-visco-plasticity and Orowan stress (dislocation mechanics). They calculated the Orowan stress τ_{or}, which acts as a resistance to the dislocation motion in γ matrices, during the rafting in the following formulation,

$$\tau_{or} = \alpha_{or} \frac{\mu b}{w} \tag{3}$$

where α_{or} is a material constant ranging from 0.238 to 2.15 [51], and μ and b are the shear modulus and the magnitude of Burgers vector, respectively. Their simulation results provided quantitative description of the rafting behavior in nickel-based single crystal superalloys.

There exists the interaction between rafting and dislocation motion. Rafting leads to the accumulation of dislocations in the γ channels, resulting in the relief of the misfit stress through the migration of the γ/γ' interface and the directional coarsening of γ' precipitates. Without mechanical loading, rafting can also prevail at high temperature if the plastic strain in superalloys reaches a threshold value [52,53]. The deformation field (plastic strain) is associated with the presence of dislocations in crystal, even though there is no external loading. The dislocation motion in superalloys at high temperature relieves the misfit stress and promotes the directional coarsening of γ' precipitates through rafting. It is the plastic deformation in the γ matrices (channels) that determines the rafting process and the microstructure evolution.

3. Crystal Plasticity Models

3.1. Crystal Plasticity Theory

Crystal plasticity theory is based on the work from Taylor and his coworkers [54]. Their work suggested that plastic deformation of crystalline metals is correlated to the change of crystallographic structure. There are two major aspects contributing to the change of crystallographic structure: one is the lattice distortion, and the other is the plastic deformation from the gliding of dislocations in slip planes. There is a significant amount of dislocations in crystalline metals, which cannot be easily described by traditional continuum mechanics [54]. The large amount of dislocations in crystalline metals ($\approx 10^7$ per cm^2 at annealed state) makes it reasonable to use the concept of continuum mechanics in the analysis of plastic deformation controlled by dislocation motion.

Assuming that dislocation gliding occurs uniformly in a grain/single crystal, Rice and Hill [55,56] proposed a kinematic formalism for the plastic deformation of crystals. They divided the deformation gradient tensor F for a crystal into F^e, representing the lattice distortion and rigid rotation, and F^p, representing dislocation gliding (Figure 7) as

$$F = F^e F^p \tag{4}$$

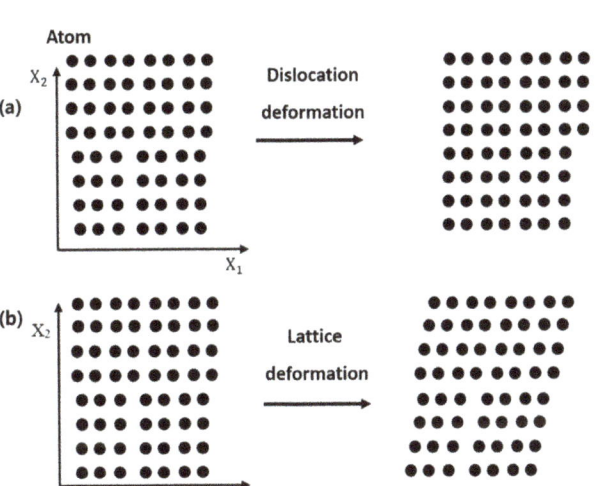

Figure 7. Deformation of a single crystal: (a) example of pure dislocation deformation $F = F^p$, and (b) example of pure lattice deformation $F = F^e$.

The plastic deformation rate tensor due to the dislocation gliding is calculated as

$$\dot{F}^p = L^p F^p \tag{5}$$

Here, L^p is the velocity gradient for plastic deformation, consisting of the contribution of the shear strain rate $\dot{\gamma}^{(\alpha)}$ on all active slip systems as

$$L^p = \sum_{\alpha=1}^{N} \dot{\gamma}^{(\alpha)} m^{(\alpha)} \otimes n^{(\alpha)} \tag{6}$$

where $m^{(\alpha)}$ and $n^{(\alpha)}$ are the unit slip direction and normal vectors of the slip plane for the α-th slip system, respectively, and N is the number of slip systems. L^p can be further divided into a symmetric part, i.e., plastic deformation rate tensor, D^p, and an antisymmetric part, i.e., spin tensor, W^p, as

$$D^p_{ij} = \frac{1}{2} \sum_{\alpha=1}^{N} \dot{\gamma}^{(\alpha)} (m^{(\alpha)}_i n^{(\alpha)}_j + m^{(\alpha)}_j n^{(\alpha)}_i) \tag{7}$$

$$W^p_{ij} = \frac{1}{2} \sum_{\alpha=1}^{N} \dot{\gamma}^{(\alpha)} (m^{(\alpha)}_i n^{(\alpha)}_j - m^{(\alpha)}_j n^{(\alpha)}_i) \tag{8}$$

The plastic deformation rate tensor, D^p, represents the incremental change in the deformation behavior of the material, and the spin tensor, W^p, represents the gliding-induced change in the crystal orientation. Equations (5)–(8) lay the foundation to establish the relationship between the deformation at macro-scale and the shear strains of individual slip systems in a grain/crystal during plastic deformation.

In addition to the kinematic relations, constitutive equations, which capture the microstructure evolution, such as the rafting and dislocation activities, and correlate stresses to strains, need to be developed in order to completely describe the deformation behavior of a grain/crystal. In the following, we present two classes of constitutive models: one is referred to as phenomenological constitutive models, and the other is referred to as physics-based constitutive models.

3.2. Phenomenological Constitutive Models

Several internal state variables are used in the development of phenomenological constitutive models. The deformation state of a material is determined by the variations of the internal state variables with thermal and mechanical loading. The phenomenological constitutive models can be categorized into power-law type (medium stress condition), hyperbolic-sine type (medium to high stress condition) and linear type or combination of linear and power-law types (low to medium stress condition). Chowdhury et al. [57] compared the applications of these constitutive models in the study of crystal plasticity. For engineering practices, the analysis of the plastic deformation of turbine blades is usually based on the power-law constitutive models. The power-law constitutive models have been widely used in the stress analysis due to that it is cost-effective in determining material parameters in the constitutive models and they are applicable in the analysis of the steady-state creep deformation of turbine blades at elevated temperature over a wide range of stresses. However, the pre-factor of the power-law constitutive models is a function of temperature due to different rate mechanisms.

For nickel-based superalloys, dislocation motion in both γ and γ' phases are the dominant mechanism for plastic deformation at elevated temperature [58,59]. In the power-law constitutive models, the ratio or difference between the resolved shear stress, $\tau^{(\alpha)}$, and the slip resistance, $g^{(\alpha)}$, determines the slipping activity of the slip systems. A general mathematic relationship between the shear strain rate, the resolved shear stress, and the slip resistance can be expressed as [60]

$$\dot{\gamma}^{\alpha} = f(\tau^{(\alpha)}, g^{(\alpha)}) \qquad (9)$$

There are three different types of power-law constitutive models for creep deformation as [57]

$$\dot{\gamma}^{(\alpha)} = \dot{\gamma}_0 \left| \frac{\tau^{(\alpha)}}{g^{(\alpha)}} \right|^n sgn(\tau^{(\alpha)}) \qquad (10)$$

$$\dot{\gamma}^{(\alpha)} = \dot{\gamma}_0 \left| \frac{\tau^{(\alpha)} - X^{(\alpha)}}{g^{(\alpha)}} \right|^n sgn(\tau^{(\alpha)} - X^{(\alpha)}) \qquad (11)$$

$$\dot{\gamma}^{(\alpha)} = \left\langle \frac{\left| \tau^{(\alpha)} - X^{(\alpha)} \right| - R^{(\alpha)}}{K} \right\rangle^n sgn(\tau^{(\alpha)} - X^{(\alpha)}) \qquad (12)$$

where $\dot{\gamma}_0$ and n are the shear rate at the reference state and the rate sensitivity parameter, respectively, $g^{(\alpha)}$ and $R^{(\alpha)}$ are two threshold stresses (both evolve with the activity of crystallographic slips [57]), $X^{(\alpha)}$ is internal stress or the back stress, and K is temperature-dependent material parameter. The angle bracket "<y>" in Equation (12) represents the positive part of y, and the "$sgn(\bullet)$" is the sign function.

For the power-law constitutive model of Equaton (10), the activity of crystallographic slips is determined by the resolved shear stress, $\tau^{(\alpha)}$. The slip resistance, $g^{(\alpha)}$, is the only internal state variable as

$$\dot{g}^{(\alpha)} = \sum_{\beta=1}^{N} h_{\alpha\beta} \left| \dot{\gamma}^{(\beta)} \right| \qquad (13)$$

with $h_{\alpha\beta}$ as the hardening matrix, which determines the hardening effect of the α-th slip system on the β-th slip system [61]. For $\alpha = \beta$, the hardening parameter, $h_{\alpha\alpha}$, represents the self-hardening modulus as [61]

$$h_{\alpha\alpha} = h(\gamma) = h_0 sech^2 \left| \frac{h_0 \gamma}{\tau_s - \tau_0} \right| \qquad (14)$$

for $\alpha \neq \beta$, the hardening parameter, $h_{\alpha\beta}$, represents the latent hardening modulus as

$$h_{\alpha\beta} = qh(\gamma), (\alpha \neq \beta) \qquad (15)$$

Here, h_0 is the initial hardening modulus, τ_s and τ_0 are the saturated shear stress and initial yield stress, respectively, and q is a constant. The cumulative shear strain, γ, is calculated as

$$\gamma = \sum_{\alpha} \int_0^t \left| \dot{\gamma}^{(\alpha)} \right| dt \qquad (16)$$

Note that the hardening matrix, $h_{\alpha\beta}$, is also reported in a power-law form as [60,62]

$$h_{\alpha\beta} = q_{\alpha\beta} h_0 \left(1 - \frac{\tau_0}{\tau_s} \right)^{n_0} \qquad (17)$$

with n_0 as a material constant. For the self-hardening modulus, $q_{\alpha\beta} = 1$, and for the latent hardening modulus, $q_{\alpha\beta} = 1.4$.

For the constitutive models of Equations (11) and (12), the activity of crystallographic slips is determined by the effective stress $\tau^{(\alpha)} - X^{(\alpha)}$ and $\left| \tau^{(\alpha)} - X^{(\alpha)} \right| - R^{(\alpha)}$, respectively. The internal stress, $X^{(\alpha)}$, can be calculated by the following equation [63]

$$X^{(\alpha)} = C^{(\alpha)} a^{(\alpha)} \qquad (18)$$

$$\dot{a}^{(\alpha)} = \phi(\nu^{(\alpha)})\dot{\gamma}^{(\alpha)} - \left|\dot{\gamma}^{(\alpha)}\right|d^{(\alpha)}a^{(\alpha)} \qquad (19)$$

$$\phi(\nu^{(\alpha)}) = \phi_0 + (1-\phi_0)e^{-\delta\nu^{(\alpha)}} \qquad (20)$$

$$\nu^{(\alpha)} = \int_0^t \left|\dot{\gamma}^{(\alpha)}\right|dt \qquad (21)$$

In Equation (18), $C^{(\alpha)}$ is the determinate internal stress at a temperature, and $a^{(\alpha)}$ is a control variable [57]. The constant d is a recovery parameter. The flow accumulation function of $\phi(\nu^{(\alpha)})$ is calculated from the accumulated shear strain of the α-th slip system of $\nu^{(\alpha)}$ and the constants of δ and ϕ_0 in Equation (20). It needs to be pointed out that some works [12,19] replaced the term of $\phi(\nu^{(\alpha)})\dot{\gamma}^{(\alpha)}$ with $|\dot{\gamma}^{(\alpha)}|$ in Equation (19) and did not consider the flow accumulation function of $\phi(\nu^{(\alpha)})$.

In general, the use of the effective stress instead of the resolved shear stress for the constitutive models of Equations (11) and (12) makes it possible to analyze the plastic deformation in materials consisting of complex microstructures involving multiphases [64,65]. This is because the internal stresses from the complex microstructures do not directly contribute to the activities of crystallographic slips and needs to be deducted. Additionally, the increments of the internal state variables in Equations (11) and (12) have provided the basis to accurately describe crystallographic slips and determine the threshold stress. All of these suggest that the power-law constitutive models of Equations (11) and (12) can likely provide better correlations between stresses and strains for the analysis of the plastic deformation of the nickel-based superalloys.

3.3. Physics-Based Constitutive Models

In the phenomenological constitutive models, the threshold stresses, $g^{(\alpha)}$ and/or $R^{(\alpha)}$, whose evolution follows a hardening rule, are used to represent the contribution of dislocation motion. However, it is very difficult, if not impossible, to experimentally determine the hardening rule and to validate the hardening rule under service-like conditions. Additionally, these constitutive models fail to capture the orientation dependence of the mechanical behavior of single crystals [66] and the hardening rule of materials at the micron scale [67]. There is a great need to develop physics-based constitutive models, which use dislocation density as an important internal state variable.

There are various physics-based constitutive models, which incorporate dislocation density in the theory of plasticity for the analysis of the plastic deformation of crystalline materials [68–71]. In the heart of the physics-based constitutive models is the evolution of statistically stored dislocation (SSD) and geometrically necessary dislocation (GND) during plastic deformation. The SSD is associated with the "homogeneous" deformation, while the GND is associated with the "inhomogeneous" deformation in single crystals only at small length scale [68].

SSDs are quantified by dislocation density, ρ, (line length per unit volume), Burgers vector, b, and unit vector of dislocation-line segment, t. The magnitude of the Burgers vector is discrete and related to the lattice constant of crystal. For simplification, the unit vector of dislocation-line segment is sometimes limited to a finite set, which makes it easy to include dislocations in numerical calculation.

The Orowan relationship instead of the phenomenological constitutive relationship is used to correlate the plastic shear rate with dislocation density in a physics-based constitutive model as [60]

$$\dot{\gamma}^{(\alpha)} = \rho_m^{(\alpha)} b \nu^{(\alpha)} \qquad (22)$$

where ρ_m is the density of mobile dislocations and is calculated from the SSD density, ρ_{SSD}, and b and ν are the magnitude of Burgers vector and average velocity of the mobile dislocations, respectively. One specific form of Equation (22) is [68,69]:

$$\dot{\gamma}^{(\alpha)} = \left(\rho_{e+}^{(\alpha)}\overline{\nu}_{e+}^{(\alpha)} + \rho_{e-}^{(\alpha)}\overline{\nu}_{e-}^{(\alpha)} + \rho_{s+}^{(\alpha)}\overline{\nu}_{s+}^{(\alpha)} + \rho_{s-}^{(\alpha)}\overline{\nu}_{s-}^{(\alpha)}\right)|b| \qquad (23)$$

where the subscripts, e and s, represent edge and screw dislocations, respectively, and the symbols, "+" and "−", represent the polarity of the SSD density. The density of mobile dislocation, ρ_m, in Equation (22) can also be divided into two portions: ρ_P for the dislocations parallel to slip planes, and ρ_F for the dislocations perpendicular to slip planes as [70,71]

$$\rho_m^{(\alpha)} = BT\sqrt{\rho_P^{(\alpha)}\rho_F^{(\alpha)}} \tag{24}$$

$$B = \frac{2k_B}{c_1\mu b^3} \tag{25}$$

$$\rho_P^{(\alpha)} = \sum_{\beta=1}^{N} \chi^{(\alpha)(\beta)} \rho_{SSD}^{(\beta)} |sin(n^{(\alpha)}, t^\beta)| \tag{26}$$

$$\rho_F^{(\alpha)} = \sum_{\beta=1}^{N} \chi^{(\alpha)(\beta)} \rho_{SSD}^{(\beta)} |cos(n^{(\alpha)}, t^\beta)| \tag{27}$$

where k_B is Boltzmann constant, c_1 is a constant for the evolution of dislocation density, and $\chi^{(\alpha)(\beta)}$ represents the interaction strength between different slip systems. The dislocation interaction plays an important role in determining the plastic deformation in single crystal superalloys. There are nucleation and annihilation of dislocations during plastic deformation, which determine the evolution of dislocation densities [68,70].

The SSD alone is not enough to describe the plastic deformation of crystalline materials. If heterogeneous two-phase microstructure exists in a material, a local strain gradient related to the activity of GNDs is always generated between the strengthening precipitates, such as γ' phases in nickel-based superalloys [72]. In crystal plasticity models, GNDs can be obtained from slip gradients, and they are further divided into the edge dislocation, $\rho_{GND,e}$, along the slip direction and the screw dislocation, $\rho_{GND,s}$, perpendicular to the slip direction as [67]

$$\rho_{GND,e}^{(\alpha)} b = -\nabla \gamma^{(\alpha)} \cdot m^{(\alpha)} \tag{28}$$

$$\rho_{GND,s}^{(\alpha)} b = \nabla \gamma^{(\alpha)} \cdot p^{(\alpha)} \tag{29}$$

where $p^{(\alpha)} = m^{(\alpha)} \times n^{(\alpha)}$, m is a unit vector parallel to the Burgers vector, and n is the unit normal to the slip plane.

With the SSD density and GND density, the internal stress or back stress, $X^{(\alpha)}$, and the slip resistance, $g^{(\alpha)}$, in phenomenological constitutive models can be calculated as [67]

$$g^{(\alpha)} = \mu b \sqrt{\sum_\beta H^{(\alpha)(\beta)} \rho^{(\beta)}} \tag{30}$$

$$X^{(\alpha)} = \frac{\mu b R^2}{8}\left[\frac{1}{(1-v)}\nabla\rho_{GND,e}^{(\alpha)} \cdot m^{(\alpha)} - 2\sum_\beta \delta_s^{(\alpha)(\beta)}(\nabla\rho_{GND,s}^{(\alpha)} \cdot p^{(\alpha)})\right] \tag{31}$$

where v is Poisson's ratio, R is a length scale in the calculation model [73], $H^{(\alpha)(\beta)}$ is the interaction matrix between two slip systems of α and β with six types of interaction in the matrix [74]. $\delta_s^{(\alpha)(\beta)}$ is the interaction coefficient as [73]

$$\delta_s^{(\alpha)(\beta)} = \begin{cases} 1 & for (\alpha,\beta) = (4,13), (6,18), (8,17), (9,15), (10,16), (11,14), \\ & (1,16), (2,17), (3,18), (5,14), (7,13), (12,15) \\ 0 & otherwise \end{cases} \tag{32}$$

Here, α (=1 to 12) represents the edge dislocation, and α (=13 to 18) represents the screw dislocation. $\rho^{(\beta)}$ in Equation (30) is the total dislocation density on the β-th slip system. Note that Tinga et al. [51] considered the contributions of the SSD density and GND density to the total dislocation density, respectively.

Some works incorporated dislocation dynamics, such as discrete dislocation dynamics (DDD) [16,17] and continuum dislocation dynamics (CDD) [18], in the analysis of plastic deformation and microstructural evolution. The DDD models are based on discrete description of dislocation motion and require sufficiently fine grid spacing and great computational cost in the simulation of dislocation activities. The CDD models are based on a continuum quantity (dislocation density) instead of individual dislocations and need much less computational cost [18]. However, the simulation with either type of dislocation dynamics models still costs more computational effort than that with the crystal plasticity models, which incorporate the dislocation activity in a phenomenological or empirical form [19,20]. Thus, crystal plasticity models are widely used to account for the microstructure evolution and plastic deformation and to provide valuable information for engineering applications [57,75]. Table 2 presents the comparisons of the plasticity models used in the rafting analysis.

Table 2. Comparisons of different plasticity models used in the rafting analysis.

Model	Pros and Cons	Application in the Rafting Analysis
Phenomenological constitutive models	Pros: Be cost-effective in determining material parameters and applicable in engineering calculations. Cons: Fail to capture the orientation dependence of the mechanical behavior of single crystals; difficult to experimentally verify the hardening rule used in the constitutive models.	Rafting with creep damage [8,19]
Physics-based constitutive models	Pros: Be able to model the microstructure evolution and include the contribution of dislocations. Cons: Fail to explicitly capture the motion of dislocations.	Coupling between rafting and crystal plasticity with dislocation densities [67]
Discrete dislocation dynamics models	Pros: Explicitly describe the dislocation distribution during microstructural evolution. Cons: Require sufficiently fine grid spacing and great computational cost in simulation.	Distribution of plastic strain in γ-channels and its effect on rafting [16]
Continuum dislocation dynamics models	Pros: Consider average distribution of dislocations and need less computational cost. Cons: Difficult to be compared with phenomenological constitutive models in engineering calculations.	Effect of initial dislocation density on rafting [18]

4. Phase-Field Models

4.1. Phase-Field Method

Phase-field method is based on the description of diffuse-interface model, and the concept of diffuse interface is derived from van der Waals [76] and Cahn and Hilliard [77]. In contrast to conventional sharp-interface models (field variables are discontinuous at interface), the interface in the diffuse-interface model is represented by a series of consecutive values [18]. Assigning different values to different phases (for example, γ' strengthening phase and γ matrix phase in nickel-based superalloys), the interface region is implicitly given [78]. Such a method makes it possible to analyze the evolution of complex microstructures, including precipitation, dissolution, coarsening, connection, and topological inversion of γ' strengthening phases [79,80]. Additionally, the physical characteristics, i.e., the γ/γ' lattice misfit, elastic constants and dislocation densities, and the mechanical behavior, i.e., heterogeneous elasticity and plastic deformation, of superalloys can be well incorporated in the mathematical framework of the phase-field method. The phase-field method has become an excellent technique to study internal stresses and plastic deformation and illustrate the mechanisms for the microstructure evolution in nickel-based superalloys.

The application of the phase-field method in the simulation of microstructural evolution was initially based on an elastic framework, where the mechanical stress, σ^A, the γ/γ' lattice misfit, δ, and the difference of mechanical properties between γ and γ' phases, ΔG, were incorporated in the model to reveal the rafting behavior [14,15,26]. The elasto-plastic frameworks, which take into account the plastic deformation in γ matrices during creep, were later developed [8,12,19].

Superalloys are generally multiphase and/or multicomponent materials, which require large number of field variables or physical quantities in the phase-field method in addition to mechanical deformation (elastic and elasto-plastic deformation). They provide a practical application of the phase-field theory in the understanding of the microstructure evolution of multiphase, polycrystal, and multicomponent materials [81]. The combination of the phase-field model and the crystal plasticity likely opens a new avenue for the study of the mechanical deformation and microstructure evolution of superalloys under concurrent action of thermal and mechanical loading.

4.2. Ni-Al Binary System

Nickel-based superalloys consist of multielements, including Ni, Al, Re, Co, etc., while Ni and Al are two major elements determining the microstructure evolution [12]. A Ni-Al binary system has been used in different phase-field models to simulate the γ/γ' microstructure evolution in the nickel-based superalloys.

$C(r,t)$ is defined as the concentration of Al, which is used to distinguish γ' phase (mainly Ni_3Al) from γ phase (mainly Ni). For a given concentration, C, it is equal to c_γ in the γ phase and $c_{\gamma'}$ in the γ' phase, and the region with the Al concentration changing from c_γ to $c_{\gamma'}$ corresponds to the γ/γ' interface. Note that the Ni-Al phase diagram is used to determine the numerical values of c_γ and $c_{\gamma'}$ at a given temperature [82]. A dimensionless parameter, C', is sometimes used instead of the concentration, C, to distinguish the γ' phase from the γ phase as [26]

$$C' = \frac{C - c_\gamma}{c_{\gamma'} - c_\gamma} \qquad (33)$$

with $C' = 1$ representing the γ' phase and $C' = 0$ representing the γ phase.

There are four different γ' variants when studying the γ/γ' microstructure evolution. This is because there are antiphase domains between different γ' variants, which affect the directional coarsening of γ' phase [79]. Three field parameters are introduced to characterize four different γ' variants: $\{\phi_1, \phi_2, \phi_3\} = \phi_0\{1,1,1\}, \phi_0\{\bar{1},\bar{1},1\}, \phi_0\{\bar{1},1,\bar{1}\}, \phi_0\{1,\bar{1},\bar{1}\}$ [12]. The concentration field, $C(r,t)$, and the field parameters, ϕ_i ($i = 1, 2, 3$), are controlled by kinetic equations, which are established on the principles of minimum free energy [12] and/or maximum entropy [83].

The principle of maximum entropy has been mainly used in the solidification analysis of alloys. It is very difficult to calculate the change of entropy during plastic deformation. The principle of minimum free energy is more common in the study of the rafting process. With the principle of minimum free energy, the concentration field, $C(r,t)$, and the field parameters, ϕ_i ($i = 1, 2, 3$), satisfy the Cahn–Hilliard and Allen–Cahn equations as

$$\frac{\partial C(r,t)}{\partial t} = \nabla \cdot \left(m_0 \nabla \frac{\delta F}{\delta C(r,t)} \right) \qquad (34)$$

$$\frac{\partial \phi_i(r,t)}{\partial t} = -l_0 \frac{\delta F}{\delta \phi_i(r,t)}, \quad (i = 1, 2, 3) \qquad (35)$$

The local volume fraction of the γ' phase, $f(r,t)$, is used sometime instead of $C(r,t)$ in Equation (34) [20,84]. Note that the local γ'-volume fraction is equivalent to the dimensionless parameter, C'. The use of $f(r,t)$ makes it easy to extend the phase-field models for the Ni-Al binary system to multicomponent systems. Here, m_0 and l_0 are the mobility coefficient and the kinetic coefficient, respectively, and F is the total free energy consisting of chemical free energy, F^{ch}, and strain energy, F^{el}. The chemical free energy can be expressed by Ginzburg–Landau functional approximation as [12]

$$F^{ch} = \int_V [f_{homo} + \kappa_1 |\nabla C|^2 + \kappa_2 \sum_{i=1,3} |\nabla \phi_i|^2] dV \qquad (36)$$

The gradients of the concentration field and field parameters in Equation (36) define the numerical resolution and interface thickness in the simulation [18]. The gradient energy coefficients of κ_1 and κ_2 are related to interfacial energy, and their values are adjusted to ensure that the interface thickness of two-phase microstructure can represent real situation [12,85]. The use of f_{homo} is to distinguish the γ phase from the γ' phase and to assure the stability of four different γ' variants at $\phi_0\{1,1,1\}$, $\phi_0\{\bar{1},\bar{1},1\}$, $\phi_0\{\bar{1},1,\bar{1}\}$, and $\phi_0\{1,\bar{1},\bar{1}\}$. The strain energy, F^{el}, is calculated as

$$F^{el} = F^a + \frac{1}{2} \int_V \bar{\bar{\lambda}} : \tilde{\varepsilon}^{el} : \tilde{\varepsilon}^{el} dV \qquad (37)$$

where $\bar{\bar{\lambda}}$ is local elasticity modulus tensor depending on the concentration field, $C(r,t)$. F^a is a homogeneous term, depending on loading condition. For strain-control condition, $F^a = 0$; for stress-control condition, $F^a = -V\sigma_{ij}^a \bar{\varepsilon}_{ij}$. Here σ_{ij}^a is the components of applied stress, and $\bar{\varepsilon}_{ij}$ is the average strain components [13].

The contribution of the hardening free energy, F^{vp}, or plastic energy, F^{pl}, to the total free energy during plastic deformation was also considered in some works [12,19]. However, the partial derivative of F^{vp} or F^{pl} to $C(r,t)$ in Equation (34) is negligible since most studies have been using the same viscoplastic parameters for both the γ phase and the γ' phase, i.e., the hardening or plastic energy function is homogeneous in two phases and independent of the concentration field, $C(r,t)$.

It is worth mentioning that calculations are performed for small deformation in the phase-field simulation. In this situation, the change of the crystal dimensions is negligible and the total strain rate tensor, $\tilde{\varepsilon}$, can be divided into three parts as [26,86]

$$\tilde{\varepsilon} = \tilde{\varepsilon}^{el} + \tilde{\varepsilon}^{0} + \tilde{\varepsilon}^{pl} \qquad (38)$$

Here, $\tilde{\varepsilon}^{el}$, $\tilde{\varepsilon}^{0}$, and $\tilde{\varepsilon}^{pl}$ are elastic strain, eigenstrain from the γ/γ' lattice misfit ξ, and plastic strain rate tensors, respectively. The plastic strain rate tensor, $\tilde{\varepsilon}^{pl}$, is obtained through the theory of crystal plasticity as [86]

$$\tilde{\varepsilon}^{pl} = \sum_{\alpha=1}^{N} \dot{\gamma}^{(\alpha)} m^{(\alpha)} \otimes n^{(\alpha)} \qquad (39)$$

Then the shear strain rate $\dot{\gamma}^{(\alpha)}$ is calculated through phenomenological or physics-based constitutive models. However, the use of the theory of crystal plasticity for small deformation indicates that the calculated strains are much less than the strains measured in experiments.

The equilibrium equations must be satisfied all the time during the microstructure evolution. The local stress equilibrium is expressed as [12]

$$\begin{cases} div \, \tilde{\sigma} = 0 \\ \bar{\sigma} = \tilde{\sigma}^a \end{cases} \in V \qquad (40)$$

Minimizing the strain energy function, F^{el}, with respect to the displacement or strain components, u_i or $\bar{\varepsilon}_{ij}$, under given boundary conditions yields [13]

$$\begin{cases} \frac{\delta F^{el}}{\delta u_i} = 0, strain-control \ condition \\ \frac{\delta F^{el}}{\delta \bar{\varepsilon}_{ij}} = \sigma_{ij}^a, stress-control \ condition \end{cases} \qquad (41)$$

The Cahn–Hilliard and Allen–Cahn equations and the equilibrium equations constitute the main framework of the phase-field models, which involve the two-way interaction between the concentration field and the energy function (stresses).

4.3. Multiphase-Field Model

Phase-field models have evolved from simple binary models with only one order parameter to multiorder or multiphase models recently. A typical example of the multiorder phase-field model is to have four different γ' variants with three ordered parameter fields, ϕ_i ($i = 1, 2, 3$), as discussed in the above section. Multiphase-field models are mainly used for polycrystalline materials [87] and single crystal with multiple components [41]. The following discussion is focused on the multicomponent models for the analysis of the rafting behavior in single crystals.

The total free energy, F, in multicomponent models can be calculated from the integration of the strain energy density, f^{el}, the interface energy, f^{it}, and the chemical free energy, f^{ch}, as [87]

$$F = \int_\Omega (f^{el} + f^{it} + f^{ch}) d\Omega \tag{42}$$

$$f^{it} = \sum_{\alpha,\beta=0}^{N} \frac{\kappa_{\alpha\beta}}{\eta_{\alpha\beta}} \left\{ \frac{\eta_{\alpha\beta}^2}{\pi^2} |\nabla\phi_\alpha \cdot \nabla\phi_\beta| + W_{\alpha\beta} \right\} \tag{43}$$

$$f^{ch} = \sum_{\alpha=0}^{N} \phi_\alpha f_\alpha(c_\alpha) + \mu\left(c - \sum_{\alpha=1}^{N} \phi_\alpha c_\alpha\right) \tag{44}$$

where ϕ_0 denotes γ matrix, ϕ_α ($\alpha = 1, 2, 3, 4$) denote different γ' variants, $\kappa_{\alpha\beta}$ is the interface energy, $\eta_{\alpha\beta}$ is the width of interface, $W_{\alpha\beta}(=\phi_\alpha\phi_\beta)$ is the repulsive potential function, and c is the concentration vector. In Equation (44), $f_\alpha(c_\alpha)$ is the bulk free energy of each phase, and $\mu\left(=\frac{\partial f^{ch}}{\partial c}\right)$ is chemical potential vector.

The strain energy density, f^{el}, in multiphase-field models has a similar expression to that in binary-field models as

$$f^{el} = \frac{1}{2} \lambda^*_{ijkl} \varepsilon^{el}_{ij} \varepsilon^{el}_{kl} \tag{45}$$

$$\varepsilon^{el}_{ij} = \varepsilon_{ij} - \varepsilon^*_{ij} - \varepsilon^{pl}_{ij} \tag{46}$$

where λ^*_{ijkl} and ε^*_{ij} are the components of effective stiffness tensor and effective eigenstrain tensor, respectively [41,87].

The kinetic equations for the microstructure evolutions are [41,87]

$$\frac{\partial \phi_\alpha}{\partial t} = -\sum_{\beta=0}^{N} \frac{\mu_{\alpha\beta}}{N} \left(\frac{\delta F}{\delta \phi_\alpha} - \frac{\delta F}{\delta \phi_\beta}\right) \tag{47}$$

$$\frac{\partial c}{\partial t} = \nabla\left(\sum_{\alpha=0}^{N} M\nabla\frac{\delta F}{\delta c}\right) \tag{48}$$

$$\frac{\partial \sigma_{ij}}{\partial r_i} = 0 \tag{49}$$

Here, $\mu_{\alpha\beta}$ is the components of the mobility-coefficient tensor, and M is the chemical mobility tensor. Comparing Equations (47)–(49) of the multiphase-field model with those in the Ni-Al binary-field model, we note similarities between both models. The driving force for the microstructure evolution in both models is the variation of the energy functions of individual phases, and the equilibrium equation needs to be satisfied all the time.

Figure 8 summarizes the basic process to numerically solve the Ni-Al binary-field models and the multiphase-field models.

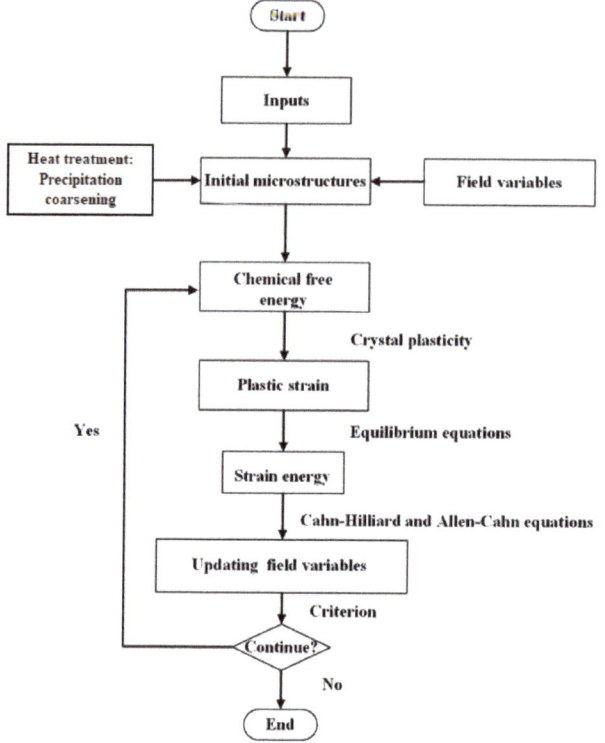

Figure 8. Flowchart showing the basic process to numerically solve the Ni-Al binary-field models and the multiphase-field models.

Step 1: Obtain initial microstructures. This can be achieved by the heat treatment of superalloys (precipitation and coarsening) at high temperature and imaging (SEM and/or TEM) [79] and/or by setting field variables from external files [8,42].

Step 2: Calculate plastic strain. The plastic strain is calculated by solving the related equations, which are based on phenomenological models or physics-based models.

Step 3: Calculate strain energy. The total strain field, $\tilde{\varepsilon}$, and the stress filed, $\tilde{\sigma}$, are firstly obtained by solving equilibrium equations [12,26]. The elastic strains are obtained by subtracting the eigenstrain and plastic strains from the total strains. Finally, the strain energy is calculated from the elastic strains and elastic constants.

Step 4: Update the field variables by solving the Allen–Cahn and Cahn–Hilliard equations.

5. Application of Crystal Plasticity and Phase-Field Method in the Rafting Analysis

5.1. Uniaxial Tension

Uniaxial tension with constant strain rates has been used to experimentally study the rafting behavior in superalloys. Using strain control in the phase-field simulation, we can achieve the tensile mode in solving equilibrium equations. Figure 9 presents the microstructure evolution and plastic strain during the tension of a nickel-based single crystal superalloy. It is evident that there is no significant change in the microstructure, and there is plastic deformation near the γ/γ′ interface and the center of the γ' phases. Comparing the numerical results with experimental results, we can determine the mechanical properties of the γ/γ' microstructure and the bulk phase (if two phase structures are not distinguishable) from the input parameters for the numerical calculation. Additionally, we can evaluate the strengthening effect of the γ' phase and determine the effect of the γ' phase on the mechanical properties of nickel-based superalloys on the macroscale.

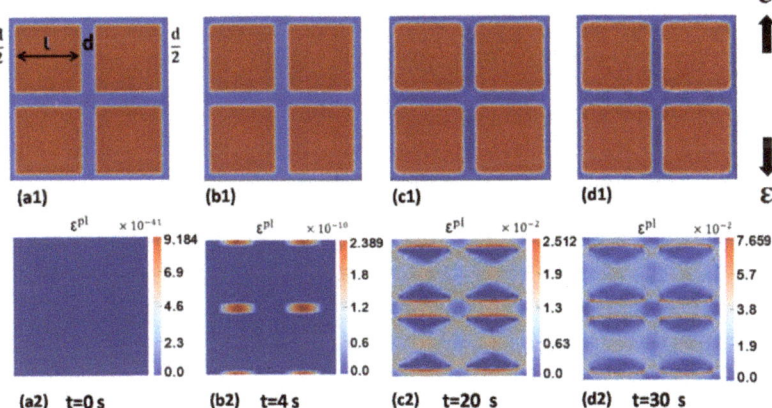

Figure 9. Numerical results of the microstructure and plastic strain at different instants during the tension of a nickel-based single crystal superalloy at a strain rate of 10^{-3} per second and 1253 K: (**a1–d1**) microstructure, and (**a2–d2**) plastic strain (Reproduced with permission from [8] © 2020 Elsevier).

Cottura et al. [67,88] introduced a characteristic plastic length, ζ, in 1D configuration to illustrate the size effect on the plastic deformation of the γ phase and the mechanical properties of the material. Their simulation results suggest that plastic strain is homogeneous in the γ phase if the characteristic plastic length is negligible. Increasing the characteristic plastic length led to the inhomogeneous distribution of the plastic strain and the decrease of the peak value of ζ. For the value of ζ being larger than the width of the γ channels, no plastic deformation would occur in the γ channels. Increasing the characteristic plastic length also increased the flow stress during the tensile deformation.

Wang et al. [8] included the Orowan stress in a visco-plasticity phase-field model to analyze the size effect in a two-dimensional system. Their results show that the flow stress during the tensile deformation decreases with the increase of the γ channel width and the flow stress remains unchanged after the γ channel width reaches a critical value. Zhang et al. [7] also reported similar results for the tension of nickel-based superalloys at high temperature. They pointed out the dependence of the yielding stage during tensile deformation on the penetration of dislocations into γ' precipitates.

5.2. Creep Deformation

5.2.1. N-Type/P-Type Rafting

There are reports on the use of the phase-field models in the analysis of the rafting during creep deformation in the frameworks of elastic deformation [14,26] and elasto-plastic deformation [8,12,19,67]. In the framework of elastic deformation, the rafting is determined by the sign of applied stress, σ^A, the γ/γ' lattice misfit, δ, and the difference of mechanical properties between γ and γ' phases, ΔG. The results from the phase-field simulation reveal the formation of N-type and P-type raft structures in consistence with the predictions shown in Figure 3. However, the simulation cannot capture the rafting features associated with plastic deformation [12,26] despite the good replication in the types of the raft structure. For example, Gaubert et al. [12] compared the simulated microstructures in elastic framework with the corresponding ones in elasto-plastic framework. They found raft structures with straight edges in elastic framework and raft structures with wavy-type edges in elasto-plastic framework. In addition, the plastic deformation increased the rafting rate and promoted the formation of raft structures. Wang et al. [8] presented raft structures and the distribution of plastic strain during creep deformation in elasto-plastic framework. As shown in Figure 10, the raft direction was not strictly perpendicular to the loading direction, and the plastic strain almost appeared in horizontal γ channels and around the γ/γ' interface. They pointed out that further plastic strain that occurred around the γ/γ' interface might result in the instability of raft structures.

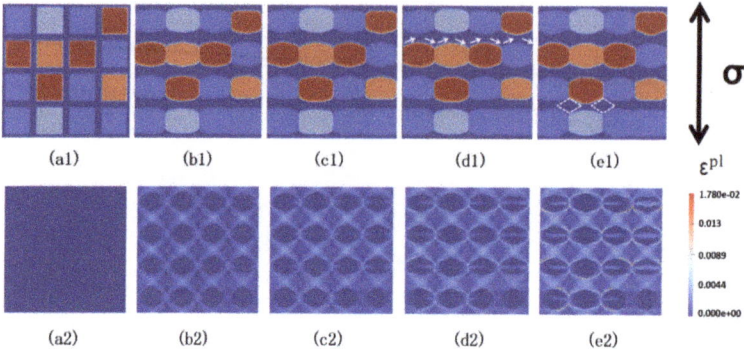

Figure 10. Microstructure evolution (top row) and the corresponding plastic strain field (bottom row) at different instants at 1253 K under 330 MPa from the phase-field simulation: (**a1,a2**) t = 0, (**b1,b2**) t = 5 h, (**c1,c2**) t = 16.83 h, (**d1,d2**) t = 32 h, and (**e1,e2**) t = 51.39 h. Dark blue in the top row denotes γ matrix, and other four colors represent four different γ' variants (Reproduced with permission from [8] © 2020 Elsevier).

Zhou et al. [89] pointed out that the kinetics for rafting are not solely determined by the difference of mechanical properties between γ and γ' phases, ΔG. They revealed that N-type raft structures can form under tensile loading even for homogeneous γ/γ' microstructures ($\Delta G = 1$) if there are dislocations present in horizontal γ channels (normal to the loading direction) at initial state. The time needed to form stable P-type raft structures is less than that to form N-type raft structures, which can be attributed to the difference in the diffusion paths of solute atoms.

According to Equations (34) and (36), the atomic migration in γ channels is controlled by the difference of chemical potentials, where forming elements of the γ' phase (mainly Al in Ni-Al binary system) diffuse from the γ channels with large chemical potential to the matrix with small chemical potential. Under compression in <001> direction, the forming atoms of the γ' phase migrate from two types of vertical γ channels (channel 1 and channel 2, see Table 1) to the horizontal γ channel (channel 3) leading to the formation of P-type raft structure. Under tension in <001> direction, atoms migrate from channel 3 to channels 1 and 2, leading to the formation N-type raft structure.

The phase-field models for Ni-Al binary system have been focused on the diffusion of Al, which is likely not enough for multielement superalloys. Multiple-component phase-field models, which contain the main chemical components of nickel-based superalloys, have been developed in order to systematically investigate the effects of alloy elements on the rafting kinetics [42]. As shown in Figure 6, the results reveal that, under tensile loading, the coalescence of γ' precipitates to form raft structures causes the diffusion of Al, Ti, and Ta from the horizontal γ channels to the vertical γ channels and removes atoms of Co, Cr, and Mo from the vertical γ channels. However, some refractory elements, such as Re and W, likely accumulate near the γ/γ' interface.

5.2.2. Complex Types of Rafting

The types of raft structures formed are largely dependent on the loading condition (direction, tension/compression, etc.), such as the 45-degree rafting in <011> direction [37] (Figure 4). Kamaraj [40] reported the formation of raft structures with 45° angle to the shear stress in a double shear creep test. Such a type of raft structure was also observed experimentally during the creep of MC2 nickel-based superalloy at high temperature and attributed to the highly localized creep strain [90]. However, the angle between the direction of the raft structure and the mechanical loading varied from the region near the fracture with highly localized creep strain to the region far away from the fracture.

Gaubert et al. [91] analyzed the microstructure evolution of <011>-oriented Ni-based superalloys using a 3D mean-field visco-plasticity model. Their results demonstrated the directional coarsening of raft structures along <100> direction. A slight deviation between the loading direction and <011> direction drove initial cubic γ' precipitates to coarsen firstly along <100> direction and extended then along <001> or <010> direction. Yang et al. [92] performed a similar study on the microstructure evolution of nickel-based superalloys with the loading direction deviated from <001> orientation. They referred to this loading mode as monoclinic loading, which can be equivalent to a shear-loading mode and a tension-loading mode. They analyzed the microstructure evolution under monoclinic, shear, and tension loadings, respectively, and found nearly no rafting behavior under the shear loading and synchronous N-type rafting for the monoclinic and tension loading under small stress. Increasing applied stress led to the formation of 45-degree rafting in the overall region under the shear loading and in the partial region under the monoclinic loading. A N-type rafting was still found under the tensile loading, and its rafting rate was smaller than that under the monoclinic loading. Increasing applied stress increased the role of the shear stress component in the rafting process under the monoclinic loading.

Ali et al. [93] combined the phase-field method with physics-based crystal plasticity model to explain the formation of a 45-degree rafting in the region with local creep strain larger than 10% under a tensile loading in <001> direction. Their results showed that creep strain in some regions was significantly higher than average creep strain. A large amount of geometrically necessary dislocations was found in these regions, which caused the change of the original direction of the raft structures and led to the formation of the 45-degree rafting.

5.2.3. Collapse and Topological Inversion

It is known that rafting usually occurs under mechanical loading at high temperature. Continuous creep deformation can cause the change of initial γ' precipitates of cubic shape to stable raft structures. However, further creep deformation may distort the stable raft structure and lead to topological inversion of γ/γ' phases, i.e., the γ' strengthening phase gradually surrounds the γ phase [94]. Incorporating damage parameters [8,19,80] and the change of misfit strain [84,95] in the phase-field models allows for the determination of the major factors contributing to the collapse and topological inversion of the γ/γ' microstructures.

The damage parameters can represent the change of the elastic constants of the γ' phase and/or the increase of the plastic shear rate during creep deformation. Some slip systems can cut into the γ' phase and distort stable raft structures [8,80] when plastic strain accumulated near the γ/γ' interface reaches a critical value. The results from the phase-field models with the variation of misfit strain reveal the decrease of misfit strain during the deformation, which makes the γ/γ' interface become semi-incoherent or incoherent. This trend destroys the stabilization mechanism in the γ channels and reduces the disjoining potential between adjacent γ' precipitates [95]. This is an essential step to achieve topological inversion, as observed experimentally.

Currently, the phase-field models with damage parameters are mainly under the framework of elasto-plastic deformation. It is a challenge to fully describe the three creep stages for the tensile creep at elevated temperature in addition to the analysis of the collapse and topological inversion of the γ/γ' microstructures and the microstructure evolution. The incorporation of the change of misfit strain in the phase-field models also enables the observation of the topological inversion of the γ/γ' microstructures, and the simulation results are in good agreement with those observed in long-term aging experiments [95].

6. Summary

Phase-field method is an excellent technique to study the rafting process in nickel-based superalloys. Incorporating the theory of crystal plasticity in the phase-field method makes it possible to analyze the rafting kinetics with internal physical characteristics (the γ/γ' lattice misfit, elastic constants, and dislocation densities) as well as the mechanical behavior (heterogeneous elasticity and plastic deformation) of superalloys. Raft structures with N-type/P-type or other complex types can be formed during creep deformation under different loading modes. The creep deformation may lead to the collapse of raft structures and the topological inversion of the γ/γ' microstructures.

Introducing damage parameters or variation of lattice misfit in the phase-field models under elasto-plastic framework makes it possible to illustrate the individual processes for the microstructure evolution. The development of multiphase-field models provides the foundation to study the contribution of solute atoms to the rafting behavior and to better design and optimize superalloys for industrial applications. Future efforts are expected to focus on complex situations for industrial applications of single crystal superalloys, which likely include orientation-dependent, temperature-dependent, and composition-dependent mechanical properties. The evolution of the mechanical properties needs to be correlated to the microstructure evolution, such as the rafting. Additionally, the interactions between raft structures and defects, such as dislocations, micro-voids, micro-cracks, etc., need to be incorporated in the models. Further study is needed to expand the capability of the phase-field simulation in the analysis of the effects of voids and cracks of millimeter size.

Author Contributions: Conceptualization, Z.Y. (Zhiyuan Yu), X.W. and F.Y.; methodology, Z.Y. (Zhiyuan Yu) and F.Y.; software, Z.Y. (Zhiyuan Yu) and F.Y.; validation, X.W., Z.Y. (Zhufeng Yue) and J.C.M.L.; formal analysis, X.W. and F.Y.; investigation, Z.Y. (Zhiyuan Yu) and F.Y.; resources, F.Y., Z.Y. (Zhufeng Yue) and J.C.M.L.; data curation, X.W. and F.Y.; writing—original draft preparation, Z.Y. (Zhiyuan Yu) and F.Y.; writing—review and editing, Z.Y. (Zhiyuan Yu), X.W., F.Y. and J.C.M.L.; visualization, Z.Y. (Zhiyuan Yu) and F.Y.; supervision, Z.Y. (Zhufeng Yue) and J.C.M.L.; project administration, X.W. and F.Y.; funding acquisition, X.W. and Z.Y. (Zhufeng Yue). All authors have read and agreed to the published version of the manuscript.

Funding: This research was funded by National Natural Science Foundation of China (Grant Nos. 51775438 and 51875461).

Conflicts of Interest: The authors declare no conflict of interest.

References

1. Murakumo, T.; Kobayashi, T.; Koizumi, Y.; Harada, H. Creep behaviour of Ni-base single-crystal superalloys with various γ' volume fraction. *Acta Mater.* **2004**, *52*, 3737–3744. [CrossRef]
2. Yue, Q.; Liu, L.; Yang, W.; He, C.; Sun, D.; Huang, T.; Zhang, J.; Fu, H. Stress dependence of the creep behaviors and mechanisms of a third-generation Ni-based single crystal superalloy. *J. Mater. Sci. Technol.* **2019**, *35*, 752–763. [CrossRef]
3. Xia, W.; Zhao, X.; Yue, L.; Zhang, Z. A review of composition evolution in Ni-based single crystal superalloys. *J. Mater. Sci. Technol.* **2020**, *44*, 76–95. [CrossRef]
4. Pearson, D.; Lemkey, F.; Kear, B. Stress Coarsening of γ' and its Influence on Creep Properties of a Single Crystal Superalloy. In Proceedings of the 4th International Symposium on Superalloys, Champion, PA, USA, 21–25 September 1980; pp. 513–520.
5. Henderson, P.J.; Mclean, M. Microstructural contributions to friction stress and recovery kinetics during creep of the nickel-base superalloy IN738LC. *Acta Metall.* **1983**, *31*, 1203–1219. [CrossRef]
6. Pollock, T.M.; Argon, A.S. Directional coarsening in nickel-base single crystals with high volume fractions of coherent precipitates. *Acta Metall. Mater.* **1994**, *42*, 1859–1874. [CrossRef]

7. Zhang, C.; Hu, W.; Wen, Z.; Tong, W.; Zhang, Y.; Yue, Z.; He, P. Creep residual life prediction of a nickel-based single crystal superalloy based on microstructure evolution. *Mater. Sci. Eng. A* **2019**, *756*, 108–118. [CrossRef]
8. Yu, Z.; Wang, X.; Yue, Z.; Sun, S. Visco-plasticity phase-field simulation of the mechanical property and rafting behavior in nickel-based superalloys. *Intermetallics* **2020**, *125*, 106884. [CrossRef]
9. Epishin, A.; Link, T.; Portella, P.D.; Bruckner, U. Evolution of the γ/γ′ microstructure during high-temperature creep of a nickel-base superalloy. *Acta Mater.* **2000**, *48*, 4169–4177. [CrossRef]
10. Wu, W.P.; Li, S.Y.; Li, Y.L. An anisotropic elastic–plastic model for predicting the rafting behavior in Ni-based single crystal superalloys. *Mech. Mater.* **2019**, *132*, 9–17. [CrossRef]
11. Fratzl, P.; Paris, O. Strain-induced morphologies during homogeneous phase separation in alloys. *Phase Transit.* **1999**, *67*, 707–724. [CrossRef]
12. Gaubert, A.; le Bouar, Y.; Finel, A. Coupling phase field and viscoplasticity to study rafting in Ni-based superalloys. *Philos. Mag.* **2010**, *90*, 375–404. [CrossRef]
13. Boussinot, G.; le Bouar, Y.; Finel, A. Phase-field simulations with inhomogeneous elasticity: Comparison with an atomic-scale method and application to superalloys. *Acta Mater.* **2010**, *58*, 4170–4181. [CrossRef]
14. Kundin, J.; Mushongera, L.; Goehler, T.; Emmerich, H. Phase-field modeling of the γ′-coarsening behavior in Ni-based superalloys. *Acta Mater.* **2012**, *60*, 3758–3772. [CrossRef]
15. Cottura, M.; le Bouar, Y.; Appolaire, B.; Finel, A. Rôle of elastic inhomogeneity in the development of cuboidal microstructures in Ni-based superalloys. *Acta Mater.* **2015**, *94*, 15–25. [CrossRef]
16. Zhou, N.; Shen, C.; Mills, M.; Wang, Y.Z. Large-scale three-dimensional phase field simulation of γ′-rafting and creep deformation. *Philos. Mag.* **2010**, *90*, 405–436. [CrossRef]
17. Huang, M.; Zhao, L.; Tong, J. Discrete dislocation dynamics modelling of mechanical deformation of nickel-based single crystal superalloys. *Int. J. Plast.* **2012**, *28*, 141–158. [CrossRef]
18. Wu, R.; Zaiser, M.; Sandfeld, S. A continuum approach to combined γ/γ′ evolution and dislocation plasticity in Nickel-based superalloys. *Int. J. Plast.* **2017**, *95*, 142–162. [CrossRef]
19. Yang, M.; Zhang, J.; Wei, H.; Gui, W.M.; Su, H.J.; Jin, T.; Liu, L. A phase-field model for creep behavior in nickel-base single-crystal superalloy: Coupled with creep damage. *Scr. Mater.* **2018**, *147*, 16–20. [CrossRef]
20. Tsukada, Y.; Koyama, T.; Kubota, F.; Murata, Y.; Kondo, Y. Phase-field simulation of rafting kinetics in a nickel-based single crystal superalloy. *Intermetallics* **2017**, *85*, 187–196. [CrossRef]
21. Jokisaari, A.M.; Naghavi, S.S.; Wolverton, C.; Voorhees, P.W.; Heinonen, O.G. Predicting the morphologies of γ′ precipitates in cobalt-based superalloys. *Acta Mater.* **2017**, *141*, 273–284. [CrossRef]
22. Wang, C.; Ali, M.A.; Gao, S.; Goerler, J.V.; Steinbach, I. Combined phase-field crystal plasticity simulation of P- and N-type rafting in Co-based superalloys. *Acta Mater.* **2019**, *175*, 21–34. [CrossRef]
23. Tsao, T.K.; Yeh, A.C.; Kuo, C.M.; Kakehi, K.; Murakami, H.; Yeh, J.W.; Jian, S.R. The High Temperature Tensile and Creep Behaviors of High Entropy Superalloy. *Sci. Rep.* **2017**, *7*, 12658. [CrossRef] [PubMed]
24. Sauza, D.J.; Dunand, D.C.; Seidman, D.N. Microstructural evolution and high-temperature strength of a γ(fcc)/γ′(L12) Co–Al–W–Ti–B superalloy. *Acta Mater.* **2019**, *174*, 427–438. [CrossRef]
25. Schmidt, I.; Gross, D. Directional coarsening in Ni–base superalloys: Analytical results for an elasticity–based model. *Proc. R. Soc. Lond. Ser. A Math. Phys. Eng. Sci.* **1999**, *455*, 3085–3106. [CrossRef]
26. Gururajan, M.P.; Abinandanan, T.A. Phase field study of precipitate rafting under a uniaxial stress. *Acta Mater.* **2007**, *55*, 5015–5026. [CrossRef]
27. Ott, M.; Mughrabi, H. Dependence of the high-temperature low-cycle fatigue behaviour of the monocrystalline nickel-base superalloys CMSX-4 and CMSX-6 on the γ/γ′-morphology. *Mater. Sci. Eng. A* **1999**, *272*, 24–30. [CrossRef]
28. Mughrabi, H.; Tetzlaff, U. Microstructure and High-Temperature Strength of Monocrystalline Nickel-Base Superalloys. *Adv. Eng. Mater.* **2000**, *2*, 319–326. [CrossRef]
29. Shui, L.; Jin, T.; Tian, S.; Hu, Z. Influence of precipitate morphology on tensile creep of a single crystal nickel-base superalloy. *Mater. Sci. Eng. A* **2007**, *454*, 461–466. [CrossRef]
30. Tian, S.G.; Wang, M.G.; Yu, H.C.; Yu, X.F.; Li, T.; Qian, B.J. Influence of element Re on lattice misfits and stress rupture properties of single crystal nickel-based superalloys. *Mater. Sci. Eng. A* **2010**, *527*, 4458–4465.
31. Mughrabi, H. The importance of sign and magnitude of γ/γ′ lattice misfit in superalloys—With special reference to the new γ′-hardened cobalt-base superalloys. *Acta Mater.* **2014**, *81*, 21–29. [CrossRef]
32. Yu, J.; Sun, Y.; Sun, X.; Jin, T.; Hu, Z. Anisotropy of high cycle fatigue behavior of a Ni-base single crystal superalloy. *Mater. Sci. Eng. A* **2013**, *566*, 90–95. [CrossRef]

33. Wen, Z.; Zhang, D.; Li, S.; Yue, Z.; Gao, J. Anisotropic creep damage and fracture mechanism of nickel-base single crystal superalloy under multiaxial stress. *J. Alloys Compd.* **2017**, *692*, 301–312. [CrossRef]
34. Fan, Y.N.; Shi, H.J.; Qiu, W.H. Constitutive modeling of creep behavior in single crystal superalloys: Effects of rafting at high temperatures. *Mater. Sci. Eng. A* **2015**, *644*, 225–233. [CrossRef]
35. Han, G.M.; Yu, J.J.; Hu, Z.Q.; Sun, X.F. Creep property and microstructure evolution of a nickel-base single crystal superalloy in orientation. *Mater. Charact.* **2013**, *86*, 177–184. [CrossRef]
36. Yu, H.C.; Su, Y.; Tian, N.; Tian, S.G.; Li, Y.; Yu, X.F.; Yu, L.L. Microstructure evolution and creep behavior of a oriented single crystal nickel-based superalloy during tensile creep. *Mater. Sci. Eng. A* **2013**, *565*, 292–300. [CrossRef]
37. Chatterjee, D.; Hazari, N.; Das, N.; Mitra, R. Microstructure and creep behavior of DMS4-type nickel based superalloy single crystals with orientations near <001> and <011>. *Mater. Sci. Eng. A* **2010**, *528*, 604–613. [CrossRef]
38. Tian, S.; Su, Y.; Yu, L.; Yu, H.; Zhang, S.; Qian, B. Microstructure evolution of a orientation single crystal nickel-base superalloy during tensile creep. *Appl. Phys. A* **2011**, *104*, 643–647. [CrossRef]
39. Li, Y.; Wang, L.; Zhang, G.; Zheng, W.; Qi, D.; Du, K.; Zhang, J.; Lou, L. Creep deformation related to γ′ phase cutting at high temperature of a oriented nickel-base single crystal superalloy. *Mater. Sci. Eng. A* **2019**, *763*, 138162. [CrossRef]
40. Kamaraj, M. Rafting in single crystal nickel-base superalloys—An overview. *Sadhana* **2003**, *28*, 115–128. [CrossRef]
41. Zhang, Y.Q.; Yang, C.; Xu, Q.Y. Numerical simulation of microstructure evolution in Ni-based superalloys during P-type rafting using multiphase-field model and crystal plasticity. *Comput. Mater. Sci.* **2020**, *172*, 109331. [CrossRef]
42. Mushongera, L.T.; Fleck, M.; Kundin, J.; Wang, Y.; Emmerich, H. Effect of Re on directional γ′-coarsening in commercial single crystal Ni-base superalloys: A phase field study. *Acta Mater.* **2015**, *93*, 60–72. [CrossRef]
43. Tian, S.G.; Zhang, S.; Liang, F.S.; Li, A.A.; Li, J.J. Microstructure evolution and analysis of a single crystal nickel-based superalloy during compressive creep. *Mater. Sci. Eng. A* **2011**, *528*, 4988–4993. [CrossRef]
44. Fedelich, B.; Künecke, G.; Epishin, A.; Link, T.; Portella, P. Constitutive modelling of creep degradation due to rafting in single-crystalline Ni-base superalloys. *Mater. Sci. Eng. A* **2009**, *510*, 273–277. [CrossRef]
45. Tinga, T.; Brekelmans, W.; Geers, M. Directional coarsening in nickel-base superalloys and its effect on the mechanical properties. *Comput. Mater. Sci.* **2009**, *47*, 471–481. [CrossRef]
46. Caron, P.; Ramusat, C.; Diologent, F. Influence of the γ′ fraction on the γ/γ′ topological inversion during high temperature creep of single crystal superalloys. In Proceedings of the 11th International Symposium on Superalloys: SUPERALLOYS 2008, Seven Springs, PA, USA, 14–18 September 2008; pp. 159–167.
47. Bergonnier, S.; Hild, F.; Roux, S. Local anisotropy analysis for non-smooth images. *Pattern Recognit.* **2007**, *40*, 544–556. [CrossRef]
48. Caccuri, V.; Desmorat, R.; Cormier, J. Tensorial nature of γ′-rafting evolution in nickel-based single crystal superalloys. *Acta Mater.* **2018**, *158*, 138–154. [CrossRef]
49. Nguyen, L.; Shi, R.P.; Wang, Y.Z.; de Graef, M. Quantification of rafting of γ′ precipitates in Ni-based superalloys. *Acta Mater.* **2016**, *103*, 322–333. [CrossRef]
50. Desmorat, R.; Mattiello, A.; Cormier, J. A tensorial thermodynamic framework to account for the γ′ rafting in nickel-based single crystal superalloys. *Int. J. Plast.* **2017**, *95*, 43–81. [CrossRef]
51. Tinga, T.; Brekelmans, W.A.M.; Geers, M.G.D. Incorporating strain gradient effects in a multiscale constitutive framework for nickel-base superalloys. *Philos. Mag.* **2008**, *88*, 3793–3825. [CrossRef]
52. Henderson, P.; Berglin, L.; Jansson, C. On rafting in a single crystal nickel-base superalloy after high and low temperature creep. *Scr. Mater.* **1998**, *40*, 229–234. [CrossRef]
53. Matan, N.; Cox, D.; Rae, C.; Reed, R. On the kinetics of rafting in CMSX-4 superalloy single crystals. *Acta Mater.* **1999**, *47*, 2031–2045. [CrossRef]
54. Taylor, C.F.E.G.I. The Distortion of an Aluminum Crystal during a Tensile Test. *Proc. R. Soc.* **1923**, *102*, 634–667.
55. Rice, J.R. Inelastic constitutive relations for solids: An internal-variable theory and its application to metal plasticity. *J. Mech. Phys. Solids* **1971**, *19*, 433–455. [CrossRef]
56. Hill, R.; Rice, J. Constitutive analysis of elastic-plastic crystals at arbitrary strain. *J. Mech. Phys. Solids* **1972**, *20*, 401–413. [CrossRef]
57. Chowdhury, H.; Naumenko, K.; Altenbach, H. Aspects of power law flow rules in crystal plasticity with glide-climb driven hardening and recovery. *Int. J. Mech. Sci.* **2018**, *146*, 486–496. [CrossRef]

58. Frost, H.J.; Ashby, M.F. *Deformation Mechanism Maps: The Plasticity and Creep of Metals and Ceramics*; Pergamon Press: Oxford, UK, 1982.
59. Kamaraj, M.; Radhakrishnan, V. First Report on the Deformation Mechanism Mapping of First and Second Generation Ni-Based Single Crystal Super Alloys. *Trans. Indian Inst. Metals* **2017**, *70*, 2485–2496. [CrossRef]
60. Roters, F.; Eisenlohr, P.; Hantcherli, L.; Tjahjanto, D.D.; Bieler, T.R.; Raabe, D. Overview of constitutive laws, kinematics, homogenization and multiscale methods in crystal plasticity finite-element modeling: Theory, experiments, applications. *Acta Mater.* **2010**, *58*, 1152–1211. [CrossRef]
61. Bassani, J.L.; Wu, T.Y. Latent hardening in single crystals. II. Analytical characterization and predictions. *Proc. R. Soc. Lond. Ser. A Math. Phys. Sci.* **1991**, *435*, 21–41.
62. Wang, X.M.; Hui, Y.Z.; Hou, Y.Y.; Yu, Z.Y.; Li, L.; Yue, Z.F.; Deng, C.H. Direct investigation on high temperature tensile and creep behavior at different regions of directional solidified cast turbine blades. *Mech. Mater.* **2019**, *136*, 103068. [CrossRef]
63. Me'ric, L.; Poubanne, P.; Cailletaud, G. Single crystal modeling for structural calculations: Part 1—Model presentation. *J. Eng. Mater. Technol.* **1991**, *113*, 162–170. [CrossRef]
64. Chen, B.; Hu, J.; Flewitt, P.; Smith, D.; Cocks, A.; Zhang, S. Quantifying internal stress and internal resistance associated with thermal ageing and creep in a polycrystalline material. *Acta Mater.* **2014**, *67*, 207–219. [CrossRef]
65. Hoppe, R.; Appel, F. Origin and magnitude of internal stresses in TiAl alloys. In *Gamma Titanium Aluminide Alloys 2014: A Collection of Research on Innovation and Commercialization of Gamma Alloy Technology*; Wiley: Hoboken, NJ, USA, 2014; pp. 159–168.
66. Kumar, A.V.; Yang, C. Study of work hardening models for single crystals using three dimensional finite element analysis. *Int. J. Plast.* **1999**, *15*, 737–754. [CrossRef]
67. Cottura, M.; Appolaire, B.; Finel, A.; le Bouar, Y. Coupling the Phase Field Method for diffusive transformations with dislocation density-based crystal plasticity: Application to Ni-based superalloys. *J. Mech. Phys. Solids* **2016**, *94*, 473–489. [CrossRef]
68. Arsenlis, A.; Parks, D.M.; Becker, R.; Bulatov, V.V. On the evolution of crystallographic dislocation density in non-homogeneously deforming crystals. *J. Mech. Phys. Solids* **2004**, *52*, 1213–1246. [CrossRef]
69. Arsenlis, A.; Parks, D.M. Modeling the evolution of crystallographic dislocation density in crystal plasticity. *J. Mech. Phys. Solids* **2002**, *50*, 1979–2009. [CrossRef]
70. Ma, A.; Roters, F.; Raabe, D. On the consideration of interactions between dislocations and grain boundaries in crystal plasticity finite element modeling—Theory, experiments, and simulations. *Acta Mater.* **2006**, *54*, 2181–2194. [CrossRef]
71. Ma, A.; Roters, F.; Raabe, D. A dislocation density based constitutive model for crystal plasticity FEM including geometrically necessary dislocations. *Acta Mater.* **2006**, *54*, 2169–2179. [CrossRef]
72. Fleck, N.; Muller, G.; Ashby, M.F.; Hutchinson, J.W. Strain gradient plasticity: Theory and experiment. *Acta Metall. Mater.* **1994**, *42*, 475–487. [CrossRef]
73. Evers, L.; Brekelmans, W.; Geers, M. Non-local crystal plasticity model with intrinsic SSD and GND effects. *J. Mech. Phys. Solids* **2004**, *52*, 2379–2401. [CrossRef]
74. Devincre, B.; Kubin, L.; Hoc, T. Physical analyses of crystal plasticity by DD simulations. *Scr. Mater.* **2006**, *54*, 741–746. [CrossRef]
75. Chaboche, J.L. A review of some plasticity and viscoplasticity constitutive theories. *Int. J. Plast.* **2008**, *24*, 1642–1693. [CrossRef]
76. Van der Waals, J.D. Thermodynamische Theorie der Kapillaritat unter voraussetzung stetiger Dichteanderung. *Z. Phys. Chem.* **1894**, *13*, 657–725.
77. Cahn, J.W.; Hilliard, J.E. Free energy of a nonuniform system. I. Interfacial free energy. *J. Chem. Phys.* **1958**, *28*, 258–267. [CrossRef]
78. Moelans, N.; Blanpain, B.; Wollants, P. An introduction to phase-field modeling of microstructure evolution. *Calphad* **2008**, *32*, 268–294. [CrossRef]
79. Wang, Y.; Banerjee, D.; Su, C.; Khachaturyan, A. Field kinetic model and computer simulation of precipitation of L12 ordered intermetallics from fcc solid solution. *Acta Mater.* **1998**, *46*, 2983–3001. [CrossRef]
80. Harikrishnan, R.; le Graverend, J.B. A creep-damage phase-field model: Predicting topological inversion in Ni-based single crystal superalloys. *Mater. Des.* **2018**, *160*, 405–416. [CrossRef]

81. Amin, W.; Ali, M.A.; Vajragupta, N.; Hartmaier, A. Studying grain boundary strengthening by dislocation-based strain gradient crystal plasticity coupled with a multi-phase-field model. *Materials* **2019**, *12*, 2977. [CrossRef]
82. Wang, J.C.; Osawa, M.; Yokokawa, T.; Harada, H.; Enomoto, M. Modeling the microstructural evolution of Ni-base superalloys by phase field method combined with CALPHAD and CVM. *Comput. Mater. Sci.* **2007**, *39*, 871–879. [CrossRef]
83. Wang, S.L.; Sekerka, R.; Wheeler, A.; Murray, B.; Coriell, S.; Braun, R.; McFadden, G. Thermodynamically-consistent phase-field models for solidification. *Phys. D Nonlinear Phenom.* **1993**, *69*, 189–200. [CrossRef]
84. Tsukada, Y.; Murata, Y.; Koyama, T.; Morinaga, M. Phase-field simulation on the formation and collapse processes of the rafted structure in Ni-based superalloys. *Mater. Trans.* **2008**, *49*, 484–488. [CrossRef]
85. Schleifer, F.; Holzinger, M.; Lin, Y.Y.; Glatzel, U.; Fleck, M. Phase-field modeling of γ/γ'' microstructure formation in Ni-based superalloys with high γ'' volume fraction. *Intermetallics* **2020**, *120*, 106745. [CrossRef]
86. Gurtin, M.E. A gradient theory of single-crystal viscoplasticity that accounts for geometrically necessary dislocations. *J. Mech. Phys. Solids* **2002**, *50*, 5–32. [CrossRef]
87. Steinbach, I.; Apel, M. Multi phase field model for solid state transformation with elastic strain. *Phys. D Nonlinear Phenom.* **2006**, *217*, 153–160. [CrossRef]
88. Cottura, M.; le Bouar, Y.; Finel, A.; Appolaire, B.; Ammar, K.; Forest, S. A phase field model incorporating strain gradient viscoplasticity: Application to rafting in Ni-base superalloys. *J. Mech. Phys. Solids* **2012**, *60*, 1243–1256. [CrossRef]
89. Zhou, N.; Shen, C.; Mills, M.J.; Wang, Y. Contributions from elastic inhomogeneity and from plasticity to γ' rafting in single-crystal Ni–Al. *Acta Mater.* **2008**, *56*, 6156–6173. [CrossRef]
90. Touratier, F.; Andrieu, E.; Poquillon, D.; Viguier, B. Rafting microstructure during creep of the MC2 nickel-based superalloy at very high temperature. *Mater. Sci. Eng. A* **2009**, *510*, 244–249. [CrossRef]
91. Gaubert, A.; Jouiad, M.; Cormier, J.; le Bouar, Y.; Ghighi, J. Three-dimensional imaging and phase-field simulations of the microstructure evolution during creep tests of <011>-oriented Ni-based superalloys. *Acta Mater.* **2015**, *84*, 237–255. [CrossRef]
92. Yang, M.; Zhang, J.; Wei, H.; Zhao, Y.; Gui, W.; Su, H.; Jin, T.; Liu, L. Study of γ' rafting under different stress states—A phase-field simulation considering viscoplasticity. *J. Alloy. Compd.* **2018**, *769*, 453–462. [CrossRef]
93. Ali, M.A.; Amin, W.; Shchyglo, O.; Steinbach, I. 45-degree rafting in Ni-based superalloys: A combined phase-field and strain gradient crystal plasticity study. *Int. J. Plast.* **2020**, *128*, 102659. [CrossRef]
94. Yu, Z.; Wang, X.; Yue, Z. The effect of stress state on rafting mechanism and cyclic creep behavior of Ni-base superalloy. *Mech. Mater.* **2020**, *149*, 103563. [CrossRef]
95. Goerler, J.V.; Lopez-Galilea, I.; Roncery, L.M.; Shchyglo, O.; Theisen, W.; Steinbach, I. Topological phase inversion after long-term thermal exposure of nickel-base superalloys: Experiment and phase-field simulation. *Acta Mater.* **2017**, *124*, 151–158. [CrossRef]

Publisher's Note: MDPI stays neutral with regard to jurisdictional claims in published maps and institutional affiliations.

© 2020 by the authors. Licensee MDPI, Basel, Switzerland. This article is an open access article distributed under the terms and conditions of the Creative Commons Attribution (CC BY) license (http://creativecommons.org/licenses/by/4.0/).

Article

Thermal Stability of Nanocrystalline Gradient Inconel 718 Alloy

Jie Ding [1,*], Yifan Zhang [1], Tongjun Niu [1], Zhongxia Shang [1], Sichuang Xue [1], Bo Yang [1], Jin Li [1], Haiyan Wang [1,2] and Xinghang Zhang [1,*]

[1] School of Materials Engineering, Purdue University, West Lafayette, IN 47907, USA; zhan2592@purdue.edu (Y.Z.); niu35@purdue.edu (T.N.); shang19@purdue.edu (Z.S.); xue97@purdue.edu (S.X.); yang837@purdue.edu (B.Y.); vincentlijin@gmail.com (J.L.); hwang00@purdue.edu (H.W.)
[2] School of Electrical and Computer Engineering, Purdue University, West Lafayette, IN 47907, USA
* Correspondence: ding173@purdue.edu (J.D.); xzhang98@purdue.edu (X.Z.)

Abstract: Gradient structures containing nanograins in the surface layer have been introduced into Inconel 718 (IN718) nickel-based alloy using the surface mechanical grinding treatment technique. The thermal stability of the gradient IN718 alloy was investigated. Annealing studies reveal that nanograins with a grain size smaller than 40 nm exhibited significantly better thermal stability than those with larger grain size. Transmission electron microscopy analyses reveal that the enhanced thermal stability was attributed to the formation of grain boundaries with low energy configurations. This study provides new insight on strategies to improve the thermal stability of nanocrystalline metals.

Keywords: nanocrystalline; thermal stability; IN718 alloy

Citation: Ding, J.; Zhang, Y.; Niu, T.; Shang, Z.; Xue, S.; Yang, B.; Li, J.; Wang, H.; Zhang, X. Thermal Stability of Nanocrystalline Gradient Inconel 718 Alloy. *Crystals* **2021**, *11*, 53. https://doi.org/10.3390/cryst11010053

Received: 12 December 2020
Accepted: 2 January 2021
Published: 11 January 2021

Publisher's Note: MDPI stays neutral with regard to jurisdictional claims in published maps and institutional affiliations.

Copyright: © 2021 by the authors. Licensee MDPI, Basel, Switzerland. This article is an open access article distributed under the terms and conditions of the Creative Commons Attribution (CC BY) license (https://creativecommons.org/licenses/by/4.0/).

1. Introduction

Nanograined (NG) metals containing a high volume fraction of grain boundaries have demonstrated much higher mechanical strength than their coarse-grained counterparts [1–3]. Severe plastic deformation (SPD) techniques such as equal channel angular pressing [4,5] and high-pressure torsion [6–8], etc., have been proven effective in grain refinement of metallic materials. However, the same grain boundaries that contribute to the high strength also lead to deterioration of the thermal stability of NG metals [9–12]. The grain boundary energy of nanograins provides a large driving force for grain coarsening. For instance, grain boundary migration takes place at 300 °C for nanocrystalline Nb (obtained by high pressure torsion) with an average grain size of 75 nm [13]. Grain growth occurs at temperatures as low as 200 °C for nanocrystalline Ni, accompanied by a substantial hardness drop [14]. In electrodeposited nanocrystalline Ni with an average grain size of 10–20 nm, grain coarsening occurs at 80 °C [15]. Similarly, grain growth takes place even at ambient temperature in nanocrystalline Cu. The poor thermal stability hinders the application of NG metallic materials at elevated temperatures [16].

Recently, surface mechanical grinding treatment (SMGT) [17], surface mechanical attrition treatment (SMAT) [18–20] and surface mechanical rolling treatment (SMRT) [18] have been applied to introduce gradient microstructures into the surface of metallic materials to improve both strength and ductility. Gradient structures containing an NG top surface layer have been introduced into several types of metals [18,21–24]. It has been reported that these surface modification techniques are more effective than conventional SPD approaches in grain refinement [22]. In contrast to the poor thermal stability of nanograins in most prior studies, it was reported that nanograins smaller than the critical values (70 nm for Cu and 43 nm for Ni) were more stable than larger grains in gradient structured pure Cu and Ni fabricated using SMGT in liquid nitrogen [25]. The surprising observation of enhanced thermal stability of nanograins was attributed to the unique grain boundaries in low energy configurations generated during low-temperature SMGT [25,26]. This evidence implies that the thermal stability of NG alloys may not necessarily be deteriorated after grain refinement.

Inconel 718 (IN718) is a common precipitation-strengthened Ni-based superalloy used for application in high-pressure turbine discs in jet engines [27–35]. However, a majority of the published works focused on wrought IN718 alloys with coarse grains [36,37]. Studies on NG IN718 are limited. Besides, the high-temperature performance of IN718 is primarily determined by the high-density γ'' phases formed after annealing [36,38–45]. However, at temperatures above 650 °C, the metastable γ'' phase transforms to stable δ phase over long-term exposure [36,37,44,46]. The application of IN718 alloy is therefore limited to temperatures below 650 °C. In this study, gradient structures containing a severely deformed NG surface layer were introduced into IN718 alloy via the SMGT technique at liquid nitrogen temperature. Studies on the NG IN718 alloy at 700 °C for up to 100 h reveal that nanograins have outstanding thermal stability. The underlying nanograin stabilization mechanisms are discussed.

2. Experimental

2.1. Materials and Processing

The IN718 Ni-based alloy with a chemical composition as listed in Table 1 was subjected to SMGT. Prior to processing, a cylindric bar was solution-treated at 1100 °C for 1 h followed by water quenching (denoted as as-processed hereafter). During processing, the bar was rotated at a velocity of 400 rpm while a WC/Co tool tip penetrated into the surface by 30 μm and slid along the axial direction at a speed of 10 mm/min. The process was repeated 8 times to generate a substantial deformation zone. Liquid nitrogen was used as the coolant during processing. Subsequently, the processed samples were annealed in a vacuum furnace at 700 °C for 5, 24 and 100 h, followed by furnace cooling. Annealing was conducted when the vacuum reached 2×10^{-6} torr.

Table 1. The chemical composition of IN718 alloy (in wt.%).

Cr	Fe	Co	Nb	Mo	Al	Ti	Ta	Ni
18.57	18.00	0.11	5.02	2.86	0.58	0.97	<0.01	Bal.

2.2. Microstructure Characterizations

The samples used for metallographic observations were ground and polished using the conventional metallographic preparation technique, and the observation was then carried out using an optical microscope. TEM samples were prepared using the focused ion beam (FIB) technique with an FEI Quanta 3D FEG Dual Beam FIB scanning electron microscope following typical protocols. The microstructure and chemical composition analyses of both the as-processed and annealed samples were performed on an FEI Talos 200X analytical transmission electron microscope operated at 200 kV, equipped with a Super-X energy-dispersive X-ray spectroscopy (EDS) detector. The grain orientation analyses were performed using the NanoMegas ASTAR (to generate an electron backscattering diffraction (EBSD)-like automated crystal orientation map with 4-nm spatial resolution) setup installed in the Talos 200X TEM microscope, and data analyses were conducted using OIM Analysis software.

3. Results

After SMGT, a severely deformed gradient structure formed on the surface, as shown in the optical microscopy image in Figure 1a. NG structures (as labeled by a dotted line) were observed on the topmost region of the gradient layer (based on TEM studies shown later). The Vickers hardness indents and the corresponding hardness along the depth direction were labeled, indicating the formation of gradient microstructures after SMGT. After annealing at 700 °C, a sharp interface formed between the topmost NG region and the deeper region of the sample, as revealed by the optical microscopy images in Figure 1b,c. It is worth mentioning that the thickness of the NG region varies with positions due to the inhomogeneous penetration depth of the gradient structure after SMGT, with a maximum depth of 30 μm.

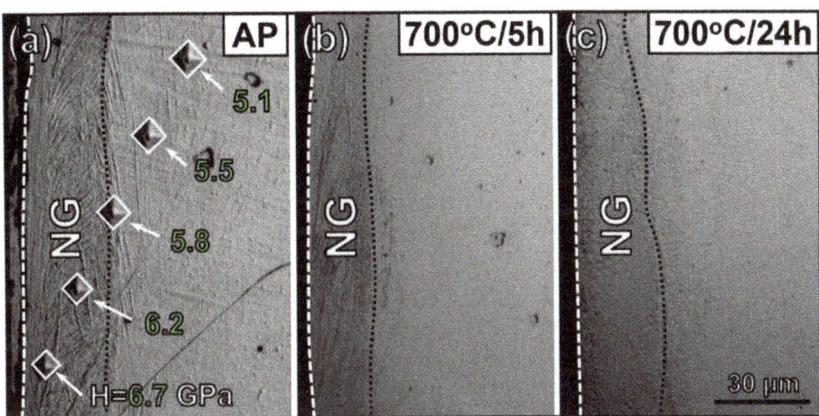

Figure 1. Optical microscopy images of (**a**) as-processed (AP) and annealed gradient IN718 alloy at 700 °C for (**b**) 5 h and (**c**) 24 h. The topmost nanograined (NG) regions are labeled by dotted lines. Vickers indentation and corresponding hardness values (in GPa) are labeled.

The TEM image of the as-processed sample in Figure 2a reveals that NG structures formed near the surface of IN718 alloy after SMGT, as confirmed by the inserted selected area diffraction (SAD) pattern. It is worth mentioning that fine and coarse nanograin layers were observed in the NG region (referred to as FNG and CNG hereafter, respectively), as labeled by the dotted lines in Figure 2a. The TEM image in Figure 2b shows the alternately distributed FNG and CNG layers in the topmost NG region (at a depth range of 2–5 μm from surface). The corresponding grain size distribution profile reveals that the average grain size is 14 and 28 nm in the alternating FNG and CNG layers, respectively. The average grain size of the CNG layers increases gradually from 25 nm at a depth of 100 nm from the surface to over 32 nm at the depth of 2200 nm from the surface (as shown in Figure 2b), further confirming the formation of complex gradient microstructures. The corresponding scanning transmission electron microscopy (STEM) image and EDS maps of the NG region in Supplementary Figure S1 (see supplementary materials) reveal that the chemical composition remains uniform. The ASTAR inverse pole figure map in Figure 2c shows the nanograins with various orientations in both FNG and CNG layers. The grain boundary map in Figure 2d shows that a majority of the grain boundaries in both the FNG and CNG layers are high-angle grain boundaries (HAGBs, indicated by blue lines). The fractions of low-angle grain boundaries (LAGBs) and coincidence site lattice (CSL) boundaries (indicated by red and yellow lines, respectively) are low.

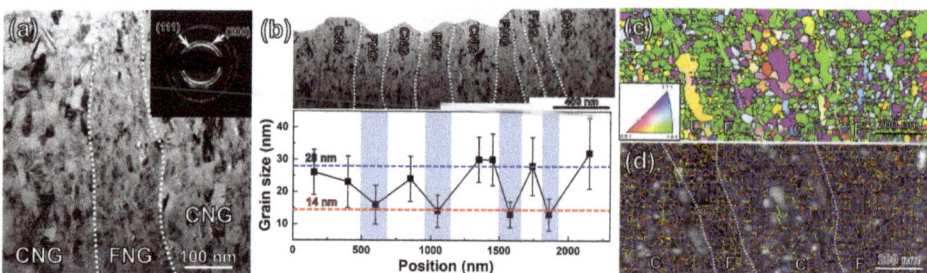

Figure 2. (**a**) TEM image showing the formation of an NG structure in the topmost region of the IN718 specimen after surface mechanical grinding treatment (SMGT). Fine nanograin (FNG) layers were sandwiched by coarse nanograin (CNG) layers. (**b**) TEM image of NG region and the corresponding grain size vs. position profile showing the grain size evolution of both FNG and CNG layers at various depth. The corresponding (**c**) ASTAR crystal orientation analyses and (**d**) grain boundary map showing the high-angle grain boundaries (blue lines), low-angle grain boundaries (red lines) and twin boundaries (yellow lines).

Upon annealing (700 °C/24 h), the nanograins of the topmost NG region retained, whereas recrystallization and grain coarsening occurred in the rest of the gradient layers. Figure 3a shows the microstructure of the distinct interface (as denoted by a dashed line) formed between the thermally stable topmost NG area and the grain coarsened areas. Furthermore, the alternatively distributed FNG/CNG structures were sustained after annealing (as labeled by dotted lines in Figure 3a). The statistic distributions of grain size in Figure 3b show that the average grain size of FNG and CNG layers in the thermally stable area is 18 and 37 nm, respectively, whereas the grain size of the adjacent area coarsened to 90 nm. The STEM images and corresponding EDS maps of both thermally stable and grain coarsened area in supplementary Figure S2a,b show that large Ni- and Nb-rich δ phases and Al-, Ni- and Nb-rich η phases formed after annealing. The difference between these two areas is that nanoscale α-Cr phases have higher density and smaller size in the thermally stable area than the grain coarsened area.

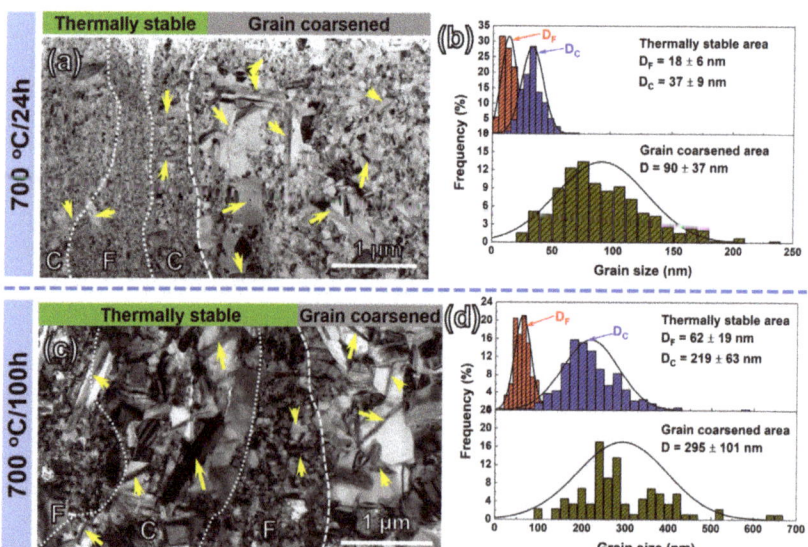

Figure 3. TEM images showing the sharp interface formed between the thermally stable area and the grain coarsened area of NG IN718 specimens after annealing at 700 °C for (**a**) 24 h and (**c**) 100 h. (**b**,**d**) The corresponding statistic distributions revealing the average grain sizes of FNG (D_F) and CNG (D_C) layers in the thermally stable and grain coarsened area.

Increasing the annealing time further to 100 h coarsened the grains in both thermally stable and grain coarsened areas of the NG region. However, the sharp interface between those two areas sustained, as labeled by the dashed line in Figure 3c. The alternatively distributed FNG/CNG structures were also observed, whereas the grain size difference between these two layers is much larger than the specimen annealed for 24 h (in Figure 3a). The statistical analyses reveal that the grain size of the FNG and CNG layers increased to 62 and 219 nm, respectively (Figure 3d). In comparison, in the grain coarsened area, the grains coarsened further to 295 nm.

4. Discussion

While grain boundaries in nanocrystalline metals improve mechanical strength, they provide a strong driving force for grain coarsening [9]. Grain coarsening of NG metals involves GB migration at a certain velocity (v), which can be expressed as [47,48]:

$$v = M_{gb} \cdot \gamma_{gb} \cdot \kappa \qquad (1)$$

where M_{gb} represents the GB mobility, γ_{gb} is the GB energy and κ is the local grain boundary curvature. This equation implies that at least two approaches can be applied to alleviate grain coarsening: the kinetics-driven stabilization approach, in which grain coarsening is suppressed by pinning grain boundaries with second-phase Zener drag or by solute drag, and a thermodynamics approach, where lowering grain boundary energy can effectively reduce the driving force for grain coarsening [49–51]. Specifically, the driving force for grain coarsening of NG metals is the excess energy stored at the grain boundaries. Zhou et al. [25] reported that the grain boundary energy of NG Cu fabricated by SMGT declined from 0.52 to 0.25 J/m² when the average grain size decreased from 125 to 50 nm and showed a thermally stable NG layer when the average grain size is smaller than 70 nm. The grain boundary energy reduction was partially attributed to the formation of low-energy grain boundaries, consisting of nanotwins and stacking faults (SFs), and partially ascribed to the grain boundary relaxation [25].

In this study, a gradient NG surface layer was introduced into IN718 alloy by SMGT. The topmost area of the NG region with smaller grain size exhibited better thermal stability than the deeper area with larger grain size at high temperatures (Figure 3), similar to the reported thermally stable NG Cu [25]. Zhang et al. [52] reported that the grain boundaries of Cu evolved from low-energy twin boundaries (TBs) possess lower grain boundaries energy than conventional HAGBs. In this study, the grain boundary map in Figure 2d reveals that the majority of grain boundaries of nanograins are HAGBs. Our previous work on the microstructure evolution of gradient structured C-22HS Ni-based alloy indicates that these HAGB-dominated NG structures may derive from deformation-induced twin structures [23]. Such transformation led to the formation of grain boundaries in low-energy configuration. Slow grain boundary migration velocity is therefore expected according to Equation (1), indicating the improvement of thermal stability.

The high resolution TEM (HRTEM) image of the nanograins in the topmost NG region in Figure 4a reveals plenty of SFs and nanotwins formed inside the nanograins. The magnified HRTEM image of area b in Figure 4a is shown in Figure 4b. Nanotwins with an average twin thickness of 2 nm were observed. These nanotwins formed inside nanograins are too thin to be detected by ASTAR due to the limited spatial resolution of the technique (4 nm), which indicates that the proportion of CSL boundaries in the NG region is higher than what was revealed in Figure 2d. These CSL boundaries are also beneficial for improving the thermal stability of GBs [52]. The {111} planes forming a twinning relationship and TBs are labeled by solid and dashed lines in Figure 4b, respectively. These observations also imply that partial dislocations rather than full dislocation activities dominated the grain refinement process at the nanoscale. The partial dislocations-dominated grain refinement process led to the formation of a thermally stable topmost NG layer because the generation of SFs or nanotwins from grain boundaries plays a critical role in grain boundary relaxation, which impacts the migration of the boundary [26]. Based on the Orowan relation [25], the resolved shear stress (τ_{RSS}) required for the expansion of a dislocation loop derived from a Frank–Reed source with a diameter of D can be expressed as:

$$\tau_{RSS} = \mu b / D \qquad (2)$$

where μ represents the shear modulus and b is the Burgers vector. Hence, the minimum grain size (D^*) required for the multiplication of full dislocations at a given yield strength (σ_y) can be calculated by:

$$D^* = \frac{\mu b M}{\sigma_y(D^*)} \qquad (3)$$

where M is the Taylor factor (3.0 for polycrystalline metals). The shear modulus and Burgers vector for Ni are 76 GPa and 0.25 nm, respectively. The yield strength $\sigma_y(D^*)$ can be calculated by the classic Hall–Petch equation as [53,54]:

$$\sigma_y(D^*) = \sigma_0 + K(D^*)^{-1/2} \qquad (4)$$

where σ_0 is the friction stress and K represents the Hall–Petch slope. Zhou et al. [25] estimated the D^* for (full dislocation multiplication mechanism to operate) pure Ni by using Equation (2) and obtained D^* = 43 nm, similar to what was calculated by Legros et al. (38 nm) [55]. Considering that the value of K for Ni alloys is higher than that for pure Ni, the calculated D^* for IN718 alloys is expected to be smaller than that of pure Ni. Hence, an average D^* value (40 nm) between those reported by Zhao et al. and Legros et al. is adopted in this story. As the grain size of IN718 is smaller than 40 nm, full dislocation activities are superseded by partial dislocation activities, leading to grain boundary relaxation. The experimental observation is consistent with the calculated results. As shown in Figure 3, the average grain size of the relative coarse-grained structures in the thermally stable area is 37 nm after annealing (700 °C/24 h), whereas prominent grain coarsening occurs in the grain coarsened area with an initial grain size greater than 40 nm.

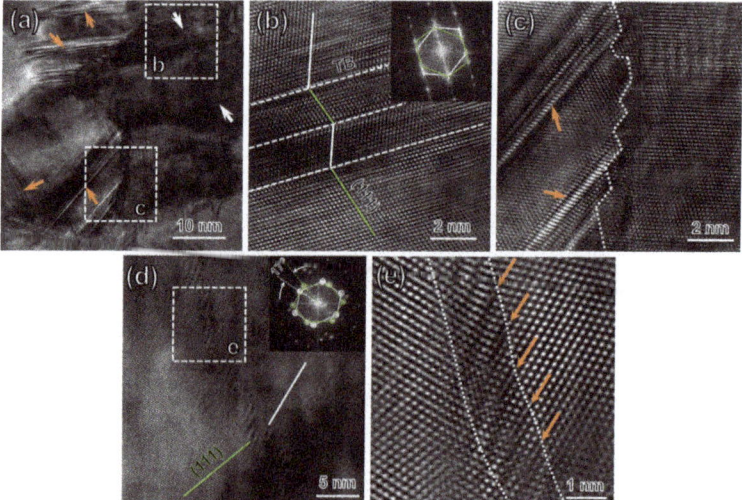

Figure 4. (**a**) HRTEM image of nanograins containing high-density stacking faults (SFs) and nanotwins (as labeled by orange and white arrows, respectively) formed after SMGT. (**b**) An atomic-resolution TEM image of the area b in (**a**) showing the formation of nanotwin structures with an average twin thickness of 2 nm. (**c**) An atomic-resolution TEM image of the area c in (**a**) showing the faceted grain boundary. (**d**) HRTEM image showing the dissociated HAGBs formed between two adjacent nanograins. (**e**) An atomic-resolution TEM image of the area e in (**d**) showing an array of SFs (as labeled by orange arrows) decorated along the dissociated grain boundary.

We also observed that grain boundaries associated with nanotwins or SFs became faceted frequently, as shown in Figure 4c, where the steps of (111) planes along the grain boundaries are labeled. A previous study reported that the emission of nanoscale twins or SFs from grain boundaries reduces the excess energy of grain boundaries [56], leading to reduced atomic diffusion along the grain boundaries. The emission of SFs and nanotwins in this study led to the formation of faceted grain boundaries. Similar grain boundaries with zig-zag configurations have been reported in nanostructured Cu with good thermal stability [57]. Generally, a given grain boundary has five crystallographic degrees of freedom (three for the misorientation of the crystallographic axes of one grain and two for the inclination of the boundary between the two adjacent grains) [58]. Grain boundaries naturally find a low-energy configuration that fixes these degrees of freedom, i.e., minimize the free energy of the system pertaining to atomic coordinates or composition by specifying the five degrees of freedom [58,59]. For a system where its boundary free energy is anisotropic with respect to the inclination of the grain boundary, the corresponding boundary may lower its free energy by forming faceted planes [58]. Faceting is the process of decomposing the grain boundary into sections with low-energy inclination. Generally, the system with low grain boundary free energy possesses geometric characteristics of low reciprocal volume density of coincidence sites, large interplanar spacing and high planar density of coincidence site, etc. [60]. The twin boundaries in Figure 4b and the most densely packed (111) plane of the zig-zagged grain boundaries in Figure 4c are examples of such low-energy grain boundaries. However, the low grain boundary free energy system is not limited to these criteria as there is no direct connection between coincidence and grain boundary energy [60]. The non-CSL 9R structure was found to minimize the grain boundary energy by forming a body-centered-cubic structured grain boundary in the face-centered-cubic matrix [60,61]. Faceting has been observed at both micrometer and nanometer scales in several cases [57,62]. These faceted grain boundaries have low-energy states and, therefore, are more thermally stable than conventional grain boundaries according to Equation (1).

Grain boundary structure may also transform to a lower-energy state (grain boundary relaxation) through the dissociation of grain boundaries during deformation. The TEM image in Figure 4d shows a dissociated HAGB in the NG IN718 specimen after SMGT. As denoted by dotted lines, broad grain boundaries decorated with plenty of SF-like structures formed. The HRTEM image (of the area e in Figure 4d) in Figure 4e reveals an array of SFs (as noted by orange arrows) emitted from the grain boundary, leading to the broad/dissociated grain boundary with 1 nm in width. The formation of these SF-decorated broad grain boundaries indicates that grain boundary dissociation may have taken place in the topmost NG region during SMGT, leading to grain boundary relaxation to low-energy states and grain boundary stabilization. Rittner et al. found that grain boundary dissociation generally occurs via the emission of SFs from one boundary and termination at a second boundary [63]. Zhou et al. [25] reported that the formation of both SFs and nanotwins from grain boundaries involves emission of partial dislocations, leading to grain boundary relaxation and stable structures. The dissociation of grain boundaries usually leads to the formation of a wider grain boundary (or three-dimensional boundaries) of up to 1 nm or more in width, similar to the structures formed in this study.

The grain size evolution of the FNG and CNG structures of the thermally stable area and the grain coarsened area with annealing time is presented in Figure 5a. The calculated minimum grain size (40 nm) is labeled by the horizontal dotted line. It is evident that the coarsening rate of grains smaller than 40 nm is lower than that with grain sizes larger than 40 nm. The grain growth kinetic, correlating the grain size (d) to the annealing time (t), can be expressed as [64]:

$$d^n - d_0^n = kt \qquad (5)$$

where d_0 is the initial grain size, n represents grain growth exponent and k is a rate constant. The evolution of grain size ($\ln(d)$) and annealing time ($\ln(t)$) in different areas of the specimen is shown in Figure 5b. The n value can be determined by the slope of linear fit lines. It reveals that the average n value of the thermally stable area with initial grain size smaller than 40 nm is 0.15, whereas the n value for the grain coarsened area is much greater, 0.59, confirming a much more sluggish grain growth behavior in the thermally stable NG structure when the initial grain size is smaller than the D^*.

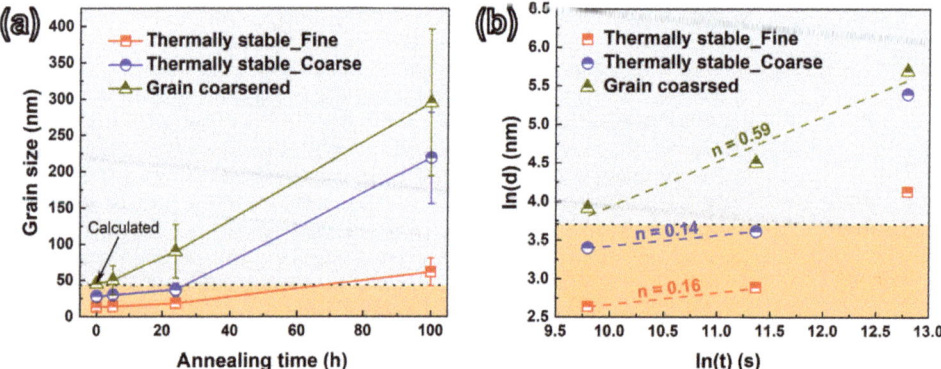

Figure 5. (a) Grain size evolution of FNG and CNG layers in the thermally stable area and the grain coarsened area of IN718 specimens with increasing annealing time. (b) The corresponding plot of ln (d) vs. ln (t) at several regions, where d and t are grain size and annealing time, respectively.

5. Conclusions

Gradient structures containing an NG surface region have been fabricated in IN718 Ni alloy using the SMGT technique. The thermal stability studies at 700 °C up to 100 h resulted in the following observations.

(1) Nanograins with a grain size smaller than 40 nm in the deformed surface exhibited significantly enhanced thermal stability compared to grains with larger grain sizes away from the surface.
(2) The average grain growth exponent of thermally stable NG structures with smaller grain sizes (<40 nm) was 0.15, in contrast to 0.59 for the larger grains.
(3) TEM studies suggest that the enhanced thermal stability of nanograins was attributed to the generation of grain boundaries in low-energy states during SMGT. The emission of SFs or nanotwins from grain boundaries leads to the dissociation of grain boundaries. The relaxation of grain boundaries to low-energy states results in their subsequent thermal stabilization.

Supplementary Materials: The following are available online at https://www.mdpi.com/2073-4352/11/1/53/s1, Figure S1: (a) STEM image and the corresponding (b–f) EDS maps of the NG layer of the as-processed IN718 alloy., Figure S2: STEM image and EDS maps of (a) thermally stable area and (b) grain coarsened areas of NG IN718 specimen after annealing at 700 °C for 24 h.

Author Contributions: J.D. and Z.S. fabricated gradient IN718 alloy. B.Y. and Y.Z. performed heat treatment. J.D. and S.X. prepared TEM samples. J.D., Y.Z., Z.S., T.N., J.L. and H.W. performed TEM imaging and results analyses. J.D. wrote the first draft of the manuscript. X.Z. and H.W. conducted data analyses and manuscript preparation. X.Z. and H.W. designed the project and secured funding. All authors have read and agreed to the published version of the manuscript.

Funding: DOE-Nuclear Energy under DE-NE0008549 and DE-NE0008787.

Institutional Review Board Statement: Not applicable.

Informed Consent Statement: Not applicable.

Data Availability Statement: The raw/processed data required to reproduce these findings cannot be shared at this time due to technical or time limitations.

Acknowledgments: J. Ding and X. Zhang acknowledge primary financial support by DOE-Nuclear Energy under DE-NE0008549. Z. Shang acknowledges support by DE-NE0008787. S. Xue and H. Wang acknowledge financial support from the U.S. Office of Naval Research (ONR) N00014-20-1-2043 (for TEM) and N0014-17-1-2087 (for sample preparation). Access to microscopy facilities at Life Science, Birck Nanotechnology Center and School of Materials Engineering at Purdue University, is also acknowledged.

Conflicts of Interest: The authors declare no conflict of interest.

References

1. Wang, Y.; Wang, K.; Pan, D.; Lu, K.; Hemker, K.; Ma, E. Microsample tensile testing of nanocrystalline copper. *Scr. Mater.* **2003**, *48*, 1581–1586. [CrossRef]
2. Youssef, K.M.; Scattergood, R.O.; Murty, K.L.; Horton, J.A.; Koch, C.C. Ultrahigh strength and high ductility of bulk nanocrystalline copper. *Appl. Phys. Lett.* **2005**, *87*, 85–88. [CrossRef]
3. Zhao, Y.; Bingert, J.F.; Liao, X.; Cui, B.; Han, K.; Sergueeva, A.V.; Mukherjee, A.K.; Valiev, R.Z.; Langdon, T.G.; Zhu, Y.T. Simultaneously Increasing the Ductility and Strength of Ultra-Fine- Grained Pure Copper. *Adv. Mater.* **2006**, *18*, 2949–2953. [CrossRef]
4. Furukawa, M.; Horita, Z.; Nemoto, M.; Langdon, T.G. Review: Processing of metals by equal-channel angular pressing. *J. Mater. Sci.* **2001**, *36*, 2835–2843. [CrossRef]
5. Iwahashi, Y.; Horita, Z.; Nemoto, M.; Langdon, T.G. The process of grain refinement in equal-channel angular pressing. *Acta Mater.* **1998**, *46*, 3317–3331. [CrossRef]
6. Zhilyaev, A.P.; Langdon, T.G. Using high-pressure torsion for metal processing: Fundamentals and applications. *Prog. Mater. Sci.* **2008**, *53*, 893–979. [CrossRef]
7. Zhilyaev, A.P.; Gimazov, A.A.; Soshnikova, E.P.; Révész, Á.; Langdon, T.G. Microstructural characteristics of nickel processed to ultrahigh strains by high-pressure torsion. *Mater. Sci. Eng. A* **2008**, *489*, 207–212. [CrossRef]
8. Kawasaki, M.; Langdon, T.G. Review: Achieving superplasticity in metals processed by high-pressure torsion. *J. Mater. Sci.* **2014**, *49*, 6487–6496. [CrossRef]
9. Ames, M.; Karos, R.; Michels, A.; Tscho, A.; Birringer, R. Unraveling the nature of room temperature grain growth in nanocrystalline materials. *Acta Mater.* **2008**, *56*, 4255–4266. [CrossRef]
10. Thuvander, M.; Abraham, M.; Cerezo, A.; Smith, G.D.W. Thermal stability of electrodeposited nanocrystalline nickel and iron–Nickel alloys Thermal stability of electrodeposited nanocrystalline nickel and iron–Nickel alloys. *Mater. Sci. Technol.* **2001**, *17*, 961–970. [CrossRef]
11. Hibbard, G.D.; Radmilovic, V.; Aust, K.T.; Erb, U. Grain boundary migration during abnormal grain growth in nanocrystalline Ni. *Mater. Sci. Eng. A* **2008**, *494*, 232–238. [CrossRef]
12. Iordache, M.; Whang, S.; Jiao, Z.; Wang, Z. Grain growth kinetics in nanostructured nickel. *Nanostruct. Mater.* **1999**, *11*, 1343–1349. [CrossRef]
13. Popov, V.V.; Popova, E.; Stolbovskiy, A.; Pilyugin, V. Thermal stability of nanocrystalline structure in niobium processed by high pressure torsion at cryogenic temperatures. *Mater. Sci. Eng. A* **2011**, *528*, 1491–1496. [CrossRef]
14. Sharma, G.; Varshney, J.; Bidaye, A.C.; Chakravartty, J.K. Grain growth characteristics and its effect on deformation behavior in nanocrystalline Ni. *Mater. Sci. Eng. A* **2012**, *539*, 324–329. [CrossRef]
15. Klement, U.; Erb, U.; El-Sherik, A.; Aust, K. Thermal stability of nanocrystalline Ni. *Mater. Sci. Eng. A* **1995**, *203*, 177–186. [CrossRef]
16. Jiang, H.; Zhu, Y.T.; Butt, D.P.; Alexandrov, I.V.; Lowe, T.C. Microstructural evolution, microhardness and thermal stability of HPT-processed Cu. *Mater. Sci. Eng. A* **2000**, *290*, 128–138. [CrossRef]
17. Fang, T.H.; Li, W.L.; Tao, N.R.; Lu, K. Revealing Extraordinary Intrinsic Tensile Plasticity in Gradient Nano-Grained Copper. *Science* **2011**, *331*, 1587–1590. [CrossRef]
18. Huang, H.W.; Wang, Z.B.; Lu, J.; Lu, K. Fatigue behaviors of AISI 316L stainless steel with a gradient nanostructured surface layer. *Acta Mater.* **2015**, *87*, 150–160. [CrossRef]
19. Wu, X.; Jiang, P.; Chen, L.; Yuan, F.; Zhu, Y. Extraordinary strain hardening by gradient structure. *Proc. Natl. Acad. Sci. USA* **2014**, *111*, 7197–7201. [CrossRef]
20. Wu, X.; Jiang, P.; Chen, L.; Zhang, J.F.; Yuan, F.P.; Zhu, Y.T. Synergetic Strengthening by Gradient Structure. *Mater. Res. Lett.* **2014**, *2*, 185–191. [CrossRef]
21. Ding, J.; Shang, Z.; Zhang, Y.F.; Su, R.; Li, J.; Wang, H.; Zhang, X. Tailoring the thermal stability of nanocrystalline Ni alloy by thick grain boundaries. *Scr. Mater.* **2020**, *182*, 21–26. [CrossRef]
22. Liu, X.; Zhang, H.; Lu, K. Formation of nano-laminated structure in nickel by means of surface mechanical grinding treatment. *Acta Mater.* **2015**, *96*, 24–36. [CrossRef]
23. Ding, J.; Li, Q.; Li, J.; Xue, S.; Fan, Z.; Wang, H.; Zhang, X. Mechanical behavior of structurally gradient nickel alloy. *Acta Mater.* **2018**, *149*, 57–67. [CrossRef]
24. Ding, J.; Neffati, D.; Li, Q.; Su, R.; Li, J.; Xue, S.; Shang, Z.; Zhang, Y.F.; Wang, H.; Kulkarni, Y.; et al. Thick grain boundary induced strengthening in nanocrystalline Ni alloy. *Nanoscale* **2019**, *11*, 23449–23458. [CrossRef] [PubMed]

25. Zhou, X.; Li, X.; Lu, K. Enhanced thermal stability of nanograined metals below a critical grain size. *Science* **2018**, *360*, 526–530. [CrossRef] [PubMed]
26. Zhou, X.; Li, X.; Lu, K. Size Dependence of Grain Boundary Migration in Metals under Mechanical Loading. *Phys. Rev. Lett.* **2019**, *122*, 126101. [CrossRef]
27. Zhang, H.J.; Li, C.; Liu, Y.C.; Guo, Q.Y.; Huang, Y.; Li, H.J.; Yu, J.X. Effect of hot deformation on γ" and δ phase precipitation of Inconel 718 alloy during deformation&isothermal treatment. *J. Alloys Compd.* **2017**, *716*, 65–72.
28. Dreler, A.; Oberwinkler, B.; Primig, S.; Turk, C.; Povoden-Karadeniz, E.; Heinemann, A.; Ecker, W.; Stockinger, M. Experimental and numerical investigations of the γ" and γ' precipitation kinetics in Alloy 718. *Mater. Sci. Eng. A* **2018**, *723*, 314–323. [CrossRef]
29. Lawitzki, R.; Hassan, S.; Karge, L.; Wagner, J.; Wang, D.; von Kobylinski, J.; Krempaszky, C.; Hofmann, M.; Gilles, R.; Schmitz, G. Differentiation of γ'- and γ"- precipitates in Inconel 718 by a complementary study with small-angle neutron scattering and analytical microscopy. *Acta Mater.* **2019**, *163*, 28–39. [CrossRef]
30. Anderson, M.; Thielin, A.L.; Bridier, F.; Bocher, P.; Savoie, J. δ Phase precipitation in Inconel 718 and mechanical properties. *Mater. Sci. Eng. A* **2017**, *679*, 48–55. [CrossRef]
31. Mei, Y.; Liu, C.; Liu, Y.; Zhou, X.; Yu, L.; Li, C.; Ma, Z.; Huang, Y. Effects of cold rolling on the precipitation and the morphology of δ-phase in Inconel 718 alloy. *J. Mater. Res.* **2016**, *31*, 443–454. [CrossRef]
32. Azadian, S.; Wei, L.-Y.; Warren, R. Delta phase precipitation in Inconel 718. *Mater. Charact.* **2004**, *53*, 7–16. [CrossRef]
33. Wang, Z.; Guan, K.; Gao, M.; Li, X.; Chen, X.; Zeng, X. The microstructure and mechanical properties of deposited-IN718 by selective laser melting. *J. Alloys Compd.* **2012**, *513*, 518–523. [CrossRef]
34. Wang, Y.; Shao, W.; Zhen, L.; Zhang, B. Hot deformation behavior of delta-processed superalloy 718. *Mater. Sci. Eng. A* **2011**, *528*, 3218–3227. [CrossRef]
35. Amato, K.; Gaytan, S.; Murr, L.; Martinez, E.C.; Shindo, P.; Hernandez, J.; Collins, S.F.; Medina, F.S. Microstructures and mechanical behavior of Inconel 718 fabricated by selective laser melting. *Acta Mater.* **2012**, *60*, 2229–2239. [CrossRef]
36. Chamanfar, A.; Sarrat, L.; Jahazi, M.; Asadi, M.; Weck, A.; Koul, A. Microstructural characteristics of forged and heat treated Inconel-718 disks. *Mater. Des.* **2013**, *52*, 791–800. [CrossRef]
37. Liu, W.; Chen, Z.; Xiao, F.; Yao, M.; Wang, S.; Liu, R. Effect of cold rolling on the kinetics of δ phase precipitation in inconel 718. *Metall. Mater. Trans. A* **1999**, *30A*, 31–40. [CrossRef]
38. Slama, C.; Servant, C.; Cizeron, G. Aging of the Inconel 718 alloy between 500 and 750 °C. *J. Mater. Res.* **1997**, *12*, 2298–2316. [CrossRef]
39. Thomas, A.; El-Wahabi, M.; Cabrera, J.; Prado, J. High temperature deformation of Inconel 718. *J. Mater. Process. Technol.* **2006**, *177*, 469–472. [CrossRef]
40. Fisk, M.R.; Andersson, J.; Du Rietz, R.; Haas, S.; Hall, S.A. Precipitate evolution in the early stages of ageing in Inconel 718 investigated using small-angle x-ray scattering. *Mater. Sci. Eng. A* **2014**, *612*, 202–207. [CrossRef]
41. Mei, Y.; Liu, Y.; Liu, C.; Li, C.; Yu, L.; Guo, Q.; Li, H. Effects of cold rolling on the precipitation kinetics and the morphology evolution of intermediate phases in Inconel 718 alloy. *J. Alloys Compd.* **2015**, *649*, 949–960. [CrossRef]
42. Kuo, Y.-L.; Horikawa, S.; Kakehi, K. The effect of interdendritic δ phase on the mechanical properties of Alloy 718 built up by additive manufacturing. *Mater. Des.* **2017**, *116*, 411–418. [CrossRef]
43. Zhang, H.; Li, C.; Guo, Q.; Ma, Z.; Huang, Y.; Li, H.; Liu, Y. Delta precipitation in wrought Inconel 718 alloy: the role of dynamic recrystallization. *Mater. Charact.* **2017**, *133*, 138–145. [CrossRef]
44. Nalawade, S.A.; Sundararaman, M.; Singh, J.B.; Verma, A.; Kishore, R. Precipitation of γ' phase in δ-precipitated Alloy 718 during deformation at elevated temperatures. *Mater. Sci. Eng. A* **2010**, *527*, 2906–2909. [CrossRef]
45. Huang, Y.; Langdon, T.G. The evolution of delta-phase in a superplastic Inconel 718 alloy. *J. Mater. Sci.* **2007**, *42*, 421–427. [CrossRef]
46. Rong, Y.H.; Chen, S.P.; Hu, G.X.; Gao, M.; Wei, R.P. Prediction and characterization of variant electron diffraction patterns for γ" and δ precipitates in an INCONEL 718 alloy. *Metall. Mater. Trans. A* **1999**, *30A*, 2297–2303. [CrossRef]
47. Burke, J.; Turnbull, D. Recrystallization and grain growth. *Prog. Met. Phys.* **1952**, *3*, 220–292. [CrossRef]
48. Schuler, J.D.; Donaldson, O.K.; Rupert, T.J. Amorphous complexions enable a new region of high temperature stability in nanocrystalline Ni-W. *Scr. Mater.* **2018**, *154*, 49–53. [CrossRef]
49. Muthaiah, V.S.; Babu, L.H.; Koch, C.C.; Mula, S. Feasibility of formation of nanocrystalline Fe-Cr-Y alloys: Mechanical properties and thermal stability. *Mater. Charact.* **2016**, *114*, 43–53. [CrossRef]
50. Li, L.; Saber, M.; Xu, W.; Zhu, Y.; Koch, C.C.; Scattergood, R.O. High-temperature grain size stabilization of nanocrystalline Fe–Cr alloys with Hf additions. *Mater. Sci. Eng. A* **2014**, *613*, 289–295. [CrossRef]
51. A Darling, K.; Kecskes, L.J.; Atwater, M.; Semones, J.; Scattergood, R.; Koch, C. Thermal stability of nanocrystalline nickel with yttrium additions. *J. Mater. Res.* **2013**, *28*, 1813–1819. [CrossRef]
52. Zhang, Y.; Tao, N.; Lu, K. Mechanical properties and rolling behaviors of nano-grained copper with embedded nano-twin bundles. *Acta Mater.* **2008**, *56*, 2429–2440. [CrossRef]
53. Hall, E.O. The Deformation and Ageing of Mild Steel. *Proc. Phys. Soc. London Sect. B* **1951**, *64*, 747–753. [CrossRef]
54. Petch, J.N. The Cleavage Strength of Polycrystals. *J. Iron Steel Inst.* **1953**, *174*, 25–28.
55. Legros, M.; Elliott, B.R.; Rittner, M.N.; Weertman, J.R.; Hemker, K.J. Microsample tensile testing of nanocrystalline metals. *Philos. Mag. A* **2000**, *80*, 1017–1026. [CrossRef]

56. Chen, K.-C.; Wu, W.-W.; Liao, C.-N.; Chen, L.-J.; Tu, K.N. Observation of Atomic Diffusion at Twin-Modified Grain Boundaries in Copper. *Science.* **2008**, *321*, 1066–1069. [CrossRef]
57. Li, X.; Zhou, X.; Lu, K. Rapid heating induced ultrahigh stability of nanograined copper. *Sci. Adv.* **2020**, *6*, eaaz8003. [CrossRef]
58. Wu, Z.; Zhang, Y.-W.; Srolovitz, D. Grain boundary finite length faceting. *Acta Mater.* **2009**, *57*, 4278–4287. [CrossRef]
59. Priedeman, J.L.; Olmsted, D.L.; Homer, E.R. The role of crystallography and the mechanisms associated with migration of incoherent twin grain boundaries. *Acta Mater.* **2017**, *131*, 553–563. [CrossRef]
60. Straumal, B.B.; Kogtenkova, O.A.; Gornakova, A.S.; Sursaeva, V.G.; Baretzky, B. Review: Grain boundary faceting-roughening phenomena. *J. Mater. Sci.* **2016**, *51*, 382–404. [CrossRef]
61. Wolf, U.; Ernst, F.; Muschik, T.; Finnis, M.W.; Fischmeister, H.F. The influence of grain boundary inclination on the structure and energy of σ = 3 grain boundaries in copper. *Philos. Mag. A.* **1992**, *66*, 991–1016. [CrossRef]
62. Bishop, G.H.; Hartt, W.; Bruggeman, G.A. Grain boundary faceting of <1010> tilt boundaries in zinc-II. *Acta Metall.* **1971**, *19*, 37–47. [CrossRef]
63. Rittner, J.; Seidman, D.; Merkle, K. Grain-boundary dissociation by the emission of stacking faults. *Phys. Rev. B Condens. Matter Mater. Phys.* **1996**, *53*, 4241–4244. [CrossRef] [PubMed]
64. Sun, C.; Yang, Y.; Liu, Y.; Hartwig, K.T.; Wang, H.; Maloy, S.A.; Allen, T.R.; Zhang, X. Thermal stability of ultrafine grained Fe-Cr-Ni alloy. *Mater. Sci. Eng. A.* **2012**, *542*, 64–70. [CrossRef]

Article

Hardness-Depth Relationship with Temperature Effect for Single Crystals—A Theoretical Analysis

Hao Liu [1], Long Yu [2] and Xiazi Xiao [3],*

[1] Department of Applied Physics, School of Physics and Electronics, Hunan University, Changsha 410082, China; haoliu@hnu.edu.cn
[2] State Key Laboratory for Turbulence and Complex System, Department of Mechanics and Engineering Science, College of Engineering, Peking University, Beijing 100871, China; longyu2014@pku.edu.cn
[3] Department of Mechanics, School of Civil Engineering, Central South University, Changsha 410075, China
* Correspondence: xxz2017@csu.edu.cn; Tel.: +86-18569499345

Received: 23 January 2020; Accepted: 11 February 2020; Published: 13 February 2020

Abstract: In this paper, a mechanistic model is developed to address the effect of temperature on the hardness-depth relationship of single crystals. Two fundamental hardening mechanisms are considered in the hardness model, including the temperature dependent lattice friction and network dislocation interaction. The rationality and accuracy of the developed model is verified by comparing with four different sets of experimental data, and a reasonable agreement is achieved. In addition, it is concluded that the moderated indentation size effect at elevated temperatures is ascribed to the accelerated expansion of the plasticity affected region that results in the decrease of the density of geometrically necessary dislocations.

Keywords: temperature effect; indentation size effect; theoretical model; nano-indentation

1. Introduction

With increasing requirements for the application of advanced structural materials under high temperature environments, it becomes necessary and vital to effectively characterize the thermal-related materials properties at elevated temperatures to ensure structural safety and device function [?]. However, for the materials with limited size and characteristic microstructures, for example, multilayer thin-films and ion-irradiated materials with defect damage, the direct application of traditional mechanical tests seems invalid, and the consideration of small-scale testing techniques becomes inevitable [?]. Among several promising candidates, the technique of nano-indentation has been well developed over the last decades, and taken as a valid method to characterize the localized materials properties at elevated temperatures due to the development of advanced high temperature indentation systems [? ? ?].

So far, plenty of experimental works have been performed in the field of high temperature nano-indentation [? ? ? ? ? ?], and some principal features observed in these experiments indicate that the well-known indentation size effect, that is, the increase of materials hardness with decreasing indentation depth, still exists even when the testing temperature T increases up to $T_m/3$ (T_m is the melting temperature) for most materials [? ? ? ? ?]. However, when compared with the test performed at room temperature, both the increasing rate of materials hardness and ultimate bulk hardness are noticed to get weakened at elevated temperatures [? ?]. For instance, Lee et al. [?] investigated the dependence of indentation size effect on T for [0 0 1] -oriented single crystalline Nb, W, Al and Au, and demonstrated that for all of them both the hardness at infinite indentation depth and intrinsical materials length scale are strong functions of T. Similar experimental phenomena [?] have also been observed in the indentation test of polycrystalline Co, Ni and Pt that the indentation size effect becomes moderated when T increases from room temperature to $T_m/3$.

In order to interpret the fundamental mechanisms related to the above observed experimental results, several theoretical models have been developed in the past years [? ? ? ? ?]. Thereinto, the most widely applied model, was proposed by Nix and Gao [?], which indicates that the intrinsic length scale is ascribed to the formation of geometrically necessary dislocations (GNDs) within the plasticity affected region. Later, Durst et al. [?] modified the Nix-Gao model by redefining the volume of the plasticity affected region, and pointing out that its radius should scale linearly with the contact radius. In addition, the contribution of intrinsic lattice friction resistance was noticed to play a dominant role in determining the material's hardness, especially for body-centered cubic (BCC) materials at low temperatures [?]. When further addressing the temperature effect on the fundamental deformation mechanisms, it is noted that increasing temperature can not only help enhance the dislocation mobility and expansion of the plasticity affected region [?], but also lead to the decrease of lattice friction at elevated temperatures for most crystalline materials [?]. However, most existing hardness models are proposed at room temperature, and the temperature effect on both microstructural evolution and lattice friction has not yet been systematically addressed [? ?].

In this work, we intend to propose a theoretical framework for the hardness-depth relationship with temperature effect for single crystals. In Section ??, a detailed derivation of the hardness model with temperature effect is presented. The dominant deformation mechanisms cover the dependence of lattice friction and network dislocation interaction on temperature. In Section ??, the experimental data of four single crystals (Cu, Al, CaF$_2$ and W) with different crystalline structures is applied to verify the accuracy and rationality of the proposed model. Moreover, corresponding deformation mechanisms and microsturcutres evolutions are discussed. Finally, we close with a conclusion in Section ??.

2. Hardness-Depth Relationship with Temperature Effect

2.1. Model Development

Intrinsically speaking, plastic deformation of pure single-crystal materials is mainly determined by the mobile ability of dislocations when there exist sufficient initial dislocations. At low temperatures, the slipping of existing dislocations, especially for BCC metals, is dominated by short-range barriers like the Peierls potential. With the increase of T, long-range barriers induced by the network dislocation interaction become the dominant role in determining materials hardness [? ?]. Therefore, the temperature dependent critical resolved shear stress $\tau_{CRSS}(T)$, indicating the impediment of slipping dislocations, can be expressed as

$$\tau_{CRSS}(T) = \tau_n(T) + \tau_f(T), \tag{1}$$

where $\tau_n(T)$ and $\tau_f(T)$, respectively, denote the dislocation hardening term and lattice friction. Therefore, $\tau_n(T)$ is influenced by both materials properties and dislocation microstructures at a given temperature. For the latter, statistically stored dislocations (SSDs) and GNDs should be simultaneously addressed when performing nano-indentation on pure single crystals. Following the Taylor law, the general form of $\tau_n(T)$ with temperature effect yields as

$$\tau_n(T) = b\mu(T)\alpha(T)[\rho_G(T) + \rho_S(T)]^{m(T)}, \tag{2}$$

where b denotes the magnitude of Burgers vector. $\mu(T)$ and $\alpha(T)$ are, respectively, the shear modulus and dislocation strength coefficient, which both decrease with the increase of T. $m(T)$ is the hardening coefficient that is generally around 0.5 at room temperature but also decreases with increasing temperature [?]. Moreover, $\rho_S(T)$ and $\rho_G(T)$ denote the density of SSDs and GNDs, respectively. Following the theory of Nix and Gao [?], GNDs are considered to generate beneath the indenter tip in order to coordinate with the gradient plastic deformation, and be stored within the plasticity affected region which is assumed to be a hemisphere, as illustrated in Figure ??. Therefore, $\rho_G(T)$ can be deduced by dividing the length λ of GNDs by the volume $V(T)$ of the plasticity affected region, that is,

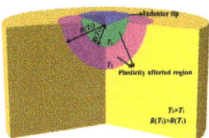

Figure 1. (Color online) Schematic of the nano-indentation of single crystal materials at two different temperatures, that is, $T_2 > T_1$. The plasticity affected region is assumed as a hemisphere with radius of R, and $R(T_2) > R(T_1)$.

$$\rho_G(T) = \frac{\lambda}{V(T)} = \frac{3}{2b\tan\theta M^3(T)} \frac{1}{h}, \qquad (3)$$

where $\lambda = \pi h^2/(b\tan\theta)$ and $V(T) = 2/3\pi R^3(T)$. Here, θ is the angle between the surface of the indenter and sample, and $R(T)$ denotes the radius of the plasticity affected region that proportionally scales with the indentation depth h with a proportional coefficient $M(T)$, that is, $R(T) = M(T)h$. At elevated temperatures, both lattice friction and dislocation impediment strength get weakened, which dramatically facilitate the expansion of the plastic region, and lead to the increase of $M(T)$ and $R(T)$ with T [?], as presented in Figure ??. Moreover, $\rho_S(T)$ is expressed as

$$\rho_S(T) = \frac{3}{2b\tan\theta} \frac{1}{h^*(T)}, \qquad (4)$$

where $h^*(T)$ represents a characteristic length that is connected to the bulk hardness [?]. According to Ashby's definition [?], SSDs are formed and accumulated in pure crystals during straining. Therefore, with higher internal strain stored in the materials at higher temperatures, more dislocations will be formed within the crystal that result in the higher density of SSDs.

Concerning the lattice friction, it is well known that $\tau_f(T)$ for most face-centered cubic (FCC) materials is negligible when compared with the dislocation hardening term, therefore, the contribution of $\tau_f(T)$ is generally ignored when addressing the temperature effect on materials' hardening [? ? ?]. Whereas, for BCC materials, the stress required to move a dislocation over the Peierls potential is a thermally activated event, and takes a dominate role in determining the materials strength at low temperatures [? ? ?]. Following the work of [?], the expression of $\tau_f(T)$ for BCC metals follows as

$$\tau_f(T) = \begin{cases} \tau_{p0}[1 - \sqrt{\frac{k_B T}{2H_k} \ln(\frac{\dot{\gamma}_{p0}}{\dot{\varepsilon}})}] & (T \leq T_0) \\ \tau_{f0}[1 - \frac{k_B T}{2H_k} \ln(\frac{\dot{\gamma}_{p0}}{\dot{\varepsilon}})]^2 & (T > T_0) \end{cases}, \qquad (5)$$

where τ_{p0} and τ_{f0} are, respectively, the reference stress for the screw dislocations when T is below and above the critical temperature T_0, which divides the deformation region into the elastic interaction and line tension regimes. k_B indicates the Boltzmann constant and $2H_k$ is the formation enthalpy of the kink pair on a screw dislocation. $\dot{\gamma}_{p0}$ and $\dot{\varepsilon}$ are the reference strain rate and loading strain rate, respectively.

By further considering the Mises flow rule [?] and Tabor's factor [?], one can connect the temperature dependent hardness $H(T)$ with $\tau_{CRSS}(T)$ as

$$H(T) = 3\sqrt{3}\tau_{CRSS}(T) = H_f(T) + H_n(T), \qquad (6)$$

where $H_f(T) = 3\sqrt{3}\tau_f(T)$ is the hardness component induced by lattice friction, and $H_n(T)$ denotes the dislocation hardening component deduced by submitting Eqsuations (??) and (??) into Equation (??), that is,

$$H_n(T) = H_0(T)[1 + \frac{\bar{h}^*(T)}{h}]^{m(T)} \qquad (7)$$

where
$$H_0(T) = 3\sqrt{3}b\mu(T)\alpha(T)\rho_S^{m(T)}(T), \qquad (8)$$

and
$$\bar{h}^*(T) = h^*(T)/M^3(T) \quad \text{with} \quad h^*(T) = \frac{9[9\sqrt{3}b\mu(T)\alpha(T)]^{\frac{1}{m(T)}}}{2b\tan\theta H_0^{\frac{1}{m(T)}}(T)}. \qquad (9)$$

Further derivation of Equation (??) indicates that $[H_n(T)/H_0(T)]^{1/m(T)}$ scales linearly with $1/h$, and the slope $\bar{h}^*(T)$ is determined by both the characteristic length $h^*(T)$ and proportional coefficient $M(T)$. On the one hand, it shows that $\rho_S(T)$ tends to increase with T, which results in the decrease of $h^*(T)$ at elevated temperatures [?]. On the other hand, as the impediment of slipping dislocations gets weakened at high temperatures, the expansion of the plasticity affected region becomes comparatively easy, which results in the increase of $M(T)$ with T [?]. Therefore, $\bar{h}^*(T)$ tends to decrease at high temperatures that results in the weakened indentation size effect at elevated temperatures. Furthermore, increasing temperature not only leads to the decrease of $H_0(T)$ and $m(T)$, but also weakens the lattice friction, thus, it becomes rational to experimentally observe that $H(T)$ decreases with the increase of T for most crystalline materials [? ? ?].

One may also note that Equation (??) offers a general law characterizing the hardness-depth relationships of pure single crystals at various temperatures. When ignoring the temperature effect, Equation (??) can be degraded into the hardness model involving lattice friction effect at room temperature. Once the hardening contribution of lattice friction is further ignored, the model is ultimately reduced to the classical Nix-Gao model [?]. This simplification is reasonable and rational for materials with small lattice friction. Whereas, when the lattice friction is relatively comparable with the dislocation hardening term, the ignoring of the former will result in the overestimation of H_0 and ρ_S.

2.2. Model Calibration

In order to parameterize the temperature dependent hardness $H(T)$ as expressed in Equation (??), the main attention turns to the determination of $H_0(T)$, $\bar{h}^*(T)$ and $m(T)$ for $H_n(T)$, given the experimental data or theoretical expression for $H_f(T)$ is generally known for the concerned materials. In the following, the calibration process is briefly introduced, that is,

(1) First, conduct nano-indentation tests for single crystals at different temperatures to obtain the hardness-depth relationships, that is, the $\tilde{H}(T) - h$ curves [?]. Hereafter, the symbol $\sim$ denotes the experimental data. It then follows from Equation (??) that the $\tilde{H}_n(T) - h$ relationships are obtained by subtracting $\tilde{H}(T)$ from $H_f(T)$.

(2) Then, transform the $\tilde{H}_n(T) - h$ relationships into the $\tilde{H}_n^{1/m(T)}(T) - 1/h$ curves. By adjusting parameter $m(T)$ to approximately obtain a straight line with the determination coefficient $r^2 \geq 0.95$. The slope and intercept with the vertical axis give $H_0^{1/m(T)}(T)\bar{h}^*(T)$ and $H_0^{1/m(T)}(T)$, respectively. It then yields the value of $H_0(T)$ and $\bar{h}^*(T)$.

(3) Next, combine Equations (??) and (??) with the fitted value of $H_0(T)$, $\bar{h}^*(T)$ and $m(T)$ as well as the previously known expression of $H_f(T)$, it finally gives the parameterized hardness model with temperature effect.

(4) Finally, compare the fitted theoretical results with corresponding experimental data at different temperatures.

3. Experimental Verifications and Results

In this section, four sets of experimental data for single crystal Cu [?], Al [?], CaF$_2$ [?] and W [?] are applied to verify the rationality and accuracy of the developed model. Before the calibration of model parameters, it should be noted that H_f for Cu and Al is so small that is generally not considered [? ?]. For CaF$_2$, H_f is informed to be 1.2 GPa at room temperature [?], and vanishes to be zero around 473 K [?]. As to W, H_f can be theoretically calculated by referring to Equation (??) with $\tau_{p0} = 1038$ MPa, $\tau_{f0} = 2035$ MPa, $T_0 = 580$ K, $k_B = 1.38 \times 10^{-23}$ J/K, $H_k = 1.65 \times 10^{-19}$ J, $\dot{\gamma}_{p0} = 3.71 \times 10^{10}$ s^{-1} and $\dot{\varepsilon} = 0.02$ s^{-1} [? ?]. Then, the experimental data can be plotted in the form as $\tilde{H}_n^{1/m} - 1/h$ for the four materials, as illustrated in Figure ??. Following the calibration procedure as mentioned in Section ??, the model parameters are obtained, as listed in Table ??.

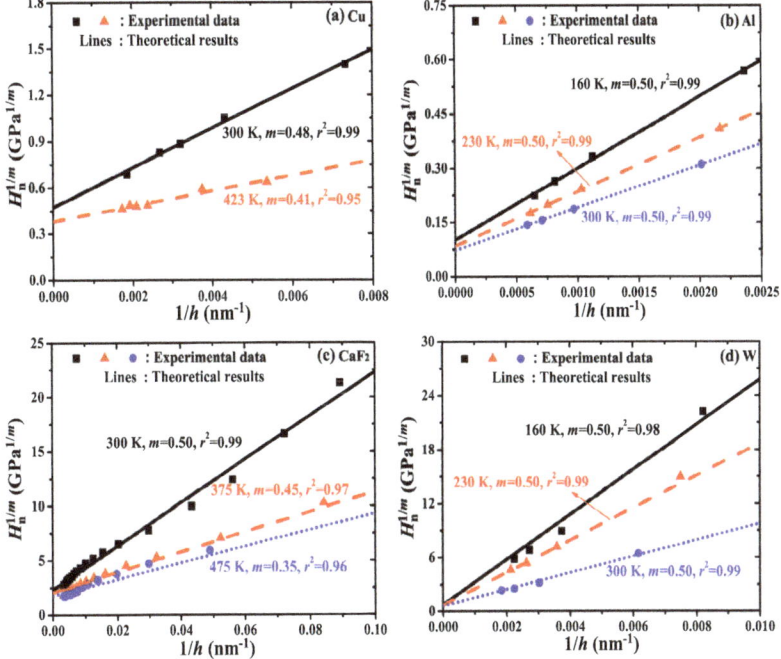

Figure 2. (Color online) Comparison of the $\tilde{H}_n^{1/m} - 1/h$ relationships at various temperatures between theoretical results (lines) and experimental data (dots) for (a) Cu [?], (b) Al [?], (c) CaF$_2$ [?] and (d) W [?].

Table 1. Parameter calibration of the proposed theoretical model for single crystal Cu [?], Al [?], CaF$_2$ [?] and W [?].

	Cu (N/A)		Al ([001[)			CaF$_2$ ([111])			W ([001])		
	300 K	423 K	160 K	230 K	300 K	300 K	375 K	475 K	160 K	230 K	300 K
H_f (GPa)	0.00	0.00	0.00	0.00	0.00	1.20	0.60	0.00	3.05	2.58	2.18
m	0.48	0.41	0.50	0.50	0.50	0.50	0.45	0.35	0.50	0.50	0.50
$H_0^{\frac{1}{m}}$ (GPa$^{\frac{1}{m}}$)	0.474	0.381	0.101	0.084	0.072	2.320	2.070	1.750	0.750	0.716	0.627
$H_0^{\frac{1}{m}} \bar{h}^*$ (GPa$^{\frac{1}{m}}$·nm)	127.6	49.6	198.5	150.3	118.3	199.8	92.4	75.9	2503	1796	915
r^2	0.99	0.95	0.99	0.99	0.99	0.99	0.97	0.96	0.98	0.99	0.99
H_0 (GPa)	0.699	0.673	0.318	0.290	0.268	1.523	1.390	1.220	0.866	0.846	0.792
$\bar{h}^*$ (nm)	268	130	1965	1789	1643	86	45	43	3337	2508	1459

With these calibrated parameters, the $H_n^{1/m} - 1/h$ relationships are compared between the theoretical results (lines) and experimental data (dots), and a reasonable agreement is achieved, as presented in Figure ??. It shows that both the intercept and slope of the fitting line decrease with increasing temperature, which mutually contribute to the decrease of H_0 and $\bar{h}^*$ at elevated temperatures. The former is mainly ascribed to the weakened elastic constants and dislocation strength coefficient with the increase of temperature, and the latter originates from the stimulated expansion of the plasticity affected region at high temperatures that results in the decrease of the intrinsical length scale and limited indentation size effect. Moreover, the hardening coefficient m of Cu and CaF$_2$ is noticed to decrease with increasing temperature when compared with that of Al and W. Similar experimental data has also been observed for OFHC copper that m decreases from 0.48 to 0.41 when T increases from 293 K to 698 K [?], which indicates the weakened work hardening behavior at elevated temperatures that resembles an elastic-ideally plastic material [?]. In order to further characterize the thermally activated deformation mechanisms resulting in the decrease of m with T, it could be addressed by the nano-indentation strain jump tests [?] or long term creep tests [?], especially in terms of the strain-rate sensitivity.

In Figure ??, we present the $H - h$ relationships obtained from the calibrated theoretical model and experimental data of single crystal Cu [?], Al [?], CaF$_2$ [?] and W [?]. As one can see the results match reasonably well for the four materials, and an obvious indentation size effect is informed at different temperatures. However, the increasing rate of H with the decrease of h tends to decrease with the increase of temperature, which is determined by the decrease of $\bar{h}^*$ at elevated temperatures, as illustrated in Figure ??. In addition, the bulk hardness at the deep indentation depth also decreases with increasing temperature as both the hardness components $H_f(T)$ and $H_0(T)$ get weakened at high temperatures.

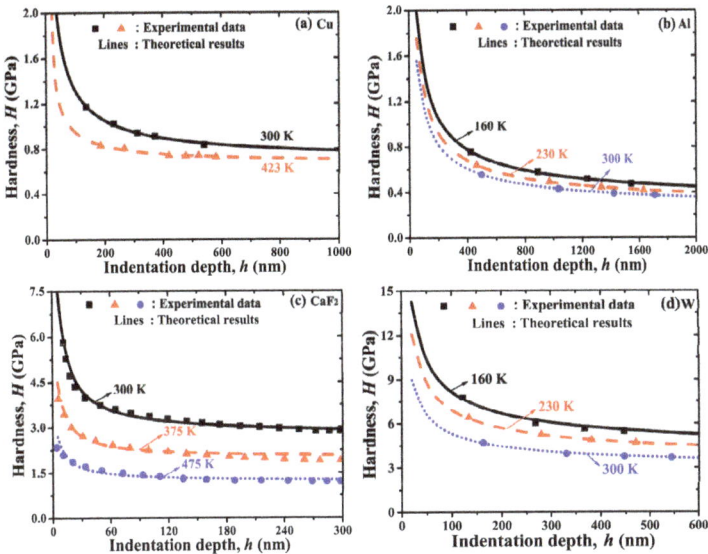

Figure 3. (Color online) Comparison of the $H-h$ relationships at different temperatures between theoretical results (lines) and experimental data (dots) for (**a**) Cu [?], (**b**) Al [?], (**c**) CaF$_2$ [?] and (**d**) W [?].

Based on the developed model, the effect of temperature on the evolution of different microstructures can be further analyzed, for example, the expansion of the plasticity affected region and evolution of dislocation density. Take W for an example, the temperature dependent shear modulus $\mu(T)$ and $\alpha(T)$ can be informed in previous works [? ?], and it is known that $b = 0.274$ nm and $\tan\theta = 0.358$ for the Berkovich nano-indentation of W [? ?], as summarized in Table ??. Therefore, according to Equation (??), one can calculate $h^*(T)$ and $M(T) = \sqrt[3]{h^*(T)/\bar{h}^*(T)}$ at 160 K, 230 K and 300 K, respectively. Figure ?? illustrates the evolution of $h^*(T)$ and $M(T)$ as a function of T for single crystal W, which indicates that $h^*(T)$ decreases while $M(T)$ increases with T. The former is rational as $\rho_S(T)$, determined by $h^*(T)$ as expressed in Equation (??), is considered to increase due to the high internal strain stored in the materials under high temperatures [?]. As a comparison, the latter is ascribed to the weakened impediment of slipping dislocations that the expansion of the plasticity affected region becomes comparatively easy at high temperatures [?].

Table 2. Material properties for single crystal W with temperature effect.

Parameter	b (nm)	μ (GPa)	α
Value	0.274	$\sqrt{C_{44}(T)[C_{11}(T)-C_{12}(T)]/2}$	$\alpha_0 - k_0 T$ with $\alpha_0 = 0.38$ and $k_0 = 4 \times 10^{-4} K^{-1}$
Ref.	[?]	[?]	[?]

b: the magnitude of Burgers vector; μ: shear modulus; α: dislocation strength coefficient; C_{11}, C_{12} and C_{44}: elastic constants; α_0: dislocation strength coefficient when the temperature equals zero; k_0: proportional coefficient; T: temperature.

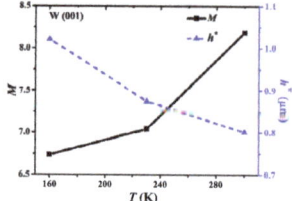

Figure 4. (Color online) Evolution of M and h^* as a function of T for single crystal W.

Last but not the least, the evolution of $\rho_G(T)$ and $\rho_S(T)$ as a function of h at different temperatures is compared for single crystal W, as illustrated in Figure ??. According to Equation (??), $\rho_G(T)$ is determined by both $M(T)$ and h. Thereinto, the inverse scaling law between $\rho_G(T)$ and h indicates the fundamental mechanism for the indentation size effect. Whereas, this scaling law tends to get weakened at high temperatures due to the increase of $M(T)$ with T. As a comparison, $\rho_S(T)$ is independent with h but only increases with T. Moreover, one should note that $\rho_G(T)$ is generally more than one order of magnitude higher than $\rho_S(T)$ at the shallow indentation region, indicating the dominant dislocation hardening mechanism originates from the contribution of GNDs.

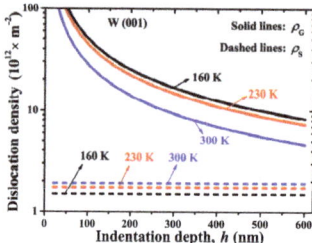

Figure 5. (Color online) Comparison of the $\rho_G - h$ and $\rho_S - h$ relationships at various temperatures for single crystal W.

4. Conclusions

In this work, a mechanistic model is proposed for the hardness-depth relationships of single crystals with temperature effect. Fundamental hardening mechanisms, including the lattice friction and network dislocation interaction, are considered in the hardness model. Four sets of experimental data are applied to verify the rationality and accuracy of the proposed model, and a reasonable agreement is achieved. Moreover, it is realized that the moderated indentation size effect at elevated temperatures is ascribed to the accelerated expansion of the plasticity affected region, which results in the decrease of the density of GNDs.

Author Contributions: Conceptualization, H.L.; Writing—Original draft preparation, H.L.; Writing—Review and editing, L.Y. and X.X.; Project administration, X.X.; Funding acquisition, H.L. and X.X. All authors have read and agreed to the published version of the manuscript.

Funding: This work is supported by the National Nature Science foundation of China (NSFC) under Contract No. 11802344, 11872379 and 11805061, and Natural Science Foundation of Hunan Province, China (Grant No. 2019JJ50809 and 2019JJ50072). H.L. thanks the Fundamental Research Funds for the Central Universities.

Acknowledgments: The author acknowledges Terentyev for the useful discussion.

Conflicts of Interest: The authors declare no conflict of interest. The funders had no role in the design of the study; in the collection, analyses, or interpretation of data; in the writing of the manuscript, or in the decision to publish the results.

References

1. Choi, I.C.; Brandl, C.; Schwaiger, R. Thermally activated dislocation plasticity in body-centered cubic chromium studied by high-temperature nanoindentation. *Acta Mater.* **2017**, *140*, 107–115. [CrossRef]
2. Chavoshi, S.Z.; Xu, S. Temperature-dependent nanoindentation response of materials. *MRS Commun.* **2018**, *8*, 15–28. [CrossRef]
3. Wheeler, J.M.; Armstrong, D.E.J.; Heinz, W.; Schwaiger, R. High temperature nanoindentation: The state of the art and future challenges. *Curr. Opin. Solid State Mater. Sci.* **2015**, *19*, 354–366. [CrossRef]
4. Duan, Z.C.; Hodge, A.M. High-temperature Nanoindentation: New Developments and Ongoing Challenges. *Jom* **2009**, *61*, 32–36. [CrossRef]
5. Schuh, C.A.; Mason, J.K.; Lund, A.C. Quantitative insight into dislocation nucleation from high-temperature nanoindentation experiments. *Nat. Mater.* **2005**, *4*, 617–621. [CrossRef]
6. Sawant, A.; Tin, S. High temperature nanoindentation of a Re-bearing single crystal Ni-base superalloy. *Scr. Mater.* **2008**, *58*, 275–278. [CrossRef]
7. Harris, A.J.; Beake, B.D.; Armstrong, D.E.J.; Davies, M.I. Development of High Temperature Nanoindentation Methodology and its Application in the Nanoindentation of Polycrystalline Tungsten in Vacuum to 950 A degrees C. *Exp. Mech.* **2017**, *57*, 1115–1126. [CrossRef]
8. Beake, B.D.; Harris, A.J.; Moghal, J.; Armstrong, D.E.J. Temperature dependence of strain rate sensitivity, indentation size effects and pile-up in polycrystalline tungsten from 25 to 950 degrees C. *Mater. Des.* **2018**, *156*, 278–286. [CrossRef]
9. Zhang, Y.; Mohanty, D.P.; Seiler, P.; Siegmund, T.; Kruzic, J.J.; Tomar, V. High temperature indentation based property measurements of IN-617. *Int. J. Plasticity* **2017**, *96*, 264–281. [CrossRef]
10. Maier, V.; Hohenwarter, A.; Pippan, R.; Kiener, D. Thermally activated deformation processes in body-centered cubic Cr-How microstructure influences strain-rate sensitivity. *Scr. Mater.* **2015**, *106*, 42–45. [CrossRef]
11. Prasitthipayong, A.; Vachhani, S.J.; Tumey, S.J.; Minor, A.M.; Hosemann, R. Indentation size effect in unirradiated and ion-irradiated 800 H steel at high temperatures. *Acta Mater.* **2018**, *144*, 896–904. [CrossRef]
12. Koch, S.; Abad, M.D.; Renhart, S.; Antrekowitsch, H.; Hosemann, P. A high temperature nanoindentation study of Al-Cu wrought alloy. *Mater. Sci. Eng. Struct. Mater. Prop. Microstruct. Process.* **2015**, *644*, 218–224. [CrossRef]
13. Gibson, J.S.K.L.; Schroeders, S.; Zehnder, C.; Korte-Kerzel, S. On extracting mechanical properties from nanoindentation at temperatures up to 1000 degrees C. *Extreme Mech. Lett.* **2017**, *17*, 43–49. [CrossRef]
14. Hangen, U.; Chen, C.L.; Richter, A. Mechanical Characterization of PM2000 Oxide-Dispersion-Strengthened Alloy by High Temperature Nanoindentation. *Adv. Eng. Mater.* **2015**, *17*, 1683–1690. [CrossRef]
15. Javaid, F.; Johanns, K.E.; Patterson, E.A.; Durst, K. Temperature dependence of indentation size effect, dislocation pile-ups, and lattice friction in (001) strontium titanate. *J. Am. Ceram. Soc.* **2018**, *101*, 356–364. [CrossRef]
16. Lee, S.W.; Meza, L.; Greer, J.R. Cryogenic nanoindentation size effect in 001-oriented face-centered cubic and body-centered cubic single crystals. *Appl. Phys. Lett.* **2013**, *103*. [CrossRef]
17. Maughan, M.R.; Leonard, A.A.; Stauffer, D.D.; Bahr, D.F. The effects of intrinsic properties and defect structures on the indentation size effect in metals. *Philos. Mag.* **2017**, *97*, 1902–1920. [CrossRef]
18. Nix, W.D.; Gao, H. Indentation size effects in crystalline materials: A law for strain gradient plasticity. *J. Mech. Phys. Solids* **1998**, *46*, 411–425. [CrossRef]
19. Karsten, D.; Bjom, B.; Mathias, G. Indentation size effect in metallic materials: Correcting for the size of the plastic zone. *Scr. Mater.* **2005**, *52*, 1093–1097.
20. Qiu, X.; Huang, Y.; Nix, W.D.; Hwang, K.C.; Gao, H. Effect of intrinsic lattice resistance in strain gradient plasticity. *Acta Mater.* **2001**, *49*, 3949–3958. [CrossRef]
21. Sadrabadi, P.; Durst, K.; Goeken, M. Study on the indentation size effect in CaF2: Dislocation structure and hardness. *Acta Mater.* **2009**, *57*, 1281–1289. [CrossRef]
22. Lim, H.; Battaile, C.C.; Carroll, J.D.; Boyce, B.L.; Weinberger, C.R. A physically based model of temperature and strain rate dependent yield in BCC metals: Implementation into crystal plasticity. *J. Mech. Phys. Solids* **2015**, *74*, 80–96. [CrossRef]

- Durst, K.; Franke, O.; Bohner, A.; Goken, M. Indentation size effect in Ni-Fe solid solutions. *Acta Mater.* **2007**, *55*, 6825–6833. [CrossRef]
- Franke, O.; Trenkle, J.C.; Schuh, C.A. Temperature dependence of the indentation size effect. *J. Mater. Res.* **2010**, *25*, 1225–1229. [CrossRef]
- Chua, J.; Zhang, R.; Chaudhari, A.; Vachhani, S.J.; Kumar, A.S.; Tu, Q.; Wang, H. High-temperature nanoindentation size effect in fluorite material. *Int. J. Mech. Sci.* **2019**, *159*, 459–466. [CrossRef]
- Xiao, X.; Song, D.; Xue, J.; Chu, H.; Duan, H. A self-consistent plasticity theory for modeling the thermo-mechanical properties of irradiated FCC metallic polycrystals. *J. Mech. Phys. Solids* **2015**, *78*, 1–16. [CrossRef]
- Lu, C.; Deng, G.Y.; Tieu, A.K.; Su, L.H.; Zhu, H.T.; Liu, X.H. Crystal plasticity modeling of texture evolution and heterogeneity in equal channel angular pressing of aluminum single crystal. *Acta Mater.* **2011**, *59*, 3581–3592. [CrossRef]
- Kiner, D.; Hosemann, P.; Maloy, S.A.; Minor, A.M. In situ nanocompression testing of irradiated copper. *Nat. Mater.* **2011**, *10*, 608–613. [CrossRef]
- Terentyev, D.; Xiao, X.; Dubinko, A.; Bakaeva, A.; Duan, H. Dislocation-mediated strain hardening in tungsten: Thermo-mechanical plasticity theory and experimental validation. *J. Mech. Phys. Solids* **2015**, *85*, 1–15. [CrossRef]
- Brunner, D.; Glebovsky, V. The plastic properties of high-purity W single crystals. *Mater. Lett.* **2000**, *42*, 290–296. [CrossRef]
- Gurson, A.L. Continuum Theory of Ductile Rupture by Void Nucleation and Growth: Part I-Yield Criteria and Flow Rules for Porous Ductile Media. *J. Eng. Mater. Technol.* **1977**, *99*, 2–15. [CrossRef]
- David, T. A simple theory of static and dynamic hardness. *Proc. R. Soc. A* **1948**, *192*, 247–274. [CrossRef]
- Oliver, W.C.; Pharr, G.M. An improved technique for determining hardness and elastic modulus using load and displacement sensing indentation experiments. *J. Mater. Res.* **1992**, *7*, 1564–1583. [CrossRef]
- Mekala, S.R. Analysis of Creep Transients in Calcium Fluoride Single Crystals following Stress Changes. Ph.D. Thesis, Friedrich-Alexander-University Erlangen-Nurnberg, Erlangen and Nuremberg, Bavaria, Germany, 2006.
- Xiao, X.; Terentyev, D.; Ruiz, A.; Zinovev, A.; Bakaev, A.; Zhurkin, E.E. High temperature nano-indentation of tungsten: Modelling and experimental validation. *Mater. Sci. Eng. Struct. Mater. Prop. Microstruct. Process.* **2019**, *743*, 106–113. [CrossRef]
- Christodoulou, N.; Jonas, J.J. Work hardening and rate sensitivity material coefficients for OFHC Cu and 99.99% Al. *Acta Metall.* **1984**, *32*, 1655–1668. [CrossRef]
- Durst, K.; Maier, V. Dynamic nanoindentation testing for studying thermally activated processes from single to nanocrystalline metals. *Curr. Opin. Solid State Mater. Sci.* **2015**, *19*, 340–353. [CrossRef]
- Lowrie, R.; Gonas, A.M. Single-Crystal Elastic Properties of Tungsten from 24 to 1800 C. *J. Appl. Phys.* **1967**, *38*, 4505. [CrossRef]

© 2020 by the authors. Licensee MDPI, Basel, Switzerland. This article is an open access article distributed under the terms and conditions of the Creative Commons Attribution (CC BY) license (http://creativecommons.org/licenses/by/4.0/).

Article

Determination of Long-Range Internal Stresses in Cyclically Deformed Copper Single Crystals Using Convergent Beam Electron Diffraction

Roya Ermagan [1], Maxime Sauzay [2], Matthew H. Mecklenburg [3] and Michael E. Kassner [1,*]

[1] Mork Family Department of Chemical Engineering and Materials Science, University of Southern California, Los Angeles, CA 90089, USA; ermagan@usc.edu
[2] CEA Paris-Saclay, DMN-SRMA, Bât. 455, 91191 Gif-sur-Yvette CEDEX, France; maxime.sauzay@yahoo.fr
[3] Core Center of Excellence in Nano Imaging (CNI), University of Southern California, Los Angeles, CA 90089, USA; matthew.mecklenburg@usc.edu
* Correspondence: kassner@usc.edu; Tel.: +1-213-740-0942

Received: 2 November 2020; Accepted: 21 November 2020; Published: 24 November 2020

Abstract: Understanding long range internal stresses (LRIS) may be crucial for elucidating the basis of the Bauschinger effect, plastic deformation in fatigued metals, and plastic deformation in general. Few studies have evaluated LRIS using convergent beam electron diffraction (CBED) in cyclically deformed single crystals oriented in single slip and there are no such studies carried out on cyclically deformed single crystals in multiple slip. In our earlier and recent study, we assessed the LRIS in a cyclically deformed copper single crystal in multiple slip via measuring the maximum dislocation dipole heights. Nearly equal maximum dipole heights in the high dislocation density walls and low dislocation density channels suggested a uniform stress state across the labyrinth microstructure. Here, we evaluate the LRIS by determining the lattice parameter in the channels and walls of the labyrinth dislocation structure using CBED. Findings of this work show that lattice parameters obtained were almost equal near the walls and within the channels. Thus, a homogenous stress state within the heterogeneous dislocation microstructure is again suggested. Although the changes in the lattice parameter in the channels are minimal (less than 10^{-4} nm), CBED chi-squared analysis suggests that the difference between the lattice parameter values of the cyclically deformed and unstrained copper are slightly higher in the proximity of the walls in comparison with the channel interior. These values are less than 6.5% of the applied stress. It can be concluded that the dominant characteristics of the Bauschinger effect may need to include the Orowan-Sleeswyk mechanism type of explanation since both the maximum dipole height measurements and the lattice parameter assessment through CBED analysis suggest a homogenous stress state. This work complements our earlier work that determined LRIS based on dipole heights by assessing LRIS through a different methodology, carried out on a cyclically deformed copper single crystal oriented for multiple slip.

Keywords: fatigue; cyclic deformation; internal stress; copper single crystal; dislocations

1. Introduction

The concept of LRIS in metals refers to the deviation of local stress from the applied stress that occurs over relatively long length scales such as that of the spacing of dislocation heterogeneities (e.g. labyrinth microstructure [1], persistent slip bands (PSB) walls, cell walls, subgrain boundaries, etc). LRIS may have been initially discussed in connection with the Bauschinger effect [2]. Understanding LRIS is essential for characterizing the Bauschinger effect and plastic deformation in cyclically deformed metals [3,4]. Metals strain harden during plastic deformation and upon reversal of the applied stress, yielding occurs at a much lower (absolute value) stress than if the material continued monotonic deformation. This effect is referred to as the Bauschinger effect and is contrary to what is expected based on isotropic hardening. When a metal is cyclically deformed, the lost strength due to the Bauschinger effect occurs with each reversal of the applied stress. This results in low hardening rates and saturation stresses compared to the (monotonic) fracture stress [3]. Different theories have been proposed for rationalizing the Bauschinger effect [5–8]. The two prominent theories are Mughrabi's composite model [5], which proposed relatively high values of LRIS, and the Orowan-type mechanism [6], which involves no internal stresses.

The Mughrabi composite model presented the heterogenous dislocation microstructure as hard (high dislocation density walls) and soft (low dislocation density channels or cell interiors) sections. In the forward direction of deformation (tension), the stress is positive in both the walls and channels (hard and soft regions). However, in the reverse direction of deformation (compression), while the stress in the walls are still positive, these regions can place the cell interiors, PSB channels, etc. in compression. Thus, the Bauschinger effect, which is the occurrence of yielding at lower stresses, is observed in this simplified model [2].

The Orowan-type mechanism, which was discussed in more details by Brown in [9,10], suggests that mobile dislocations in the forward direction of straining encounter an increasing lineal density of obstacles (forest dislocations or dislocation walls). However, in the reverse direction, there is a lower lineal density of obstacles (channels or cells interiors). Therefore, plastic strains are accumulated at a much lower stresses on reversal [6,7].

Historic LRIS assessment studies on high dislocation density heterogeneities of cyclically deformed single crystals oriented for single slip suggest internal stresses in the so-called hard phase varying from a factor of 1.0 (no LRIS) to 3, or more (larger than the applied stress). These studies were based on measurements of dislocation dipole height, dislocation loop radii, asymmetry in X-ray peaks, and lattice parameter measurements through CBED analysis [5,11–19]. In an earlier recent study by the authors [1], the maximum dipole heights were discovered to be approximately independent of location, being almost identical in the walls and channels of the labyrinth dislocation microstructure in <001> Cu single crystals oriented for multiple slip. Since the maximum dipole height strength values may be indicative of the local stresses, nearly equal maximum dipole heights in the walls and channels support a uniform stress state and low LRIS. However, the maximum value for dipole heights suggest dipole strengths that were about a factor of 2.4 higher than the applied stress based on the usual athermal equation to separate the dislocations of a dipole. Extra stress at the dipoles may be provided by tripoles or small dislocation pile-ups [20]. A nearly homogenous stress distribution with only small internal stresses were suggested by the authors in an earlier study [1] to be present based on the maximum dipole separation stress values. This is consistent with the observation of uniform dipole height across the heterogeneous dislocation microstructure. Other studies reported similar behaviors, observing homogenous dipole heights and higher dipole separation stresses for cyclically deformed metal single crystals oriented in single slip (except aluminum, which has a similar dipole separation stress to the applied stress but homogenous dipole heights as usual) [11–13]. It should be noted that accounting on either the anisotropy of cubic elasticity or the finite elongation ratios of dipoles did not allow the authors to explain the aluminum specificity. As stated earlier, since the maximum dipole heights are the upper limit of stable dipoles under the imposed local stress, they can predict the local stresses in cyclically deformed materials. The local stresses may be more accurately measured by determining the lattice parameters using convergent beam electron diffraction. This method involves using a small convergent electron probe to generate a diffraction pattern containing the higher order Laue zone (HOLZ) lines that are very sensitive to small elastic distortions in the lattice.

There have been few studies on internal stress assessment using the CBED technique in creep deformed and fatigued polycrystals and single crystals oriented in single slip. Such studies on cyclically deformed single crystals oriented in multiple slip are missing in the literature. Straub et al. [21] and Maier et al. [22] examined internal stresses using CBED analysis in polycrystalline copper specimens experiencing either creep or cyclic deformation. They did not quantify the internal stresses but suggested that internal stresses exist. It should be noted that both of these studies used kinematical simulations for deriving the position of the HOLZ lines, but dynamical effects may be important. Kassner et al. observed a homogenous stress distribution with no internal stresses in an unloaded monotonically (creep deformed) polycrystalline copper using CBED analysis [23]. In another CBED study by Kassner et al., an absence of internal stresses in creep deformed aluminum single crystal was reported [24]. In the most recent study by Kassner et al. on a cyclically deformed copper single crystal oriented for single slip at ambient temperature, lattice parameter measurements in the channels and close to the vein bundles showed no evidence of LRIS. The uncertainty of these measurements was ±30 MPa, which is 80% of their applied stress [25]. Furthermore, it was not determined whether relaxation occurred leading to a reduction in the LRIS. Legros et al. [26] assessed the internal stresses in a cyclically deformed silicon single crystal oriented for single slip at 1078 K using chi-squared analysis on CBED patterns. Chi-squared is the typical refinement method for producing the best match between the simulated and experimental CBED patterns [27]. Legros et al. suggested small internal stresses closer to the dislocation wall (7 MPa or about 14% of their applied stress) and negligible internal stresses within the cell interior exist in the cyclically deformed silicon single crystal [26]. This is basically consistent with the earlier work by the authors of this paper on the structures without PSBs. Again, all of the studies in the literature that use CBED for strain measurements were performed on unloaded material, and, of course, examined on the thin regions of the foil. Thus, LRIS relaxation is possible.

In this study, we evaluated the lattice parameters using CBED in the channels and close to the walls of the labyrinth microstructure of a cyclically deformed copper <001> single crystal oriented in multiple slip which complements the dipole study of our earlier work [1].

2. Materials and Methods

Copper single crystals of 99.999% purity oriented in [001], were cyclically deformed in tension/compression at room temperature to 157 cycles at a strain amplitude of 4.0×10^{-3} and a strain rate of 2×10^{-3} s^{-1}. This single crystal copper was cyclically deformed in multiple slip with 8 active slip systems of identical Schmid factor of 0.408. Figure 1a illustrates the stress versus strain behavior of the cyclically deformed specimen. Figure 1b illustrates that the copper single crystal was fatigued to saturation to an axial stress of 275 MPa (a resolved shear stress of 112 MPa in the <110> direction on a {111} plane). A deceleration in the cyclic hardening rate up to the maximum peak stress at the 108th cycle followed by a very slow softening until the 157th cycle was observed. The specimens were stored in a liquid nitrogen container to suppress recovery and recrystallization subsequent to deformation, which has been observed in other high purity metals [28,29]. TEM disks from the (100), (001), and (010) planes were prepared using conventional jet electropolishing with a Fischione twin jet (Fischione Instruments Inc., Export, PA, USA). Foil preparation details can be found in earlier publication by the authors [1]. Figure 1a shows that the macroscopic back stress is about one half of the maximum stress, independent from the number of cycles (a few cycles: no labyrinth structure versus about 80 cycles close to the quasi-saturation stage, considered as the labyrinth domain).

CBED studies were done on both the cyclically deformed copper and a 99.999% pure unstrained copper using the JEOL JEM-2100F transmission electron microscope (TEM) (JEOL Inc., Peabody, MA, USA) at the University of Southern California at an accelerating voltage of 200 kV and a beam diameter of about 40 nm. In order to obtain the lattice parameter from CBED patterns, comparisons between the experimental CBED patterns with simulated ones were made [27]. In this study the EMSOFT codes that consider a dynamical behavior of the electrons within the specimen were used for CBED patterns simulations [30].

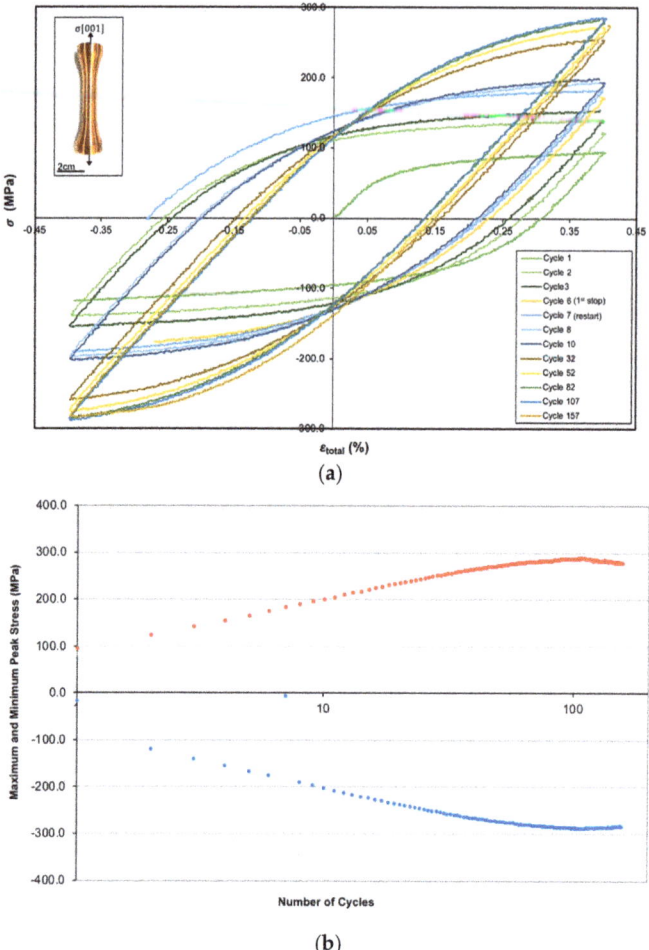

Figure 1. (a) The cyclic deformation of [001]-oriented copper single crystal at 298 K. Strains are plastic and elastic. (b) The evolution of the maximum and minimum peak stress versus the number of cycles confirms saturation is reached.

Figure 2 illustrates the 3-dimensional microstructure based on TEM images taken from the (100), (010), and (001) planes of the cyclically deformed copper [1]. The stress axis is parallel to the vertical [001] direction. The heterogeneous labyrinth dislocation microstructure consisting of orthogonal high dislocation density walls and low dislocation density channels is observed. Dislocation density in the walls and channels of the labyrinth structure was 8.6×10^{14} m/m^3 and 1.55×10^{13} m/m^3 respectively [1].

Figure 3 illustrates TEM micrographs of the labyrinth structure viewing from the [010] direction. All the dipole height measurements [1] and CBED analysis were performed on the labyrinth structure on the (010) planes.

Figure 2. A transmission electron microscope "cube" based on images taken from the (100), (010), and (001) planes of a copper specimen cyclically deformed to saturation as shown in Figure 1a. The stress axis is parallel to the [001] direction (Taken from Ref. [1]).

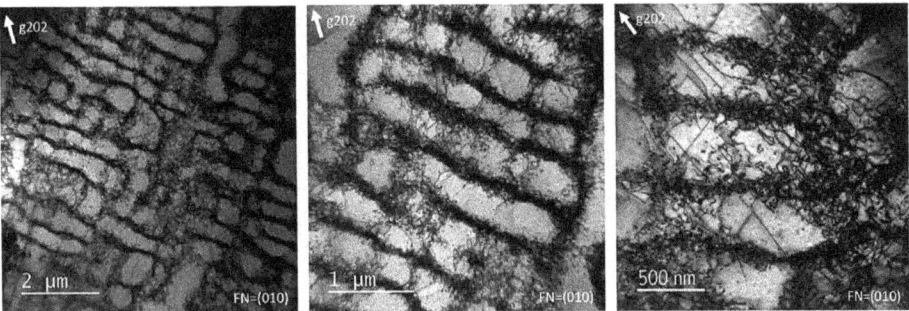

Figure 3. Transition electron microscope (TEM) micrographs of the labyrinth structure from the {010} planes of the cyclically deformed copper.

3. Results and Discussion

Strain determination is based on the shifts of the HOLZ lines in the strained specimen relative to the unstrained pattern. Specific orientations are more sensitive to the changes in the strain state. The <411> zone axis has been shown to be highly sensitive to changes in strain of face centered cubic (FCC) crystals [31]. The <411> CBED patterns were acquired in small volumes of the cyclically deformed copper very close to and remote from a dislocation heterogeneity (dislocation walls) and in the direction of applied stress ([001]). A channel with five locations where a CBED pattern was acquired is illustrated in Figure 4. The closest a CBED pattern could be acquired from a dislocation heterogeneity was approximately 30 nm. Below 30 nm, the dislocation tangles within the walls are too close to the electron probe, causing perturbations within the CBED pattern.

Lattice parameter determination was performed using the normalized distance between different HOLZ line intersections. Normalization by using the ratios of the distance between different intersections was used to adjust for differences in magnification. The comparison of simulated and experimental patterns is achieved by using the chi-squared equation, which is the typical refinement method for producing the best match between the simulated and experimental patterns [27]. The chi-squared equation is defined as:

$$\chi^2 = \sum_i^N \frac{1}{d_{is}}(d_{is} - d_{ix})^2 \tag{1}$$

where N is the number of data points, d_s is the normalized distance between two intersections in the simulated pattern, and d_x is the normalized distance between the same intersections of the experimental pattern. As stated earlier, the CBED pattern of an unstrained copper single crystal has also been recorded and compared with the simulated patterns to determine the observed lattice parameter of the undeformed copper. Consequently, the strain can be evaluated by comparing the lattice parameter of the cyclically deformed copper with the unstrained value. The strain was converted into stress using the elastic modulus along the [001] direction E [001] = 66.6 ± 0.5 GPa [32], accounting for the cubic elasticity anisotropy of copper.

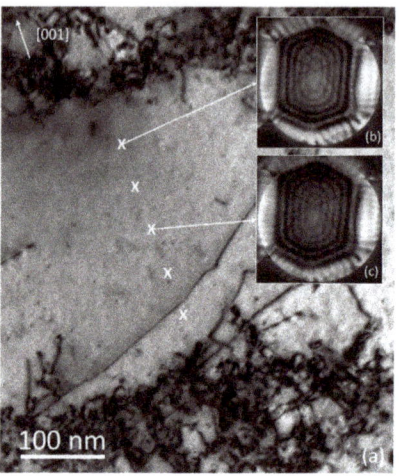

Figure 4. Convergent beam electron diffraction (CBED) patterns were recorded across channels to assess the internal stresses in the direction of applied stress [001]. (**a**) A channel illustrating the locations where the CBED patterns were acquired. (**b**) The <411> CBED pattern that was recorded closer to the wall and (**c**) in the middle of the channel.

Figure 5 shows the lattice parameter measurements and stress calculations corresponding to different positions within the channel. The data are obtained from four channels in two TEM foils. The horizontal axis shows the distance from the walls normalized by the channel width. The average channel width is 0.36 µm and the wall width is about 0.12 µm. Minor scatterings exist in the lattice parameters of different channels that is ±2 × 10^{-4} nm. Identical values of lattice parameters in each channel show that the internal stresses are homogenous in the channel and close to the walls of the labyrinth dislocation microstructure. Comparing the lattice parameter of cyclically deformed copper single crystal with that of unstrained copper indicates that the internal stresses are minimal. Considering Brown's note on permanent softening, this weak backstress is in line with the fact that the measured cyclic softening in our test was weak [9,10]. Of course, it is possible that internal stresses exist and are less than the measurement error. The accuracy of lattice parameter measurements is about ±1 × 10^{-4} nm. The error in stress measurements is then approximately ±18 MPa that is 6.5% of the applied axial stress of 275 MPa.

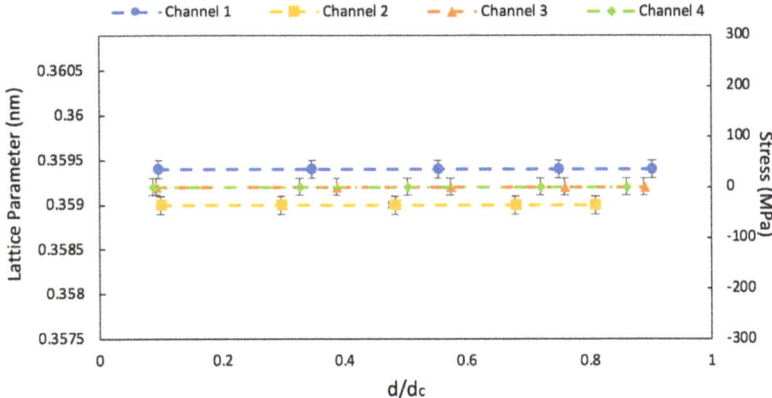

Figure 5. The obtained lattice parameters and corresponding axial stress values in different locations within a dislocation channel. The lattice parameter of an unloaded copper was found to be 0.3592 nm.

Following the Legros et al. [26] method of chi-squared analysis, an effort was made to assess the local changes of the lattice parameter within a single channel. Since the lattice parameter of the unstrained copper was observed to be 0.3592 nm, channel four with the lattice parameter of 0.3592 nm in the cyclically deformed copper was chosen for this analysis. HOLZ lines of simulated CBED patterns corresponding to the lattice parameter of 0.35920 nm were considered as the reference point for chi-squared analysis. The data illustrated in Figure 6 is a chi-squared fit between the aforementioned HOLZ lines intersection ratios for cyclically deformed (yellow) and undeformed (red) copper. Chi-squared analysis can "refine" the strain measurement below 10^{-4} and provide increased resolution of the elastic strain, although precise values of the strain within in the 10^{-5} range is not possible. The chi-squared results plotted in Figure 6 are somewhat qualitative. Although the changes in the lattice parameter in a channel are minimal and less than 1×10^{-4} nm, the data shown in Figure 6 can show that the difference between the lattice parameter values of the cyclically deformed copper and the unstrained copper are slightly higher in the proximity of the walls in comparison with the values in the channel interior. This result is consistent with the composite model but with much lower values of internal stresses (less than 6.5% of the applied stress). It should also be noted that the aforementioned differences in the chi-squared values might be due to the higher quality of HOLZ lines in the middle of the channels as opposed to the vicinity of the walls. Considering the two renowned theories that rationalize the Bauschinger effect (Composite model and Orowan-Sleeswyk mechanism), it appears that the dominant characteristics of the Bauschinger effect may need to include the Orowan–Sleeswyk [6] mechanism type of explanation since both the maximum dipole height measurements and the lattice parameter assessment through CBED analysis suggest a relatively homogenous stress state. As stated earlier, no internal stresses are involved in the Orowan-Sleeswyk mechanism where the Bauschinger effect is rationalized by the lower lineal density of obstacles in reverse direction of straining.

It should be noted that dislocations may eject out of the surface in the thin areas of the TEM foil. This will result in stress relaxation and can subsequently alter the values of the internal stress. This is rather challenging since thin areas of the specimen are to be used for acquiring high quality HOLZ lines in a CBED pattern. It must be emphasized that these relaxations may be negligible as the labyrinth pattern with similar characteristics such as dislocation density, channel, and wall width were observed both in the thin regions as well as thicker areas. The dislocation density in the thinner regions (approximately 130 ± 10 nm) where CBED patterns were recorded was 8.2×10^{14} m/m^3 in the walls and 1.8×10^{13} m/m^3 in the channels. This is very close to the wall dislocation density of 8.6×10^{14} m/m^3 and channel dislocation density of 1.5×10^{13} m/m^3 in relatively thicker regions (approximately 0.23 µm) where dislocation densities were measured [1]. Although copper has a fairly low stacking fault energy of 60 mJ/m² and the labyrinth microstructure characteristics including dislocation densities are consistent in the thick and thin regions, relaxations caused by dislocations ejecting the thin regions of the foil cannot be completely neglected.

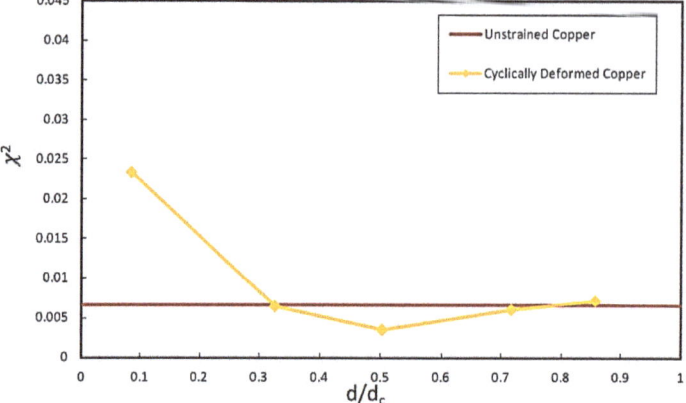

Figure 6. Change in the [411] CBED pattern higher order Laue zone (HOLZ) lines of cyclically deformed copper in a channel (and near the walls) and undeformed copper relative to simulated pattern with lattice parameter of 0.35920 nm. The data is shown as a χ^2 analysis fit between the HOLZ lines of the aforementioned CBED patterns.

As stated earlier, there have been few studies on internal stress assessment using the CBED technique in creep deformed and fatigued polycrystals and single crystals oriented in single slip. Such studies on cyclically deformed single crystals oriented in multiple slip are missing in the literature. The current study shows a uniform stress state within the crystal since the lattice parameters are almost identical near the dislocation walls and within the channel. This is similar to the findings of earlier studies on fatigued copper and silicon oriented for single slip and creep deformed aluminum and copper [21–26]. All of the deformation conditions of the current study along with other studies that used CBED to assess LRIS are summarized in Table 1.

Table 1. Summary of studies on creep and cyclically deformed materials that utilized CBED to assess long range internal stresses.

CBED Studies	This Study	Kassner [25]	Legros [26]	Kassner [24]	Kassner [23]		Straub [21]	Maier [22]
Material	Copper [001] Single Crystal	Copper [123] Single Crystal	Silicon [231] Single Crystal	Aluminum Single Crystal	Copper Polycrystal		Copper Polycrystal	Copper Polycrystal
Deformation Type	Cyclic	Cyclic	Cyclic	Creep Steady State	Creep Steady State		Creep	Cyclic
Applied Stress (MPa)	112 (Shear) Saturation	19 (Shear) Presaturation	49 (Shear) Presaturation	7 (Axial)	20 (Axial)	40	162 (Axial)	144 (Axial)
Strain	0.40%	0.125% Plastic only	6×10^{-4} Plastic only	-	0.04	0.05	3.6	5×10^{-3} Plastic only
Strain Rate (s⁻¹)	2×10^{-3}	2.5×10^{-3}	3×10^{-4}	-	2.5×10^{-5}	5.6×10^{-7}	1.9×10^{-3}	-
Slip	Polyslip	Single Slip	Single Slip	-	-		-	-
Number of Cycles	157	200	-	-	-		-	2000
Temperature (K)	293	293	1078	663	823		573	293
LRIS	None	None	None	None	None		Observed	Observed

4. Conclusions

The lattice parameter assessment through convergent beam electron diffraction patterns recorded in the channels and close to the walls of the labyrinth dislocation structure suggest very low long range internal stresses (LRIS). Our earlier dipole height work on the same cyclically deformed copper also suggests low internal stresses. The stress to separate dipoles with the widest height was independent of the location. Therefore, the results of the current CBED analysis study is consistent with our earlier maximum dipole height measurement work.

Minimal changes (less than 10^{-4} nm) were observed in the lattice parameters recorded throughout a single channel. These values are less than 6.5% of the applied stress. Hence, negligible internal stresses in the channel interior and near the dislocation walls were observed. The Kassner et al. X-ray synchrotron study on monotonically deformed (to 30% strain at ambient temperature) copper single crystals [32] suggest that long range internal stresses were nearly 10% of the applied stress. Thus, the outcome of the present cyclic deformation study is consistent with the earlier monotonic deformation work. Although the changes in the lattice parameter in a channel are very minimal (less than 10^{-4} nm), chi-squared analysis suggest that the difference between the lattice parameter values of the cyclically deformed copper and the unstrained copper are slightly higher in the proximity of the walls in comparison with the channel interior. These internal stresses are less than 6.5% of the applied stress. This is consistent with the composite model originally suggested by Mughrabi but with perhaps lower values of internal stresses. Therefore, it appears that a low proportion of the Bauschinger effect may be influenced by the existence of long range internal stresses. The dominant feature of the Bauschinger effect may include the Orowan–Sleeswyk [6] mechanism type of explanation, since both the maximum dipole height measurements and the lattice parameter assessment through CBED analysis suggest a relatively homogenous stress state within the heterogeneous dislocation microstructure.

Author Contributions: M.E.K. conceived of and designed the project; R.E., M.S., and M.E.K. contributed to theoretical analysis and conceptualization; M.S. prepared the fatigued samples and was involved in fundamental discussions with M.E.K.; R.E. prepared the TEM samples; R.E. and M.H.M. performed TEM studies and data analysis; Literature searches were performed by R.E.; Manuscript preparation and proof-reading were performed by R.E., M.E.K., M.S., and M.H.M. All authors have read and agreed to the published version of the manuscript.

Funding: This research was funded by the National Science Foundation under grant DMR-1401194.

Acknowledgments: The authors greatly appreciate the assistance of Professor Marc Legros and Tannaz Sattari Tabrizi.

Conflicts of Interest: The authors declare no conflict of interest.

References

1. Ermagan, R.; Sauzay, M.; Kassner, M.E. Assessment of Internal Stresses Using Dislocation Dipole Heights in Cyclically Deformed [001] Copper Single Crystals. *Metals* **2020**, *10*, 512. [CrossRef]
2. Mughrabi, H. Dislocation Clustering and Long-Range Internal Stresses in Monotonically and Cyclically Deformed Metal Crystals. *Rev. Phys. Appliquée* **1988**, *23*, 367–379. [CrossRef]
3. Levine, L.E.; Stoudt, M.R.; Creuziger, A.; Phan, T.Q.; Xu, R.; Kassner, M.E. Basis for the Bauschinger Effect in Copper Single Crystals: Changes in the Long-Range Internal Stress with Reverse Deformation. *J. Mater. Sci.* **2019**, *54*, 6579–6585. [CrossRef]
4. Kassner, M.E.; Geantil, P.; Levine, L.E. Long Range Internal Stresses in Single-Phase Crystalline Materials. *Int. J. Plast.* **2013**, *45*, 44–60. [CrossRef]
5. Mughrabi, H. Dislocation Wall and Cell Structures and Long-Range Internal Stresses in Deformed Metal Crystals. *Acta Metall.* **1983**, *31*, 1367–1379. [CrossRef]
6. Orowan, E. Causes and Effects of Internal Stresses. In *Internal Stresses and Fatigue in Metals*; General Motors; Elsevier: Amsterdam, The Netherlands, 1959; pp. 59–80, General Motors.
7. Sleeswyk, A.W.; James, M.R.; Plantinga, D.H.; Maathuis, W.S.T. Reversible Strain in Cyclic Plastic Deformation. *Acta Metall.* **1978**, *26*, 1265–1271. [CrossRef]

8. Seeger, A.; Diehl, J.; Mader, S.; Rebstock, H. Work-Hardening and Work-Softening of Face-Centred Cubic Metal Crystals. *Philos. Mag. J. Theor. Exp. Appl. Phys.* **1957**, *2*, 323–350. [CrossRef]
9. Brown, L.M. Orowan's Explanation of the Bauschinger Effect. *Scr. Metall.* **1977**, *11*, 127–131. [CrossRef]
10. Pedersen, O.B.; Brown, L.M. Equivalence of Stress and Energy Calculations of Mean Stress. *Acta Metall.* **1977**, *25*, 1303–1305. [CrossRef]
11. Tippelt, B.; Breitschneider, J.; Hähner, P. The Dislocation Microstructure of Cyclically Deformed Nickel Single Crystals at Different Temperatures. *Phys. Status Solidi A* **1997**, *163*, 11–26. [CrossRef]
12. Kassner, M.E.; Wall, M.A. Microstructure and Mechanisms of Cyclic Deformation in Aluminum Single Crystals at 77 K: Part II. Edge Dislocation Dipole Heights. *Metall. Mater. Trans. A* **1999**, *30*, 777–779. [CrossRef]
13. Kassner, M.E.; Wall, M.A.; Delos-Reyes, M.A. Primary and Secondary Dislocation Dipole Heights in Cyclically Deformed Copper Single Crystals. *Mater. Sci. Eng. A* **2001**, *317*, 28–31. [CrossRef]
14. Paterson, M.S. X-Ray Line Broadening from Metals Deformed at Low Temperatures. *Acta Metall.* **1954**, *2*, 823–830. [CrossRef]
15. Ungar, T.; Mughrabi, H.; Rönnpagel, D.; Wilkens, M. X-Ray Line-Broadening Study of the Dislocation Cell Structure in Deformed [001]-Orientated Copper Single Crystals. *Acta Metall.* **1984**, *32*, 333–342. [CrossRef]
16. Mughrabi, H. Self-Consistent Experimental Determination of the Dislocation Line Tension and Long-Range Internal Stresses in Deformed Copper Crystals by Analysis of Dislocation Curvatures. *Mater. Sci. Eng. A* **2001**, *309–310*, 237–245. [CrossRef]
17. Brown, L.M. On the Shape of a Dislocation Shear Loop in Stable Equilibrium. *Philos. Mag. Lett.* **2001**, *81*, 617–621. [CrossRef]
18. Brown, L.M. Dislocation Bowing and Passing in Persistent Slip Bands. *Philos. Mag.* **2006**, *86*, 4055–4068. [CrossRef]
19. Kubin, L.; Sauzay, M. Persistent Slip Bands: The Bowing and Passing Model Revisited. *Acta Mater.* **2017**, *132*, 517–524. [CrossRef]
20. Neumann, P.D. The Interactions between Dislocations and Dislocation Dipoles. *Acta Metall.* **1971**, *19*, 1233–1241. [CrossRef]
21. Straub, S.; Blum, W.; Maier, H.J.; Ungar, T.; Borbély, A.; Renner, H. Long-Range Internal Stresses in Cell and Subgrain Structures of Copper during Deformation at Constant Stress. *Acta Mater.* **1996**, *44*, 4337–4350. [CrossRef]
22. Maier, H.J.; Renner, H.; Mughrabi, H. Local Lattice Parameter Measurements in Cyclically Deformed Copper by Convergent-Beam Electron Diffraction. *Ultramicroscopy* **1993**, *51*, 136–145. [CrossRef]
23. Kassner, M.E.; Pérez-Prado, M.T.; Long, M.; Vecchio, K.S. Dislocation Microstructure and Internal-Stress Measurements by Convergent-Beam Electron Diffraction on Creep-Deformed Cu and Al. *Metall. Mater. Trans. A* **2002**, *33*, 311–317. [CrossRef]
24. Kassner, M.E.; Perez-Prado, M.T.; Vecchio, K.S. Internal Stress Measurements by Convergent Beam Electron Diffraction on Creep-Deformed Al Single Crystals. *Mater. Sci. Eng. A* **2001**, *319–321*, 730–734. [CrossRef]
25. Kassner, M.E.; Pérez-Prado, M.-T.; Vecchio, K.S.; Wall, M.A. Determination of Internal Stresses in Cyclically Deformed Copper Single Crystals Using Convergent-Beam Electron Diffraction and Dislocation Dipole Separation Measurements. *Acta Mater.* **2000**, *48*, 4247–4254. [CrossRef]
26. Legros, M.; Ferry, O.; Houdellier, F.; Jacques, A.; George, A. Fatigue of Single Crystalline Silicon: Mechanical Behaviour and TEM Observations. *Mater. Sci. Eng. A* **2008**, *483–484*, 353–364. [CrossRef]
27. Zuo, J.M. Automated Lattice Parameter Measurement from HOLZ Lines and Their Use for the Measurement of Oxygen Content in $YBa_2Cu_3O_{7-\delta}$ from Nanometer-Sized Region. *Ultramicroscopy* **1992**, *41*, 211–223. [CrossRef]
28. Kassner, M.E.; Pollard, J.; Evangelista, E.; Cerri, E. Restoration Mechanisms in Large-Strain Deformation of High Purity Aluminum at Ambient Temperature and the Determination of the Existence of "Steady-State". *Acta Metall. Mater.* **1994**, *42*, 3223–3230. [CrossRef]
29. Kassner, M.E.; McQueen, H.J.; Pollard, J.; Evangelista, E.; Cerri, E. Restoration Mechanisms in Large-Strain Deformation of High Purity Aluminum at Ambient Temperature. *Scr. Metall. Mater.* **1994**, *31*, 1331–1336. [CrossRef]
30. Singh, S.; Ram, F.; Graef, M.D. EMsoft: Open Source Software for Electron Diffraction/Image Simulations. *Microsc. Microanal.* **2017**, *23*, 212–213. [CrossRef]

31. Rozeveld, S.J.; Howe, J.M. Determination of Multiple Lattice Parameters from Convergent-Beam Electron Diffraction Patterns. *Ultramicroscopy* **1993**, *50*, 41–56. [CrossRef]
32. Levine, L.E.; Larson, B.C.; Yang, W.; Kassner, M.E.; Tischler, J.Z.; Delos-Reyes, M.A.; Fields, R.J.; Liu, W. X-Ray Microbeam Measurements of Individual Dislocation Cell Elastic Strains in Deformed Single-Crystal Copper. *Nat. Mater.* **2006**, *5*, 619–622. [CrossRef] [PubMed]

Publisher's Note: MDPI stays neutral with regard to jurisdictional claims in published maps and institutional affiliations.

© 2020 by the authors. Licensee MDPI, Basel, Switzerland. This article is an open access article distributed under the terms and conditions of the Creative Commons Attribution (CC BY) license (http://creativecommons.org/licenses/by/4.0/).

Article

Micromechanics of Void Nucleation and Early Growth at Incoherent Precipitates: Lattice-Trapped and Dislocation-Mediated Delamination Modes

Qian Qian Zhao [1], Brad L. Boyce [2] and Ryan B. Sills [1,*]

[1] Department of Materials Science and Engineering, Rutgers University, Piscataway, NJ 08854, USA; qz161@scarletmail.rutgers.edu
[2] Sandia National Laboratories, Albuquerque, NM 87185, USA; blboyce@sandia.gov
* Correspondence: ryan.sills@rutgers.edu

Abstract: The initial stages of debonding at hard-particle interfaces during rupture is relevant to the fracture of most structural alloys, yet details of the mechanistic process for rupture at the atomic scale are poorly understood. In this study, we employ molecular dynamics simulation of a spherical Al_2Cu θ precipitate in an aluminum matrix to examine the earliest stages of void formation and nanocrack growth at the particle-matrix interface, at temperatures ranging from 200–400 K and stresses ranging from 5.7–7.2 GPa. The simulations revealed a three-stage process involving (1) stochastic instantaneous or delayed nucleation of excess free volume at the particle-matrix interface involving only tens of atoms, followed by (2) steady time-dependent crack growth in the absence of dislocation activity, followed by (3) dramatically accelerated crack growth facilitated by crack-tip dislocation emission. While not all three stages were present for all stresses and temperatures, the second stage, termed *lattice-trapped delamination*, was consistently the rate-limiting process. This lattice-trapped delamination process was determined to be a thermally activated brittle fracture mode with an unambiguous Arrhenius activation energy of 1.37 eV and an activation area of 1.17 $Å^2$. The role of lattice-trapped delamination in the early stages of particle delamination is not only relevant at the high strain-rates and stresses associated with shock spallation, but Arrhenius extrapolation suggests that the mechanism also operates during quasi-static rupture at micrometer-scale particles.

Keywords: fracture; interfacial delamination; nucleation; void formation; cracking; alloys

1. Introduction

Void nucleation is the first step towards fracture in many different contexts, including quasi-static tearing, dynamic spall, creep rupture, irradiation creep, and wear debris generation. Voids are predominantly thought to nucleate at second phase particles, either when the particles crack or when the interface between the particle and matrix debonds [1]. Here, we focus on the earliest stages of void nucleation via particle delamination. Subsequent to nucleation, these voids grow until they induce fracture. While many studies have employed continuum models, such as finite element modeling [2–5], to evaluate the process of void nucleation, the atomistic mechanisms governing nucleation are less well studied. And yet, since void nucleation begins at the nanoscale, it is intrinsically an atomistic process [6]. Our goal in this work is to evaluate void nucleation in a model system with an incoherent, second-phase particle (θ-particle in Al) in an effort to reveal the underlying micromechanics and kinetically limiting processes.

Given its central role in fracture, continuum damage models commonly invoke void nucleation in their underlying formalisms. Perhaps the most popular approach is the porous plasticity model of Gurson, Tvergaard, and Needleman [7] which utilizes a yield

criterion that is a function of the void volume fraction, f. In this model, the void volume fraction is governed by a differential equation of the form:

$$\dot{f} = \dot{f}_{\text{nuc}} + \dot{f}_{\text{growth}}$$

where $\dot{f}_{\text{nuc}}$ and $\dot{f}_{\text{growth}}$ account for the contributions from nucleation and growth events, respectively. Ideally, the term $\dot{f}_{\text{nuc}}$ would be derived from a fundamental, micromechanics-based understanding of void nucleation. Given that this understanding is lacking, a more phenomenological approach is commonly utilized. For example, it is commonly assumed that void nucleation occurs at particles when a critical plastic strain, ε_c^P, is reached, and that ε_c^P varies from particle to particle according to a probability density function $F(\varepsilon^P/\varepsilon_c^P)$ where ε^P is the equivalent plastic strain [8,9]. The void volume fraction then increases in time as a result of void nucleation according to the expression:

$$\dot{f}_{\text{nuc}} = F(\varepsilon^P/\varepsilon_c^P)\dot{\varepsilon}^P$$

where $\dot{\varepsilon}^P$ is the equivalent plastic strain rate. Usually, $F(\varepsilon^P/\varepsilon_c^P)$ is assumed to be a normal distribution with mean $\bar{\varepsilon}_c^P$ and standard deviation s_{ε_c} [8], however there is no direct evidence which justifies this choice. Furthermore, $\bar{\varepsilon}_c^P$ and s_{ε_c} are treated as empirical parameters that are fitted against experimental data (e.g., stress-strain curves). While this and other similar phenomenological approaches have been applied pervasively across the literature, the Sandia Fracture Challenges have recently shown that these models often fare poorly when making blind fracture predictions [10–12]. This motivates a deeper look at the micromechanics of void nucleation, so that strong assumptions about what governs fracture (a critical strain?) and how the propensity for fracture varies across the population of particles (normally distributed?) can be lifted.

Unfortunately, the critical strain ε_c^P is not easily studied using micromechanical simulations because plastic strain is really a homogenized, macroscale concept; at the microscale where discrete dislocations interact with particles, plastic strain is not a very relevant concept. On the other hand, some damage mechanics models employ a critical stress σ_c at which nucleation occurs [9], which is more consistent with micromechanics modeling (the stress state can be specified in molecular dynamics, for example). Here, we argue, however, that rather than focusing on a "critical stress" at which void nucleation occurs, it makes more sense to consider how the *nucleation rate* varies with stress state, temperature, etc. In other words, void nucleation can occur over a range of stresses, with the nucleation rate increasing as the stress is increased. This view is more consistent with other works on crack nucleation, which focus on the nucleation rate [13,14]. Within this view, the critical stress is the stress at which the nucleation rate goes to infinity, meaning that nucleation occurs instantaneously. We note that the nucleation rate for a given state is likely only well-defined in an average sense, because nucleation is a stochastic phenomenon. This means that at each state, there is a distribution of nucleation rates (which could be interpreted in probabilistic terms). We argue that the possibility of "subcritical" nucleation, i.e., with $\sigma < \sigma_c$, and the statistical aspects of nucleation could be important to the overall nucleation process. For these reasons, our focus here is on the stress and temperature dependence of the void nucleation rate.

An important nuance to the study of void nucleation is deciding when exactly a void is said to "nucleate." As soon as a crack appears within the particle-matrix interface? Or after a significant fraction of the interface has delaminated? We may expect that a clear nucleation event occurs whereby a crack "suddenly" appears along the interface, allowing us to disentangle this terminological ambiguity, although the appearance of a crack is often limited by the spatial and temporal resolution of the techniques employed. In the present approach, with atomic-scale and picosecond resolution, the initial emergence of a crack is still difficult to define: we observed steady growth of an interfacial crack starting from a vacancy-sized nucleus. We were unable to determine the precise mechanism by which the vacancy-sized nucleus appeared, however. Furthermore, the appearance of the vacancy-sized nucleus did not control the kinetics of void nucleation. Instead, we found that it was the subsequent growth of the crack that governed the overall delamination (e.g., void nucleation) process. Hence, we find that it is the delamination rate, controlled by the growth of a crack, which governs the void nucleation rate. For this reason, we refer to our simulations as studying void nucleation and "early growth."

Void nucleation has been studied in perfect crystals [15–22], at grain boundaries [23,24], ahead of crack tips [25], and at second-phase particles [26–29] using molecular dynamics. In most cases, void nucleation results from interactions between several crystallographic defects, such as grain boundaries and twins/dislocations [23,24], pairs of intersecting stacking faults [21], and particles and dislocations [26,27,30]. The previous work on particle-mediated void nucleation is most relevant here. Coffman et al. [31] studied void nucleation in Si under uniaxial tension with a cubic nanograin "particle" that delaminated from the surrounding matrix. They first performed atomistic simulations to calibrate a continuum fracture model (a cohesive zone model), and then compared atomistic and continuum predictions of void nucleation. In general, they found poor agreement between the models, motivating the need for further studies of void nucleation with atomistic resolution. Pogorelko and Mayer [26–29] and Cui and Chen [30] studied void nucleation at spherical particles in a variety of material systems, considering the influence of strain rate, temperature, simulation box size, and particle volume fraction on the delamination behavior under a fixed uniaxial strain rate. In simulations with face-centered cubic (FCC) matrices [26,27,29,30], nucleation was observed to occur in two stages: first a crack nucleated at the top and bottom poles along the loading axis (similar to the behavior predicted by continuum models [2]), and then after some subsequent growth dislocations were emitted from the crack tips. On the other hand, nucleation with body-centered cubic and hexagonal close-packed (HCP) matrices seemed to initiate from defects in the matrix rather than at the particle-matrix interface [28]. While the tensile strength of these systems has been characterized extensively using these simulation results, the nucleation rate could not be estimated because of the fixed-strain-rate boundary conditions.

Our study here had two goals: (1) to assess the stress and temperature dependence of the void nucleation and early growth rate with MD and (2) identify the micromechanical processes which govern the kinetics. In contrast to previous work [26–30], we perform simulations here with a fixed stress state (and temperature), so that our results can be used to estimate the stress and temperature-dependent rates. Our findings indicate that void nucleation may be rate limited by the kinetics of crack growth processes rather than the kinetics of crack nucleation (e.g., the time it takes for a crack to appear). Furthermore, we show two distinct delamination modes with drastically different growth kinetics. Finally, we conclude that a brittle delamination mode which we term *lattice-trapped delamination* may be an important contributor to void nucleation. While the stress range employed here (5.7 to 7.2 GPa) is high relative to quasi-static loading, it is in the range where shock spallation is observed under high loading rates [32,33]. Furthermore, through a thermal activation analysis of our data we are able to extrapolate our results down to lower stress conditions. The remainder of the work is organized as follows. In Section 2 we discuss our simulation setup and analysis methods, in Section 3 we present our results, in Section 4 we discuss the implications of our findings and compare results with existing theories, and finally conclude the manuscript in Section 5.

2. Materials and Methods

As a model system, we consider void nucleation at θ-particles in an FCC Al matrix. θ is the thermodynamically stable intermetallic phase of the Al-Cu system and is commonly observed in Al-Cu-copper alloys (e.g., 2xxx series) in the overaged condition [34]. θ-particles have a composition of Al_2Cu and a body-centered tetragonal C16 crystal structure. They are incoherent with the matrix and typically adopt plate-like geometries [35]. However, for simplicity in this work we will use a spherical precipitate geometry. While there is evidence that voids may nucleate at θ-particles [36], we emphasize that we are using the θ-Al system as a model incoherent precipitate system with the goal of gaining general insight into the micromechanics and kinetics of void nucleation.

MD simulations were performed using LAMMPS [37] with the Al-Cu angular-dependent interatomic potential of Apostol and Mishin [38]. The angular-dependent potential framework is an enriched version of the embedded atom method that enables incorporation of angular-dependent interactions. These interactions enable the potential to capture the lattice constants, anisotropic elastic moduli, and formation energy of the θ-phase with reasonable accuracy. Our simulation cell geometry is shown in Figure 1a; we initially inserted an incoherent spherical particle with a radius of $R = 50$ Å into a pure Al lattice using the zero-temperature lattice parameters predicted by the potentials. The Al lattice and θ-particle are oriented so that their unit cell axes are aligned with the simulation box. The c-axis of the θ-lattice is oriented in the z-direction of the simulation cell. Periodic boundary conditions were used in all directions with a 200 Å cubic simulation cell. The sequence of each simulation is shown in Figure 1b. During the relaxation stage, we used an NPT (constant number of atoms N, pressure P, and temperature T) ensemble and simulated 2 ps at the chosen temperature and zero hydrostatic stress. Subsequently, during the ramping stage we ramped the hydrostatic stress up to the target value σ^H over a duration of 23 ps. This duration was chosen empirically; if the stress was ramped too quickly we observed "premature" void nucleation, likely because of stress spikes resulting from imperfect performance of the barostat. See Appendix A for additional information. Finally, during the holding stage, the hydrostatic stress was held constant for the duration of the simulation until the particle completely debonded from the matrix or the simulation terminated after 1-week of wall time. All simulations used a thermostat damping parameter value of 0.01 ps, barostat damping parameter value of 1 ps, and a time step size of 0.001 ps.

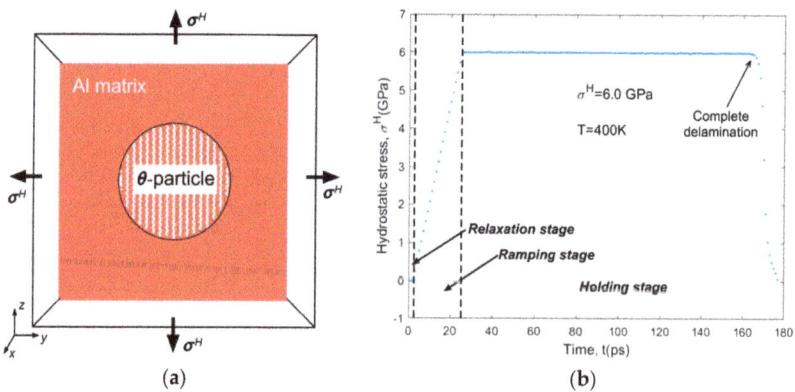

Figure 1. Simulation details. (**a**) Snapshot showing periodic simulation cell with a spherical θ-particle loaded hydrostatically; (**b**) time history of hydrostatic stress from a sample simulation at T = 400 K and σ^H = 6.0 GPa with simulation stages marked. When the particle completely delaminates the applied stress can no longer be sustained, causing the precipitous drop at the end.

We note that while the MD barostat controls the average (virial) stress state in the simulation cell, the local stress state may vary. In fact, we expect there to be variation because the elastic constants between the matrix and particle differ, i.e., this is an Eshelby inhomogeneity problem [39]. Furthermore, the particle images resulting from periodic boundary conditions will interact with each other, further complicating the stress field. These effects are quite complex, especially given the anisotropic nature of the C16 θ-phase. For simplicity, in our analysis of the data we assume that the applied hydrostatic stress σ^H

dominates the delamination behavior; our successful thermal activation analysis below justifies this assumption. We note that Pogorelko and Mayer have analyzed the spatially varying stress field near a second-phase particle under uniaxial loading and its influence on the delamination process [26,27,29].

Results from a total of 290 MD simulations are reported in this study at temperatures ranging from 200 to 400 K and stresses in the range of 5.7 to 7.2 GPa (the precise stress range differed for each temperature). In most cases, 10 simulations with different thermalization histories (initial atomic velocities) were performed at each stress-temperature condition. In many simulations, we observed nucleation of dislocations at the particle interface. To enable efficient detection of the appearance of Shockley partial dislocations in our simulation cell, we exploited the fact that atoms situated in stacking faults (e.g., produced by a Shockley partial dislocation) appear as HCP atoms when analyzed via common neighbor analysis (CNA) [40]. Hence, by simply monitoring the number of "HCP" atoms in the simulation cell N_{HCP}, we could identify when a dislocation appeared. We note that there was always a small, non-zero number of HCP atoms detected, due to thermal noise in the lattice and the imperfect detection capacity of CNA. To prevent false detection of a dislocation, we established a threshold value for the appearance of a dislocation, N_{HCP}^d, based on empirical analysis of our data. We set this threshold at $N_{HCP}^d = 40$ HCP atoms, so if $N_{HCP} > 40$ we "detected" appearance of a dislocation. Furthermore, we applied a moving average to the raw N_{HCP} vs. time data using a window of width 0.005 ps. This served to smooth out the data a bit and remove spurious spikes in N_{HCP} which did not lead to a sustained increase in N_{HCP} over time (as was expected if a dislocation had nucleated and remained in the system). This approach to dislocation detection was validated by manually analyzing several datasets in OVITO Pro [41]. All simulation snapshots were produced using OVITO Pro [41].

3. Results

We begin the Results section by discussing the overall behaviors observed in our simulations, showing that two modes of delamination were observed which we call *lattice-trapped delamination* and *dislocation-mediated delamination*.

3.1. Void Nucleation Process

In our simulations, we generally observed the sequence of events depicted in Figure 2. At the start of the load holding step there was no evidence of cracking or voiding at the particle interface. After some time, one or more clusters of atoms began exhibiting relatively large atomic volumes, indicating the nucleation of a crack with the size of a small vacancy cluster. For example, in Figure 2a we show atoms in blue whose atomic volume exceeds 30 Å^3 based on Voronoi analysis. For reference, the atomic volume of aluminum is 16.7 Å^3. The specific mechanism by which this vacancy clustered appeared could not be determined. Over time this crack grew, leading to a larger patch of atoms with volumes exceeding 30 Å^3 as shown in Figure 2b. Importantly, this crack growth was not accompanied by any dislocation activity. Instead, the crack grew steadily in time; in the Discussion we demonstrate that this delamination rate is governed by the lattice trapping phenomenon [42], and hence refer to this as *lattice-trapped delamination*. Eventually, once the crack reached a critical size, Shockley partial dislocations nucleated at the tip of the crack approximately in the plane of the crack, as shown in Figure 2c. These dislocations then rapidly glided away from the particle into the bulk and began to multiply, joined by additional dislocations nucleating from the crack tip. The crack growth rate increased rapidly upon appearance of the dislocations, leading to total delamination of the particle from the matrix. We call this process *dislocation-mediated delamination*.

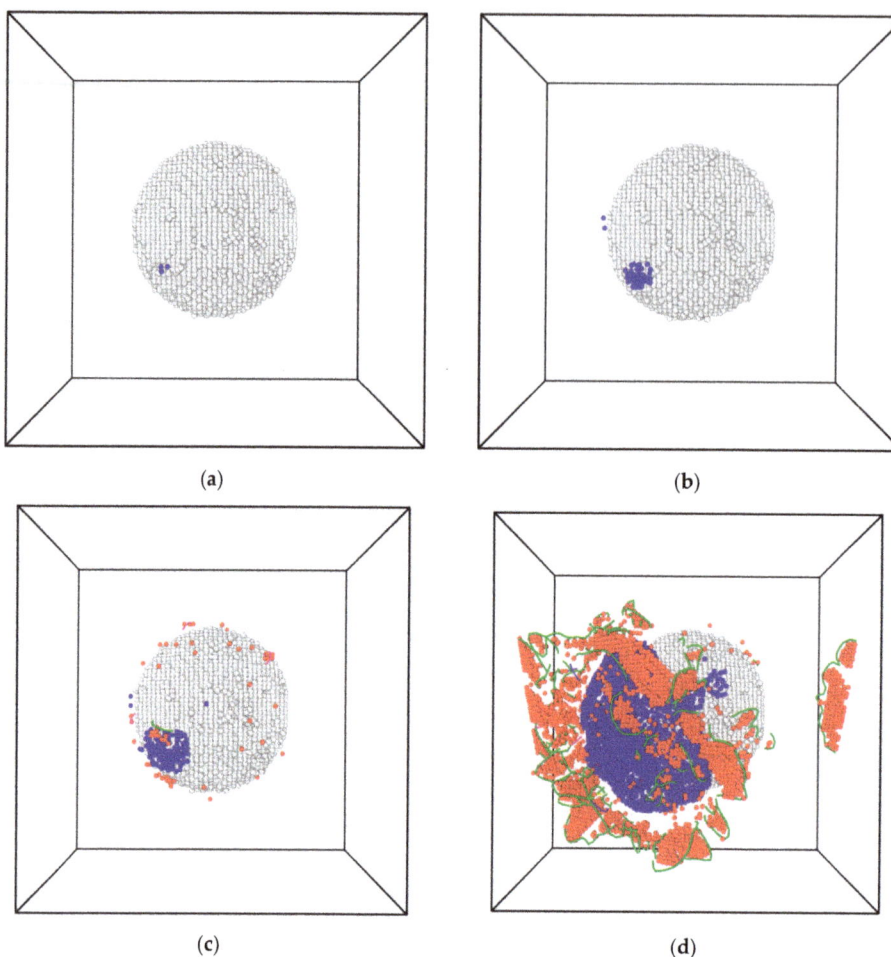

Figure 2. Simulation snapshots from a simulation with T = 400 K and σ^H = 6.0 GPa. Blue atoms are associated with a crack (have an atomic volume > 30 Å^3). (**a**) Initial appearance of a small crack involving a few atoms at t = 37 ps (relative to the start of the holding stage). (**b**) Subsequent growth of the crack via lattice-trapped delamination at t = 105 ps. (**c**) Initial appearance of dislocations at the crack tips at t = 126 ps. (**d**) Rapid crack growth via dislocation-mediated delamination at t = 139 ps. Green lines are Shockley-Read partial dislocations. Red atoms are associated with stacking faults.

Figure 3 shows atomic displacements associated with the two delamination modes. Figure 3a shows the atomic displacement vectors as black arrows over an 18 ps time period from the snapshot in Figure 2b to the moment just before dislocation nucleation Figure 2c. As shown, most of the crack growth is accommodated by displacements of atoms at the crack tip along the circumference of the particle. These displacements under lattice-trapped delamination are rather incoherent, in the sense that their direction and magnitude vary along the crack front in an uncoordinated manner. For example, several atoms experience large displacements while their immediate neighbors do not displace much at all (see yellow circled atoms). Displacements associated with dislocation-mediated crack growth over a 2 ps time period between the snapshots shown in Figure 2c,d are shown in Figure 3b. Once again, the displacements largely occur at the crack tip and in the circumferential direction, however these displacements are more coordinated in their direction and magnitude. These displacements appear to be due to nucleation and glide of the dislocations visible in the figure, i.e., they are generally aligned with the Burgers vectors.

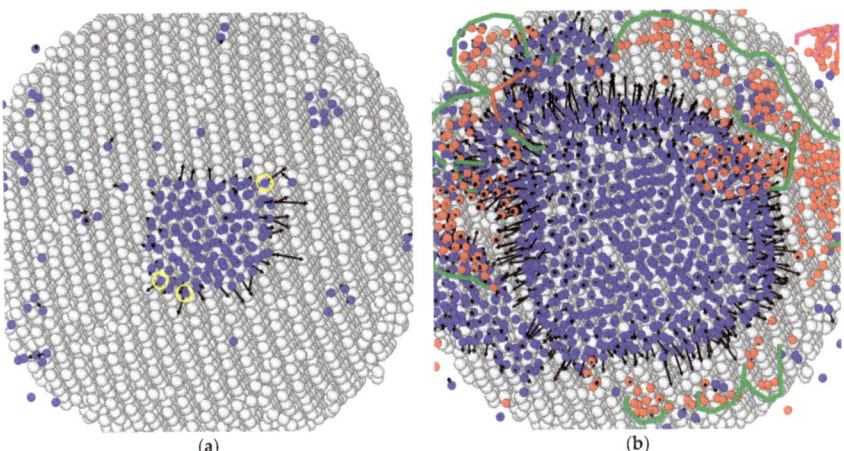

Figure 3. Snapshots from the same simulation as Figure 2 showing atomic displacements (black arrows) associated with (**a**) lattice-trapped delamination over an 18 ps time window and (**b**) dislocation-mediated delamination during a 2 ps time window. Blue atoms are associated with a crack (have an atomic volume > 30 Å^3). Green lines are Shockley-Read partial dislocations. Red atoms are associated with stacking faults. Yellow circles denote atoms whose atomic trajectories differ significantly from their immediate neighbors.

To determine whether there were preferential nucleation sites on the particle's surface, we extracted the approximate crack nucleation location from 40 simulations and plot these locations in Figure 4 as a point cloud projected onto the *x-z* plane. This figure shows that while there may be a slight preference for nucleation at the negative *z*-axis pole of the precipitate (since there is a small cluster there), nucleation was also common at other points around the surface. Similarly, there is a lack of data points at the positive *z*-axis pole, indicating nucleation there is unfavorable. These results imply that our boundary conditions, simulation cell size, and precipitate orientation did not significantly influence the simulation behaviors (i.e., they did not introduce strong preferential sites).

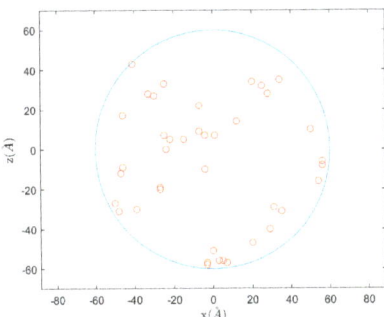

Figure 4. Approximate crack nucleation locations projected onto the *x-z* plane from 40 simulations with various stresses and temperatures.

To assess the delamination behavior in the absence of thermal fluctuations, we performed molecular statics simulations of hydrostatic straining. We progressively increased the hydrostatic strain of the box by increasing the volume in 0.003% increments and minimizing the energy of the system after each strain increment. Each minimization step iterated until the change in energy during a minimization step was less than 10^{-6}% or the norm of the global force vector was less than 10^{-8} eV/Å. The resulting stress-strain curve is shown in Figure 5. We observe that at a hydrostatic stress of around 10 GPa, the particle catastrophically delaminates from the matrix. Hence, 10 GPa can be regarded as the athermal critical stress for void nucleation. The first peak in Figure 5 corresponds to the nucleation of dislocations at the poles of the particle along the z-axis. The subsequent peaks correspond to nucleation of dislocations around the entire circumference of the particle. Hence, it seems that athermal void nucleation is governed by the athermal nucleation of dislocations.

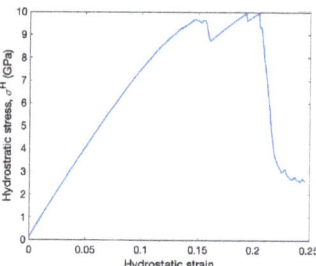

Figure 5. Hydrostatic stress-strain curve from a molecular statics simulation of delamination at T = 0 K. The peak stress is the stress required to delaminate the particle without the aid of thermal fluctuations.

To get information about the crack growth, we estimate the crack volume as $V_{crack}(t) = V(t) - V_0$, where $V(t)$ is the volume of the simulation cell at time t and V_0 is the volume at the start of the holding phase. In Figure 6 we show a few examples of how the crack volume evolves over time during the load holding phase. In some cases, there appears to be an "incubation period" where the volume does not increase at all, followed by a gradual increase over time indicating the nucleation and growth of a crack. This gradual increase corresponds to the lattice-trapped delamination mode. In other cases, the volume appears to increase from the start of the load holding phase, with no obvious incubation period. In most cases it was difficult to unambiguously identify a clear "nucleation" event which correlated with a local increase in atomic volume at the void's surface. For this reason, we were unable to analyze any sort of "nucleation" rate directly from the incubation time. Regardless, lattice-trapped delamination was always observed in our simulations. We also mark in Figure 6 the time where the first dislocation nucleated. Upon nucleation of one or more dislocations, the system volume increases precipitously as the crack growth rate accelerates to complete delamination.

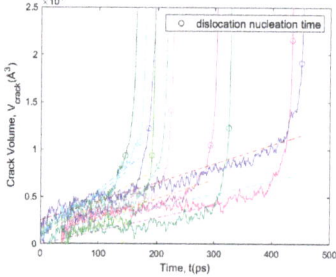

Figure 6. Crack volume as a function of time for simulations with T = 400 K and $\sigma^H = 6.0$ GPa, each with a different random seed for atomic velocity initialization. Dashed lines denote the lattice-trapped delamination rates and circles mark the appearance of a dislocation.

3.2. Kinetics of Lattice-Trapped Delamination

To characterize the kinetics of lattice-trapped delamination, we extracted the crack growth rate in terms of the rate of increase of the crack volume as shown in Figure 6. Since no other defects were present in the simulation cell, it is expected that essentially all increases in volume must be due to growth of interfacial cracks. Visual inspection of simulations and identification of volume "hot spots" where atomic volumes were elevated confirmed this assumption; we never observed significant volume increases in the bulk, only at the interface near cracks. It is more conventional, of course, to characterize a crack in terms of its area or radius. However, since cracks typically adopted complex shapes and morphologies, it was difficult to directly extract their area and/or radius. For these reasons, we use changes in simulation cell volume to quantify the delamination rate.

We determined the lattice-trapped delamination rate by taking a linear regression of the volume vs. time curve from the start of the load holding phase to the time of dislocation nucleation, shown by the dashed lines in Figure 6. We find that under the same nominal conditions (stress and temperature), the delamination rate was sensitive to the initial conditions (e.g., initial random atomic velocities). For this reason, 10 replicate simulations were performed at each condition. Figure 7 shows histograms of the lattice-trapped delamination rate at a few conditions with 30 replica simulations, demonstrating the spread of the data. Interestingly, the delamination rates are systematically biased towards lower rates (rather than being symmetrical about the mean), qualitatively taking the form of exponential distributions. Figure 8 shows the mean delamination rate as a function of σ^H at temperatures of 200, 250, 300, 350, and 400 K. There is clearly a strong sensitivity to both stress and temperature. We will further analyze these data in the Discussion.

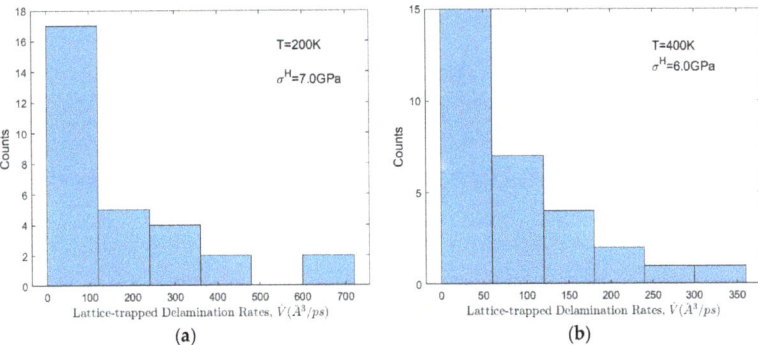

Figure 7. Histograms of lattice-trapped delamination rates from 30 replicas at (**a**) T = 200 K and σ^H = 7.0 GPa and (**b**) T = 400 K and σ^H = 6.0 GPa.

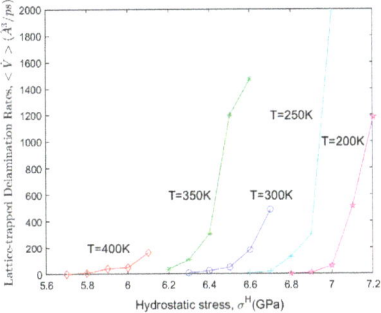

Figure 8. Average lattice-trapped delamination rate among 10 replicas, $\langle \dot{V} \rangle$, as a function of hydrostatic stress over temperatures ranging from 200 to 400 K.

3.3. Analysis of Dislocation-Mediated Delamination

While we always observed lattice-trapped delamination in our simulations, dislocation nucleation was only observed in "high stress" simulations. Furthermore, if the stress was too high then dislocations would nucleate at the beginning of the simulation, much like the behavior observed in our molecular statics simulation. One important question is: what is the interplay between lattice-trapped and dislocation-mediated delamination? What governs the transition from one mode to the other? To gain insight into this question, we extracted the crack volume at the moment of dislocation nucleation from our simulations and present the average values in Figure 9. At lower stresses, the crack volume increases to a peak value and then subsequently decays to zero, indicating immediate nucleation of dislocations. We believe that these trends result from changes in the lattice-trapped delamination and dislocation nucleation rates with stress and temperature, and changes in the driving force for dislocation nucleation as the crack grows. We defer further analysis to the Discussion.

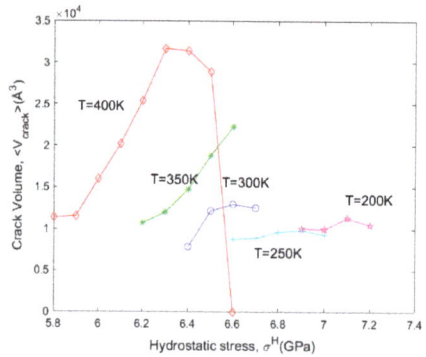

Figure 9. Average crack volume among 10 replicas at the moment of dislocation nucleation as a function of hydrostatic stress over temperatures ranging from 200 to 400 K.

In all simulations, as soon as the first dislocation nucleation event occurred, the delamination rate increased dramatically. The delamination rate was so large that a well-defined rate could not be determined. In some cases, dislocations nucleated from the "crack tip" near the particle interface, but not in all cases. For example, dislocation nucleation in the middle of the crack face in the Al matrix was also observed.

4. Discussion

4.1. Thermal Activation Analysis

Here we further analyze the lattice-trapped delamination rate data and demonstrate that lattice-trapped delamination is thermally activated. For a system containing a crack of length a loaded by hydrostatic stress σ^H, the theory of thermally activated crack growth says that the growth rate is [43]

$$\dot{a}\left(\sigma^H, a\right) = \dot{a}_0 \exp\left(\frac{-E_a + A^* K\left(\sigma^H, a\right)^2 / E}{k_B T}\right) \quad (1)$$

where E_a is the activation energy for bond breaking, K is the stress intensity factor, A^* is the activation area, E is the modulus of elasticity, $\dot{a}_0$ is the exponential prefactor related to the attempt frequency, T is the absolute temperature, and k_B is Boltzmann's constant. It is important to note that Equation (1) is only valid when $K > K_c$, the critical stress intensity factor. If $K < K_c$, then the free energy of the system increases when the crack grows, violating the second law of thermodynamics [44]. The stress intensity factor can always be written in the form

$$K = Y\sigma^H\sqrt{\pi a} \qquad (2)$$

where Y is a geometric factor dictated by the geometry of the problem. To relate the crack volume to the crack length, we approximate the crack as an ellipsoid with two axes of radius a and the other of radius $a/2$ (consistent with cracks observed in our simulations). The volumetric crack growth rate is then $\dot{V} = 4\pi a^2 \dot{a}$. In our simulations where lattice-trapped delamination occurs, the crack length does not increase significantly (cracks remain relatively small during lattice-trapped delamination, see Figure 2). Hence, for simplicity, we neglect changes in a and assume that $K = K(\sigma^H, \bar{a})$ and $\dot{V} = 4\pi \bar{a}^2 \dot{a}$, where $\bar{a}$ is the average crack length during the simulation. With this assumption and using Equation (1), we obtain that the activation enthalpy (numerator in the exponential) is

$$\Delta H_a = E_a - Ck_B\left(\sigma^H\right)^2 \qquad (3)$$

where $C = A^*Y^2\pi\bar{a}/(Ek_B)$ and simple algebraic manipulation further shows that

$$\ln \dot{V} = \ln \dot{V}_0 - \frac{E_a}{k_B T} + C\frac{\left(\sigma^H\right)^2}{T} \qquad (4)$$

where $\dot{V}_0 = 4\pi \bar{a}^2 \dot{a}_0$. Hence, if we plot $\ln\langle\dot{V}\rangle$ from our simulations as a function of $(\sigma^H)^2/T$, the dataset for each temperature should form a straight line with slope C if growth is thermally activated. In Figure 10 we plot the data in this way and see a consistent linear trend across all datasets. Specifically, we find that the same slope fits all datasets, indicating that $C = 258$ K/GPa^2. Next, we extract the y-intercept from each of these linear fits, and Equation (4) indicates that these intercepts should scale with $1/T$. Figure 11 plots the y-intercepts from Figure 10 as a function of $1/T$ and once again a linear behavior is recovered as expected. The slope of Figure 10 is $-E_a/k_B$ and the y-intercept is $\ln \dot{V}_0$; we obtain values of $E_a = 1.37$ eV and $\dot{V}_0 = 1.23 \times 10^9$ Å^3/ps. The fact that our data so strongly reproduces the behaviors predicted by Equation (4) indicates that the lattice-trapped delamination observed in our MD simulations is indeed thermally activated, and that our neglect of the crack length dependence of K does not introduce any significant errors into our analysis.

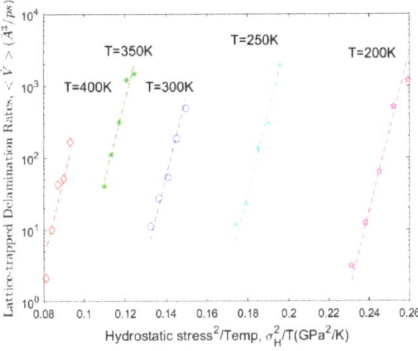

Figure 10. Lattice-trapped delamination rate data from Figure 8 plotted based on the theory of thermally activated crack growth. Dashed lines are linear fits with identical slopes.

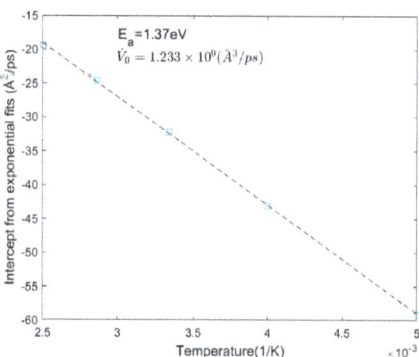

Figure 11. Arrhenius plot of ordinate intercepts from linear curve fits in Figure 10, showing Arrhenius behavior consistent with the theory of thermally activated crack growth.

The thermally activated delamination mode that we observe is likely governed by the so-called lattice trapping phenomenon [42]. Under lattice trapping, a crack which is loaded supercritically (i.e., with $K > K_c$) grows in a step-wise manner as atomic bonds at the crack tip are sequentially broken, often by a kink-pair mechanism [43–46]. As the crack grows, it experiences an oscillating potential energy landscape due to the periodicity of the lattice, and the height of these oscillations dictates the activation energy for growth. To our knowledge this crack growth mode has only been observed in brittle materials like Si [45] and glass [47], but not in ductile materials like Al considered here. This is nonetheless reasonable, since in effect the Al system is acting in a brittle manner in the absence of dislocation activity.

To further analyze our extracted parameters, we need to estimate the stress intensity factor. Tan and Gao [48] numerically determined the stress intensity factor for an axisymmetric interfacial crack on a sphere which forms an angle ϕ from the radial axis, as shown in the inset in Figure 12a, with various modulus ratios E_p/E_m where m stands for matrix and p for precipitate. We can express their results in the form

$$K_0 = Y(\phi, E_p/E_m)\sigma^H \sqrt{\pi R} \tag{5}$$

where $K_0 = \sqrt{K_I^2 + K_{II}^2}$ is the "overall" stress intensity factor for the mixed-mode loading (the crack is generally mixed mode) and $Y(\phi, E_p/E_m)$ was determined by Tan and Gao via numerical boundary integral methods. Using Equation (5) enables us to obtain the activation area as

$$A^* = C \frac{k_B \overline{E}}{\pi R Y(\phi, E_p/E_m)^2} \tag{6}$$

where $\overline{E} = 2E_m E_p/(E_m + E_p)$ is the bimaterial modulus for interfacial fracture [49]. For approximation purposes, we employ the typical Young's modulus of untextured polycrystals instead of the anisotropic single crystal elastic constants. Estimating $E_m = 70$ GPa (experimental value for pure Al) and $E_p \approx 120$ GPa [50] gives $\overline{E} \approx 88$ GPa and $E_p/E_m \approx 1.7$. Tan and Gao found that for cracks varying from $\phi = 22.5$ to 67.5 degrees with a modulus ratio of $E_p/E_m = 2$, $Y(\phi, E_p/E_m)$ varied from 1.31 to 1.79. Unfortunately, cracks observed in our simulations typically had angles $\phi < 22.5°$, so it is difficult to apply Tan and Gao's solution. Note that according to Equation (5), $Y \to 0$ as $\phi \to 0$ since K_0 must go to zero when the crack length is zero (i.e., $\phi = 0$). Hence, we expect the Y values in our simulations to be less than 1.31. Nonetheless, to gain insight into orders of magnitude for the thermal activation parameters we assume $Y \approx 1.31$ in the analysis below. Using these parameter values with $R = 50$ Å in Equation (6) leads to $A^* \approx 1.17$ A^2. According to Schoeck [43]

$$A^* = (1+\beta)\Delta A \tag{7}$$

where ΔA is the atomistic area of crack advance between the equilibrium and saddle position of the crack front and β is a factor in the range between 0 and 2. Schoeck argued that for crack advance by breaking of individual bonds, $\Delta A \approx 1$ A^2; hence, our A^* value gives the correct order of magnitude, further bolstering our conclusion that lattice-trapped delamination is thermally activated.

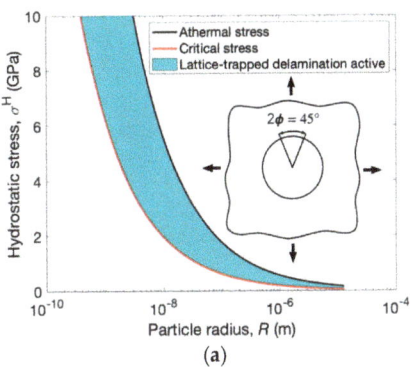
(a)

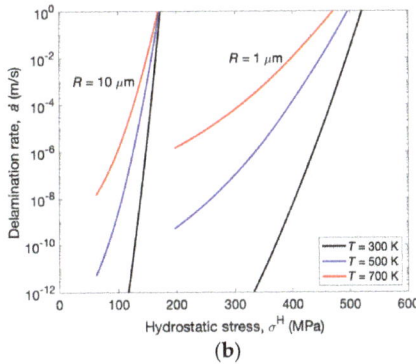
(b)

Figure 12. Predictions of lattice-trapped delamination for a crack with $\phi = 22.5°$. (**a**) Stress range over which lattice-trapped delamination may operate. (**b**) Lattice-trapped delamination rates for particles of radius $R = 1$ µm and 10 µm at three temperatures.

Finally, we note that an important aspect of thermally activated crack growth is the athermal stress intensity factor, K_{ath}, at which the activation enthalpy goes to zero. At and above this load, the crack growth rate is no longer governed by thermally activated bond breaking. According to Equation (1) and accounting for the modulus mismatch, the athermal stress intensity factor is

$$K_{ath} = \sqrt{\frac{E_a \bar{E}}{A^*}}. \tag{8}$$

Using our estimates for parameters above, we obtain that $K_{ath} \approx 1.29$ MPa·m$^{1/2}$. According to Equation (5), K_{ath} is reached when

$$\sigma_{ath}^H = \frac{K_{ath}}{Y\sqrt{\pi R}} \tag{9}$$

and using the estimates above we obtain $\sigma_{ath}^H \approx 7.8$ GPa. This value is consistent with our data since we did not observe lattice-trapped delamination above 7.2 GPa (although we did not attempt to obtain a maximum stress where lattice-trapped delamination rates could be obtained).

4.2. Dislocation Nucleation

Delamination and crack growth via dislocation nucleation have been observed by other researchers in the past [30,51]. Similar to the thermally activated lattice-trapped delamination mode discussed above, dislocation nucleation is thermally activated. Above some critical load, the activation barrier for dislocation nucleation goes to zero and spontaneous nucleation occurs [51]. Hence, we have a race between two thermally activated processes—lattice-trapped delamination and dislocation nucleation—each characterized by their own activation enthalpy which varies with the stress intensity factor. This race gives rise to the crack volume trends exhibited in Figure 9. At lower stress, lattice-trapped delamination is slow so that the crack does not grow much before a dislocation is nucleated. As the stress is increased, the lattice-trapped delamination rate increases and there is a time lag before the dislocation nucleates, causing the crack volume at dislocation nucleation to increase. Finally, at very high stresses the athermal load is reached for dislocation nucleation so that a dislocation nucleates immediately during the simulations, causing the crack volume at dislocation nucleation to go to zero. This qualitatively explains the trends observed in Figure 9. A quantitative explanation would require a detailed understanding of the nucleation parameters for a dislocation at the crack tip, which are non-trivial to compute and beyond the scope of this work [51].

Unfortunately, the conditions of the simulations here were such that as soon as a dislocation nucleated, the particle delaminated almost instantaneously. This indicates that our systems were overdriven with the respect to dislocation-mediated delamination. It is possible that quantitatively useful delamination rates could be obtained under dislocation-mediated delamination from simulations at lower stresses if a large enough pre-crack is manually inserted into the system. Here, if our applied stress was too low, the lattice-trapped delamination rate was so small that a crack never became large enough to enable dislocation nucleation. Future work should focus on quantifying the dislocation-mediated delamination rate.

4.3. Implications for Damage and Rupture of Materials

The present simulations bear direct relevance to spall formation under shock loading conditions; for example, spall by microvoid coalescence has been observed in 1100-O aluminum at shock stresses as high as 6 GPa [33]. However, it is reasonable to ask if the present observations also relate to quasi-static ductile rupture of aluminum alloys. The main gaps between our simulation conditions and those of typical experiments are that our applied stresses are much higher (for example, the yield strength of 2xxx series aluminum alloys is in the range of 150–350 MPa depending on alloy and heat treatment) and our particles sizes are smaller.

First, we address dislocation-mediated growth. The reality is that aside from high rate loading scenarios where high stresses are attained [52] and highly localized loading (e.g., under a nano-indenter), dislocation nucleation is typically far too slow to meaningfully impact material behaviors [51,53]. Hence, we do not expect that dislocation-mediated growth, as observed here through nucleation of dislocations, will be relevant under most loading scenarios. Indeed, even within the present spectrum of simulations, the dislocation-mediated growth only occurred at the highest stresses approaching the athermal critical stress. On the other hand, Sills and Boyce recently showed that dislocation-mediated growth processes are greatly accelerated when dislocations are already present in the material, and thus do not require nucleation [54]. Kinematically speaking, it is equivalent to nucleate a dislocation at a crack tip or to adsorb an existing dislocation of opposite sign. We believe that dislocation-mediated growth via adsorption of *existing* dislocations is likely to play an important role in particle delamination, just as Sills and Boyce argue it does for void growth.

In terms of the relevance of lattice-trapped delamination, we are fortunate to have a fully characterized thermal activation law for the lattice-trapped delamination rate. This means that we can extrapolate the model to quasi-static conditions since the physics of thermal activation is equally valid at lower stresses and larger particle sizes. Importantly, the process of lattice-trapped delamination is bookended by two conditions on the stress intensity factor; namely, $K_c < K < K_{at}$. Below the critical stress intensity factor, K_c, delamination is not energetically favorable. And above the athermal stress intensity factor, K_{at}, thermal activation is not operable. Equation (9) provides an estimate for the athermal hydrostatic stress σ_{ath}^H at which K_{at} is reached. For brittle fracture, $K_c = \sqrt{(\gamma_{sm} + \gamma_{sp})\bar{E}}$

where γ_{sm} and γ_{sp} are the surface energies for the matrix and precipitate, respectively. The surface energy for Al is known experimentally to be $\gamma_{sm} = 0.98$ J/m^2 [55]. To estimate the surface energy for θ-phase, we use the average value for the interatomic potential used here (computed for (100) and (110) surfaces), giving $\gamma_{sp} = 1.38$ J/m^2 [38]. These values give a critical stress intensity factor of $K_c = 0.457$ MPa·m$^{1/2}$. Using these results, we estimate the stress and particle size range where lattice-trapped delamination may operate (i.e., where $K_c < K < K_{at}$) when $\phi = 22.5°$ in Figure 12a. The plot shows that in the particle size range considered here, stresses must exceed 2.8 GPa for lattice-trapped delamination to operate. On the other hand, for particles with a radius of 1 μm, lattice-trapped delamination can occur in the range of 200 to 550 MPa, which falls within the stress range relevant to quasi-static loading conditions, especially considering that inhomogeneous microstructural stresses can far exceed the homogeneous far-field applied stresses. Using this stress range, we plot the predicted lattice-trapped delamination rate $\dot{a}$ as a function of stress for particle radii of 1 and 10 μm at temperatures of 300, 500, and 700 K. Interestingly, the resulting delamination rates are large enough that they may be relevant to applications. For example, full delamination of a 1 μm particle requires ~3 μm of crack growth (half of the circumference), and for this to occur in 1 years' time would require a delamination rate of just $\dot{a} \approx 10^{-13}$ m/s. We emphasize that because we have made a number of approximations in our analysis, these delamination rates should be interpreted only in terms of their rough order of magnitude. Nonetheless, these results indicate that lattice-trapped delamination via thermally activated brittle crack growth may be broadly relevant to void nucleation.

While we have only considered the influence of hydrostatic loading on the delamination behavior here, shear stresses are also expected to affect delamination. For example, experimental work by Croom et al. [56] and Achouri et al. [57] has demonstrated void nucleation under shear-dominated conditions in pure Cu and particle-containing Al, respectively. In the case of particle-mediated nucleation, in addition to the applied shear stress the non-uniform plastic strain accumulation around the particle provides a driving force for delamination. In the context of lattice-trapped delamination, this driving force is difficult to quantify since it requires an elastic-plastic analysis of deformation with a significant accumulation of plastic strain. None-the-less, lattice-trapped delamination could also be operative under shear-dominated loading, but additional research is necessary to quantify the influence of shear stresses and shear deformation.

Perhaps the most confounding aspect of our simulations is that we set out to study crack nucleation and instead wound-up studying crack growth. One might have expected that the rate controlling step of void nucleation would be the appearance of an interfacial crack, in the sense that a crack spontaneously appeared along the interface. However, this was not the behavior we observed. Instead, we found that excess free volume in the form of vacancy-type defect clusters at the interface appeared quickly and then grew by lattice-trapped delamination. The majority of the simulation time was then spent growing the crack by lattice-trapped delamination until dislocation nucleation occurred. Hence, for our simulations, the rate controlling step for void nucleation was actually lattice-trapped delamination. Once the first dislocations appeared, it is true that the void fully nucleated in a catastrophic manner more consistent with a true nucleation event, but we believe that this may be an artifact of our high stresses; in our lower stress simulations, the dislocation-mediated delamination stage never occurred. We note that additional research is necessary to determine the mechanism for vacancy-type cluster nucleation so that a comprehensive picture for void nucleation can be assembled. While the present observations are directly applicable to the strain-rates associated with shock-induced spallation, extrapolation of the thermally activated process of lattice-trapped delamination suggests that the process is also relevant at quasi-static timescales for micrometer-sized particles.

5. Conclusions

We observed that early stage nanoscale delamination at a particle-matrix interface was a three stage process: (1) a nuclei of excess free volume is formed at the particle-matrix interface, either immediately upon loading or after a brief delay; (2) subsequently, the nuclei grows in the absence of dislocation activity in a process we term *lattice-trapped delamination*; (3) when the crack reaches sufficient size and the crack tip stress intensity is sufficient, Shockley partial dislocations are emitted causing a rapid acceleration of crack growth and ultimately complete debonding of the particle-matrix interface. The second stage, lattice-trapped delamination, as described for the first time herein, limits the rate of early-stage growth of the nanocrack, and is not only relevant to shock spallation, but extrapolation of the kinetics of this process suggest that it is also relevant at quasi-static strain rates for micrometer-scale particles.

Author Contributions: Conceptualization, R.B.S.; methodology, R.B.S. and Q.Q.Z.; analysis, R.B.S., Q.Q.Z., and B.L.B.; writing, R.B.S., Q.Q.Z., and B.L.B. All authors have read and agreed to the published version of the manuscript.

Funding: This research received no external funding.

Institutional Review Board Statement: Not applicable.

Informed Consent Statement: Not applicable.

Data Availability Statement: The data presented in this study are openly available through Zenodo at 10.5281/zenodo.4420413.

Acknowledgments: B.L.B. was supported by the Center for Integrated Nanotechnologies funded by the DOE Office of Basic Energy Science. Sandia National Laboratories is a multimission laboratory managed and operated by National Technology and Engineering Solutions of Sandia LLC, a wholly owned subsidiary of Honeywell International Inc. for the U.S. Department of Energy's National Nuclear Security Administration under contract DE-NA0003525. The views expressed in the article do not necessarily represent the views of the U.S. DOE or the United States Government.

Conflicts of Interest: The authors declare no conflict of interest.

Appendix A. Selection of Ramping Duration

One aspect of these simulations which we found to be challenging was that the hydrostatic stress had to be applied carefully to the atomistic system. Otherwise, artifacts in the form of unwanted and uncontrolled defects would be introduced into the system prior to the holding phase. After trying out several strategies for applying stress to the atomistic system, we found that a simple linear ramp via an *NPT* ensemble was sufficient as long as the loading duration was sufficiently long. If the ramping duration were too short, we observed that void nucleation would occur prematurely via nucleation of dislocations. Figure A1 shows the time at which a void nucleated relative to the start of the holding phase (indicated by a 10% increase in the simulation cell volume relative to the start of the holding phase). Each data point is an average over five replicas. To ensure that the load ramping does not introduce artifacts, we need to confirm that the nucleation time is insensitive to the ramping duration. Figure A1 shows, within the scatter of the data, that 12 ps is a sufficiently long ramping duration. When ramped over durations less than 12 ps, the nucleation time decreases. We believe that this behavior results from stress waves and spikes introduced into the cell by the *NPT* barostat during rapid loading, which nucleate defects during ramping.

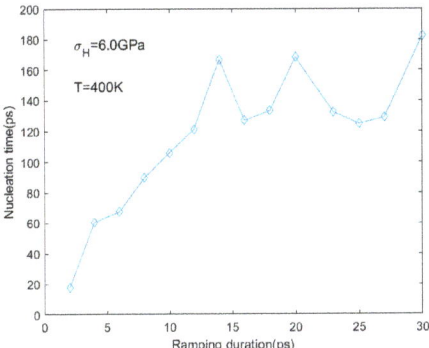

Figure A1. Void nucleation time, measured as time required for the simulation cell volume to increase by 10% during the holding phase, as a function of ramping duration.

References

1. Goods, S.H.; Brown, L.M. The Nucleation of Cavities by Plastic Deformation. *Acta Met.* **1979**, *27*, 1–15. [CrossRef]
2. Needleman, A. A Continuum Model for Void Nucleation by Inclusion Debonding. *J. Appl. Mech.* **1987**, *54*, 525–531. [CrossRef]
3. Shabrov, M.N.; Needleman, A. An analysis of inclusion morphology effects on void nucleation. *Model. Simul. Mater. Sci. Eng.* **2002**, *10*, 163–183. [CrossRef]
4. Besson, J. Continuum Models of Ductile Fracture: A Review. *Int. J. Damage Mech.* **2010**, *19*, 3–52. [CrossRef]
5. Shakoor, M.; Navas, V.M.T.; Munõz, D.P.; Bernacki, M.; Bouchard, P.-O. Computational Methods for Ductile Fracture Modeling at the Microscale. *Arch. Comput. Methods Eng.* **2019**, *26*, 1153–1192. [CrossRef]
6. Bitzek, E.; Kermode, J.R.; Gumbsch, P. Atomistic aspects of fracture. *Int. J. Fract.* **2015**, *191*, 13–30. [CrossRef]
7. Tvergaard, V.; Needleman, A. The Modified Gurson Model. In *Handbook of Materials Behavior Models*; Elsevier: Amsterdam, The Netherlands, 2001; pp. 430–435.
8. Chu, C.C.; Needleman, A. Void Nucleation Effects in Biaxially Stretched Sheets. *J. Eng. Mater. Technol.* **1980**, *102*, 249–256. [CrossRef]
9. Tvergaard, V. Material Failure by Void Growth to Coalescence. In *Advances in Applied Mechanics*; Elsevier: Amsterdam, The Netherlands, 1989; Volume 27, pp. 83–151.
10. Boyce, B.L.; Kramer, S.L.B.; Fang, H.E.; Cordova, T.E.; Neilsen, M.K.; Dion, K.; Kaczmarowski, A.K.; Karasz, E.; Xue, L.; Gross, A.J.; et al. The Sandia Fracture Challenge: Blind round robin predictions of ductile tearing. *Int. J. Fract.* **2014**, *186*, 5–68. [CrossRef]
11. Boyce, B.L.; Kramer, S.L.B.; Bosiljevac, T.R.; Corona, E.; Moore, J.A.; Elkhodary, K.; Simha, C.H.M.; Williams, B.W.; Cerrone, A.R.; Nonn, A.; et al. The second Sandia Fracture Challenge: Predictions of ductile failure under quasi-static and moderate-rate dynamic loading. *Int. J. Fract.* **2016**, *198*, 5–100. [CrossRef]
12. Kramer, S.L.B.; Jones, A.; Mostafa, A.; Ravaji, B.; Tancogne-Dejean, T.; Roth, C.C.; Bandpay, M.G.; Pack, K.; Foster, J.T.; Behzadinasab, M.; et al. The third Sandia fracture challenge: Predictions of ductile fracture in additively manufactured metal. *Int. J. Fract.* **2019**, *218*, 5–61. [CrossRef]
13. Berdichevsky, V.; Le, K.C. On The Microcrack Nucleation In Brittle Solids. *Int. J. Fract.* **2005**, *133*, L47–L54. [CrossRef]
14. Dias, C.L.; Kröger, J.; Vernon, D.; Grant, M. Nucleation of cracks in a brittle sheet. *Phys. Rev. E* **2009**, *80*, 066109. [CrossRef] [PubMed]
15. Rawat, S.; Raole, P.M. Molecular dynamics investigation of void evolution dynamics in single crystal iron at extreme strain rates. *Comput. Mater. Sci.* **2018**, *154*, 393–404. [CrossRef]
16. Yang, X.; Zeng, X.; Wang, J.; Wang, J.; Wang, F.; Ding, J. Atomic-scale modeling of the void nucleation, growth, and coalescence in Al at high strain rates. *Mech. Mater.* **2019**, *135*, 98–113. [CrossRef]
17. Agarwal, G.; Dongare, A.M. Defect and damage evolution during spallation of single crystal Al: Comparison between molecular dynamics and quasi-coarse-grained dynamics simulations. *Comput. Mater. Sci.* **2018**, *145*, 68–79. [CrossRef]
18. Shao, J.-L.; Wang, P.; He, A.-M.; Zhang, R.; Qin, C.-S. Spall strength of aluminium single crystals under high strain rates: Molecular dynamics study. *J. Appl. Phys.* **2013**, *114*, 173501. [CrossRef]
19. Liao, Y.; Xiang, M.; Zeng, X.; Chen, J. Molecular dynamics study of the micro-spallation of single crystal tin. *Comput. Mater. Sci.* **2014**, *95*, 89–98. [CrossRef]
20. Rawat, S.; Warrier, M.; Chaturvedi, S.; Chavan, V.M. Temperature sensitivity of void nucleation and growth parameters for single crystal copper: A molecular dynamics study. *Model. Simul. Mater. Sci. Eng.* **2011**, *19*, 025007. [CrossRef]

21. Pang, W.-W.; Zhang, P.; Zhang, G.-C.; Xu, A.-G.; Zhao, X.-G. Dislocation creation and void nucleation in FCC ductile metals under tensile loading: A general microscopic picture. *Sci. Rep.* **2015**, *4*, 6981. [CrossRef]
22. Mayer, A.E.; Mayer, P.N. Strain rate dependence of spall strength for solid and molten lead and tin. *Int. J. Fract.* **2020**, *222*, 171–195.
23. Paul, S.K.; Kumar, S.; Tarafder, S. Effect of loading conditions on nucleation of nano void and failure of nanocrystalline aluminum: An atomistic investigation. *Eng. Fract. Mech.* **2017**, *176*, 257–262. [CrossRef]
24. Fensin, S.J.; Cerreta, E.K.; Iii, G.T.G.; Valone, S.M. Why are some Interfaces in Materials Stronger than others? *Sci. Rep.* **2015**, *4*, 5461. [CrossRef]
25. Xu, S.; Deng, X. Nanoscale void nucleation and growth and crack tip stress evolution ahead of a growing crack in a single crystal. *Nanotechnology* **2008**, *19*, 115705. [CrossRef] [PubMed]
26. Pogorelko, V.V.; Mayer, A.E. Influence of copper inclusions on the strength of aluminum matrix at high-rate tension. *Mater. Sci. Eng. A* **2015**, *642*, 351–359. [CrossRef]
27. Pogorelko, V.V.; Mayer, A.E. Influence of titanium and magnesium nanoinclusions on the strength of aluminum at high-rate tension: Molecular dynamics simulations. *Mater. Sci. Eng. A* **2016**, *662*, 227–240. [CrossRef]
28. Pogorelko, V.V.; Mayer, A.E. Tensile strength of Fe–Ni and Mg–Al nanocomposites: Molecular dynamic simulations. *J. Phys. Conf. Ser.* **2018**, *946*, 012043. [CrossRef]
29. Pogorelko, V.V.; Mayer, A.E. Tensile strength of Al matrix with nanoscale Cu, Ti and Mg inclusions. *J. Phys. Conf. Ser.* **2016**, *774*, 012034. [CrossRef]
30. Cui, Y.; Chen, Z. Void initiation from interfacial debonding of spherical silicon particles inside a silicon-copper nanocomposite: A molecular dynamics study. *Model. Simul. Mater. Sci. Eng.* **2017**, *25*, 025007. [CrossRef]
31. Coffman, V.R.; Sethna, J.P.; Heber, G.; Liu, M.; Ingraffea, A.; Bailey, N.P.; Barker, E.I. A comparison of finite element and atomistic modelling of fracture. *Model. Simul. Mater. Sci. Eng.* **2008**, *16*, 065008. [CrossRef]
32. Ramesh, K.T. High Rates and Impact Experiments. In *Springer Handbook of Experimental Solid Mechanics*; Sharpe, W.N., Ed.; Springer: Boston, MA, USA, 2008; pp. 929–960.
33. Williams, C.L.; Chen, C.Q.; Ramesh, K.T.; Dandekar, D.P. On the shock stress, substructure evolution, and spall response of commercially pure 1100-O aluminum. *Mater. Sci. Eng. A* **2014**, *618*, 596–604. [CrossRef]
34. Nie, J.-F. 20 Physical Metallurgy of Light Alloys. In *Physical Metallurgy*; Laughlin, D.E., Hono, K., Eds.; Elsevier: Oxford, UK, 2014; pp. 2009–2156.
35. Vaughan, D.; Silcock, J.M. The Orientation and Shape of θ Precipitates Formed in an Al-Cu Alloy. *Phys. Status Solidi B* **1967**, *20*, 725–736. [CrossRef]
36. Wisner, B.; Kontsos, A. Investigation of particle fracture during fatigue of aluminum 2024. *Int. J. Fatigue* **2018**, *111*, 33–43. [CrossRef]
37. Plimpton, S. Fast Parallel Algorithms for Short-Range Molecular Dynamics. *J. Comput. Phys.* **1995**, *117*, 1–19. [CrossRef]
38. Apostol, F.; Mishin, Y. Interatomic potential for the Al-Cu system. *Phys. Rev. B* **2011**, *83*, 054116. [CrossRef]
39. Mura, T. *Micromechanics of Defects in Solids*; Springer: Dordrecht, The Netherlands, 1987; Volume 3.
40. Stukowski, A. Structure identification methods for atomistic simulations of crystalline materials. *Model. Simul. Mater. Sci. Eng.* **2012**, *20*, 045021. [CrossRef]
41. Stukowski, A. Visualization and analysis of atomistic simulation data with OVITO–the Open Visualization Tool. *Model. Simul. Mater. Sci. Eng.* **2010**, *18*, 015012. [CrossRef]
42. Thomson, R.; Hsieh, C.; Rana, V. Lattice Trapping of Fracture Cracks. *J. Appl. Phys.* **1971**, *42*, 3154–3160. [CrossRef]
43. Schoeck, G. Thermally activated crack-propagation in brittle materials. *Int. J. Fract.* **1990**, *44*, 1–14. [CrossRef]
44. Argon, A.S. Thermally activated crack growth in brittle solids? *Scr. Metall.* **1982**, *16*, 259–264. [CrossRef]
45. Zhu, T.; Li, J.; Yip, S. Atomistic Configurations and Energetics of Crack Extension in Silicon. *Phys. Rev. Lett.* **2004**, *93*, 205504. [CrossRef]
46. Zhu, T.; Li, J.; Yip, S. Atomistic characterization of three-dimensional lattice trapping barriers to brittle fracture. *Proc. R. Soc. Math. Phys. Eng. Sci.* **2006**, *462*, 1741–1761. [CrossRef]
47. Wiederhorn, S.M.; Johnson, H.; Diness, A.M.; Heuer, A.H. Fracture of Glass in Vacuum. *J. Am. Ceram. Soc.* **1974**, *57*, 6. [CrossRef]
48. Tan, C.L.; Gao, Y.L. Stress intensity factors for cracks at spherical inclusions by the boundary integral equation method. *J. Strain Anal. Eng. Des.* **1990**, *25*, 197–206. [CrossRef]
49. Rice, J.R. Elastic Fracture Mechanics Concepts for Interfacial Cracks. *J. Appl. Mech.* **1988**, *55*, 98–103. [CrossRef]
50. Zhang, J.; Huang, Y.N.; Mao, C.; Peng, P. Structural, elastic and electronic properties of θ (Al_2Cu) and S (Al_2CuMg) strengthening precipitates in Al–Cu–Mg series alloys: First-principles calculations. *Solid State Commun.* **2012**, *152*, 2100–2104. [CrossRef]
51. Zhu, T.; Li, J.; Yip, S. Atomistic Study of Dislocation Loop Emission from a Crack Tip. *Phys. Rev. Lett.* **2004**, *93*, 025503. [CrossRef]
52. Gray, G.T., III. High-Strain-Rate Deformation: Mechanical Behavior and Deformation Substructures Induced. *Annu. Rev. Mater. Res.* **2012**, *42*, 285–303. [CrossRef]
53. Nguyen, L.D.; Warner, D.H. Improbability of Void Growth in Aluminum via Dislocation Nucleation under Typical Laboratory Conditions. *Phys. Rev. Lett.* **2012**, *108*, 035501. [CrossRef]
54. Sills, R.B.; Boyce, B.L. Void growth by dislocation adsorption. *Mater. Res. Lett.* **2020**, *8*, 103–109. [CrossRef]
55. Murr, L.E. *Interfacial Phenomena in Metals and Alloys*; Addison-Wesley: London, UK, 1975.

56. Croom, B.P.; Jin, H.; Noell, P.J.; Boyce, B.L.; Li, X. Collaborative ductile rupture mechanisms of high-purity copper identified by in situ X-ray computed tomography. *Acta Mater.* **2019**, *181*, 377–384. [CrossRef]
57. Achouri, M.; Germain, G.; Santo, P.D.; Saidane, D. Experimental characterization and numerical modeling of micromechanical damage under different stress states. *Mater. Des.* **2013**, *50*, 207–222. [CrossRef]

Article

Analysis of the Crack Initiation and Growth in Crystalline Materials Using Discrete Dislocations and the Modified Kitagawa–Takahashi Diagram

Kuntimaddi Sadananda [1,*], Ilaksh Adlakha [2], Kiran N. Solanki [3] and A.K. Vasudevan [1]

1 Technical Data Analysis, Falls Church, VA 22046, USA; akruva@gmail.com
2 Department of Applied Mechanics, Indian Institute of Technology-Madras, Chennai 600036, India; ilaksh.adlakha@iitm.ac.in
3 School for Engineering of Matter, Transport, and Energy, Arizona State University, 501 Tyler Mall, Tempe, AZ 85287, USA; kiran.solanki@asu.edu
* Correspondence: kuntimaddisada@yahoo.com

Received: 19 March 2020; Accepted: 26 April 2020; Published: 1 May 2020

Abstract: Crack growth kinetics in crystalline materials is examined both from the point of continuum mechanics and discrete dislocation dynamics. Kinetics ranging from the Griffith crack to continuous elastic-plastic cracks are analyzed. Initiation and propagation of incipient cracks require very high stresses and appropriate stress gradients. These can be obtained either by pre-existing notches, as is done in a typical American Society of Testing and Materials (ASTM) fatigue and fracture tests, or by in situ generated stress concentrations via dislocation pile-ups. Crack growth kinetics are also examined using the modified Kitagawa–Takahashi diagram to show the role of internal stresses and their gradients needed to sustain continuous crack growth. Incipient crack initiation and growth are also examined using discrete dislocation modeling. The analysis is supported by the experimental data available in the literature.

Keywords: crack growth; dislocation models; pile-ups; kitagawa-takahashi diagram; fracture mechanics; internal stresses

1. Introduction

Mechanical failure of materials occurs by crack initiation and growth. Griffith [1] provided the first crack growth analysis using the change in the strain energy of the cracked body in comparison to uncracked bodies, in generating new surfaces by crack initiation and the growth process. In his classical analysis, interestingly, the crack initiation and its growth are considered the same. Most metallic materials, however, yield by plastic flow before cracks form and grow. In addition, if the material deforms uniformly across the specimen cross-section, then cracks may not be initiated. The specimen may fail by the necking process. Because cracks are high energy defects, they need localized stress concentrations for their initiation and growth. Simple necking to failure can be seen in some superplastic materials [2].

Most engineering materials, however, are polycrystalline with distributed grain sizes and orientations. This results in inhomogeneous deformations across the specimen cross-section, leading to stress concentrations. The origin of the famous Hall–Petch equation that describes the grain size dependence for deformation and fracture is intrinsically related to the inhomogeneous deformations [3]. In addition, surface grains are likely to deform first due to reduced constraints near the surface. In the following, we examine the crack initiation and growth process in crystalline materials to extract some of the basic concepts involved. Discrete dislocation analysis [4] and elastic-plastic fracture mechanics methods [5] are used to understand how localized deformation and stress concentration help in the formation and growth of an incipient crack leading to specimen failure.

2. Crack Growth Analysis

It is important to look at the problem from the basics. Figure 1 gives a schematic illustration describing the kinetics of crack initiation and the growth process, from an elastic (Griffith) crack to elastic-plastic crack. The problem is discussed both from the point of continuum mechanics principles and from the discrete dislocation modeling.

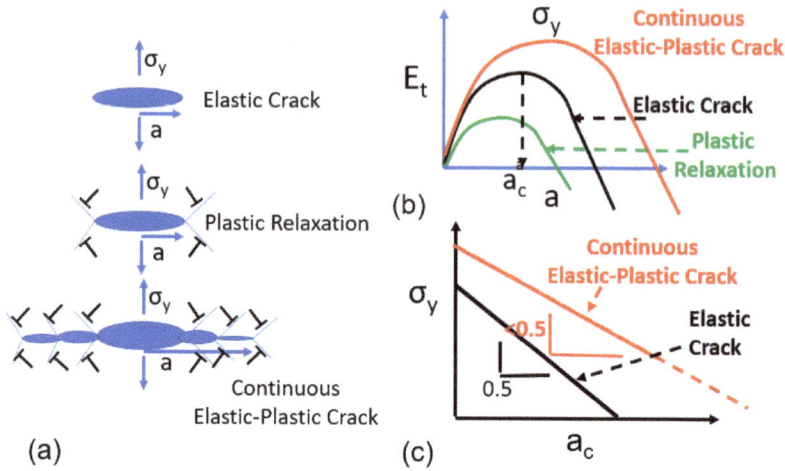

Figure 1. Analysis of crack initiation and growth for elastic to elastic-plastic cracks: (**a**) crack configurations (**b**) variation of total energy with crack length (**c**) log of stress vs. log of crack length for elastic and elastic-plastic cracks.

In Figure 1a, Griffith's elastic crack [1] is described. The total energy of the system involves the change in the elastic energy due to the presence of a crack under applied stress σ_y, work done by the applied stress, and work expended in creating two new surfaces. The total energy reaches a peak, where crack length longer than the critical size a_c will expand continuously with the reduction in the total energy, Figure 1b. The energy gradient provides the crack-tip driving force. If the energy to nucleate crystal dislocations from the crack tip is lower than the energy needed for the crack to expand further as an elastic crack, then the crack undergoes plastic relaxation, causing a reduction in the total energy of the system.

Starting from Bilby, Cottrell, and Swindon [6], continuum dislocation models are used to characterize the growth of a crack with the dislocation-plasticity. For a given crack size, the crack and associated plastic zone are analyzed using continuum dislocations. The conditions for a critical crack size for its continuous expansion are determined as a function of applied stress and lattice friction stress. In this model for mathematical simplicity, each crack with its plastic zone is treated separately. Hence, inherently the history dependence of a growing incipient crack with its plastic zone was not considered. A similar approach has been adopted by Orawan [7]. There have been many attempts to characterize tensile elastic-plastic cracks with crystal plasticity on inclined slip planes [8–10]. For a valid reason, the problem becomes mathematically intractable, and there are many attempted numerical solutions to the problem.

Figure 1a also shows the continuously expanding elastic-plastic crack. The crack growth alternates between glide and cleavage modes of crack growth. The total energy of such a crack also reaches a maximum as a function of crack length, Figure 1b. Analytically it is difficult to formulate the growth of such a crack. Figure 1c shows the log(stress) vs. log(crack length) plot for both elastic and continuous elastic-plastic cracks. For an elastic crack, which is the same as the Griffith crack, the slope is 0.5, depicting square root singularity of stress with crack length. For the continuous elastic-plastic crack, the slope is less than 0.5, and depends on the relative ratio of friction stress (or ~yield stress) to applied stress, μ.

Since it is difficult to formulate the tensile elastic-plastic cracks analytically, Marcinkoski and his group [4,11–13] analyzed the problem using discrete dislocations. Here, we present some recent results using such models [14], correlated with the results derived from continuum elastic-plastic fracture mechanics calculations [15,16], and also bring in our modified Kitagawa–Takahashi diagram [15] to extract some basic physical principles. In comparing the continuum models with the dislocation models, caution is exercised by recognizing that the scale of applications is different. Dislocation models are at the micron size level, while the continuum models are more relevant at the continuum level. Nevertheless, the fundamental concepts remain the same, as will be shown here.

2.1. Discrete Dislocation Models

Figure 2 shows schematics of discrete dislocation models. The cracks are modeled using crack dislocations, meaning the Burgers vectors of the dislocations can be variable and need not correspond to the translation vectors of the crystals. The crack is packed with a sufficient number of dislocations and their equilibrium positions are then found under the imposed applied tensile stress. The equilibrium position corresponds to the zero-force on the dislocations. Therefore, it corresponds to the traction-free condition at the crack surfaces. The total energy of the system involves an algebraic sum of the self-energies, mutual interaction energies, the work done by the applied stress in moving the dislocation from the origin (center of the crack) to their equilibrium positions, and the surface energy in creating the two crack surfaces. For glide dislocations, one has to consider also the work done against the lattice frictional stress. The total energy reaches a peak that corresponds to the Griffith condition. Beyond the peak, the total energy decreases with the crack accelerating as it grows, contributing to specimen failure. The calculations are valid if, for the elastic crack, the results match with that of the Griffith crack.

The crack can emit crystal-lattice dislocations on the slip plane (Burgers Vectors of these correspond to the crystal in question) that make an angle with the crack plane (cleavage plane) [17–19]. These dislocations move against the lattice frictional stress (which can be equated to the yield stress of the material) depending on the relative ratio of frictional stress to the applied stress (μ). If the frictional stress reaches zero, the emitted dislocations can go to infinity. On the other hand, if the stress is infinite, the crack reduces to an elastic crack. Thus, a range of material behavior can be obtained by changing this relative ratio, μ.

2.2. Continuous Elastic-Plastic Crack

By emitting crystal lattice dislocations, a growing elastic crack can reduce its energy. As the dislocations accumulate on the slip plane against the frictional stresses and mutual repulsive forces, a stage will be reached where the back stress from the emitted dislocations in the plastic zone prevents further emission of the dislocations at the crack tip. The configuration corresponds closely to that of the Bilby, Cottrell, and Swindon (BCS) crack other than the fact that the dislocations are on an inclined plane.

The continuous elastic-plastic crack, on the other hand, starts emitting crystal lattice dislocations as it grows. Hence, during the calculations, at each crack increment, the total energy is compared for dislocation emission vs. further crack extension as a cleavage crack. If the energy for emission is lower, then the crystal dislocation on the glide plane is allowed. The same test is again done for the emission of the second dislocation on the glide plane, in contrast to further crack extension as an elastic crack. For most cases, depending on the value of μ, the emission of the second dislocation is prevented due to the back stress from the previously emitted dislocation. Hence, the calculations are somewhat tedious and become intensive as the crack grows with changing glide to cleavage components depending on μ and the surface energy of the material. However, the case represents a more realistic situation with the plastic zone accumulating in the wake of the growing crack. The continuous elastic-plastic crack also captures crack growth history. Thus, depending on the μ and γ (surface energy) values, the relative components of glide vs. cleavage components change. The total energy of the incipient crack increases with an increase in the length of the crack until it reaches a peak value. Further increase in crack size only reduces the total energy, resulting in the acceleration of the crack, contributing to the total failure, Figure 2b. The material can harden as the crack grows (thus changing the μ value), thereby altering the energy-crack length curve or contributing to crack growth toughness. Figure 1c shows the log of stress vs. the log of critical crack size (at the peck energy value). Calculations show that in the log–log coordinates, the stress vs. crack length for the continuous elastic-plastic crack follows a straight line but with the slope less that of the elastic Griffith crack, which is 0.5. The slope decreases with a decrease in μ. This is similar to the effect of the decrease of the yield stress of the material. Conversely, as the yield stress increases, the crack growth behavior approaches that of the elastic crack.

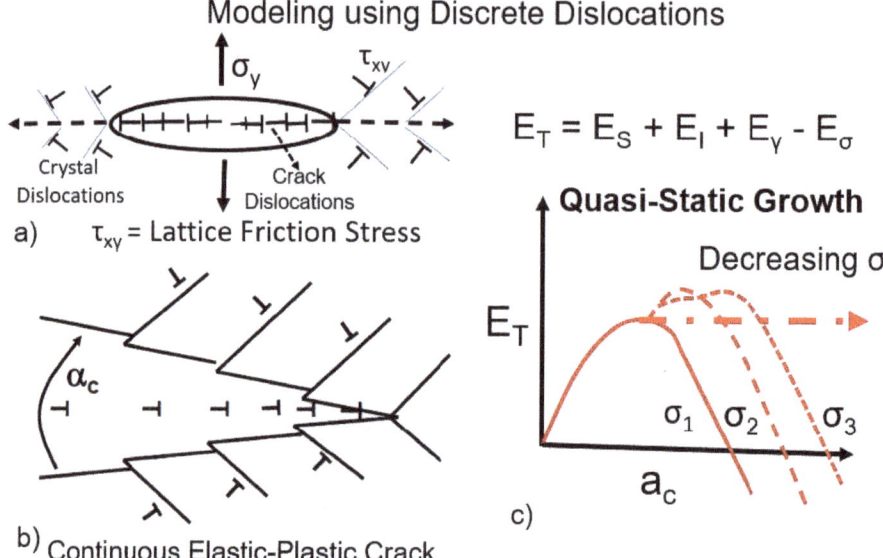

Figure 2. Discrete dislocation modeling of a continuously growing elastic-plastic crack. (a) Crack with crack and crystal dislocations with applied and lattice frictional forces. (b) Continuously expanding elastic-plastic crack with inbuilt history. Based on the total energy of the system, the crack can expand elastically or emit crystal dislocations. The crack mouth angle depends on the relative ratio of glide and cleavage planes. (c) Quasi-static growth of the crack under continuously decreasing stress holding the total energy constant after its first nucleation.

Figure 2c also depicts a thought experiment. After reaching the peak-energy for given stress as the energy starts decreasing, if the applied stress is reduced, then the energy has to again peak at a larger a_c value. One can think of an infinitesimal change in the applied stress to keep the energy at the same peak level, without it increasing or decreasing. This process can be continued with a continuous decrease in the applied stress as the crack length slowly increases to maintain the growth in equilibrium. Since the total energy remains constant, such a crack grows at a quasi-steady state. For an elastic crack, the applied stress has to be reduced, maintaining the Griffith stress with the increasing crack length. Hence in the log–log plot, the stress vs. crack length line represents the quasi-steady crack growth condition for continuously decreasing stress. If the stress is higher than the Griffith line, then the growing crack accelerates. On the other hand, if the stress falls below the line, the growing crack is arrested. This forms the condition for the crack arrest of an incipient growing crack due to a sharp decrease of applied or internal stresses that are contributing to the growth of a crack. It also leads to the Kitagawa–Takahashi type of diagram [20], as will be discussed below.

2.3. Crack Initiation at Pre-Existing Stress Concentrations

Because cracks are high energy defects, it is difficult for them to be initiated in otherwise defect-free crystals. Most fracture mechanics tests using the American Society of Testing and Materials (ASTM) criteria [21] are conducted using notched specimens and, sometimes, notched and pre-cracked specimens. The stress field of a notch is characterized by the elastic stress concentration factor K_t and the notch tip radius, ρ. Figure 3a shows an incipient crack initiated at the notch tip. The stress at the notch tip corresponds to $K_t \sigma$ but decreases with distance depending on the notch tip radius, ρ, approaching the remote stress, σ. For sharp notches, the stress gradient is sharp, while for blunt notches, the rate of decrease is slower. We have analyzed the growth of a short crack at the notch tip using elastic-plastic fracture mechanics [16]. The results are shown schematically in Figure 3c. The stress intensity factor for the short crack increases sharply from zero, decreases to some minimum, and then increases slowly with a further increase in the crack length. When the short crack length is zero, K for the short crack is also zero. The sharp increase is due to the very high notch tip stresses. Hence the initial sharp increase can be considered as within the process zone or from the point of dislocations within the core region of the notch. The decrease of K as the short crack grows is due to the gradient in the notch tip stress field. Further increase in the K value arises as the crack grows due to the remote applied stress since K increases with the crack length for a given stress. Hence, the depicted behavior of K_{sc} is expected due to the notch tip stress gradient. It may be noted that for just purely elastic calculation, K_{sc} monotonically increases and does not show the observed minimum [22].

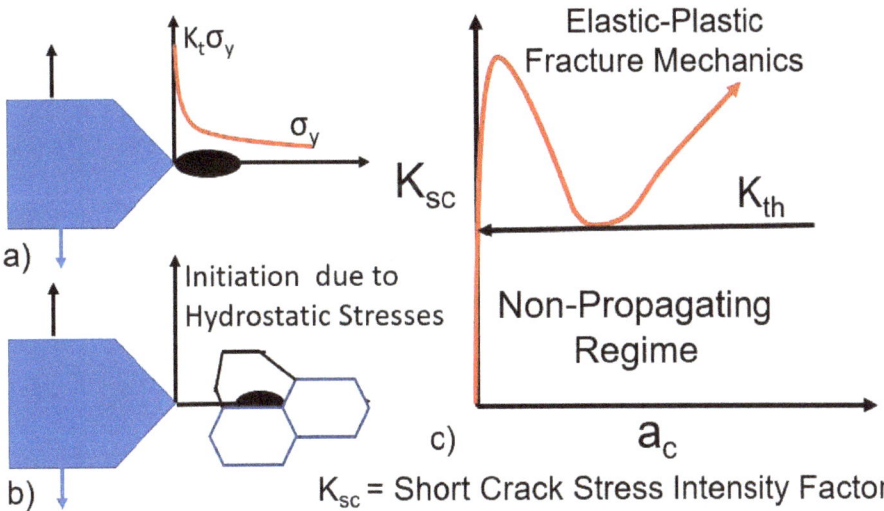

Figure 3. Crack initiation at a notch tip. (**a**) The stress field ahead of the notch tip. (**b**) Crack initiation at grain boundaries ahead of the notch tip due to hydrostatic stresses. (**c**) The variation of the stress intensity factor K_{sc} of the incipient crack growing nearing a notch tip.

For the continuous growth of the initiated crack at the notch tip, the minimum must exceed the threshold stress intensity factor, K_{th} for crack growth. Otherwise, the incipient crack that is growing in the high-stress field of the notch is arrested when K_{sc} drops below K_{th}. The minimum of the K_{sc} value is related to the internal stress (notch tip stress) magnitude and its gradient. For very sharp notches ($\rho \sim 0$), the stress gradient can be sharp, leading to arrest of the growing short crack leading to non-propagating cracks at sharp notches. This is observed, particularly under fatigue, leading to fatigue stress concentration factor, K_{FC}, differing from the elastic stress concentration factor, K_t. The magnitude of the stress at the notch tip also depends on the applied stress, σ_{apl}. We have shown [16] that the minimum applied stress needed for the continuous growth of incipient crack near the stress concentration can be expressed as:

$$\sigma_{apl} = \frac{2K_{th}}{(K_t)^{1.3} \times \sqrt{\rho}} \qquad (1)$$

where K_{th} corresponds to the threshold for crack growth. It can be a threshold for any subcritical crack growth (thresholds for fatigue, stress corrosion, corrosion-fatigue, sustained load, or even for a fracture, such as K_{IC}). K_t and ρ are elastic stress concentration factor and notch-tip root radius, respectively. The equation has been successfully applied to the extensive notch-fatigue data available in the open literature. Recently, the equation has been applied to determine the pit to crack transition under corrosion fatigue [23].

Figure 3b also shows the crack initiation ahead of the notch tip. This can occur at grain boundary junctions (accentuated by the presence of carbide particles or inclusions) due to high hydrostatic stresses present. Several models have been developed assuming such nucleation [24–26]. We show later a similar problem was analyzed using discrete dislocations where short crack nucleation ahead of a blunt crack is considered. Figure 4 shows typical results of short crack initiation and growth near notches with different K_t values but for a fixed ρ, showing how minimums in K_{sc} values become sharper with increasing K_t value.

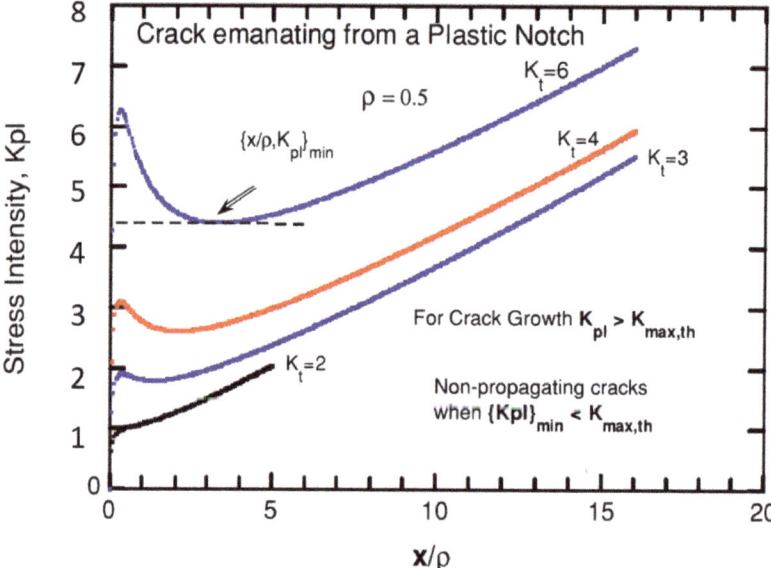

Figure 4. Typical results of actual calculations of K_{sc} (called K_{pl} due to elastic-plastic stress fields) for different K_t values but for a fixed ρ, showing the minimum in K_{sc}.

2.4. Crack Initiation at the In Situ Generated Stress-Concentrations

Since stress concentrations are essential for the initiation of cracks in the material, if there are no pre-existing cracks or stress concentrations, then the inhomogeneous deformations produce localized stress concentrations via the formation of dislocation pile-ups, as discussed by several authors, beginning with Eshelby [27]. The typical internal stress profile ahead of a dislocation pile-up deduced by Eshelby is shown in Figure 5, which was experimentally confirmed by X-ray analysis by Gao et al. [28]. The internal stress profile is not much different from that of a notch tip stress field. In fact, it was shown that most of the stress fields at stress concentrations follow a similar pattern [29]. The implication is that incipient cracks that form at stress concentrations will also have typical profiles for their stress intensity factor, as shown in Figures 3 and 4, with minima that must exceed the governing threshold for continuous crack growth. Otherwise, initiated growing short cracks are arrested. Experimentally, one finds many short cracks are initiated near the surface under, for example, corrosion, but only one or two grow and thus contribute to specimen failure.

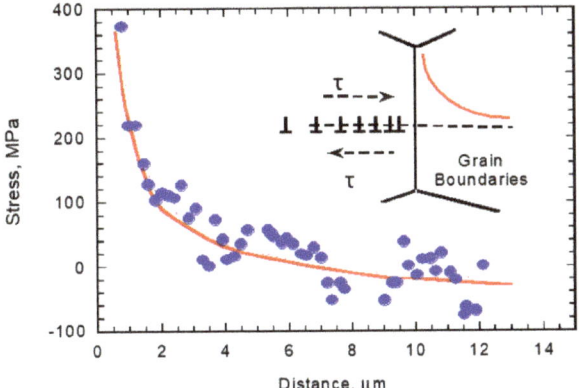

Figure 5. Stress field in the next grain due to dislocation pile-up, from Eshelby, and X-ray data from Gao et al. (2014).

2.5. Role of Internal Stresses and the Modified Kitagawa-Takahashi Diagram

From Griffith's equation, we note that the stress for crack growth increases with decreasing crack length according to the relation:

$$\sigma_{apl} = \frac{\sqrt{(2E\gamma)}}{\sqrt{\pi a(1-\nu)}} \qquad (2)$$

Where the required applied stress increases with decreasing size of the crack approaching infinity when the crack length goes to zero. Figure 6 shows this in a log–log plot. For ductile materials, with the increase in the applied stress, the material yields before a crack is initiated. From Figure 6, we can define the minimum internal stresses, and the gradients required for crack initiation can be defined by what we call an internal stress triangle. In some sense, we are generating the local internal stresses, by way of dislocation pile-ups, to augment the applied stress for the crack initiation and growth, until the remote applied stress is sufficient to sustain further growth of the initiated crack. The internal stress triangle thus defines both the magnitude and gradient required for incipient crack initiation and its growth. Based on this figure, the following points can be made: (a) In addition to the magnitude of the stress, its gradient is also involved in sustaining the growth of the incipient crack. If the gradient is too sharp, as in the case of sharp notches, the initiated crack may be arrested if the stress falls below Griffith's line. (b) Furthermore, it is difficult to separate the initiation vs. growth as both are simultaneously involved since Griffith's condition corresponds to the maxima in the total energy (unstable equilibrium). It may be possible that the initiated crack can be stabilized due to oxidation of the mating crack surfaces, but this is a separate issue.

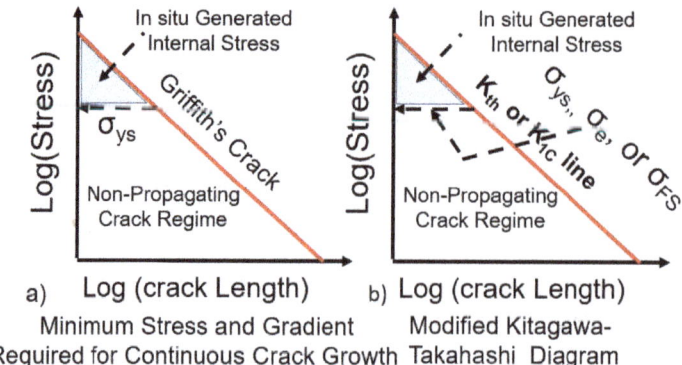

Figure 6. (a) Griffith crack representation with yield stress defining the required minimum internal stress magnitude and gradient. (b) Parallel representation of the modified Kitagawa–Takahashi diagram for subcritical crack growth.

Figure 6b shows a similar behavior that can be extracted using the modified Kitagawa–Takahashi diagram, which was initially developed for fatigue failure. The crack-growth threshold stress-intensity factor replaces Griffith's criterion, and the endurance limit replaces the yield stress. Our modification involves extending the threshold line beyond the endurance limit, thereby defining the internal stress triangle. At the endurance of a smooth specimen, close to 10^7 cycles are needed for the crack to initiate and grow. These cycles are required for the development of the needed internal stresses and their gradients for an incipient crack to form and grow. The initiation and growth of the short crack in the endurance have been accounted for by the fracture mechanics community by invoking the similitude break down and proposing that the short crack threshold is different from that of long crack thresholds due to crack closure. We have shown using the dislocation theory that the crack-closure concept is inherently faulty in the plane strain regim e, and no similitude break down is needed to account for the short crack growth behavior. The short crack grows due to the presence of both applied and in situ generated internal stresses arising from inhomogeneous deformations in the polycrystalline materials. The thresholds do not depend on the crack size, and one has to properly account for the local build-up of internal stresses and their gradients due to dislocation pile-ups. A detailed review of short crack growth was provided recently [30].

Internal stresses are difficult to determine. One can compute them based on some physical models such as dislocation pile-ups. The Kitagawa diagram provides some way to estimate the required minimum internal stresses and their gradients for the incipient crack to grow at the threshold condition. The stresses are higher than those causing acceleration of the crack growth while the stresses are lower than those causing crack arrest. Overloads and underloads, for example, can change local internal stresses (sometimes referred to as residual stresses, which are only a subset of the internal stresses) and contribute to changes in the crack growth kinetics. From equilibrium consideration, the internal stresses are self-equilibrating. That is, there will always be a plus/minus type with the net result of maintaining the specimen in equilibrium. The fact is internal stresses resulting from inhomogeneous deformations are involved in the nucleation and propagation of the cracks in specimens, even though they are difficult to measure.

2.6. Role of Chemical Forces

Figure 7 provides a simple case where chemical forces manifest in terms of reduction in the surface energy of the crack surfaces, thereby reducing the required applied stress needed for a Griffith crack, for example, to be initiated and grow (Equation (2)). The total energy as a function of the crack length is reduced (Figure 7a), and, correspondingly, the applied stress needed for a given crack length is reduced (Figure 7b). The micromechanisms involved in the reduction of crack tip driving force can be complex. Nonetheless, the net result from the point of engineering considerations is that there is a reduction in the required applied stress to contribute to the same crack length or crack growth rate. Hence, we can formally define the mechanical equivalent of chemical forces based on the reduction in the required stress to cause the same crack growth rate. This is shown in Figure 7b. To compute the chemical stresses involved, we need, therefore, crack growth in the inert medium as a reference state. For corrosion fatigue, fatigue in an inert medium can be used as a reference state. However, care should be exercised since fatigue is a two-load parametric problem requiring σ_{max} and $\Delta\sigma$ for Stress vs. number of cycles for failure (S-N fatigue) fatigue or two stress intensity factors (K_{max}, ΔK) for fatigue crack growth [30]. However, for stress corrosion or sustained load crack growth, there is no subcritical crack growth in an inert environment for a reference state. Only the fracture toughness value in an inert medium provides the reference.

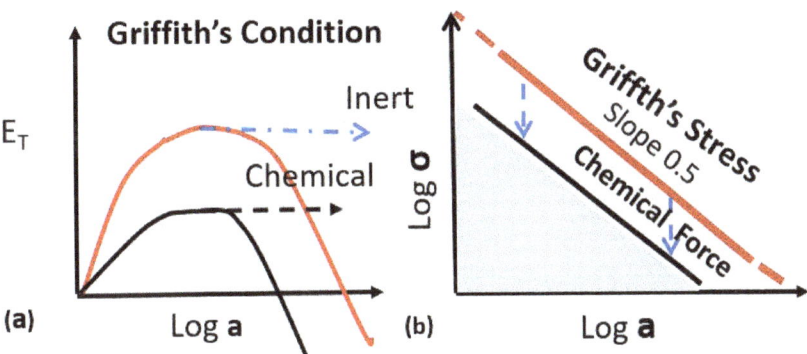

Figure 7. Role of chemical forces and their quantification. (**a**) Total energy as a function of crack length and (**b**) log(stress) vs. log(crack length) plot.

3. Experimental Support for the Above Concepts

Figure 8 shows experimental threshold data under corrosion fatigue for steels with different heat treatments resulting in different yields stresses (from Usami [31]). Interestingly the change in the yield stress affects only the endurance limit but not long crack growth thresholds. Why this is so is not clear. With increasing yield stress, the material's behavior is increasingly elastic, as one should expect. Correspondingly, the internal stress triangle becomes smaller with increasing yield stress, indicating that the needed contribution from the local internal stress decreases. When the yield stress exceeds the fracture stress, the material becomes brittle. Figure 9 shows another example based on Hiroshi–Mura data [32] on stress corrosion of 4340 steel in $H_2(SO)_4$. The data are plotted in the form of the Kitagawa–Takahashi diagram. The endurance stress is similar to the minimum failure stress, σ_{th}, of a smooth specimen loaded in the corrosive environment. The mechanical equivalent of chemical internal stress is defined in the figure. The extent of experimental data on smooth and fracture mechanics specimens in corrosive media available in the open literature is limited. Nevertheless, the analysis shows that transition from short cracks to long cracks and the role of internal stresses in accentuating the crack initiation and growth process are general for all subcritical crack growth processes in materials.

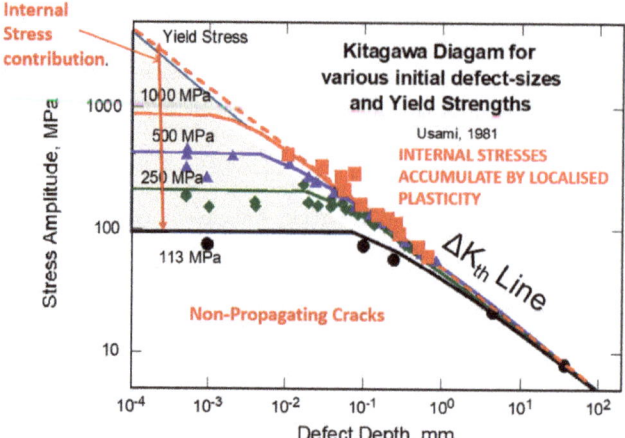

Figure 8. Corrosion-fatigue crack growth in alloy steels with varying yield stress, from Usami, 1981.

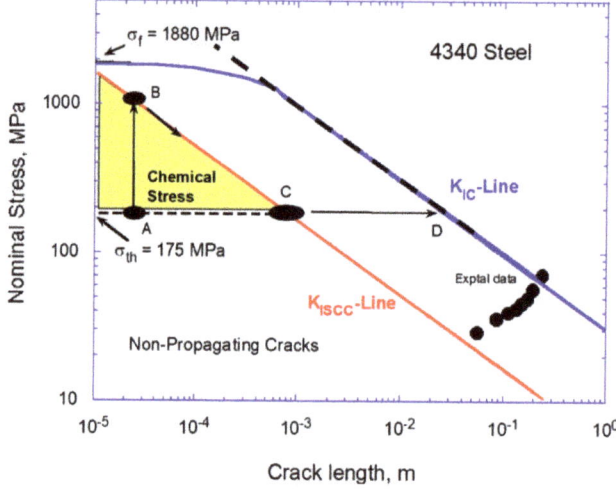

Figure 9. Modified Kitagawa–Takahashi diagram for stress corrosion crack growth in 4340 steel in H2SO4 solution, extracted from Hiroshi–Mura data.

3.1. Application to Fracture Toughness

The above concepts are relevant not only for subcritical crack growth, but also to fracture toughness (Figure 10). Experimental data from Bucci (1996; [33]) are shown in Figure 10a, and the data are plotted in terms of the modified Kitagawa–Takahashi diagram for two selected alloys (Figure 10b), one with low and the other with high fracture stresses. The horizontal portions of the plot correspond to the tensile fracture stress, while the inclined lines represent the K_{1C} lines. The extension of the K_{1C} lines defines the internal stresses needed to initiate a crack in a smooth specimen. The internal stress triangle is large for the low yield stress 2024-T3 alloy in comparison to the high yield stress 7075-T6 alloy, as expected.

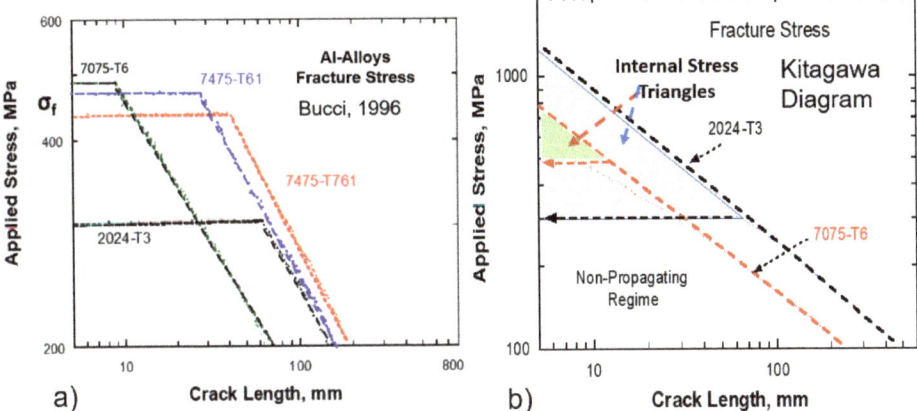

Figure 10. Application to K_{1C}, fracture data. (**a**) Experimental data from Bucci, 1996. (**b**) Representation of the data in terms of the modified Kitagawa–Takahashi diagram showing the internal stress triangle for the two cases, 7075-T6 and 2024-T3 Al-alloys.

3.2. Discrete Dislocation Models

The formulation of the discrete dislocation models for continuous elastic-plastic cracks was discussed earlier (Section 2.2). Here we present some results of our calculations [14]. Since the computations are quite involved, the analysis could only be undertaken for microscopically sized cracks. In Figure 11, the effect of μ, the ratio of friction stress to applied stress on the growing elastic-plastic cracks is shown. With the decrease of μ, the glide component increases with the cleavage component. Figure 11 shows the slope of the log(applied stress) vs. the log(crack length) decreases with the decrease of μ. Nevertheless, the exponential relation remains with the exponent of σ vs. a decrease from 0.5. Figure 12 shows the total energy of the growing crack as a function of crack size for two μ values, 0.6 and 0.8, based on the calculations reported in ref. [14]. The crack grows along the path that has lower energy. The figure shows that glide and cleavage components fluctuate until the crack becomes unstable. The relative proportions of the two components vary depending on the μ value. Experiments undertaken with varying H concentrations support these results.

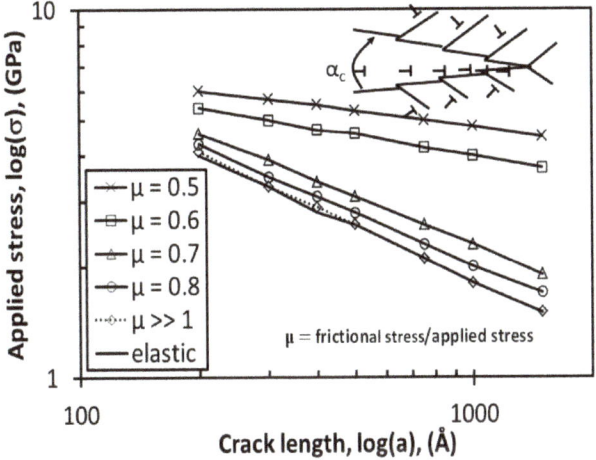

Figure 11. Discrete dislocation analysis of continuous elastic-plastic cracks for various μ values.

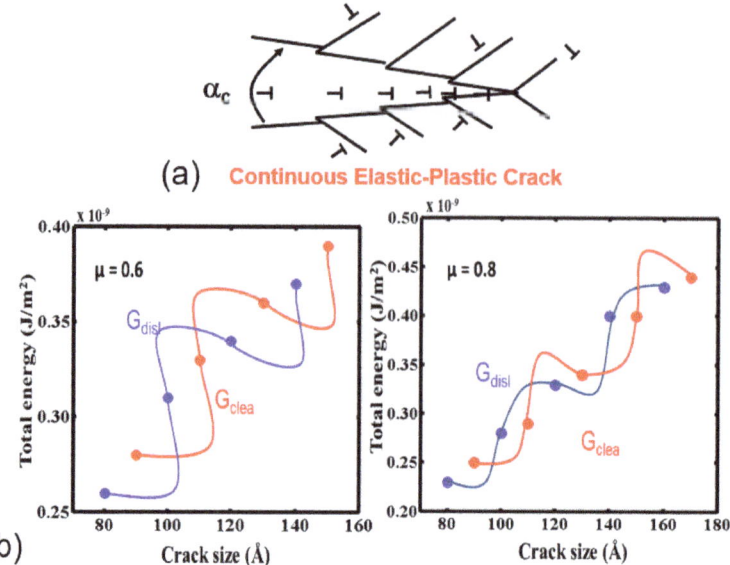

Figure 12. Changes in the relative glide and cleavage components with the change in μ values for continuous elastic-plastic cracks.

3.3. Effect of Hydrogen Pressure

The effect of hydrogen on the embrittlement of steels has received significant attention [34]. Recently, Adlakha and Solanki [35] analyzed this problem extensively using both atomic simulations and continuum-based models. They show that hydrogen segregation around the crack tip enhanced both dislocation emission and cleavage behavior. Figure 13 shows the discrete dislocation analysis of elastic-plastic cracks in the presence of hydrogen [14]. The results indicate that for a given μ, the ratio of lattice friction to applied stress, the slope of log(stress) vs. log(crack length) increases as the hydrogen pressure increases, due to the increased cleavage component, and the curve moves closer to the Griffith crack.

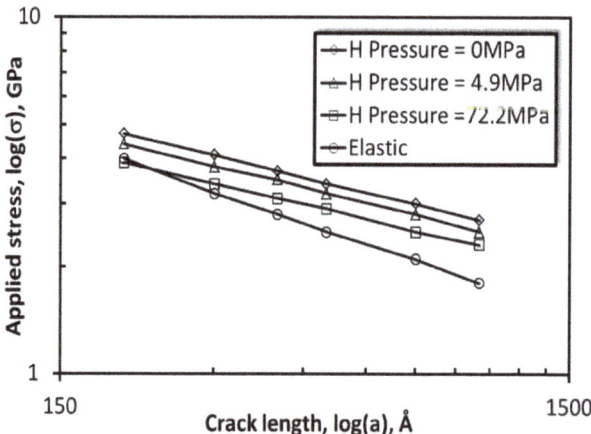

Figure 13. Effect of hydrogen pressure on crack growth for a given μ ratio. With the increase in H pressure, the lines move closer to the elastic Griffith crack behavior.

Experimental support for the analysis is provided in Figure 14, which is taken from Vehoff and Rothe [36], and shows that the cleavage component increases with increasing H-pressure, thereby reducing the crack mouth angle, α. Attached micrography shows the reduction in the crack mouth angle with increasing H-pressure. Further analysis (Figure 15) shows that the hydrogen effect saturates at high pressures. Thus, we have a negligible effect at shallow pressures and a saturation effect at high pressures. In the intermediate range, the presence of hydrogen reduces the mechanical driving force for crack growth, making the material more brittle. Experimental data are shown in Figure 16, taken from ref. [37], on the hydrogen embrittlement in alloy steels, which supports the above results.

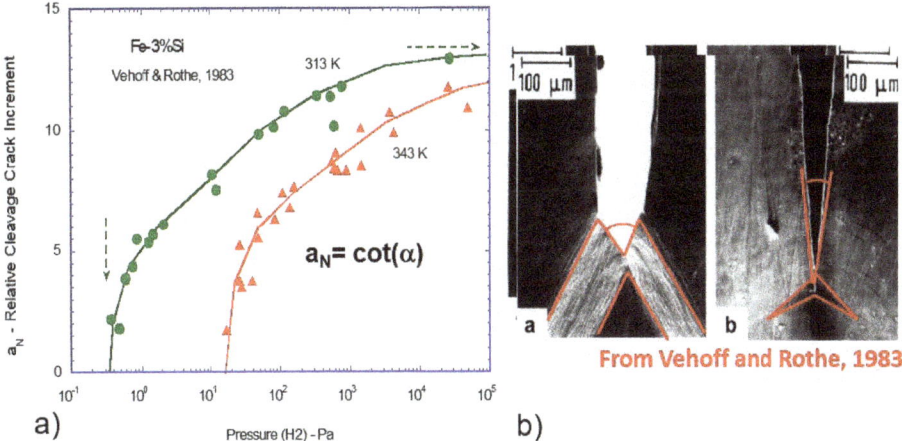

Figure 14. Experimental results from Vehoff and Roth, 1983, on Fe-3%Si showing the effect of hydrogen pressure on crack growth. (**a**) With the increase in the hydrogen pressure cot(α) increases (α decreases) due to the increase in the cleavage component. (**b**) The associated micrograph indicates the crack mouth angle decreases when hydrogen pressure increases.

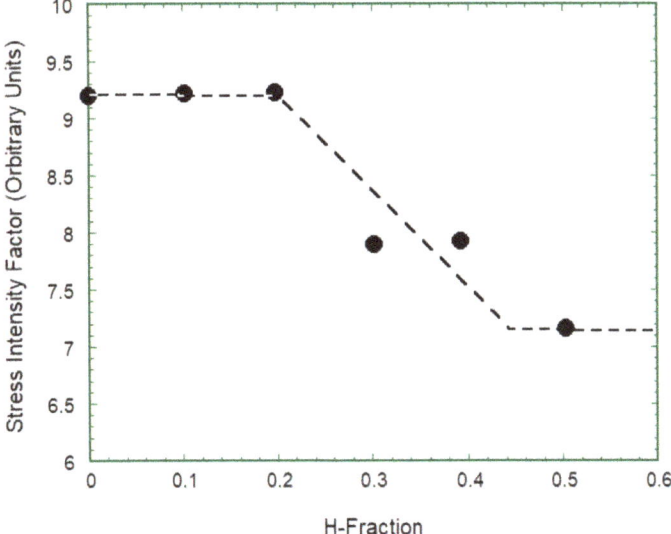

Figure 15. Results of discrete dislocation analysis showing that the stress-intensity factor of an incipient crack shows a saturation effect with H coverage.

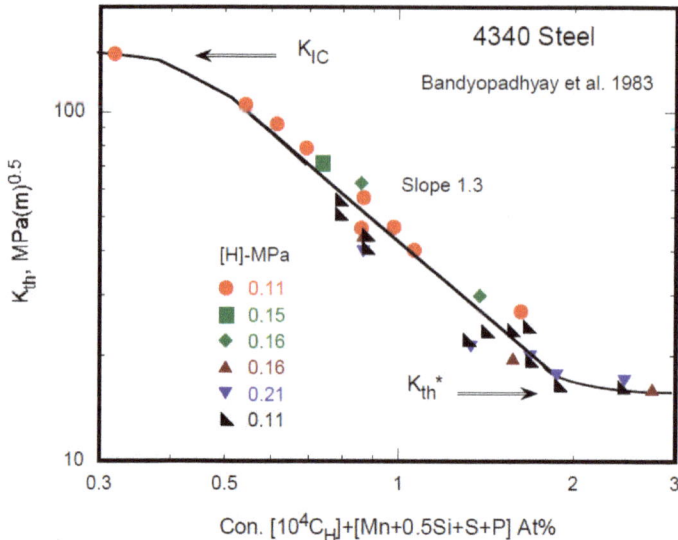

Figure 16. Experimental results showing the saturation effect with the embrittling species on the crack growth threshold.

3.4. Crack Initiation Ahead of the Main Crack

Finally, we treat the growth of a short crack initiated ahead of the main crack, similar to that shown in Figure 3b. The equilibrium configuration of both the glide and cleavage dislocations are shown for selected fractions of hydrogen coverage. With an increase in hydrogen coverage, the material becomes more brittle, or from the dislocation point, the cleavage crack becomes narrower, while the glide part of the crack becomes wider, reflecting crack opening displacements. Figure 17 shows (a) the short crack grows more towards the main crack, (b) the crack tip plasticity or glide component is more for the segment facing the main crack, and (c) with increasing hydrogen coverage, the glide components at both ends decrease and approach the cleavage mode of crack growth. In these calculations, the blunting radius of the main crack is held constant. Future calculations will involve consideration of the plasticity at both the main crack and the initiated short crack, although such calculations will be time-consuming since the number of dislocations involved becomes large.

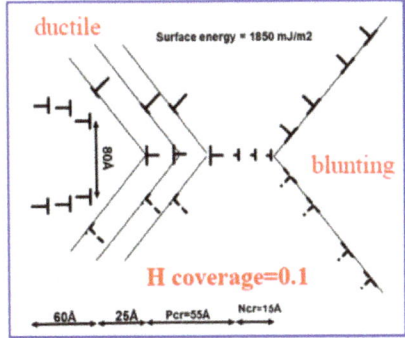
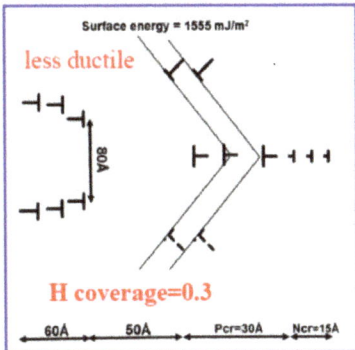
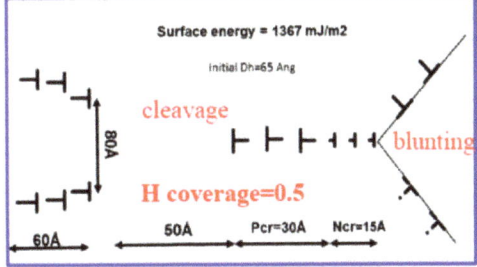

Figure 17. Discrete dislocation models of growth of an incipient crack from ahead of the main crack for various hydrogen coverages.

4. Summary

A detailed analysis of crack initiation and growth in crystalline materials is provided both from the perspective of continuum mechanics and dislocation dynamics. Basic principles underlying crack initiation and growth are highlighted. The role of pre-existing or in situ generated stress concentrations in the initiation of the incipient cracks in crystalline materials is outlined. It is shown that stress concentrations are essential for the nucleation of cracks. They provide the local internal stresses and gradients needed for crack initiation and growth. The application of the modified Kitagawa–Takahashi diagram in accounting for the role of internal stresses for the initiation and growth of short cracks, for both subcritical crack growth and fracture, is discussed. Discrete dislocation models are presented, and the effect of hydrogen on crack growth kinetics is analyzed.

Author Contributions: All authors have read and agreed to the published version of the manuscript. Conceptualization; K.S., I.A., K.N.S. Methodology K.S., I.A., K.N.S., A.K.V.; Software I.A., K.N.S., Validation, I.A., K.N.S.; Formal Analysis, K.S., I.A., K.N.S.; Investigation, I.A., K.N.S.; resources K.S., I.A., K.N.S., A.K.V.; data curation, K.S., I.A., K.N.S.; Writing-original draft preparation, K.S., I.A.; writing review and editing, K.S., I.A.; Visualization, K.S.; Supervision, K.S., K.N.S.; project administration, Nagaraja Iyer, Technical Data Analysis, funding acquisition.

Funding: This research is supported by funds received from the Office of Naval Research, USA, under Technical Data Analysis Contract # N68335-16-C-0135 with Anisur Rahman as the Project Monitor.

Conflicts of Interest: The authors declare no conflict of interest.

Symbols Used

ASTM—American Society of Testing and Materials
KT diagram—Kitagawa Takahashi
μ—Ratio of frictional stress to applied stress.
BCS model—Bilby, Cottrell, and Swindon model
a—Crack Length

a_c—Critical Crack length
σ_y—Normal Stress
E_T—Total Energy of the System
E_S—Self Energy of all dislocations
E_I—Interaction energy of all dislocation
E_γ—Surface energy
E_σ—Work done by applied stress
τ_{xy}—Lattice frictional stress
α_c—Crack mouth angle
K_t—Elastic Stress Concentration Factor
ρ—Notch tip radius
K_{SC}—Stress intensity factor for short crack
K_{th}—Threshold stress intensity factor for crack growth
K_{IC}—Fracture Toughness
K_{ISCC}—Stress Corrosion Crack Growth Threshold
K_{pl}—Stress intensity factor for short crack in the elastic plastic notch tip field.
ΔK_{th}—Fatigue crack growth threshold Stress intensity range.
σ_{apl}—Applied Stress
σ_{ys}—Yield Stress
σ_e—Endurance Stress
σ_{FS}—Fracture Stress
E—Elastic Modulus
γ—Surface Energy
ν—Poisson ratio

References and Note

1. Griffith, A.A. The phenomena of rupture and flow in solids. *Philos. Trans. R. Soc. Lond.* **1921**, *221*, 163–198.
2. Dieter, G.E. *Mechanical Metallurgy*, 3rd ed.; McGraw-Hill Inc.: New York, NY, USA, 1986; pp. 299–301, 452–453. ISBN 0-07-016893-8.
3. Armstrong, R.W. Material grain size and crack size influences on cleavage fracturing. *Philos. Trans. R. Soc. A* **2015**, *373*, 20140124. [CrossRef] [PubMed]
4. Jagannadham, K.; Marcinkowski, M.J. *Unified Theory of Fracture*; Trans Tech Publications Ltd.: Zurich, Switzerland, 1983.
5. Rice, J.R. Elastic-plastic fracture mechanics. *Eng. Fract. Mech.* **1973**, *5*, 1019–1022. [CrossRef]
6. Bilby, B.A.; Cottrell, A.H.; Swinden, K.H. The spread of plastic yield from a notch. *Proc. R. Soc. Lond. A* **1963**, *272*, 304–314.
7. Orowan, E. Fracture and strength of solids. *Rep. Prog. Phys. XII* **1948**, *12*, 185–232. [CrossRef]
8. Zhao, R.H.; Dai, S.-H.; Li, J.C.M. Dynamic emission of dislocations from a crack tip, a computer simulation. *Int. J. Fract.* **1985**, *29*, 3. [CrossRef]
9. Rice, J.R. Tensile crack tip fields in elastic-ideally plastic crystals. *Mech. Mater.* **1987**, *6*, 317–335. [CrossRef]
10. Weertman, J. *Mathematical Theory of Dislocations*; Mura, T., Ed.; American Society of Mechanical Engineers: New York, NY, USA, 1969; p. 178.
11. Marcinkowski, M.J. Dislocation model of a plastic tensile crack. *J. Appl. Phys.* **1975**, *46*, 496. [CrossRef]
12. Marcinkowski, M.J.; Das, E.S.P. Relationship between crack dislocations and crystal lattice dislocations. *Phys. Status Solidi A* **1971**, *8*, 249. [CrossRef]
13. Sadananda, K.; Jagannadham, K.; Marcinkowski, M.J. Discrete dislocation analysis of a plastic tensile crack. *Phys. Status Solidi A* **1977**, *44*, 633–642. [CrossRef]
14. Adlakha, I.; Sadananda, K.; Solanki, K.N. Discrete dislocation modeling of stress corrosion cracking in an iron. *Corros. Rev.* **2015**, *33*, 467–475. [CrossRef]
15. Sadananda, K.; Sakar, S. Modified Kitagawa Diagram and Transition from Crack Nucleation to Crack Propagation. *Metall. Mater. Trans.* **2012**, *44*, 1175–1189. [CrossRef]
16. Sadananda, K.; Arcari, A.; Vasudevan, A.K. Does a nucleated crack propagate? *Eng. Fract. Mech.* **2017**, *176*, 144–160. [CrossRef]

17. Rice, J.R.; Thomson, R.M. Ductile versus brittle behavior of crystals. *Philos. Mag.* **1974**, *29*, 73. [CrossRef]
18. Rice, J.R. Dislocation nucleation from a crack tip: An analysis based on the Peierls concept. *J. Mech. Phys. Solid.* **1992**, *40*, 239–271. [CrossRef]
19. Beltz, G.E.; Rice, J.R.; Shih, C.F.; Xia, L. A self-consistent model for cleavage in the presence of plastic flow. *Acta Mater.* **1996**, *44*, 3943–3954. [CrossRef]
20. Kitagawa, H.; Takahashi, S. Applicability of fracture mechanics to a very small crack or the cracks in the early stage. In Proceedings of the Second International Conference on Mechanical Behavior of Materials, Cleveland, OH, USA, 16–20 August 1976; pp. 627–631.
21. Active Standard ASTM E399, Standard Test Method for Linear-Elastic Plane-Strain Fracture Toughness K_{IC} of Metallic Materials.
22. Raju, I.S.; Atluri, S.N.; Newman, J.C., Jr. Stress intensity factors for small surface and corner cracks in plates. In *Fracture Mechanics: Perspectives and Directions (Twentieth Symposium)*; Wei, R.P., Gangloff, R.P., Eds.; American Society for Testing and Materials (ASTM): Philadelphia, PA, USA, 1989; pp. 297–316.
23. Sadananda, K.; Vasudevan, A.K. Analysis of pit to crack transition under corrosion fatigue & the safe-life approach using the modified Kitagawa-Takahashi. *Int. J. Fatigue* **2020**, *134*, 105471.
24. Rice, J.R.; Tracy, D.M. On the ductile enlargement of voids in the triaxial stress fields. *J. Mech. Phys. Solids* **1969**, *17*, 201–217. [CrossRef]
25. Ritchie, R.O.; Knott, J.F.; Rice, J.R. Relationship between critical tensile stress and fracture toughness in mild steel. *J. Mech. Phys. Solids* **1973**, *21*, 395–410. [CrossRef]
26. McCoy, R.A.; Gerberich, W.W. Hydrogen Embrittlement Studies of a Trip Steel. *Metall. Trans.* **1973**, *4*, 539–547. [CrossRef]
27. Eshelby, J.D.; Frank, F.C.; Nabarro, F.R.N. The equilibrium array of Dislocations. *Philos. Mag.* **1951**, *42*, 351–364. [CrossRef]
28. Guo, Y.; Britton, T.B.; Wilkinson, A.J. Slip band grain boundary interactions in commercial pure Titanium. *Acta Mater.* **2014**, *76*, 1–12. [CrossRef]
29. Glinka, G.; Newport, A. Universal function of elastic notch tip stress fields. *Int. J. Fatigue* **1987**, *9*, 144–150. [CrossRef]
30. Sadananda, K.; Nani Babu, M.; Vasudevan, A.K. A Review of Fatigue crack growth resistance in the short crack growth regime. *Mater. Sci. Eng. A* **2019**, *754*, 674–701. [CrossRef]
31. Usami, S. Applications of threshold cyclic-plastic-zone-size criterion to some limit problems. In *Fatigue Thresholds*; Blacklund, I., Bloom, A., Beevers, C.J., Eds.; EMAS: Stockholm, Sweden, 1981; pp. 205–238.
32. Hirose, Y.; Mura, T. Nucleation mechanism of stress corrosion cracking from notches. *Eng. Fract. Mech.* **1984**, *19*, 317–329. [CrossRef]
33. Bucci, R.J.; Nordmark, G.; Starke, E.A., Jr. Selecting Aluminum alloys to resist failure by fracture mechanisms. In *ASM Handbook 19, Fatigue and Fracture*; Materials Park, OH 44073-0002; ASM Handbook Committee: Oxfordshire, UK, 1996; pp. 771–812.
34. Gangloff, R.P. Environmental Cracking-Corrosion Fatigue, Ch.26. In *Corrosion Tests and Standard Manual*, 2nd ed.; Baboian, R., Ed.; MIL-HDBK 729; ASTM International: West Conshohocken, PA, USA, 2005.
35. Adlakha, I.; Solanki, K.N. Critical assessment of hydrogen effects on the slip transmission across grain boundaries in α-Fe. *Proc. R. Soc. A* **2016**, *472*, 617. [CrossRef] [PubMed]
36. Vehoff, H.; Rothe, W. Gaseous hydrogen embrittlement in *FeSi* and Ni single crystals. *Acta Metall.* **1983**, *31*, 1781–1793. [CrossRef]
37. Bandyopadhyay, N.; Kameda, J.; McMahon, C.J., Jr. Hydrogen-induced cracking in 4340-type steel: Effects of composition, yield strength and hydrogen pressure. *Metall. Trans. A* **1983**, *14*, 881–888. [CrossRef]

© 2020 by the authors. Licensee MDPI, Basel, Switzerland. This article is an open access article distributed under the terms and conditions of the Creative Commons Attribution (CC BY) license (http://creativecommons.org/licenses/by/4.0/).

Article

Tensile Deformation of Ultrafine-Grained Fe-Mn-Al-Ni-C Alloy Studied by In Situ Synchrotron Radiation X-ray Diffraction

Si Gao [1,*], Takuma Yoshimura [1], Wenqi Mao [1], Yu Bai [1], Wu Gong [2,3], Myeong-heom Park [2], Akinobu Shibata [1,2], Hiroki Adachi [4], Masugu Sato [5] and Nobuhiro Tsuji [1,2]

1. Department Materials Science and Engineering, Kyoto University, Kyoto 606-8501, Japan; yoshimura.takuma.34n@st.kyoto-u.ac.jp (T.Y.); mao.wenqi.26m@st.kyoto-u.ac.jp (W.M.); bai.yu.6m@kyoto-u.ac.jp (Y.B.); shibata.akinobu.5x@kyoto-u.ac.jp (A.S.); nobuhiro-tsuji@mtl.kyoto-u.ac.jp (N.T.)
2. Elements Strategy Initiative for Structural Materials (ESISM), Kyoto University, Kyoto 606-8501, Japan; gong.wu.3x@kyoto-u.ac.jp (W.G.); park.myeongheom.8r@kyoto-u.ac.jp (M.-h.P.)
3. J-PARC Center, Japan Atomic Energy Agency, Ibaraki 319-1195, Japan
4. Department of Materials and Synchrotron Radiation Engineering, Graduate School of Engineering, University of Hyogo, Himeji 671-2280, Japan; adachi@eng.u-hyogo.ac.jp
5. Japan Synchrotron Radiation Research Institute (JARSI), Sayo-gun, Hyogo 679-5198, Japan; msato@spring8.or.jp
* Correspondence: gao.si.8x@kyoto-u.ac.jp

Received: 6 October 2020; Accepted: 3 December 2020; Published: 7 December 2020

Abstract: Intermetallic compounds are usually considered as deleterious phase in alloy designing and processing since their brittleness leads to poor ductility and premature failure during deformation of the alloys. However, several studies recently found that some alloys containing large amounts of NiAl-type intermetallic particles exhibited not only high strength but also good tensile ductility. To clarify the role of the intermetallic particles in the excellent tensile properties of such alloys, the tensile deformation behavior of an ultrafine-grained Fe-Mn-Al-Ni-C alloy containing austenite matrix and B2 intermetallic particles was investigated by using in situ synchrotron radiation X-ray diffraction in the present study. The elastic stress partitioning behavior of two constituent phases during tensile deformation were quantitively measured, and it was suggested that B2 particles played an important role in the high strength and large tensile ductility of the material.

Keywords: ultrafine-grained materials; intermetallic compounds; B2 phase; strain hardening behavior; synchrotron radiation X-ray diffraction

1. Introduction

Intermetallic compounds with coarse particle size are usually avoided in the alloy designing, because their brittleness often leads to poor deformability at room temperature. However, recent studies showed that some steels and alloys containing intermetallic compounds as second phase exhibited excellent combination of tensile strength and ductility, even at room temperature. Furuta et al. [1] reported that a heavily deformed Fe-Ni-Al-C alloy containing NiAl-type B2 intermetallic compound particles showed a yield strength of 2.2 GPa, whilst still keeping a 25% tensile elongation. Kim et al. [2] developed an Fe-Mn-Al-Ni-C light-weight steel composed of ultrafine-grained austenite and B2 intermetallic compounds phase that exhibited quite high specific strength and good tensile elongation. In addition, B2 phase has also been frequently observed in high entropy or multi-component alloys having high strength and good tensile elongation [2–6]. It was believed that the B2 phase played an important role in the excellent mechanical properties of those materials. Yang et al. [7] studied the strain hardening behavior of Fe-Mn-Al-Ni-C steel and suggested that the high back stress, rising from the incompatibility between the matrix and B2 phase, accounted for the high strain hardening rate and the excellent tensile properties of the material. The present author studied tensile properties of an ultrafine-grained dual-phase Fe-24Ni-6Al-0.4C alloy composed of ultra-fine grained (UFG) austenite and B2 phase and suggested that B2 phase was somehow important to the high strength of the material [8]. More recently, Kim et al. [9] argued that at least in the Fe-Mn-Al-Ni-C system, the high strain hardening rate was largely attributed to the intensive planar slip of dislocations in the austenite matrix that was enhanced by the short-range ordering of the Mn-C, rather than the existence of B2 particles. Those studies, mostly using post-mortem microstructure characterization, are either qualitative or indirect. The role of B2 phase in the tensile properties of the materials has not yet been dynamically evaluated during deformation. In situ diffraction measurement has been proven as an appropriate tool to study the deformation behavior of alloys consisting of multiple phases for its capability to follow and distinguish the evolution of the stress state and phase volume fraction of different constituent phases during mechanical test [8–15]. In the present study, a tensile test with in situ synchrotron radiation X-ray diffraction measurement was carried out on an ultrafine-grained Fe-Mn-Al-Ni-C alloy containing B2 particles, in order to elucidate the contribution of B2 phases to the tensile properties of the material.

2. Materials and Methods

An Fe-20%Mn-8%Al-5%Ni-0.8%C (mass%) was used in the present study, which has a chemical composition similar to that used in previous studies by Kim et al. and Yang et al. [16,17]. An as-received ingot having dimension of 100 mm (length) × 50 mm (width) × 20 mm (thickness) was cut into small blocks with 10 mm in thickness, and then solution-treated at 1280 °C for 1 h in argon atmosphere followed by water quench. After that, the plate was cold-rolled from 10 to 1.2 mm by multiple rolling passes, which corresponded to an 88% rolling reduction in thickness. The cold-rolled specimen was annealed at 850 °C for 120 min in a salt bath and then quenched into water. Tensile test specimens having 7.5 mm in gauge length, 2.5 mm in gauge width, and 1.2 mm thick were cut from the annealed sheet, with the tensile axis parallel to the rolling direction (RD) of the sheets. The surface perpendicular to the normal direction (ND) of the sheet specimen was decorated by white and black paint, and a CCD camera was used to record the images of the surface during the tensile test. Those images were analyzed afterwards by a digital image correlation (DIC) software (VIC2D) to obtain the precise tensile elongation and strain localization behaviors during the tensile deformation. The details about the DIC system used in the present experiments were described in Reference [18]. Tensile tests were carried out at room temperature at an initial strain rate of 8.3×10^{-4} s^{-1}. Microstructures of the specimen were observed from the transverse direction (TD) of the specimens by a scanning electron microscope (SEM) equipped with an electron backscatter diffraction (EBSD) system (JSM-7100F, JOEL Ltd., Tokyo, Japan), using an acceleration voltage of 15 kV and a scan step size of 50 nm.

To investigate the individual deformation behavior of constituent phases in the material during tension, another tensile test with in situ X-ray diffraction (XRD) was conducted at the beam line BL46XU of SPring-8 in Harima, Japan. The experimental setup of the in situ XRD measurement is schematically illustrated in Figure 1. The energy of the X-ray was 30 keV, corresponding to a wavelength of 0.0413 nm. The incident X-ray beam having a size of 0.5 × 0.3 mm was irradiated perpendicular to the tensile test specimen during tensile deformation, and the diffracted X-ray in the diffraction angle (2θ) from 9.8° to 40.3° was detected by a one-dimensional detector composed of six MYTHEN detectors (DECTRIS Ltd., Baden-Daettwil, Switzerland) arranged in a line. The exposure time for each X-ray diffraction profile was set to 1 s. The tensile test specimen described earlier was further polished to 0.5 mm in thickness to ensure the penetration of the X-ray beam, and a strain rate of 8.3×10^{-4} s^{-1} was applied for the tensile test. After the tensile test, the diffraction peaks in a profile were fitted using the pseudo-Voigt function. For more details of the in situ XRD measurements in BL46XU, one can refer to References [19–21].

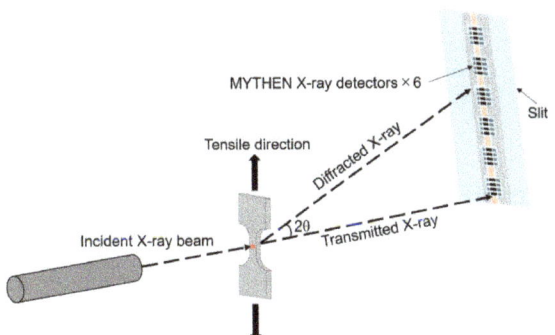

Figure 1. Schematic illustration of the in situ X-ray diffraction (XRD) measurement system in SPring-8.

3. Results and Discussions

Figure 2 shows the EBSD grain boundary map superimposed on the phase map of the specimen. It is seen that after annealing, the microstructures were composed of fully recrystallized equiaxed austenite grains with face centered cubic (FCC) structure (green) and fine yellow particles with body centered cubic (BCC) structure that primarily located at the austenite grain boundaries and grain boundary triple junctions. Although EBSD mapping could not identify an ordered structure, it was later confirmed by in situ XRD measurement that the yellow BCC particles were B2 phase. The area fraction of the B2 phase was measured as 0.09 (9%) and the average grain size of the austenite matrix and B2 phase was 1.2 μm (including annealing twin boundary) and 0.3 μm, respectively. Although some very fine B2 particles could be observed in the interior of austenite grains, as was also observed in similar alloys [16], the majority of the B2 particles had coarse sizes. It has been shown that the B2 particles in the alloy were not brittle but plastically deformable [22,23]. Therefore, in the present study, the specimen was considered as a dual-phase alloy having ultrafine-grained microstructures rather than a precipitation strengthened alloy, as was also suggested in Reference [17].

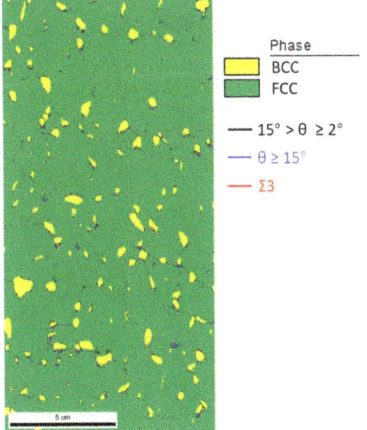

Figure 2. Grain boundary map superimposed with phase map of the dual-phase ultrafine-grained (UFG) Fe-20Mn-8Al-5Ni-0.8C alloy, observed by SEM-EBSD. Low-angle grain boundaries ($15° > \theta \geq 2°$), high-angle grain boundaries ($\theta \geq 15°$), and twin boundaries ($\Sigma 3$) are respectively indicated by black, blue, and red lines. The grain and yellow background represent austenite phase and B2 phase, respectively. The B2 phase in the material was indexed as BCC (α) phase by EBSD because they have quite similar Kikuchi patterns.

The tensile stress-strain curve of the annealed specimen is shown in Figure 3. The specimen exhibited an excellent combination of strength and tensile ductility, with upper yield stress of 932 MPa, ultimate tensile strength (UTS) of 1174 MPa, uniform elongation of 31%, and total elongation of 42%. It should be noted that, as shown in the insets of Figure 3, the stress-strain curve at the beginning of the tensile test showed a yield plateau, and the strain contour maps of the specimen surface obtained by the DIC analysis confirmed that the yield plateau corresponded to the initiation and propagation of two Lüders bands initiated from upper and lower sides of the specimen gauge. The Lüders band deformation is well-known to appear in low carbon ferritic steels showing discontinuous yielding due to the Cottrell atmosphere formed by impurity atoms around dislocations. In recent years, it has been shown that the Lüders bands' deformation can appear in most of the polycrystalline metals and alloys when their grain sizes are decreased down to an ultra-fine range below 1–2 µm [24–32]. The discontinuous yielding in UFG metals is probably because the number of mobile dislocations within each grain becomes too small to initiate the plastic deformation of the specimen in a continuous manner [28,33]. This seems to be the case in the present observation, since both the austenite matrix and B2 particles have quite fine mean grain sizes. The strain in the area swept by the Lüders band reached to 0.02 measured from the DIC local strain mapping, which agreed with the magnitude of the Lüders strain measured on the stress-strain curve. After the Lüders band deformation, the specimen showed continuous strain hardening behavior until the UTS was reached, and then macroscopic necking occurred followed by tensile failure.

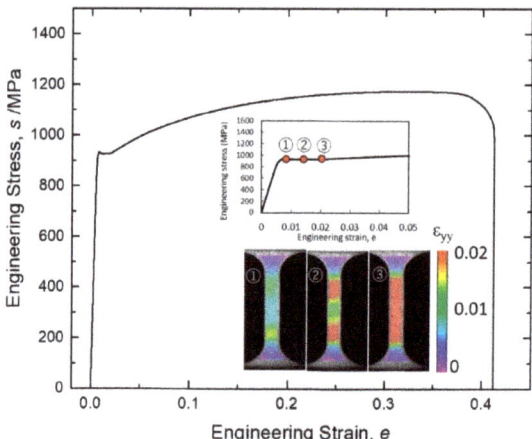

Figure 3. Nominal stress-strain curve of the UFG dual-phase Fe-20Mn-8Al-5Ni-0.8C alloy. The insets show the enlarged stress-strain curve at the beginning of tensile deformation and the corresponding strain contour on the surface of the specimen measured by the DIC method. The Lüders plateau can be clearly seen in the enlarged stress-strain curve, and the three strain contour maps corresponding to the points ①, ②, and ③ in the stress-strain curve show the initiation and propagation of the Lüders bands on the specimen.

The tensile stress-strain curve obtained during the in situ X-ray diffraction experiments is shown in Figure 4. The curve showed quite similar tensile properties to those shown in Figure 3, except for a smaller total tensile elongation of 32% and a slightly lower UTS of 1130 MPa. This was probably due to the specimen size effect on the total elongation and tensile strength [34], given that the thickness of the tensile specimen gauge was decreased to 0.6 mm for the in situ XRD experiment. XRD profiles obtained by the in situ X-ray diffraction measurement are shown in Figure 5. For the diffraction profile of the specimen before the tensile test, diffraction peaks of (hkl) planes of austenite and B2 phase were indexed, including the (100) superlattice peak of B2 phase. The volume fraction of the B2 phase was calculated, using integrated intensity of the diffraction peaks according to the following equation [35]:

$$f_{B2} = \frac{\frac{1}{m}\sum_{i=1}^{m}\frac{I_{i,B2}}{R_{i,B2}}}{\frac{1}{n}\sum_{j=1}^{n}\frac{I_{j,\gamma}}{R_{j,\gamma}} + \frac{1}{m}\sum_{j=1}^{m}\frac{I_{i,B2}}{R_{i,B2}}} \quad (1)$$

where I is the integrated intensity of the diffraction peak, R is the material scattering factor, and m and n are the numbers of diffraction peaks used for B2 and austenite phases. The volume fraction of B2 phase calculated was 0.089 (8.9%), which was quite close to the area fraction (9%) measured from the EBSD map (Figure 2). Changes of $(111)_\gamma$ diffraction peak during tensile deformation are exhibited in the inset of Figure 5. In elastic deformation under a stress of 500 MPa, the $(111)_\gamma$ peak shifted to smaller diffraction angle, i.e., the lattice spacing of $(111)_\gamma$ planes increased, in response to the external tensile stress. As the tensile deformation continued to the plastic region (ε = 5%), the peak broadening was recognized in addition to the peak shift, which was due to the inhomogeneous micro-strains caused mainly by increasing dislocation densities.

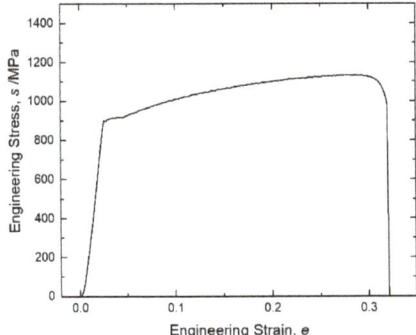

Figure 4. Nominal stress-strain curve of the UFG dual-phase Fe-20Mn-8Al-5Ni-0.8C alloy measured during the tensile test in situ XRD measurement system in SPring-8.

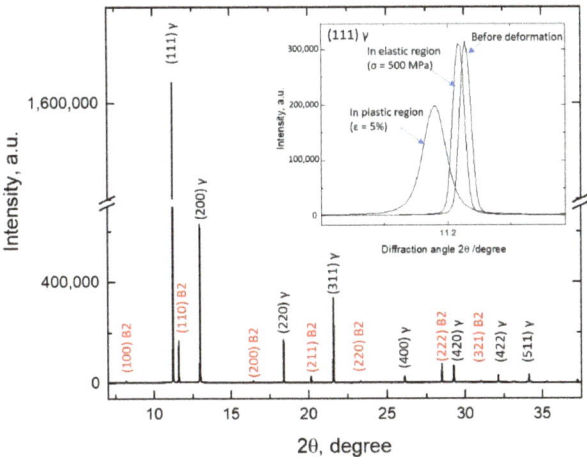

Figure 5. A diffraction profile measured before loading of the UFG Fe-20Mn-8Al-5Ni-0.8C alloy during the tensile test with in situ XRD measurements. The changing of the (111) peak of austenite (γ phase) during the tensile test is shown in the insets, in which the peak shifting and peak broadening during the tensile test are illustrated.

The stress partitioning behavior during tensile deformation between different phases can be analyzed by measuring the lattice strain evolution of the phases. The lattice spacing d of (hkl) planes during tensile deformation can be estimated from the peak shift by the use of the Bragg's law, $2d\sin\theta = \lambda$, and the lattice strain of (hkl) plane of a constituent phase i during tensile loading, ε_i^{hkl}, is calculated by the following equation:

$$\varepsilon_i^{hkl} = \frac{d_i^{hkl} - d_{i,0}^{hkl}}{d_{i,0}^{hkl}} \qquad (2)$$

where d_i^{hkl} is the lattice spacing of the (hkl) lattice plane of a constituent phase i measured during the tensile test, and $d_{i,0}^{hkl}$ is the reference lattice spacing corresponding to its stress-free state. The lattice spacing before the tensile test was regarded as its stress-free state, although upon quenching, residual stress might arise because of the different coefficient of the thermal expansion of the two phases. In the present study, the angle between the scattering vector and the tensile axis, namely θ, was generally small, owing to the short wavelength of the X-ray and the transmission geometry of the measurement. Therefore, the measured lattice strains of the (hkl) planes were regarded approximately equal to the elastic strains in the crystal family grains whose <hkl> directions are oriented to the tensile direction, and this approximation was more accurate for the (hkl) planes having smaller diffraction angles, such as for $(111)_\gamma$ planes and $(110)_{B2}$. The changes of lattice strains in austenite and B2 phase are plotted as a function of the tensile true stress in Figure 6, with the superimposition of the true stress-strain curve of the specimen. It could be seen that the lattice strains of both phases increased linearly with the true stress in the elastic deformation region. The measured slope of ε_γ^{111} and ε_γ^{311}, i.e., the diffraction elastic moduli E_γ^{111} and E_γ^{311}, are 216 and 166 GPa, which were comparable with those of another austenitic steel (245 and 187 GPa, respectively) reported in a previous study [10]. The E_{B2}^{110} and E_{B2}^{211} were measured to be 194 and 185 GPa, which were about 30 GPa lower than those reported for BCC iron (221 and 221GPa, respectively) [36]. Such a difference in the elastic modulus between different grain families is attributed to the elastic anisotropy of crystalline materials. When the yield stress was achieved, the lattice strains of two phases exhibited a dramatic separation, where ε_γ^{111} and ε_γ^{311} decreased, while ε_{B2}^{110} and ε_γ^{211} rapidly increased. It has been well-established that such a separation of the lattice strains of different phases or different grain families of single phase in the plastic region indicates the occurrence of stress partitioning between different phases or grain families [9,37,38]. In such a case, the internal stress was transferred from the soft domain (phase or grain families) to the hard domain, due to larger amounts of plastic deformation in the softer domain. However, such a dramatic lattice strain partitioning at the beginning of tensile deformation observed in Figure 6 has not commonly been reported in other dual-phase alloys [37,39]. It was noteworthy that the rapid partitioning of lattice strains coincided with the Lüders plateau on the true stress-strain curve. Considering that the Lüders band deformation occurred in a manner of propagating localized deformation region in the specimen gauge, as shown in the DIC contour map inserted in Figure 3, it was suggested that the dramatic stress partitioning behavior between the austenite and B2 phases at the beginning of plastic deformation was associated with the rapid sweeping of the plastic-strain localized band over the region on which the X-ray beam was irradiated. In addition, it should be noted that the lattice strains of austenite phase decreased, while those of B2 phase increased during the Lüders deformation, suggesting that the stress was transferred from the soft austenite grains to the hard B2 particles during the discontinuous yielding. After the Lüders band deformation, the lattice strains of two phases started to increase with the tensile true stress. Meanwhile, the separation in the lattice strains increased between different grain families within each phase, which indicated the occurrence of stress partitioning and therefore the plastic deformation progressing not only in the austenite phase but also in the B2 phase. These results, along with the observations on the deformed microstructures of B2 phase in similar alloys [22,23], suggested that the B2 phase in the present alloy was essentially not brittle and was capable for plastic deformation.

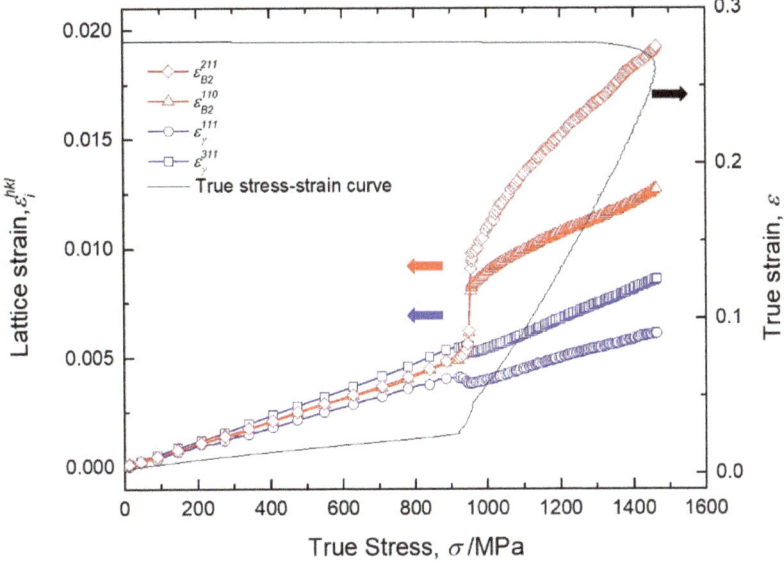

Figure 6. Changes in the (111) and (311) lattice strains of austenite phase, and the (110) and (211) lattice strains of B2 phase as a function of tensile true stress. The true stress-strain curve is superimposed in the figure. The lattice strain after tensile fracture is not shown in the figure for the sake of simplicity.

The elastic stress in each constituent phase, i.e., the so-called phase stress [37,40,41], can be evaluated from the phase strains using Hook's law and Poisson's ratios. A simplified estimation is often used, under the assumption that the phase strain can be represented by the lattice strain of certain (hkl) planes, for evaluating the phase stress when the strain in the transverse direction is not available [41]:

$$\sigma_i = E_i^{hkl} \varepsilon_i^{hkl} \tag{3}$$

where i represents austenite or B2 phase in the present case. In the present study, the $(111)_\gamma$ and $(110)_{B2}$ were used to calculate the phase stress of the austenite and B2 phases. The calculated values are plotted as a function of the tensile true strain of the specimen in Figure 7. A dramatic separation of phase stresses was observed at the beginning of plastic deformation, which corresponded with the Lüders deformation mentioned earlier. After that, the phase stresses in both phases increased with increasing tensile true strain, and the B2 phase bore significantly higher phase stress, nearly twice that in the austenite in the entire plastic region, presumably because the B2 phase was plastically much harder than the austenite phase. These results clearly demonstrated that the present alloy should be understood as a dual-phase alloy rather than a precipitation/dispersion hardened alloy with the matric involving finely dispersed second phase. It was also interesting to note that the increasing rate of σ_{B2} was higher than that of σ_γ, especially in the later part of the tensile deformation.

In order to evaluate the contribution from each constituent phase to the total tensile flow stress, the fraction-weighted phase stress was calculated by the following equation [41]:

$$\sigma_{cont,i} = \sigma_i f_i \tag{4}$$

where σ_i and f_i are the phase stress and volume fraction of phase i. In addition, the fraction-weighted average flow stress of the specimen σ_F can be calculated by summing up the contributed stress of the two phases as a composite model using the following equation [41]:

$$\sigma_F = \sigma_{cont,\gamma} + \sigma_{cont,B2} = \sigma_\gamma f_\gamma + \sigma_{B2} f_{B2} \tag{5}$$

The obtained $\sigma_{cont,\gamma}$, $\sigma_{cont,B2}$, and σ_F are plotted as a function of tensile true strain in Figure 8, with the experimental tensile true stress-strain curve superimposed. The calculated flow stress (σ_F) showed a good agreement with the experimentally obtained global true stress of the specimen, and a slight difference between them was probably associated with the fact that the lattice strain measured by diffraction was not exactly parallel to the tensile direction, which caused an underestimation of the phase stress along the tensile direction. It was obvious that the austenite phase contributed to the large majority of the tensile flow stress in the entire stages of the tensile test, owing to its high volume fraction of 0.91 (91%) and essentially good strain hardening ability. However, it should also be noted that the B2 phase with a small volume fraction of only 0.09 (9%) withstood more than 15% of the flow stress of the specimen during the tensile deformation.

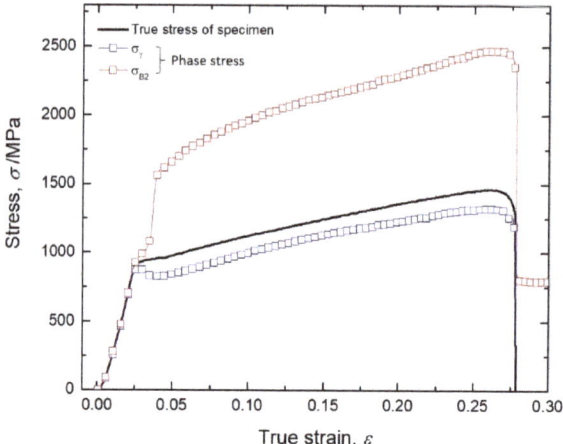

Figure 7. Calculated phase stresses of austenite and B2 phase are plotted as a function of tensile true strain. The true stress-strain curve of the specimen is also plotted. A significant stress partitioning between the austenite phase and the B2 phase during plastic deformation can be readily observed.

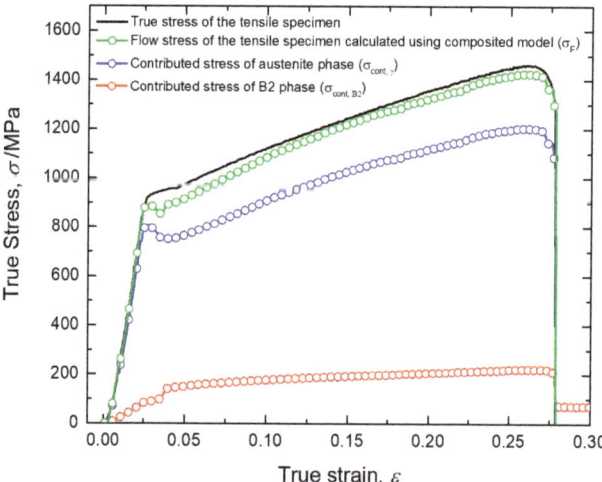

Figure 8. The contributed flow stress of the austenite phase and B2 phase, and the flow stress calculated using a composited model, are plotted as a function of the tensile true strain. The experimental tensile true stress-strain curve is also plotted. A good agreement is noticed between the calculated flow stress and the experimental flow stress.

The uniform tensile ductility of the material, i.e., the onset of necking, is determined by the Considère plastic instability criterion:

$$\left(\frac{d\sigma}{d\epsilon}\right) \leq \sigma \tag{6}$$

where σ is the flow stress, and $d\sigma/d\epsilon$ is the strain hardening rate which is critical to the plastic instability. To further understand the role of B2 phase during the tensile deformation, the slope of $\sigma_{cont,\gamma}$ and $\sigma_{cont,B2}$, namely $d\sigma_{cont,\gamma}/d\epsilon$ and $d\sigma_{cont,B2}/d\epsilon$, are plotted as a function of tensile true strain in Figure 9, together with the experimental true stress-strain curve and the strain hardening rate ($d\sigma/d\epsilon$) of the specimen. It should be noted that $d\sigma_{cont,\gamma}/d\epsilon$ and $d\sigma_{cont,B2}/d\epsilon$ did not represent the strain hardening behavior of the individual constituent phase, since partitioning of plastic strain usually takes place between the constituent phases during deformation and the exact strain in each phase cannot be directly measured. Nevertheless, the slope can be regarded as the hardening rate contributed by a constituent phase to the whole specimen at a given global strain. It could be seen in Figure 9 that $d\sigma/d\epsilon$ started to decrease with the true strain after the Lüders band deformation, meanwhile the $d\sigma_{cont,\gamma}/d\epsilon$ and $d\sigma_{cont,B2}/d\epsilon$ also decreased with the true strain, and the decreasing rate was similar to that of $d\sigma/d\epsilon$. However, the decreasing rate of $d\sigma/d\epsilon$ notably slowed down when the tensile true strain increased from 0.13 to 0.23 (indicated by the black arrow), which was found to interestingly coincide with the level off of the $d\sigma_{cont,B2}/d\epsilon$ (indicated by the red arrow) in the same region of tensile strain; meanwhile, the hardening rate of austenite phase was still in a deacceleration at high values. This result implies that the level off in the hardening rate of B2 phase slowed down the decreasing of the strain hardening rate of the whole specimen in the same tensile strain region, and therefore delayed the onset of plastic instability (necking) of the specimen afterwards, as readily exhibited by the black dashed line. These results suggested that although the B2 phase withstood a small portion of the total flow stress in the whole material, B2 provided a proper hardening rate in deformation of the specimen, especially at the later stage of deformation, which effectively delayed the onset of plastic instability (macroscopic necking) and led to a large tensile ductility of the specimen. The reason for this unique hardening behavior of the B2 phase is not yet understood, but it is considered to associate with the plastic deformation in the B2 particles during tensile deformation.

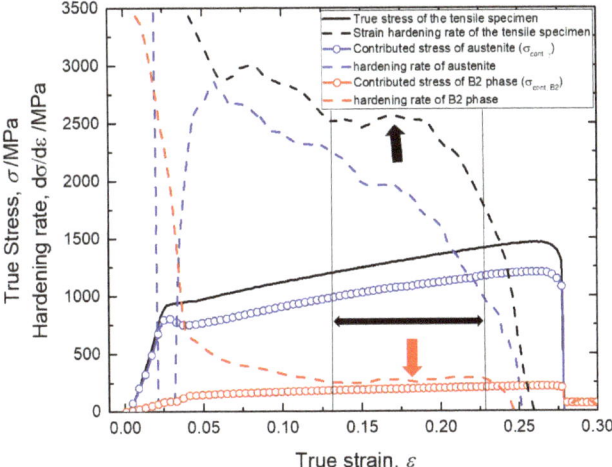

Figure 9. The contributed flow stress of austenite phase and B2 phase, the experimental tensile flow stress, and their slopes are plotted as a function of the tensile true strain. The region indicated by the double arrow corresponds to where the decreasing of the hardening rate of B2 phase (red dashed line) slowed down, so that the decreasing of strain hardening rate of the tensile test specimen (black dashed line) in the region was slowed down.

The diffraction line profile analysis was carried out in order to reveal the plastic deformation of each constituent phase during the tensile test. The evolution of full width at half maximum (FWHM) of the diffraction peaks during the tensile test is illustrated in Figure 10. It is seen that the FWHM of the diffraction peaks in both phases increased rapidly during the Lüders deformation, suggesting a strain broadening and/or a size broadening caused by a rapid generation of defects and/or a reduction of crystallite size. It should be noted that the synchrotron X-ray beam was irradiated at a particular region in the tensile specimen, so that such a rapid increase in the FWHM during the Lüders deformation was due to the quick sweeping of the Lüders front, where plastic strain was localized, on the X-ray irradiated region. After the Lüders deformation, the increasing rate of FWHM gradually slowed down with increasing the tensile strain.

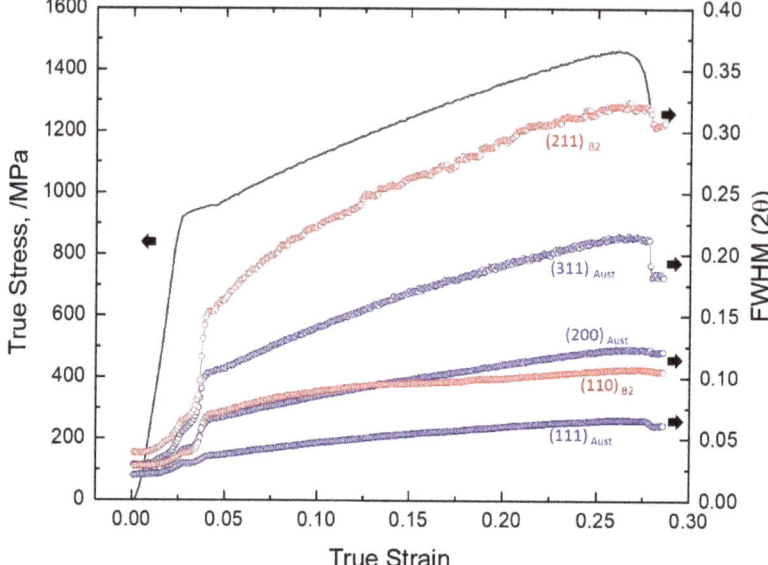

Figure 10. The full width at half maximum (FWHM) of the diffraction peaks of austenite and B2 phase are plotted as a function of the tensile true strain. The global true stress-strain curve is also superimposed.

The dislocation density of each constituent phase was estimated by the classical Williamson-Hall method [42] shown as:

$$\frac{\Delta 2\theta \cos\theta}{\lambda} = \frac{0.9}{D} + 2\varepsilon \frac{\sin\theta}{\lambda} \qquad (7)$$

where λ is the wavelength of the incident X-ray, θ is the diffraction peak angle, $\Delta 2\theta$ is the FWHM, D is the crystallite size, and ε is the inhomogeneous strain. The ε and D are the slope and intercept of the linear relationship by plotting $\Delta 2\theta \cos\theta/\lambda$ against $2\sin\theta/\lambda$ for each diffraction peak. The dislocation density, ρ, was then estimated from the average values of the crystallite size and inhomogeneous strain by the following equation [43,44]:

$$\rho = \frac{3\sqrt{2\pi}\varepsilon}{Db} \qquad (8)$$

where b is the Burgers vector of the material. The Burgers vectors of 0.258 and 0.25 nm were respectively used for austenite and B2 phase, assuming a/2<110> dislocations for austenite and a/2<111> dislocations for B2 phase. The estimated dislocation densities in austenite phase and B2 phase during tensile deformation are plotted in Figure 11. Before the tensile test, the dislocation densities in austenite phase and B2 phase were 6.0×10^{13} m^{-2} and 4.5×10^{13} m^{-2}, which were close to the values in fully recrystallized metals previously reported [21,45]. During the Lüders deformation, the dislocation densities in both phases rapidly increased, which was similar to the tendency of the FWHM evolution. After the Lüders deformation, the dislocation density in austenite phase almost linearly increased with increasing the tensile strain until tensile fracture occurred. Such linear relationship between the dislocation density and tensile strain in austenite phase has been reported by Dini et al. in an Fe–31Mn–3Al–3Si austenitic steel [46]. On the other hand, the dislocation density in the B2 phase increased at a similar rate to that in the austenite phase after the Lüders deformation, while the increasing rate was notably enhanced when the tensile strain reached 0.13 until tensile fracture. This enhanced dislocation accumulation rate in B2 phase interestingly coincided with the slowing down of the decreasing rate in the hardening rate of B2 phase in the same strain range shown in Figure 9, suggesting that the dislocation activities in B2 phase played an important role in hardening of B2 phase, especially in the later stage of tensile deformation. The reason for the enhanced increasing rate of dislocation density in B2 phase might result from mechanical interaction between B2 and austenite phases at their interfaces, which needs to be further clarified through microstructures' observations. It should be noted that the value of average dislocation densities in austenite and B2 phases are not directly related to the amount of plastic strain in each phase, because the grain size of the two phases are different and the increasing rate of geometrically necessary dislocations with plastic strain can be significantly different [47].

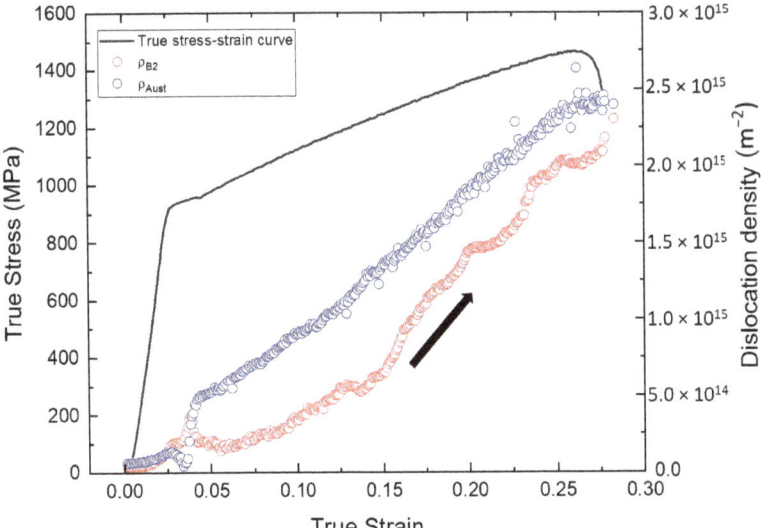

Figure 11. The estimated dislocation densities of austenite phase (blue circle) and B2 phase (red circle) during tensile deformation are plotted as a function of the tensile true strain. The increasing of dislocation accumulation rate in the B2 phase from a tensile strain of 0.13 is indicated by the arrow.

The dislocation density can be related to the flow stress, σ_F, according to the Bailey-Hirsh equation [48]:

$$\sigma_F = \sigma_0 + M\alpha Gb\sqrt{\rho} \qquad (9)$$

where G is the shear modulus, b is the Burgers vector, M is the average Taylor factor, and α is a constant depending on the dislocation interaction in the material. σ_0 is generally considered to be the additivity of the stresses associated with other strengthening mechanisms, such as friction stress, grain boundary strengthening, and precipitation strengthening. In the present analysis, the phase stress of each constituent phase in the range from $\varepsilon = 0.05$ (after Lüders deformation) to $\varepsilon = 0.26$ during tensile deformation is plotted against the square root of their respective dislocation densities in Figure 12. Good linear relationships were realized in both phases, suggesting that the increasing of dislocation density could account for the phase stress increment in both austenite and B2 phase during tensile deformation. By extrapolating the fitted linear relationships, the value of σ_0 was determined to be 487 MPa for the austenite and 1576 MPa for the B2 phase. The σ_0 of the austenite is in a reasonable agreement with the estimated yield strength of a Fe-22Mn-0.6C austenitic steel, having a similar mean grain size ($\sigma_y = 563$ MPa, $d = 1.2$ µm) estimated from its Hall-Petch relationship (σ_y (MPa) $= 133 + 472 \cdot d^{-1/2}$) [49]. The σ_0 of the B2 phase in the present study was comparable with the estimated yield strength of a B2 Fe-Al alloy having a similar mean grain size ($\sigma_y = 1747$ MPa, $d = 0.3$ µm, σ_y (MPa) $= 386 + 745 \cdot d^{-1/2}$) [50]. These results support that the σ_0 obtained through the extrapolation of the Bailey-Hirsh relationship can be regarded as the additivity of the lattice friction stress and the grain size refinement strengthening in the present alloy. Considering the values of lattice friction stress of austenitic steels and other FCC alloys [49,51] as well as those of B2 alloys [50,52], significant grain refinement strengthening is expected in the austenite and B2 phase having ultrafine grain sizes in the present specimen, although the exact values of grain size refinement strengthening are difficult to separate from the σ_0.

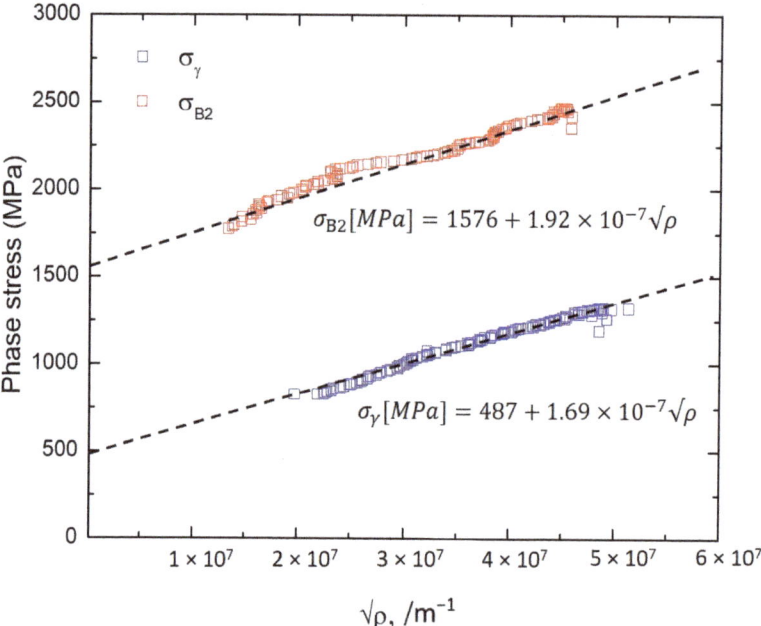

Figure 12. The phase stresses of austenite phase (blue square) and B2 phase (red square) during tensile deformation are plotted as a function of the square root of their dislocation densities. Good linear relationships between the phase stress and dislocation density are recognized, suggesting that the incremental phase stress in each constituent phase during tensile deformation can be explained by dislocation accumulation. /m^{-1}.

4. Conclusions

In conclusion, an excellent combination of strength and tensile ductility was achieved in the ultrafine-grained Fe-Mn-Ni-Al-C alloy containing B2 intermetallic compounds as second phase particles. The tensile test with in situ X-ray diffraction measurement revealed a rapid stress partitioning between austenite phase and B2 phase at the very beginning of plastic deformation, which was associated with the Lüders band deformation of the specimen. In addition, through the stress partitioning analysis, it was found that the B2 particles, although they took only 0.09 (9%) volume fraction of the material, withstood very high phase stress during tensile deformation. More importantly, the B2 phase exhibited a unique hardening behavior that could effectively slow down the decrease of the strain hardening rate of the whole specimen and delayed the onset of plastic instability, suggesting an importance of the hard B2 phase in the strength and ductility synergy of the material. Through the Williamson-Hall analysis, it was found that the dislocation accumulation rate in the B2 phase was enhanced during tensile deformation, which seemed to interpret the unique hardening rate of the B2 phase. Further studies should be focused on clarifying the interaction between B2 particles and austenite grains through microstructures' observations.

Author Contributions: Conceptualization, S.G. and N.T.; methodology, T.Y., W.M., Y.B., W.G., M.-h.P., H.A., M.S. and A.S.; writing—original draft preparation, S.G. and N.T.; writing—review and editing, N.T. All authors have read and agreed to the published version of the manuscript.

Funding: This research received no external funding.

Acknowledgments: The present study was financially supported by JST CREST (JPMJCR1994), Elements Strategy Initiative for Structural Materials (ESISM, No. JPMXP0112101000), and KAKENHI (Grant-in-Aid for Scientific Research from JSPS; No.15H05767 and No.20H00306) all through the Ministry of Education, Culture, Sports, Science and Technology (MEXT), Japan. The synchrotron radiation experiments (beam-time No. 2018B1760) at SPring-8 were performed with the approval of the Japan Synchrotron Radiation Research Institute (JASRI). The alloy used in this study was kindly provided by Dmitri LOUZGUINE from Advanced Institute for Materials Research in Tohoku University. The authors gratefully appreciate all the support.

Conflicts of Interest: The authors declare no conflict of interest.

References

1. Furuta, T.; Kuramoto, S.; Ohsuna, T.; Oh-Ishi, K.; Horibuchi, K. Die-hard plastic deformation behavior in an ultrahigh-strength Fe-Ni-Al-C alloy. *Scr. Mater.* **2015**, *101*, 87–90. [CrossRef]
2. Wang, Z.; Baker, I. Effects of annealing and thermo-mechanical treatment on the microstructures and mechanical properties of a carbon-doped FeNiMnAl multi-component alloy. *Mater. Sci. Eng. A* **2017**, *693*, 101–110. [CrossRef]
3. Wang, Z.; Baker, I.; Cai, Z.; Chen, S.; Poplawsky, J.D.; Guo, W. The effect of interstitial carbon on the mechanical properties and dislocation substructure evolution in $Fe_{40.4}Ni_{11.3}Mn_{34.8}Al_{7.5}Cr_6$ high entropy alloys. *Acta Mater.* **2016**, *120*, 228–239. [CrossRef]
4. Wang, Z.; Wu, M.; Cai, Z.; Chen, S.; Baker, I. Effect of Ti content on the microstructure and mechanical behavior of $(Fe_{36}Ni_{18}Mn_{33}Al_{13})_{100-x}Ti_x$ high entropy alloys. *Intermet* **2016**, *75*, 79–87. [CrossRef]
5. Wani, I.S.; Bhattacharjee, T.; Sheikh, S.; Lu, Y.P.; Chatterjee, S.; Bhattacharjee, P.; Guo, S.; Tsuji, N. Ultrafine-Grained $AlCoCrFeNi_{2.1}$ Eutectic High-Entropy Alloy. *Mater. Res. Lett.* **2016**, *4*, 174–179. [CrossRef]
6. Bhattacharjee, T.; Wani, I.S.; Sheikh, S.; Clark, I.T.; Okawa, T.; Guo, S.; Bhattacharjee, P.; Tsuji, N. Simultaneous Strength-Ductility Enhancement of a Nano-Lamellar $AlCoCrFeNi_{2.1}$ Eutectic High Entropy Alloy by Cryo-Rolling and Annealing. *Sci. Rep.* **2018**, *8*, 1–8. [CrossRef]
7. Tomota, Y.; Lukas, P.; Harjo, S.; Park, J.-H.; Tsuchida, N.; Neov, D. In situ neutron diffraction study of IF and ultra low carbon steels upon tensile deformation. *Acta Mater.* **2003**, *51*, 819–830. [CrossRef]
8. Gao, S.; Bai, Y.; Shibata, A.; Tsuji, N. Microstructures and mechanical property of a Fe-Ni-Al-C alloy containing B2 intermetallic compounds. *Mater. Sci. Eng.* **2017**, *219*, 012020. [CrossRef]
9. Asoo, K.; Tomota, Y.; Harjo, S.; Okitsu, Y. Tensile Behavior of a TRIP-aided Ultra-fine Grained Steel Studied by Neutron Diffraction. *ISIJ Int.* **2011**, *51*, 145–150. [CrossRef]
10. Harjo, S.; Abe, J.; Aizawa, K.; Gong, W.; Iwahashi, T. Deformation Behavior of An Austenitic Steel by Neutron Diffraction. In Proceedings of the Proceedings of the 12th Asia Pacific Physics Conference (APPC12), Makuhari, Japan, 14–19 July 2013; Volume 014017, pp. 1–6.
11. Fu, S.; Bei, H.; Chen, Y.; Liu, T.; Yu, D.; An, K. Deformation mechanisms and work-hardening behavior of transformation-induced plasticity high entropy alloys by in -situ neutron diffraction. *Mater. Res. Lett.* **2018**, *6*, 620–626. [CrossRef]
12. Sun, Z.; Song, G.; Sisneros, T.A.; Clausen, B.; Pu, C.; Li, L.; Gao, Y.; Liaw, P.K. Load partitioning between the bcc-iron matrix and NiAl-type precipitates in a ferritic alloy on multiple length scales. *Sci. Rep.* **2016**, *6*, 23137. [CrossRef]

13. Saleh, A.A.; Brown, D.W.; Pereloma, E.V.; Clausen, B.; Davies, C.; Tomé, C.N.; Gazder, A.A. An in-situ neutron diffraction study of a multi-phase transformation and twinning-induced plasticity steel during cyclic loading. *Appl. Phys. Lett.* **2015**, *106*, 171–911. [CrossRef]
14. Shibata, A.; Takeda, Y.; Park, N.; Zhao, L.; Harjo, S.; Kawasaki, T.; Gong, W.; Tsuji, N. Nature of dynamic ferrite transformation revealed by in-situ neutron diffraction analysis during thermomechanical processing. *Scr. Mater.* **2019**, *165*, 44–49. [CrossRef]
15. Bae, J.W.; Kim, J.G.; Park, J.M.; Woo, W.; Harjo, S.; Kim, H.S. In situ neutron diffraction study of phase stress evolution in a ferrous medium-entropy alloy under low-temperature tensile loading. *Scr. Mater.* **2019**, *165*, 60–63. [CrossRef]
16. Kim, S.-H.; Kim, H.; Kim, N.J. Brittle intermetallic compound makes ultrastrong low-density steel with large ductility. *Nat. Cell Biol.* **2015**, *518*, 77–79. [CrossRef] [PubMed]
17. Yang, M.; Yuan, F.; Xie, Q.; Wang, Y.; Ma, E.; Wu, X. Strain hardening in Fe–16Mn–10Al–0.86C–5Ni high specific strength steel. *Acta Mater.* **2016**, *109*, 213–222. [CrossRef]
18. Zhao, L.; Park, N.; Tian, Y.; Chen, S.; Shibata, A.; Tsuji, N. Novel thermomechanical processing methods for achieving ultragrain refinement of low-carbon steel without heavy plastic deformation. *Mater. Res. Lett.* **2016**, *5*, 61–68. [CrossRef]
19. Miyajima, Y.; Okubo, S.; Miyazawa, T.; Adachi, H.; Fujii, T. In-situ X-ray diffraction during tensile deformation of ultrafine-grained copper using synchrotron radiation. *Philos. Mag. Lett.* **2016**, *96*, 294–304. [CrossRef]
20. Adachi, H.; Miyajima, Y.; Sato, M.; Tsuji, N. Evaluation of dislocation density for 1100 aluminum with different grain size during tensile deformation by using In-situ X-ray diffraction technique. *J. Jpn. Inst. Light Met.* **2014**, *64*, 463–469. [CrossRef]
21. Adachi, H.; Karamatsu, Y.; Nakayama, S.; Miyazawa, T.; Sato, M.; Yamasaki, T. Elastic and Plastic Deformation Behavior Studied by In-Situ Synchrotron X-ray Diffraction in Nanocrystalline Nickel. *Mater. Trans.* **2016**, *57*, 1447–1453. [CrossRef]
22. Hwang, J.; Trang, T.; Lee, O.; Park, G.; Zargaran, A.; Kim, N.J. Improvement of strength – ductility balance of B2-strengthened lightweight steel. *Acta Mater.* **2020**, *191*, 1–12. [CrossRef]
23. Liu, X.; Xue, Q.; Wang, W.; Zhou, L.; Jiang, P.; Ma, H.; Yuan, F.; Wei, Y.; Wu, X. Back-stress-induced strengthening and strain hardening in dual-phase steel. *Materialia* **2019**, *7*, 100–376. [CrossRef]
24. Gao, S.; Chen, M.; Joshi, M.; Shibata, A.; Tsuji, N. Yielding behavior and its effect on uniform elongation in IF steel with various grain sizes. *J. Mater. Sci.* **2014**, *49*, 6536–6542. [CrossRef]
25. Tian, Y.; Gao, S.; Zhao, L.; Lu, S.; Pippan, R.; Zhang, Z.; Tsuji, N. Remarkable transitions of yield behavior and Lüders deformation in pure Cu by changing grain sizes. *Scr. Mater.* **2018**, *142*, 88–91. [CrossRef]
26. Tsuji, N.; Ito, Y.; Saito, Y.; Minamino, Y. Strength and ductility of ultrafine grained aluminum and iron produced by ARB and annealing. *Scr. Mater.* **2002**, *47*, 893–899. [CrossRef]
27. Zheng, R.; Bhattacharjee, T.; Shibata, A.; Sasaki, T.; Hono, K.; Joshi, M.; Tsuji, N. Simultaneously enhanced strength and ductility of Mg-Zn-Zr-Ca alloy with fully recrystallized ultrafine grained structures. *Scr. Mater.* **2017**, *131*, 1–5. [CrossRef]
28. Kamikawa, N.; Huang, X.; Tsuji, N.; Hansen, N. Strengthening mechanisms in nanostructured high-purity aluminium deformed to high strain and annealed. *Acta Mater.* **2009**, *57*, 4198–4208. [CrossRef]
29. Terada, D.; Inoue, S.; Tsuji, N. Microstructure and mechanical properties of commercial purity titanium severely deformed by ARB process. *J. Mater. Sci.* **2007**, *42*, 1673–1681. [CrossRef]
30. Zheng, R.; Du, J.-P.; Gao, S.; Somekawa, H.; Ogata, S.; Tsuji, N. Transition of dominant deformation mode in bulk polycrystalline pure Mg by ultra-grain refinement down to sub-micrometer. *Acta Mater.* **2020**, *198*, 35–46. [CrossRef]
31. Schwab, R. Understanding the complete loss of uniform plastic deformation of some ultrafine-grained metallic materials in tensile straining. *Int. J. Plast.* **2019**, *113*, 218–235. [CrossRef]
32. Schwab, R.; Ruff, V. On the nature of the yield point phenomenon. *Acta Mater.* **2013**, *61*, 1798–1808. [CrossRef]
33. Tsuji, N.; Ogata, S.; Inui, H.; Tanaka, I.; Kishida, K.; Gao, S.; Mao, W.; Bai, Y.; Zheng, R.; Du, J.-P. Strategy for managing both high strength and large ductility in structural materials–sequential nucleation of different deformation modes based on a concept of plaston. *Scr. Mater.* **2020**, *181*, 35–42. [CrossRef]
34. Zhao, Y.; Guo, Y.; Wei, Q.; Topping, T.; Dangelewicz, A.; Zhu, Y.; Langdon, T.; Lavernia, E. Influence of specimen dimensions and strain measurement methods on tensile stress–strain curves. *Mater. Sci. Eng. A* **2009**, *525*, 68–77. [CrossRef]

35. De, A.K.; Murdock, D.C.; Mataya, M.C.; Speer, J.G.; Matlock, D.K. Quantitative measurement of deformation-induced martensite in 304 stainless steel by X-ray diffraction. *Scr. Mater.* **2004**, *50*, 1445–1449. [CrossRef]
36. Takaki, S.; Akama, D.; Jiang, F.; Tsuchiyama, T. Correction of the Williamson-Hall Plots by the Diffraction Young's Modulus. *J. Soc. Mater. Sci. Jpn.* **2018**, *67*, 383–388. [CrossRef]
37. Tomota, Y.; Lukas, P.; Neov, D.; Harjo, S.; Abe, Y. In situ neutron diffraction during tensile deformation of a ferrite-cementite steel. *Acta Mater.* **2003**, *51*, 805–817. [CrossRef]
38. Morooka, S.; Tomota, Y.; Kamiyama, T. Heterogeneous Deformation Behavior Studied by in Situ Neutron Diffraction during Tensile Deformation for Ferrite, Martensite and Pearlite Steels. *ISIJ Int.* **2008**, *48*, 525–530. [CrossRef]
39. Jia, N.; Cong, Z.; Sun, X.; Cheng, S.; Nie, Z.; Ren, Y.; Liaw, P.; Wang, Y.D. An in situ high-energy X-ray diffraction study of micromechanical behavior of multiple phases in advanced high-strength steels. *Acta Mater.* **2009**, *57*, 3965–3977. [CrossRef]
40. Oliver, E.; Daymond, M.R.; Withers, P.J. Interphase and intergranular stress generation in carbon steels. *Acta Mater.* **2004**, *52*, 1937–1951. [CrossRef]
41. Harjo, S.; Tsuchida, N.; Abe, J.; Gong, W. Martensite phase stress and the strengthening mechanism in TRIP steel by neutron diffraction. *Sci. Rep.* **2017**, *7*, 1–11. [CrossRef]
42. Williamson, G.; Hall, W. X-ray line broadening from filed aluminium and wolfram. *Acta Met.* **1953**, *1*, 22–31. [CrossRef]
43. Smallman, R.E.; Westmacott, K.H. Stacking faults in face-centred cubic metals and alloys. *Philos. Mag.* **1957**, *2*, 669–683. [CrossRef]
44. Williamson, G.K.; Smallman, R.E. III. Dislocation densities in some annealed and cold-worked metals from measurements on the X-ray debye-scherrer spectrum. *Philos. Mag.* **1956**, *1*, 34–46. [CrossRef]
45. Zhong, Z.Y.; Brokmeier, H.-G.; Gan, W.M.; Maawad, E.; Schwebke, B.; Schell, N. Dislocation density evolution of AA 7020-T6 investigated by in-situ synchrotron diffraction under tensile load. *Mater. Charact.* **2015**, *108*, 124–131. [CrossRef]
46. Dini, G.; Ueji, R.; Najafizadeh, A.; Monir-Vaghefi, S. Flow stress analysis of TWIP steel via the XRD measurement of dislocation density. *Mater. Sci. Eng. A* **2010**, *527*, 2759–2763. [CrossRef]
47. Ashby, M.F. The deformation of plastically non-homogeneous materials. *Philos. Mag.* **1970**, *21*, 399–424. [CrossRef]
48. Bailey, J.E.; Hirsch, P.B. The dislocation distribution, flow stress, and stored energy in cold-worked polycrystalline silver. *Philos. Mag.* **1960**, *5*, 485–497. [CrossRef]
49. Tian, Y.; Bai, Y.; Zhao, L.; Gao, S.; Yang, H.; Shibata, A.; Zhang, Z.; Tsuji, N. A novel ultrafine-grained Fe 22Mn 0.6C TWIP steel with superior strength and ductility. *Mater. Charact.* **2017**, *126*, 74–80. [CrossRef]
50. Crimp, M.A.; Vedula, K.M. The relationship between cooling rate, grain size and the mechanical behavior of B2 Fe-Al alloys. *Mater. Sci. Eng. A* **1993**, *165*, 29–34. [CrossRef]
51. Yoshida, S.; Bhattacharjee, T.; Bai, Y.; Tsuji, N. Friction stress and Hall-Petch relationship in CoCrNi equi-atomic medium entropy alloy processed by severe plastic deformation and subsequent annealing. *Scr. Mater.* **2017**, *134*, 33–36. [CrossRef]
52. Baker, I.; Nagpal, P.; Liu, F.; Munroe, P. The effect of grain size on the yield strength of FeAl and NiAl. *Acta Met. Mater.* **1991**, *39*, 1637–1644. [CrossRef]

Publisher's Note: MDPI stays neutral with regard to jurisdictional claims in published maps and institutional affiliations.

© 2020 by the authors. Licensee MDPI, Basel, Switzerland. This article is an open access article distributed under the terms and conditions of the Creative Commons Attribution (CC BY) license (http://creativecommons.org/licenses/by/4.0/).

Article

Elastic Coefficients of β-HMX as Functions of Pressure and Temperature from Molecular Dynamics

Andrey Pereverzev * and Tommy Sewell *

Department of Chemistry, University of Missouri, Columbia, MO 65211, USA
* Correspondence: pereverzeva@missouri.edu (A.P.); sewellt@missouri.edu (T.S.)

Received: 15 November 2020; Accepted: 5 December 2020; Published: 10 December 2020

Abstract: The isothermal second-order elastic stiffness tensor and isotropic moduli of β-1,3,5,7-tetranitro-1,3,5,7-tetrazoctane (β-HMX) were calculated, using the $P2_1/n$ space group convention, from molecular dynamics for hydrostatic pressures ranging from 10^{-4} to 30 GPa and temperatures ranging from 300 to 1100 K using a validated all-atom flexible-molecule force field. The elastic stiffness tensor components were calculated as derivatives of the Cauchy stress tensor components with respect to linear strain components. These derivatives were evaluated numerically by imposing small, prescribed finite strains on the equilibrated β-HMX crystal at a given pressure and temperature and using the equilibrium stress tensors of the strained cells to obtain the derivatives of stress with respect to strain. For a fixed temperature, the elastic coefficients increase substantially with increasing pressure, whereas, for a fixed pressure, the elastic coefficients decrease as temperature increases, in accordance with physical expectations. Comparisons to previous experimental and computational results are provided where possible.

Keywords: HMX; elastic properties; molecular dynamics

1. Introduction

The high explosive 1,3,5,7-tetranitro-1,3,5,7-tetrazoctane, also known as octahydro-1,3,5,7-tetranitro-1,3,5,7-tetrazocine (HMX), is an important energetic material which is used in a number of high performance military explosive and propellant formulations [1]. Several polymorphs of HMX are known [2–5], among which β-HMX is the thermodynamically stable form at standard ambient conditions. The elastic properties of β-HMX are important for understanding processes such as shock propagation and the resulting hot-spot formation that ultimately lead to ignition and detonation initiation [6].

Accurate experimental determination of elastic coefficients in molecular high explosives is challenging due to the low crystal symmetries characteristic of many energetic substances. This is particularly evident in the case of β-HMX for which there exist substantial differences among measured values of the elastic tensor obtained using different experimental techniques [7–9]. This disparity in the experimental data has been ascribed to sample purity and processing variations and is further complicated with questions about how to interpret or process the results [6,10].

In part because of these complexities, there are virtually no experimental data for elastic properties of β-HMX under the high temperatures and pressures relevant to high-explosive initiation and detonation. In the absence of practical experimental alternatives, molecular dynamics (MD) provides a viable path to obtaining some of the needed information. In the present study, we use MD to obtain elastic coefficients of β-HMX for wide intervals of temperature and hydrostatic pressure. The results obtained herein can be used both as a general reference and as input data for mesoscale continuum models of shock propagation and detonation initiation in β-HMX [11,12].

2. Computational Details

2.1. Theoretical Background and System Preparation

The elastic stiffness tensor C_{ij} (written here in Voigt notation) was calculated using the direct method, which is based on the following equation

$$C_{ij} = \frac{\partial \sigma_i}{\partial \varepsilon_j}. \tag{1}$$

Here, ε_j denotes one of six linear strain components and σ_i is one of six Cauchy stress components. For monoclinic crystals such as β-HMX, the stiffness tensor contains 13 independent non-zero elastic coefficients. We chose linear strain rather than the often-used Lagrange strain because it is work-conjugate to the Cauchy stress (see [13], Section 2.2.4) and it is the Cauchy stress that is typically provided as output from MD simulations. For the small strains used in this work we can expect any differences arising due to the use of linear rather than Lagrange strains to be negligible. The partial derivatives in Equation (1) were calculated numerically using finite differences. To achieve this, we adopted the following four-step procedure.

First, a crystal cell of β-HMX (P2$_1$/n space group setting) was equilibrated for 1 ns at the temperature and pressure of interest using the isobaric-isothermal (NPT) ensemble. Next, 100 ps of isochoric-isothermal (NVT) simulation was performed using the time-averaged crystal cell parameters from the NPT trajectory. In the third step, small linear strains were imposed on the NVT-equilibrated crystal cell and NVT simulations were performed on the strained cells. Finally, the elastic coefficients C_{ij} were obtained by numerical differentiation of the stresses with respect to strains.

We used a 3D-periodic crystal cell consisting of $6 \times 4 \times 5$ unit-cell replications along the a, b, and c crystallographic axes, respectively. This crystal cell contains 240 HMX molecules (6720 atoms). Similar cell sizes for both β-HMX and other molecular-crystal high explosives are thought to be large enough to produce size-converged elastic properties of bulk crystals [6,14]. Temperatures $T = 300, 500, 700, 900$, and 1100 K and pressures $P = 10^{-4}, 5, 10, 20$, and 30 GPa were considered. These pressures were chosen to match the interval considered in Ref. [15]. At atmospheric pressure (10^{-4} GPa), β-HMX is known to undergo a phase transition to the α polymorph (orthorhombic) at approximately 420 K and the δ polymorph (hexagonal) when heated above approximately 450 K [16,17]; δ-HMX melts at about 550 K. However, with the force field that we use in this work the crystal remains stable (in the sense that melting does not occur) at 700 K and 1 atm for simulations lasting several nanoseconds. The crystal does melt at 1 atm somewhere between 700 and 900 K. Therefore, for the pressure $P = 1$ atm, we only considered temperatures up to 700 K. However, we observed that β-HMX remains crystalline all the way to 1100 K at 5 GPa and higher pressures. This is consistent with the MD-based melt curve of β-HMX reported recently by Kroonblawd and Austin [18]. They predicted that at 5 GPa β-HMX melts at 1320 K and the melting temperatures get higher for pressures above 5 GPa. Thus, overall, we studied 23 temperature and pressure combinations. We emphasize that the results below for the higher temperatures should be interpreted with caution because HMX may decompose on relevant time scales at those temperatures, at least at 1 atm [16].

When choosing the strain size, one has to consider the following arguments. On the one hand, one wants the strain to be sufficiently small to yield a reliable finite-difference approximation for the numerical evaluation of derivatives. On the other hand, if the strain is too small, the resulting change in stress is also small and, because there is a significant thermal noise in stress calculations, long simulation times are required to obtain converged results. Considering these factors, strains ε_i were chosen to be ± 0.004 when $i = 1, 2$ or 3 and ± 0.008 when $i = 4, 5$ or 6. The six strains required to obtain the full set of elastic coefficients were applied by distorting the simulation box in the following way. If the edge vectors of an unperturbed triclinic cell are given by vectors **a**, **b**, and **c**, then the geometry of the simulation box can, with no loss of generality, be described by the upper triangular matrix

$$h_0 = \begin{pmatrix} a_x & b_x & c_x \\ 0 & b_y & c_y \\ 0 & 0 & c_z \end{pmatrix} \tag{2}$$

by aligning **a** with the positive x axis and requiring **b** to lie in the xy plane [19]. Then, the upper triangular matrix h for the the strained cell is given by

$$h = (I \pm \epsilon_i)h_0, \tag{3}$$

where I is the 3×3 identity matrix and ϵ_i is one of the following six upper triangular matrices

$$\epsilon_1 = \begin{pmatrix} \epsilon_1 & 0 & 0 \\ 0 & 0 & 0 \\ 0 & 0 & 0 \end{pmatrix}, \quad \epsilon_2 = \begin{pmatrix} 0 & 0 & 0 \\ 0 & \epsilon_2 & 0 \\ 0 & 0 & 0 \end{pmatrix}, \quad \epsilon_3 = \begin{pmatrix} 0 & 0 & 0 \\ 0 & 0 & 0 \\ 0 & 0 & \epsilon_3 \end{pmatrix}, \tag{4}$$

$$\epsilon_4 = \begin{pmatrix} 0 & 0 & 0 \\ 0 & 0 & \epsilon_4 \\ 0 & 0 & 0 \end{pmatrix}, \quad \epsilon_5 = \begin{pmatrix} 0 & 0 & \epsilon_5 \\ 0 & 0 & 0 \\ 0 & 0 & 0 \end{pmatrix}, \quad \epsilon_6 = \begin{pmatrix} 0 & \epsilon_6 & 0 \\ 0 & 0 & 0 \\ 0 & 0 & 0 \end{pmatrix}. \tag{5}$$

Each of the strained simulation cells was equilibrated in the NVT ensemble for 100 ps after which the average stress tensor over a 4 ns NVT production trajectory was accumulated. For the pressure of 1 atm and temperatures of 300, 500, and 700 K the changes in stress tensor were small because the crystal is less rigid at the low pressure. Therefore, for those three cases, the trajectory was run for 12 ns to achieve better convergence of the elastic coefficients. Once the stress tensor components were obtained, the numerical derivatives in Equation (1) were evaluated as follows

$$C_{ij} \approx \frac{\sigma_i(\epsilon_j) - \sigma_i(-\epsilon_j)}{2\epsilon_j}, \tag{6}$$

where $\sigma_i(\epsilon_j)$ denotes the time-averaged strain tensor component for the equilibrated crystal cell with strain ϵ_j applied. Because of the elastic tensor symmetry, we expect $C_{ij} = C_{ji}$ when $i \neq j$. This was indeed the case, with differences between C_{ij} and C_{ji} typically less than 0.2 GPa. For these off-diagonal elements of the elastic tensor, we report the average of C_{ij} and C_{ji}. As part of our analysis we also obtained the eight elastic coefficients that are expected to be zero in a monoclinic crystal. These coefficients were, indeed, very small, never exceeding 0.08 GPa in magnitude. Three examples of the full 6×6 C_{ij} matrices obtained in this work are given in the Supplementary Materials. In addition to the elastic coefficients, the Voigt, Reuss, and Voigt–Reuss–Hill average bulk and shear moduli were calculated using standard expressions [8,20].

2.2. Force Field and MD Simulation Details

All MD simulations were performed using the LAMMPS package [19]. The nonreactive, fully flexible molecular potential for nitramines proposed by Smith and Bharadwaj [21] and further developed by Bedrov et al. [22] was employed in all of the MD simulations. This force field is well-validated and has been used in numerous previous studies of HMX [14,15,22–26]. In the current study, we modified the original nitramine force field used in [22] by adjusting the N-O and C-H harmonic bond stretching force constants to better reproduce the corresponding experimental vibrational mode frequencies (see [25] for more details). We further modified the original force field by adding a $1/r^{12}$ repulsive-core term to the non-bonded pair interaction potential. The reason for this latter modification is to eliminate the unphysical very-short-range attractive well in the Buckingham pair potential used in the original force field. The parameters of the $1/r^{12}$ repulsive core are chosen in such a way that the system dynamics is practically unaffected under the conditions we study in this paper. More details can be found in the supporting information for the work by Zhao et al. [27]. A cut-off distance of 11 Å was used for repulsion, dispersion, and short-range Coulomb interactions. Long-range electrostatic interactions were calculated using the PPPM method with the relative error in the forces set to 1×10^{-6}. A time step of 0.2 fs was used. Sample input decks including all force-field parameters and the crystal cell description are included in the Supplementary Materials.

2.3. Error Analysis

The output of the MD simulations is the stress tensor as a function of time for a given thermodynamic state. Generally, the observations in this sequence are correlated for time differences that are shorter than some intrinsic correlation time. Therefore, we cannot use the sample standard deviation for error analysis. Instead, we need to apply a more sophisticated analysis applicable to correlated data. We follow the approach of Flyvbjerg and Petersen [28], which requires calculation of $c(t)$, the time correlation function for each dataset (stress component in our case), for data obtained after n time steps ($n + 1$ data points). The correlation function is given by

$$c(t) = \frac{1}{n-t} \sum_{k=1}^{n-t} (\sigma_i(t) - \langle \sigma_i \rangle)(\sigma_i(k+t) - \langle \sigma_i \rangle), \tag{7}$$

where angular brackets denote the time average. Then, the error s is calculated using the following equation:

$$s^2 = \frac{c(0) + 2\sum_{t=1}^{t'}(1 - \frac{t}{n})c(t)}{n - 2t' - 1 + \frac{t'(t'+1)}{n}}. \tag{8}$$

Here, t' is a cut-off time chosen to be much larger than the correlation time of $c(t)$ but much smaller than n. We observed typical relaxation times for $c(t)$ of about 5000 time steps (1 ps) so we used $t' = 50{,}000$ (10 ps). The errors for the elastic coefficient values reported in the next section are reported as $\pm s$.

3. Results and Discussion

3.1. Isotherms and Isobars

Although the main topic of the present study is calculation of the elastic coefficients, we also obtain predictions of β-HMX unit-cell volume as functions of pressure and temperature. Figure 1 shows unit-cell volume V as a function of pressure P at 300 K obtained in this work, along with MD results due to Sewell et al. [14] and experimental data of Yoo and Cynn [29], Gump and Peiris [30], and Olinger et al. [31]. The inset in Figure 1 shows the lower pressure part of the same data. The red curve in Figure 1 shows the third-order Birch–Murnaghan isotherm [32] fitted to our isotherm data. Fitting compression data to isotherms is subtle [10]. Here, we applied simple, unweighted fits of the Birch–Murnaghan fitting form to the experimental and simulated $V = V(P)$ data. Our results are close to those of Sewell et al. This is not surprising as they used practically the same force field but with all covalent bonds fixed at constant values. The experimental results of Yoo and Cynn [29] are in overall qualitative agreement with our data but show slightly higher compressibility of the material. In addition, our simulations do not predict the subtle phase transition reported in [29] at approximately 27 GPa. Although we have limited results below 10 GPa, the changes in volume that we predict in this region are also similar to the volume changes observed experimentally in [30,31]. We provide in the Supplementary Materials the full set of lattice parameters and unit-cell volumes for all pressures and temperatures studied. The bulk moduli extracted from the experimental isotherm data are compared to our MD results in the next subsection.

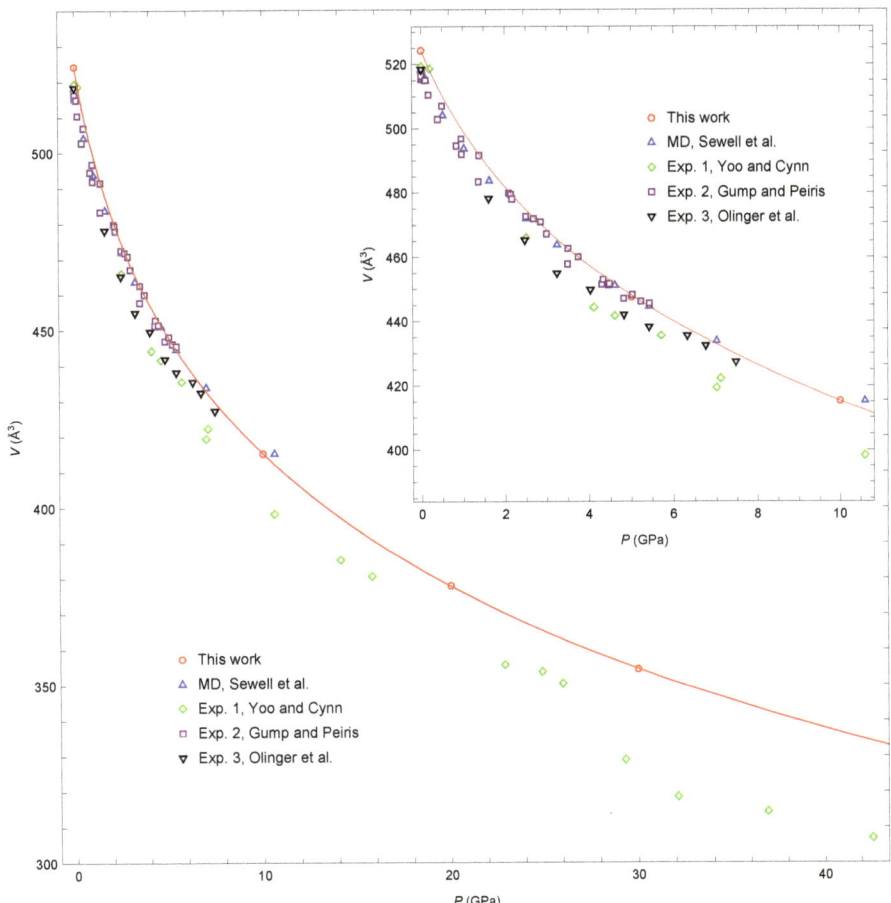

Figure 1. Comparison of the 300 K isotherm from the present simulations to previous MD [14] and experimental results [29–31] at standard ambient temperature. V is the unit-cell volume. The red curve is the third-order Birch–Murnaghan isotherm [32] fitted to our data. The two fitting parameters of the Birch–Murnaghan equation are $K_0 = 16.3$ GPa and $K_0' = 9.1$. The inset shows the lower pressure part of the same data.

Figure 2 shows the five isotherms calculated in the present study. The higher temperature isotherms lie above the lower temperature ones. This behavior is physically reasonable: unit-cell volume increases as the temperature increases due to thermal expansion. The relative volume increase with increasing temperature becomes less pronounced for higher pressures. Otherwise, the isotherms are all quite similar.

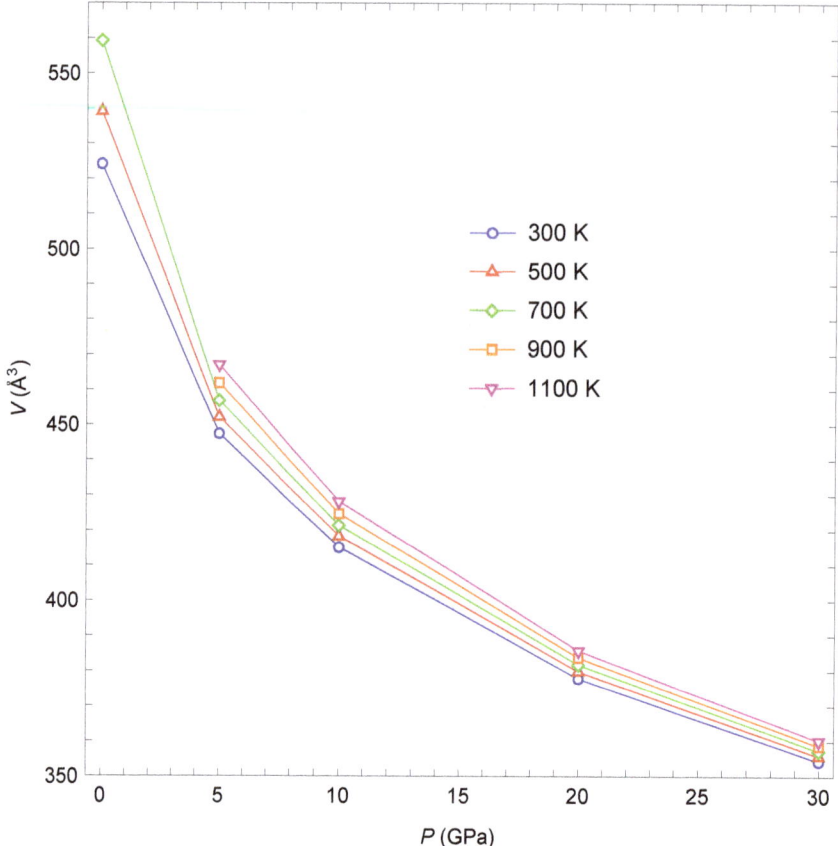

Figure 2. Isotherms for 300, 500, 700, 900, and 1100 K. V is the unit-cell volume. Lines are added to guide the eye.

Figure 3a shows isobars for the five pressures considered. As expected, the unit-cell volume increases as the temperature increases. The volume increase becomes less pronounced for higher pressure isobars. While the increase in unit-cell volume with increasing temperature is not surprising, the changes in the unit-cell geometry exhibit some interesting features. Figure 3b–d show the lengths of the unit-cell vectors **a**, **b**, and **c**, respectively. Surprisingly, at 5 and 10 GPa, the length of vector **a** *decreases* as the temperature increases. Somewhat similarly, at 20 and 30 GPa, vector **c** shows almost no change in length as the temperature changes. Similar counterintuitive behavior of some unit-cell lattice vectors was observed experimentally but not emphasized by Gump and Peiris [30]. Figure 3e shows how angle β (the angle between vectors **a** and **c**) changes with temperature and pressure. For a given temperature, β decreases with increasing pressure (the crystal becomes "more orthorhombic"). For a given pressure, the angle increases slightly as the temperature increases.

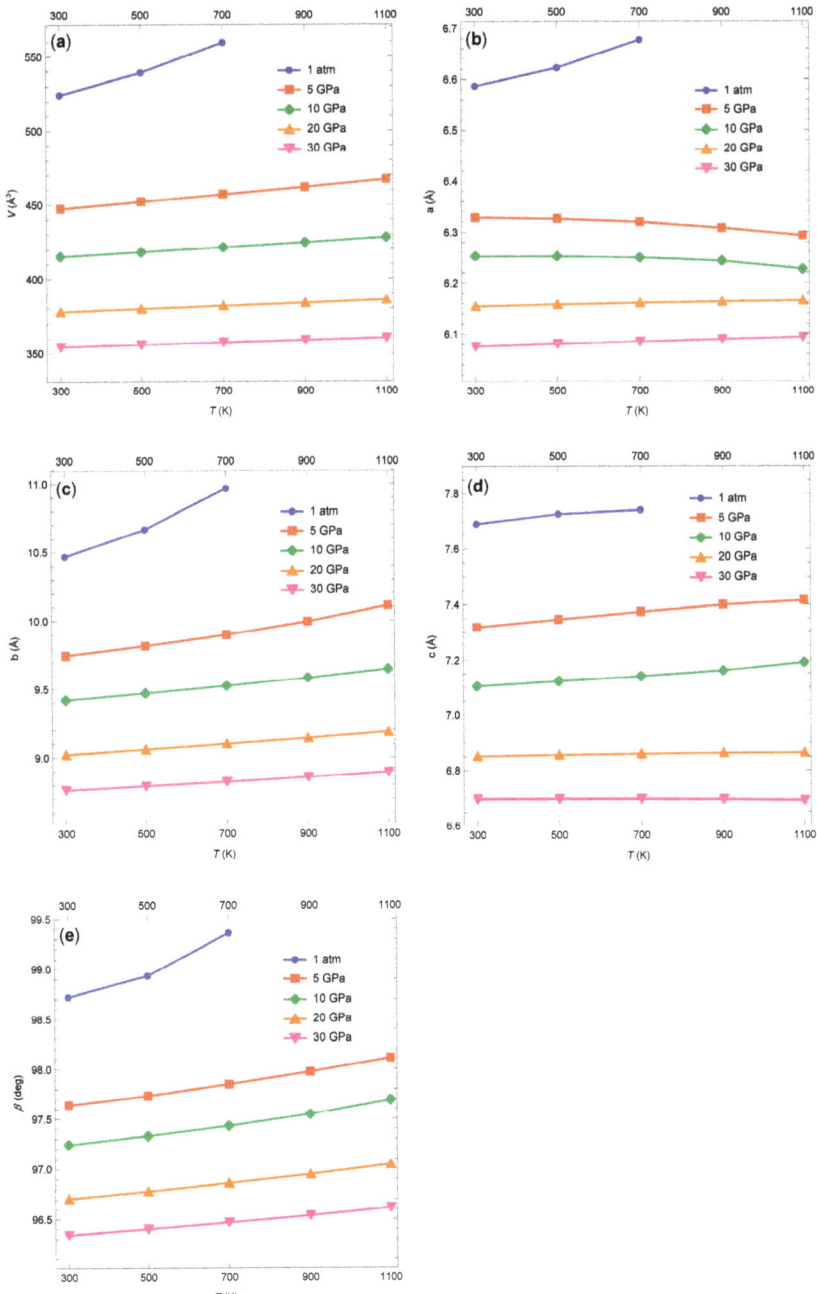

Figure 3. (a) Unit-cell volume as a function of temperature for five different pressures. (b) The length of unit-cell vector **a** as a function of temperature for five different pressures. (c) The same as (b) but for vector **b**. (d) The same as (b) but for vector **c**. (e) The same as (b) but for monoclinic lattice angle β. Lines are added to guide the eye.

3.2. Elastic Coefficients

The calculated elastic coefficients and corresponding bulk and shear moduli on the 300, 500, 700, 900, and 1100 K isotherms are listed, respectively, in Tables 1–5. There are two clear trends in the results. For a given temperature, magnitudes of the elastic coefficients increase substantially with increasing pressure. For a given pressure, magnitudes of the elastic coefficients decrease with increasing temperature. Both trends are physically plausible: As the crystal is compressed at a given temperature, it is expected to become more stiff and so the elastic coefficients become larger. On the other hand, as the crystal temperature increases at constant pressure, the crystal expands and in doing so becomes less stiff.

Table 1. Pressure-dependent elastic coefficients and isotropic moduli of β-HMX at 300 K in units of GPa.

P	10^{-4} GPa	5 GPa	10 GPa	20 GPa	30 GPa
C_{11}	22.97 ± 0.04	87.71 ± 0.05	136.23 ± 0.05	212.62 ± 0.07	277.01 ± 0.05
C_{22}	22.62 ± 0.05	67.08 ± 0.07	101.66 ± 0.07	160.72 ± 0.06	214.17 ± 0.06
C_{33}	21.67 ± 0.04	62.11 ± 0.06	93.95 ± 0.05	153.45 ± 0.04	208.63 ± 0.04
C_{44}	8.645 ± 0.004	19.461 ± 0.007	26.790 ± 0.008	37.463 ± 0.008	46.138 ± 0.009
C_{55}	10.407 ± 0.006	34.08 ± 0.01	49.280 ± 0.009	70.034 ± 0.007	84.259 ± 0.008
C_{66}	9.527 ± 0.005	19.662 ± 0.008	25.694 ± 0.008	33.804 ± 0.006	39.855 ± 0.007
C_{12}	9.20 ± 0.03	36.93 ± 0.04	59.88 ± 0.04	101.23 ± 0.05	139.99 ± 0.04
C_{13}	12.32 ± 0.03	52.95 ± 0.04	81.07 ± 0.04	126.08 ± 0.04	165.53 ± 0.03
C_{23}	12.37 ± 0.03	46.49 ± 0.05	73.71 ± 0.04	123.13 ± 0.04	169.11 ± 0.03
C_{15}	−0.43 ± 0.01	−11.32 ± 0.02	−17.00 ± 0.02	−22.33 ± 0.02	−25.23 ± 0.02
C_{25}	4.47 ± 0.01	11.1 ± 0.02	16.05 ± 0.02	22.02 ± 0.02	25.63 ± 0.02
C_{35}	1.84 ± 0.01	2.48 ± 0.02	4.37 ± 0.02	7.22 ± 0.01	9.39 ± 0.01
C_{46}	2.248 ± 0.003	6.06 ± 0.005	8.517 ± 0.005	12.455 ± 0.006	15.391 ± 0.006
K_V	15.00 ± 0.02	54.40 ± 0.02	84.57 ± 0.02	136.41 ± 0.02	183.23 ± 0.02
K_R	14.6 ± 0.4	53 ± 3	82 ± 4	134 ± 4	180 ± 4
K_{VRH}	14.8 ± 0.2	54 ± 1	83 ± 2	135 ± 2	182 ± 2
G_V	7.941 ± 0.006	20.010 ± 0.008	28.164 ± 0.009	40.017 ± 0.009	49.062 ± 0.007
G_R	6.79 ± 0.03	13.70 ± 0.06	19.04 ± 0.07	28.10 ± 0.07	35.45 ± 0.06
G_{VRH}	7.37 ± 0.02	16.86 ± 0.03	23.6 ± 0.03	34.06 ± 0.03	42.25 ± 0.03

Table 2. Pressure-dependent elastic coefficients and isotropic moduli of β-HMX at 500 K in units of GPa.

P	10^{-4} GPa	5 GPa	10 GPa	20 GPa	30 GPa
C_{11}	18.13 ± 0.05	82.57 ± 0.07	130.73 ± 0.07	207.17 ± 0.06	271.05 ± 0.06
C_{22}	17.86 ± 0.06	63.7 ± 0.1	98.52 ± 0.09	157.49 ± 0.08	210.88 ± 0.07
C_{33}	18.11 ± 0.04	59.69 ± 0.09	90.79 ± 0.08	149.96 ± 0.08	204.88 ± 0.05
C_{44}	7.421 ± 0.006	18.07 ± 0.01	25.473 ± 0.009	36.35 ± 0.01	44.75 ± 0.01
C_{55}	8.474 ± 0.008	32.25 ± 0.01	47.30 ± 0.01	68.13 ± 0.01	82.48 ± 0.01
C_{66}	8.076 ± 0.006	18.887 ± 0.009	25.21 ± 0.01	33.44 ± 0.01	39.47 ± 0.01
C_{12}	7.17 ± 0.04	35.38 ± 0.06	58.46 ± 0.05	99.99 ± 0.05	138.53 ± 0.05
C_{13}	9.83 ± 0.03	51.81 ± 0.06	80.23 ± 0.05	125.67 ± 0.05	165.20 ± 0.04
C_{23}	9.26 ± 0.04	43.70 ± 0.07	70.94 ± 0.06	120.28 ± 0.05	166.16 ± 0.04
C_{15}	−0.16 ± 0.01	−10.98 ± 0.02	−16.66 ± 0.02	−22.08 ± 0.02	−25.00 ± 0.02
C_{25}	3.78 ± 0.02	10.57 ± 0.02	15.54 ± 0.03	21.81 ± 0.02	25.53 ± 0.02
C_{35}	1.19 ± 0.01	1.47 ± 0.03	3.64 ± 0.03	6.65 ± 0.02	8.91 ± 0.02
C_{46}	1.779 ± 0.004	5.732 ± 0.007	8.071 ± 0.006	11.830 ± 0.007	14.779 ± 0.008
K_V	11.85 ± 0.02	51.97 ± 0.03	82.14 ± 0.03	134.06 ± 0.02	180.73 ± 0.02
K_R	11.4 ± 0.4	51 ± 5	80 ± 6	131 ± 5	178 ± 5
K_{VRH}	11.6 ± 0.2	51 ± 3	81 ± 3	133 ± 3	179 ± 3
G_V	6.650 ± 0.007	18.85 ± 0.01	27.0 ± 0.01	38.83 ± 0.01	47.803 ± 0.009
G_R	5.74 ± 0.04	12.85 ± 0.1	18.02 ± 0.09	27.19 ± 0.08	34.52 ± 0.07
G_{VRH}	6.20 ± 0.02	15.85 ± 0.05	22.49 ± 0.05	33.01 ± 0.04	41.16 ± 0.04

Table 3. Pressure-dependent elastic coefficients and isotropic moduli of β-HMX at 700 K in units of GPa.

P	10^{-4} GPa	5 GPa	10 GPa	20 GPa	30 GPa
C_{11}	12.78 ± 0.06	77.56 ± 0.07	125.53 ± 0.06	201.65 ± 0.07	265.28 ± 0.09
C_{22}	12.09 ± 0.07	60.1 ± 0.1	95.0 ± 0.1	154.48 ± 0.09	207.48 ± 0.08
C_{33}	14.05 ± 0.06	57.6 ± 0.1	87.6 ± 0.1	146.64 ± 0.07	201.32 ± 0.07
C_{44}	6.051 ± 0.007	16.60 ± 0.01	23.84 ± 0.02	35.04 ± 0.01	43.44 ± 0.01
C_{55}	6.147 ± 0.009	30.50 ± 0.02	45.25 ± 0.013	66.17 ± 0.01	80.61 ± 0.01
C_{66}	6.316 ± 0.008	17.89 ± 0.01	24.63 ± 0.01	33.09 ± 0.01	39.07 ± 0.01
C_{12}	4.87 ± 0.05	34.02 ± 0.06	57.05 ± 0.06	98.76 ± 0.06	137.35 ± 0.06
C_{13}	7.09 ± 0.04	51.09 ± 0.07	79.52 ± 0.06	125.28 ± 0.05	165.20 ± 0.05
C_{23}	6.43 ± 0.05	41.23 ± 0.08	68.23 ± 0.07	117.79 ± 0.06	163.43 ± 0.06
C_{15}	0.09 ± 0.02	−10.86 ± 0.03	−16.35 ± 0.02	−21.93 ± 0.03	−24.75 ± 0.02
C_{25}	2.94 ± 0.02	10.03 ± 0.04	15.16 ± 0.03	21.52 ± 0.03	25.47 ± 0.03
C_{35}	0.75 ± 0.02	0.29 ± 0.04	2.93 ± 0.03	6.02 ± 0.02	8.46 ± 0.02
C_{46}	1.272 ± 0.006	5.485 ± 0.007	7.73 ± 0.01	11.31 ± 0.01	14.22 ± 0.01
K_V	8.41 ± 0.02	49.77 ± 0.03	79.75 ± 0.03	131.82 ± 0.03	178.45 ± 0.03
K_R	7.9 ± 0.4	48 ± 6	77 ± 8	129 ± 7	175 ± 7
K_{VRH}	8.1 ± 0.2	49 ± 3	78 ± 4	130 ± 4	177 ± 4
G_V	5.072 ± 0.009	17.59 ± 0.02	25.64 ± 0.01	37.6 ± 0.01	46.50 ± 0.01
G_R	4.32 ± 0.04	11.7 ± 0.1	16.8 ± 0.1	26.1 ± 0.1	33.47 ± 0.09
G_{VRH}	4.70 ± 0.02	14.67 ± 0.06	21.20 ± 0.06	31.86 ± 0.05	39.98 ± 0.05

Table 4. Pressure-dependent elastic coefficients and isotropic moduli of β-HMX at 900 K in units of GPa.

P	5 GPa	10 GPa	20 GPa	30 GPa
C_{11}	72.71 ± 0.08	120.59 ± 0.08	196.33 ± 0.07	259.59 ± 0.07
C_{22}	56.03 ± 0.02	91.5 ± 0.1	150.98 ± 0.09	204.1 ± 0.1
C_{33}	56.5 ± 0.1	84.3 ± 0.1	143.0 ± 0.1	197.67 ± 0.07
C_{44}	15.23 ± 0.02	22.01 ± 0.02	33.48 ± 0.02	42.01 ± 0.02
C_{55}	28.94 ± 0.02	43.11 ± 0.02	64.13 ± 0.02	78.73 ± 0.02
C_{66}	16.63 ± 0.02	23.86 ± 0.02	32.70 ± 0.02	38.67 ± 0.01
C_{12}	32.93 ± 0.08	55.99 ± 0.08	97.40 ± 0.06	136.14 ± 0.06
C_{13}	50.54 ± 0.07	79.25 ± 0.07	124.88 ± 0.06	165.18 ± 0.05
C_{23}	39.11 ± 0.09	65.75 ± 0.09	115.00 ± 0.07	160.90 ± 0.06
C_{15}	−10.91 ± 0.03	−16.33 ± 0.03	−21.70 ± 0.03	−24.55 ± 0.03
C_{25}	9.36 ± 0.04	14.71 ± 0.03	21.39 ± 0.04	25.34 ± 0.03
C_{35}	−1.43 ± 0.03	2.15 ± 0.04	5.44 ± 0.03	7.95 ± 0.03
C_{46}	5.22 ± 0.01	7.54 ± 0.01	10.80 ± 0.01	13.66 ± 0.01
K_V	47.83 ± 0.04	77.60 ± 0.04	129.43 ± 0.03	176.20 ± 0.03
K_R	46 ± 9	75 ± 13	126 ± 8	173 ± 9
K_{VRH}	47 ± 5	76 ± 6	128 ± 4	175 ± 5
G_V	16.34 ± 0.02	24.16 ± 0.02	36.3 ± 0.01	45.16 ± 0.01
G_R	10.7 ± 0.2	15.1 ± 0.2	24.9 ± 0.1	32.3 ± 0.1
G_{VRH}	13.53 ± 0.08	19.62 ± 0.08	30.58 ± 0.05	38.72 ± 0.06

Table 5. Pressure-dependent elastic coefficients and isotropic moduli of β-HMX at 1100 K in units of GPa.

P	5 GPa	10 GPa	20 GPa	30 GPa
C_{11}	67.54 ± 0.09	115.6 ± 0.1	190.9 ± 0.1	253.4 ± 0.1
C_{22}	51.8 ± 0.2	87.3 ± 0.2	147.7 ± 0.1	200.82 ± 0.09
C_{33}	56.5 ± 0.1	80.7 ± 0.2	139.39 ± 0.08	193.90 ± 0.06
C_{44}	14.16 ± 0.02	19.98 ± 0.02	31.76 ± 0.03	40.52 ± 0.02
C_{55}	27.44 ± 0.02	40.98 ± 0.03	62.00 ± 0.02	76.64 ± 0.02
C_{66}	15.38 ± 0.02	22.77 ± 0.02	32.27 ± 0.02	38.21 ± 0.02
C_{12}	32.1 ± 0.1	55.07 ± 0.09	96.25 ± 0.09	135.02 ± 0.08
C_{13}	49.66 ± 0.08	79.58 ± 0.09	124.56 ± 0.07	165.05 ± 0.06
C_{23}	37.5 ± 0.1	63.1 ± 0.1	112.78 ± 0.07	158.41 ± 0.06
C_{15}	−10.79 ± 0.03	−16.53 ± 0.03	−21.57 ± 0.03	−24.35 ± 0.03
C_{25}	8.34 ± 0.04	14.36 ± 0.04	21.12 ± 0.04	25.19 ± 0.04
C_{35}	−3.60 ± 0.04	1.40 ± 0.04	4.87 ± 0.03	7.44 ± 0.03
C_{46}	4.9 ± 0.01	7.40 ± 0.01	10.45 ± 0.01	13.16 ± 0.01
K_V	46.03 ± 0.04	75.46 ± 0.05	127.24 ± 0.04	173.90 ± 0.03
K_R	44 ± 11	72 ± 24	124 ± 13	171 ± 10
K_{VRH}	45 ± 6	74 ± 12	126 ± 6	172 ± 5
G_V	15.17 ± 0.02	22.48 ± 0.02	34.83 ± 0.02	43.72 ± 0.02
G_R	9.9 ± 0.2	12.8 ± 0.2	23.4 ± 0.2	31.0 ± 0.1
G_{VRH}	12.55 ± 0.09	17.6 ± 0.1	29.10 ± 0.07	37.34 ± 0.06

As an example, Figure 4 shows the pressure dependence of the elastic coefficients and isotropic moduli at 300 K (using data from Table 1). As the pressure changes from 1 atm to 30 GPa, the magnitudes of the elastic coefficients increase by a minimum of about four-fold for C_{66} to about 60-fold for C_{15}. The Voigt–Reuss–Hill bulk and shear moduli increase by about 12- and 6-fold, respectively. Similar behavior is observed at the higher temperatures (not shown).

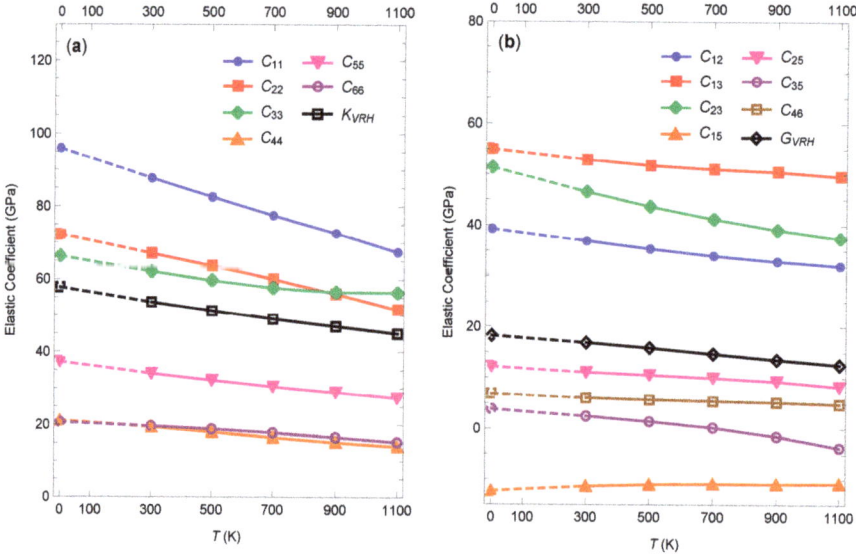

Figure 4. Components of the elastic stiffness tensor at 300 K as functions of hydrostatic pressure. (**a**) Diagonal components and the Voigt–Reuss–Hill bulk modulus. (**b**) Off-diagonal components and the Voigt–Reuss–Hill shear modulus. Lines are added to guide the eye.

The typical temperature dependence of the elastic coefficients is shown in Figure 5, where the results for pressure 5 GPa are presented. This is the lowest pressure among the ones we studied for which β-HMX remains crystalline for all temperatures on the time scale considered. The elastic coefficients decrease with increasing temperature approximately linearly. The rate at which the magnitude of the elastic coefficients decreases with temperature ranges from about 2.5 GPa per 100 K for C_{11} to about 0.07 GPa per 100 K for C_{15}. The Voigt–Reuss–Hill bulk and shear moduli decrease with temperature at a rate of about 1 GPa per 100 K and 0.5 GPa per 100 K, respectively. Qualitatively similar temperature dependence of the elastic coefficients was observed at higher pressures but the rates at which the coefficients decrease become smaller.

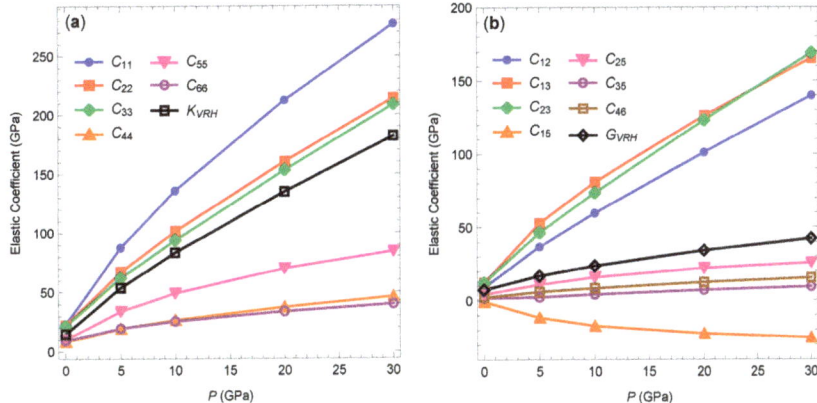

Figure 5. Components of the elastic stiffness tensor at 5 GPa as functions of temperature. (a) Diagonal components and the Voigt–Reuss–Hill bulk modulus. (b) Off-diagonal components and the Voigt–Reuss–Hill shear modulus. Data for 0 K are taken from Mathew and Sewell [15], who used practically the same force field as here and a similar methodology. Lines are added to guide the eye.

In Figure 6, we compare our results at 1 atm and 300 K to the published theoretical and experimental data for the β-HMX elastic coefficients variously at ambient conditions [7–9,14] or zero temperature and pressure [15,33,34]. Our results are, in general, consistent with the MD simulations of Sewell et al. [14] for which practically the same force field was used but with all the covalent bonds frozen. As a result of the bond constraints, most of the elastic coefficients in [14] are slightly larger in magnitude compared to our values. Similarly, there is an overall consistency with the MD/energy-minimization results of Mathew and Sewell [15], obtained with practically the same force field used here but at zero temperature. As can be expected, their values are higher than ours because the crystal stiffens with decreasing temperature, as discussed above. The DFT results in [33,34] give higher values of elastic coefficients and elastic moduli compared to ours. This is consistent with the fact that those results correspond to zero temperature and pressure. Note that, although DFT calculations have been used to calculate lattice parameters and unit-cell volumes on the cold curve (see, e.g., [33]), to our knowledge, they have not been used to predict elastic coefficients at elevated pressures. Moreover, incorporating temperature dependence into such calculations using, for example, the quasi-harmonic approximation would be unreliable due to the need to account for (anisotropic) thermal expansion across hundreds of kelvins; and explicit simulations analogous to, and on the scale of, those studied here are practically infeasible. There is a significant disparity among the various experimental results, which has been attributed to variations in measurement techniques and sample purity. Reanalysis [6,14] of the experimental data has shown that this can also result from lack of redundancy in the acoustic velocity measurements and sensitivity of the numerical solution to initial conditions of the multivariate minimization used to extract the elastic coefficients. For example, it is known that only five of the thirteen nonzero elastic coefficients reported in [7] were accurately determined [14]. Our results agree reasonably well with the experimental data of Sun et al. [9], with more pronounced differences compared to the other experimental data.

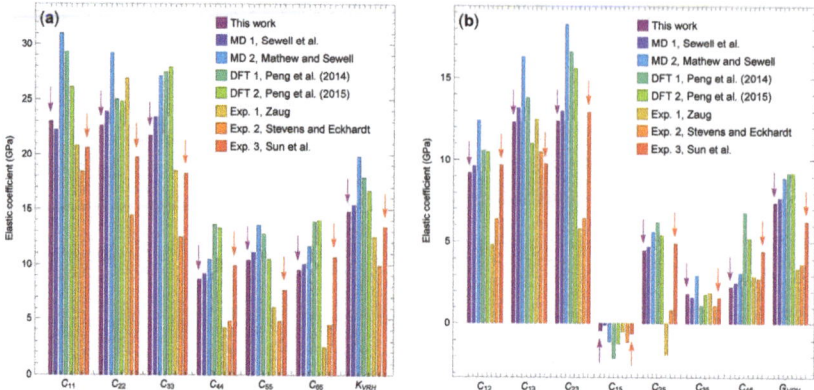

Figure 6. Components of the elastic stiffness tensor at 1 atm and 300 K from the present study, previous theoretical results [14,15,33,34], and experimental values [7–9]. (**a**) Diagonal components and the Voigt–Reuss–Hill bulk modulus. (**b**) Off-diagonal components and the Voigt–Reuss–Hill shear modulus. The purple and red arrows emphasize the results from this work and the experimental results due to Sun et al. [9], respectively.

To the best of our knowledge, there are no experimental results for the elastic coefficients of β-HMX at higher pressures even for 300 K. However, the room-temperature bulk modulus as a function of pressure can be obtained from fits of experimental isothermal compression data to an equation-of-state fitting form (here, the third-order Birch–Murnaghan equation of state), which yields pressure P as a function of volume V. The volume-dependent bulk modulus $K(V)$ can be calculated as $K(V) = -V(\partial P/\partial V)_T$. The pressure-dependent bulk modulus $K(P)$ can then be obtained by expressing the volume in $K(V)$ as a function of pressure using the equation of state. This approach is less precise than obtaining K directly from the elastic tensor at a given T and P due both to the assumed form of the equation of state and the typically small numbers of data points on the isotherm available for fitting. Table 6 lists the room-temperature bulk moduli calculated at five pressures using this approach, for three sets of experimental data [29–31] and our calculated 300 K isotherm in Figure 1. The table also includes the pressure-dependent Voigt–Reuss–Hill bulk moduli reported in Table 1. Note that the results for [30,31] at 10, 20, and 30 GPa represent extrapolations of the fitting form to pressures higher than those for which the corresponding isotherms were measured.

Table 6. Room-temperature bulk moduli of β-HMX for five different pressures. Units are GPa.

P	10^{-4} GPa	5 GPa	10 GPa	20 GPa	30 GPa
This work (direct, K_{VRH})	14.8 ± 0.2	54 ± 1	83 ± 2	135 ± 2	182 ± 2
This work (isotherm fit)	16.3	52.3	81.1	133.9	183.9
Yoo and Cynn [29]	12.4	49.6	79.3	132.8	182.7
Gump and Peiris [30]	21.0	52.7	80.0	130.2	177.4
Olinger et al. [31]	8.4	64.1	102.1	167.8	227.4

In the table, one can see that the MD-based results (top two rows) are self-consistent in that the bulk moduli computed directly from the elastic tensor and via the equation-of-state fit are in good agreement for all pressures considered. Our values obtained by the direct approach are slightly larger than those obtained using the data from Yoo and Cynn [29] for all pressures except 30 GPa, for which they are about the same. Our estimates based on the data from [30,31] are less reliable because, as noted above, the pressure states considered in those references were below 8 GPa. Nevertheless, there is good agreement between our results and those obtained based on the work of Gump and Peiris [30] with the exception of $P = 1$ atm. Our values are lower than those obtained from the data in [31] for all pressures except $P = 1$ atm, for which our value is larger.

4. Conclusions

We calculated second-order elastic coefficients of β-HMX ($P2_1/n$ space group setting) as derivatives of the Cauchy stress with respect to the linear strain using MD with a well-validated fully flexible force field. A set of pressures between 1 atm (10^{-4} GPa) and 30 GPa and temperatures between 300 and 1100 K was considered. The resulting elastic coefficients and bulk and shear elastic moduli are reported in tabular form. The following dominant trends are observed. For a given temperature, magnitudes of the elastic coefficients increase substantially with increasing pressure. For a given pressure, magnitudes of the elastic coefficients decrease with increasing temperature. The strong dependence of the elastic coefficients on pressure should be taken into account in mesoscale continuum modeling and simulations.

Supplementary Materials: The following are available online at http://www.mdpi.com/2073-4352/10/12/1123/s1, Table S1: Equilibrium unit-cell volumes and lattice parameters for β-HMX as functions of temperature and pressure. Table S2: Full 6 × 6 matrix of elastic coefficients C_{ij} of β-HMX in units of GPa at 300 K and 1 atm (10^{-4} GPa). The first column on the left and the top row are indices i and j. The matrix is very close to being symmetric. Sixteen coefficients are close to zero as expected for the monoclinic crystal. Table S3: Full 6 × 6 matrix of elastic coefficients C_{ij} of β-HMX in units of GPa at 700 K and 1 atm (10^{-4} GPa). The first column on the left and the top row are indices i and j. The matrix is very close to being symmetric. Sixteen coefficients are close to zero as expected for the monoclinic crystal. Table S4: Full 6 × 6 matrix of elastic coefficients C_{ij} of β-HMX in units of GPa at 1100 K and 30 GPa. The first column on the left and the top row are indices i and j. The matrix is very close to being symmetric. Sixteen coefficients are close to zero as expected for the monoclinic crystal.

Author Contributions: Conceptualization, T.S. and A.P.; methodology, A.P. and T.S.; software, A.P.; validation, A.P. and T.S.; formal analysis, A.P.; investigation, A.P.; resources, T.S.; writing—original draft preparation, A.P.; writing—review and editing, A.P. and T.S.; visualization, A.P.; project administration, T.S.; and funding acquisition, T.S. All authors have read and agreed to the published version of the manuscript.

Funding: This research was funded by Air Force Office of Scientific Research grant number FA9550-19-1-0318.

Conflicts of Interest: The authors declare no conflict of interest.

Abbreviations

The following abbreviations are used in this manuscript:

HMX	1,3,5,7-tetranitro-1,3,5,7-tetrazoctane
MD	Molecular Dynamics
NPT	isobaric-isothermal
NVT	isochoric-isothermal
DFT	density functional theory

References

1. Gibbs, T.; Popolato, A. *LASL Explosive Property Data*; University of California: Berkeley, CA, USA, 1980.
2. Choi, C.S.; Boutin, H.P. A study of the crystal structure of β-cyclotetramethylene tetranitramine by neutron diffraction. *Acta Crystallogr. B Struct. Cryst. Cryst. Chem.* **1970**, *26*, 1235–1240. [CrossRef]
3. Kohno, Y.; Maekawa, K.; Azuma, N.; Tsuchioka, T.; Hashizume, T.; Imamura, A. Ab initio calculations for a relationship between impact sensitivity and molecular structure in HMX polymorphs. *Kogyo Kayako* **1992**, *53*, 227–237.
4. Cady, H.H.; Larson, A.C.; Cromer, D.T. The crystal structure of α-HMX and a refinement of the structure of β-HMX. *Acta Crystallogr.* **1963**, *16*, 617–623. [CrossRef]
5. Cobbledick, R.E.; Small, R.W.H. The crystal structure of the δ-form of 1,3,5,7-tetranitro-1,3,5,7-tetraazacyclooctane (δ-HMX). *Acta Crystallogr. B Struct. Cryst. Cryst. Chem.* **1974**, *30*, 1918–1922. [CrossRef]

6. Hooks, D.E.; Ramos, K.J.; Bolme, C.A.; Cawkwell, M.J. Elasticity of crystalline molecular explosives. *Propellants Explos. Pyrotech.* **2015**, *40*, 333–350. [CrossRef]
7. Zaug, J.M. Elastic constants of β-HMX and Tantalum, equations of state of supercritical fluids and fluid mixtures and thermal transport determinations. In Proceedings of the Eleventh Detonation Symposium, Snowmass, CO, USA, 31 August–4 September 1998; p. 498.
8. Stevens, L.L.; Eckhardt, C.J. The elastic constants and related properties of β-HMX determined by Brillouin scattering. *J. Chem. Phys.* **2005**, *122*, 174701. [CrossRef]
9. Sun, B.; Winey, J.M.; Gupta, Y.M.; Hooks, D.E. Determination of second-order elastic constants of cyclotetramethylene tetranitramine (β-HMX) using impulsive stimulated thermal scattering. *J. Appl. Phys.* **2009**, *106*, 053505. [CrossRef]
10. Menikoff, R.; Sewell, T.D. Fitting forms for isothermal data. *High Press. Res.* **2001**, *21*, 121–138. [CrossRef]
11. Zaug, J.M.; Austin, R.A.; Armstrong, M.R.; Crowhurst, J.C.; Goldman, N.; Ferranti, L.; Saw, C.K.; Swan, R.A.; Gross, R.; Fried, L.E. Ultrafast dynamic response of single-crystal β-HMX (octahydro-1,3,5, 7-tetranitro-1,3,5,7-tetrazocine). *J. Appl. Phys.* **2018**, *123*, 205902. [CrossRef]
12. Drouet, D.; Picart, D.; Bailly, P.; Bruneton, E. Plastic dissipation and hot-spot formation in HMX-based PBXs subjected to low velocity impact. *Propellants Explos. Pyrotech.* **2020**, *45*, 1–11. [CrossRef]
13. Bower, A.F. *Applied Mechanics of Solids*; Taylor & Francis: Boca Raton, FL, USA, 2010. Available online: http://solidmechanics.org (accessed on 8 December 2020).
14. Sewell, T.D.; Menikoff, R.; Bedrov, D.; Smith, G.D. A molecular dynamics simulation study of elastic properties of HMX. *J. Chem. Phys.* **2003**, *119*, 7417–7426. [CrossRef]
15. Mathew, N.; Sewell, T. Pressure-dependent elastic coefficients of β-HMX from molecular simulations. *Propellants Explos. Pyrotech.* **2018**, *43*, 223–227. [CrossRef]
16. Hall, P.G. Thermal decomposition and phase transitions in solid nitramines. *Trans. Faraday Soc.* **1971**, *67*, 556–562. [CrossRef]
17. Myint, P.C.; Nichols, A.L. Thermodynamics of HMX polymorphs and HMX/RDX mixtures. *Ind. Eng. Chem. Res.* **2017**, *56*, 387–403. [CrossRef]
18. Kroonblawd, M.P.; Austin, R.A. Sensitivity of pore collapse heating to the melting temperature and shear viscosity of HMX. *Mech. Mater.* **2021**, *152*, 103644. [CrossRef]
19. Plimpton, S. Fast parallel algorithms for short-range molecular dynamics. *J. Comput. Phys.* **1995**, *117*, 1–19. Available online: http://lammps.sandia.gov/ (accessed on 8 December 2020). [CrossRef]
20. Hill, R. The elastic behaviour of a crystalline aggregate. *Proc. Phys. Soc. A* **1952**, *65*, 349–354. [CrossRef]
21. Smith, G.D.; Bharadwaj, R.K. Quantum chemistry based force field for simulations of HMX. *J. Phys. Chem. B* **1999**, *103*, 3570–3575. [CrossRef]
22. Bedrov, D.; Ayyagari, C.; Smith, G.D.; Sewell, T.D.; Menikoff, R.; Zaug, J.M. Molecular dynamics simulations of HMX crystal polymorphs using a flexible molecule force field. *J. Comput. Aid. Mater. Des.* **2001**, *8*, 77–85. [CrossRef]
23. Bedrov, D.; Smith, G.D.; Sewell, T.D. Thermal conductivity of liquid octahydro 1,3,5,7-tetranitro-1,3, 5,7-tetrazocine (HMX) from molecular dynamics simulations. *Chem. Phys. Lett.* **2000**, *324*, 64–68. [CrossRef]
24. Bedrov, D.; Smith, G.D.; Sewell, T.D. Temperature-dependent shear viscosity coefficient of octahydro-1,3, 5,7-tetranitro-1,3,5,7-tetrazocine (HMX): A molecular dynamics simulation study. *J. Chem. Phys.* **2000**, *112*, 7203–7208. [CrossRef]
25. Kroonblawd, M.P.; Mathew, N.; Jiang, S.; Sewell, T.D. A generalized crystal-cutting method for modeling arbitrarily oriented crystals in 3D periodic simulation cells with applications to crystal-crystal interfaces. *Comput. Phys. Commun.* **2016**, *207*, 232–242. [CrossRef]
26. Chitsazi, R.; Kroonblawd, M.P.; Pereverzev, A.; Sewell, T. A molecular dynamics simulation study of thermal conductivity anisotropy in β-octahydro-1,3,5,7-tetranitro-1,3,5,7-tetrazocine (β-HMX). *Modell. Simul. Mater. Sci. Eng.* **2020**, *28*, 025008. [CrossRef]
27. Zhao, P.; Lee, S.; Sewell, T.; Udaykumar, H.S. Tandem molecular dynamics and continuum studies of shock-induced pore collapse in TATB. *Propellants Explos. Pyrotech.* **2020**, *45*, 196–222. [CrossRef]
28. Flyvbjerg, H.; Petersen, H.G. Error estimates on averages of correlated data. *J. Chem. Phys.* **1989**, *91*, 461–466. [CrossRef]
29. Yoo, C.S.; Cynn, H. Equation of state, phase transition, decomposition of β-HMX (octahydro-1,3, 5,7-tetranitro-1,3,5,7-tetrazocine) at high pressures. *J. Chem. Phys.* **1999**, *111*, 10229–10235. [CrossRef]

30. Gump, J.C.; Peiris, S.M. Isothermal equations of state of beta octahydro-1,3,5,7-tetranitro-1,3,5,7-tetrazocine at high temperatures. *J. Appl. Phys.* **2005**, *97*, 53513. [CrossRef]
31. Olinger, B.; Roof, B.; Cady, H. The linear and volume compression of β-HMX and RDX to 9 GPa (90 kilobar). In *Proceedings of the Symposium (International) on High Dynamic Pressures*; Commissariat a l'Energie Atomique: Paris, France, 1978; pp. 3–8.
32. Birch, F. Finite strain isotherm and velocities for single-crystal and polycrystalline NaCl at high pressures and 300 K. *J. Geophys. Res.* **1978**, *83*, 1257–1268. [CrossRef]
33. Peng, Q.; Wang, G.; Liu, G.R.; De, S. Structures, mechanical properties, equations of state, and electronic properties of β-HMX under hydrostatic pressures: A DFT-D2 study. *Phys. Chem. Chem. Phys.* **2014**, *16*, 19972–19983. [CrossRef]
34. Peng, Q.; Wang, G.; Liu, G.R.; Grimme, S.; De, S. Predicting elastic properties of β-HMX from first-principles calculations. *J. Phys. Chem. B* **2015**, *119*, 5896–5903. [CrossRef]

Publisher's Note: MDPI stays neutral with regard to jurisdictional claims in published maps and institutional affiliations.

© 2020 by the authors. Licensee MDPI, Basel, Switzerland. This article is an open access article distributed under the terms and conditions of the Creative Commons Attribution (CC BY) license (http://creativecommons.org/licenses/by/4.0/).

Article

Investigating the Interaction between Persistent Slip Bands and Surface Hard Coatings via Crystal Plasticity Simulations

Mohammad S. Dodaran [1,2], Jian Wang [3,4], Nima Shamsaei [1,2] and Shuai Shao [1,2,*]

1. Department of Mechanical Engineering, Auburn University, Auburn, AL 36849, USA; mzs0189@auburn.edu (M.S.D.); shamsaei@auburn.edu (N.S.)
2. National Center for Additive Manufacturing Excellence (NCAME), Auburn University, Auburn, AL 36849, USA
3. Mechanical and Materials Engineering, University of Nebraska-Lincoln, Lincoln, NE 68588, USA; jianwang@unl.edu
4. Nebraska Center for Materials and Nanoscience, University of Nebraska-Lincoln, Lincoln, NE 68588, USA
* Correspondence: sshao@auburn.edu; Tel.: +1-334-844-4867

Received: 21 September 2020; Accepted: 4 November 2020; Published: 6 November 2020

Abstract: Fatigue cracks often initiate from the surface extrusion/intrusions formed due to the operation of persistent slip bands (PSBs). Suppression of these surface topographical features by hard surface coatings can significantly extend fatigue lives under lower stress amplitudes (i.e., high cycle fatigue), while cracks initiate early in the coating or in the coating–substrate interface under higher stress amplitudes (i.e., low cycle fatigue), deteriorating the fatigue performance. However, both beneficial and detrimental effects of the coatings appear to be affected by the coating–substrate material combination and coating thickness. A quantitative understanding of the role of these factors in the fatigue performance of materials is still lacking. In this study, crystal plasticity simulations were employed to elucidate the dependence of the coating's effects on two factors—i.e., the coating thickness and loading amplitudes. The results revealed that the thicker coatings more effectively suppress the operation of the PSBs, but generate higher tensile and shear stresses, normal and parallel to the interfaces, respectively, promoting interfacial delamination. The tensile stresses parallel to the interface within the coating, which favors coating fracture, are not sensitive to the coating thickness.

Keywords: crystal plasticity simulations; persistent slip band; surface hard coating; fatigue crack initiation

1. Introduction

Fatigue cracks for many metals often initiate from surface markings (i.e., intrusions and extrusions) formed due to cyclic slip localization [1,2]. This localized slip activity is associated with permanent changes in the microstructure of the material developed during the cyclic loading, and the slip markings can re-emerge at the same locations upon reapplication of the load even after surface polishing [3]. Due to their persistent nature, these surface markings are commonly referred to as persistent slip markings (PSMs), and the localized deformation volume is referred to as persistent slip bands (PSBs). Cyclic plastic deformation in the PSBs is typically accommodated by the motion of dislocations on the primary slip system [4–6].

For FCC materials, these dislocations have the same Burgers vectors and form very organized bundles known as veins (see Figure 1a). When these bundles collapse along the primary slip direction, PSBs are formed [7]. A typical dislocation microstructure (i.e., the ladder structure) of the PSBs constitutes regularly spaced dislocation walls separating dislocation channels, in which screw "runner" dislocation propagates in a to-and-fro manner. The width of the dislocation channels within PSBs are on the order of 1 µm and is significantly larger than that of the one between dislocation veins. The cyclic plastic deformation therefore localizes at the PSBs [4].

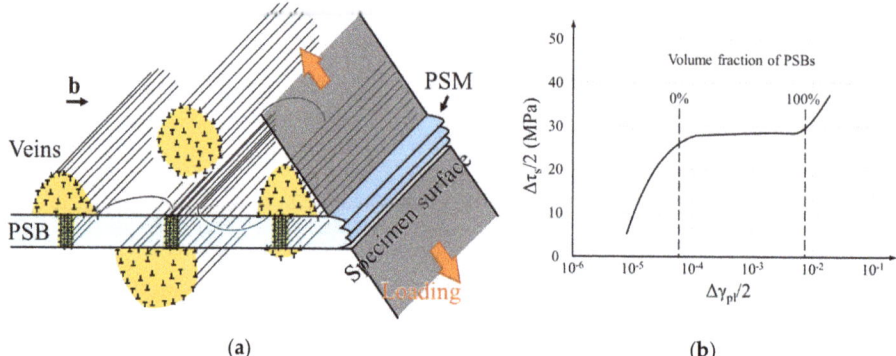

Figure 1. (**a**) A schematic illustration of a persistent slip band (PSB), vein structure, specimen surface, and the surface persistent slip marking (PSM). (**b**) The cyclic stress–strain curve of single-crystal Cu [8].

PSBs typically form within a specific range of resolved plastic shear strain amplitude of $\Delta\gamma_{pl}/2 \approx 0.0001 \sim 0.01$, within which the volume fraction of the PSBs linearly varies from 0% to 100%. The precise values of this range are also material-dependent. This behavior for single-crystal Cu is shown by the cyclic stress–strain curve obtained by Mughrabi [8] (see Figure 1b). Accordingly, the plastic strain amplitude within the PSB is constant at around 1% [9]. In addition, the width of the PSBs is slightly material-dependent but generally invariant with respect to the loading amplitude. As a result, the change in loading amplitude only alters the PSB density—i.e., the higher the loading amplitude is, the higher the PSB density becomes [10].

In polycrystalline metals, the location of the PSBs may shift from the surface to subsurface under very low stress or strain amplitudes (i.e., in the very high cycle fatigue regime) owing to the absence of global plasticity and the dominance of the subsurface, localized plastic deformations due to the elastic incompatibility among grains or between grains and defects [10]. Thus, the corresponding PSB–grain boundary interaction may lead to subsurface fatigue crack initiation [11]. Nevertheless, the PSM-induced surface crack initiation is still the governing mechanism in the low cycle fatigue (LCF) and high cycles fatigue (HCF) regimes encountered in many engineering applications [12–14]. Therefore, any surface treatments that act to suppress or interfere with the formation and/or operation of the PSBs may delay the initiation of fatigue cracks and substantially extend fatigue life. Well-known methods in this regard include intermittent surface repolishing, shot/sand surface peening, and surface hard coatings [3,15–19].

Surface repolishing aims to completely remove the PSMs—together with any other roughness created due to cyclic loading—at appropriate intervals mid-service, eliminating any stress concentrations (even crack embryos). Fatigue life had been shown to be extended almost indefinitely using this method, as long as sufficient serviceable material remained [15]. However, this method, due to its subtractive nature, is not suitable for part surfaces that require tight tolerances. Shot/sand peening, on the other hand, is performed prior to service and induces plastic deformation on the surface layer driven by the impact of shot or sand particles, which leaves a compressive residual stress and a refined microstructure on the surface [16,18]. However, this method results in a relatively rough surface finish and is not suitable for mating surfaces. Lastly, surface hard coatings can circumvent some disadvantages of the other methods because they are generally very thin (a few hundred nm to a µm) and create a smooth surface [19].

There are extensive works focused on examining the effect of various types of coatings—including metallic mono-/multi-layers, ceramic, metallic glass, diamond-like carbon—on the fatigue resistance of metallic specimens [17,20–26]. It has been generally observed that the coatings can improve the HCF performance of the test specimens, while they are often detrimental to the LCF performance. In the LCF regime, higher loading amplitudes tend to induce fracture within the coatings and/or lead to delamination at the coating–substrate interfaces. The precise effect of coatings on the fatigue resistance of a material greatly depends on several factors, including the coating–substrate material combination, mechanical properties and thickness of the coating, and the binding and shear strengths of the coating–substrate interface, etc. [17,19,27]. However, due to the time-intensive nature of the fatigue data generation, systemic evaluations of the influence of the aforementioned factors are still lacking.

Using crystal plasticity (CP) simulations, we aim to elucidate the effect of the coating thickness and the applied cyclic plastic strain amplitude on several characteristics of PSB–coating interaction that tend to influence the fatigue performance of coated materials. These characteristics include (1) the suppressive effects of coatings on the operation of PSBs, (2) the normal stress developed in the coating layer along the loading direction, (3) the interfacial normal stress developed perpendicular to the interface, and (4) the interfacial shear stress developed parallel to the interface. The first characteristic retards, while the rest accelerate, the initiation of fatigue cracks in the substrate. This work by no means attempts to perform direct CP simulations of fatigue failure of coated metallic materials. Instead, it focuses on the interaction between the coating and the PSBs under different loading amplitudes and aspires to generate an understanding that may benefit the geometrical design and material selection of the coatings. Indeed, cyclic damage (in the forms of dislocations and vacancy) accumulates during each cycle via cyclic plasticity and, upon reaching a critical level, leads to the initiation of fatigue cracks [28,29]. The surface coatings' suppression of the plastic deformation per cycle is therefore indicative of the coatings' beneficial effects on a part's overall fatigue resistance. As such, instead of simulating the accumulation of cyclic damage over the entire fatigue life, this work only considers the loading portion of a single cyclic loading period after the PSBs has formed.

2. Computational Methods

CP simulations were performed using the Düsseldorf advanced material simulation kit (DAMASK) developed by the Max-Planck-Institut für Eisenforschung [30]. Spectral solver based on fast Fourier transform (FFT) implemented in the Portable, Extensible Toolkit for Scientific Computation (PETSc) was utilized to solve for the displacement field [31,32]. The simulation model is composed of substrates made of a single crystalline austenitic stainless steel (SS) 316 and physical vapor-deposited (PVD) thin Cr coatings. Prior literature has indicated that the common PVD Cr coatings are nanocrystalline and exhibit isotropic mechanical behavior [19].

2.1. Model Setup

Figure 2 illustrates the geometries of the simulation model. To investigate the characteristics of the interaction between PSBs and surface coatings, only the tensile loading portion of one fatigue cycle is applied. The intended boundary conditions (BCs) on the lateral coating surfaces are free—i.e., the surface tractions are zero. A rate-controlled uniaxial tension was applied by enforcing only the zz component of the deformation gradient rate tensor ($\dot{F}_{zz}$), while ensuring that $\sigma_{xx} = 0$ and $\sigma_{yy} = 0$. Since this work employed the spectral solver implemented in PETSc, which imposed a full periodic BC on the computational cells [30], the presence of free surfaces was mimicked by adding two soft ~1-μm thick buffers layers on both sides of the sample (see Figure 2a), which is similar to the approach by [4,33,34]. A strong contrast in elastic moduli and strengths existed between the buffer layers and the samples (the elastic constants were at least one order of magnitude lower and the strengths were at least three orders of magnitudes lower for the buffer layer). The effective BCs are therefore free, periodic, and periodic in the x-, y-, and z-directions, respectively.

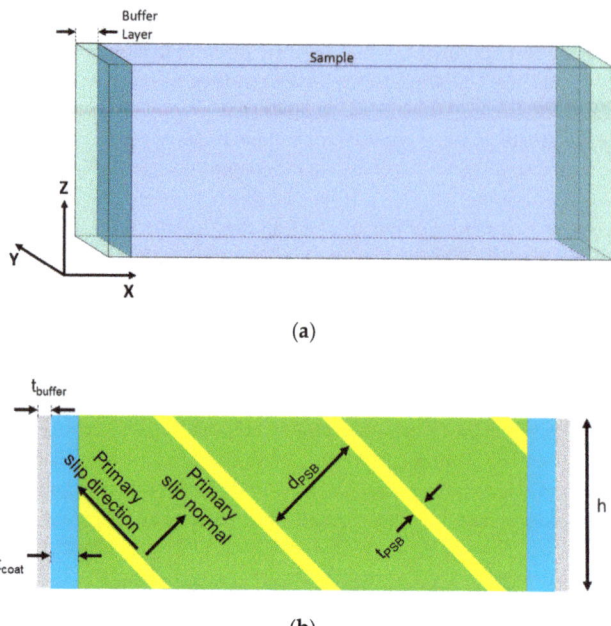

Figure 2. Schematic representation of model geometries: (**a**) a three-dimensional view of the overall geometry of the models omitting the details of the sample, and (**b**) the front view of the simulation cells highlighting the interior geometry and dimensions of the sample.

To conserve computational resources, the y-dimension of the simulation cells was minimized and kept constant at ~0.2 μm. Thus, the models were thin slab-shaped. The width of the substrate along the x-direction was also kept constant at ~34 μm. The PSBs were modeled to be 45° off the loading axis (z) and to have a constant thickness of t_{PSB} = 1 μm in accordance with direct experimental observations in the open literature. Indeed, for metallic materials such as Cu, Ni, and SS 316, the thickness of the PSBs was ~1 μm [12,35]. Constrained by the periodic BC, the height of the models along the z-direction was dictated by the thickness (t_{PSB}) of and spacing (d_{PSB}) between the PSBs—i.e., h = $\sqrt{2}$ (t_{PSB} + d_{PSB}). The spacing d_{PSB} varied between 1 and 8 μm, which corresponds to a PSB volume fraction of 50%~11%, and an overall shear plastic strain range of approximately $\Delta \gamma_{pl}$ = 0.01~0.002 assuming a 0.01 plastic strain amplitude in the PSBs. Three different coating thicknesses (t_{coat}), namely 0.5, 1.0, and 2.0 μm, were considered here. These parameters of the models are listed in Table 1.

Table 1. Design of simulations performed in the current study. The meaning of t_{PSB} and d_{PSB} are shown in Figure 2b.

t_{PSB} (μm)	d_{PSB} (μm)			
0.5	1	2	4	8
1	1	2	4	8
2	1	2	4	8

Both isotropic and anisotropic plastic flow rules have been utilized in the models. The buffer layers and the coatings were treated by isotropic plasticity. This assumption is sound since the buffer layers only have marginal resistance to deformation and the nanocrystalline Cr coatings exhibit isotropic mechanical behavior. Correspondingly, the plastic flow rule is written as [30]

$$\dot{\gamma}_p = \dot{\gamma}_0 \left(\frac{\sqrt{3J_2}}{M\xi} \right)^n, \tag{1}$$

where $\dot{\gamma}_p$ is the plastic shear strain rate, $\dot{\gamma}_0$ is a reference strain rate, n is the stress exponent, J_2 is the second invariant of the deviatoric stress tensor, and M is the Taylor factor. The ξ term in the denominator is the resistance to plastic flow. The rate of ξ is given as

$$\dot{\xi} = \dot{\gamma}_p h_0 \left| 1 - \frac{\xi}{\xi_\infty} \right|^a \mathrm{sgn}\left(1 - \frac{\xi}{\xi_\infty}\right), \tag{2}$$

where $\dot{\gamma}_p$ is the plastic shear strain rate, h_0 is the strain hardening coefficient, ξ_∞ is the saturation resistance to plastic flow, and a is a material-dependent exponent.

Anisotropic plasticity was used for the substrate (including both the matrix and the PSBs) and a phenomenological hardening law was used. The flow rule is written as

$$\dot{\gamma}_p^\alpha = \dot{\gamma}_0^\alpha \left| \frac{\tau^\alpha}{\xi^\alpha} \right|^n \mathrm{sgn}(\tau^\alpha), \tag{3}$$

where τ^α is the resolved shear stress on the slip system α, ξ^α is the slip resistance on the slip system, $\dot{\gamma}_0^\alpha$ is the reference strain rate, and n is the stress exponent. Since only the loading portion of a cyclic loading period was modeled, the back-stress term—which is necessary to capture the kinematic hardening effect in cyclic loading—is not included in the present study. The rate form of the resistance ξ^α is given as

$$\dot{\xi}^\alpha = h_0^{s-s} \sum_{\alpha'=1}^{N_s} \left| \dot{\gamma}^{\alpha'} \right| \left| 1 - \frac{\xi^{\alpha'}}{\xi_\infty^{\alpha'}} \right|^a \mathrm{sgn}\left(1 - \frac{\xi^{\alpha'}}{\xi_\infty^{\alpha'}}\right) h^{\alpha\alpha'}, \tag{4}$$

where $\dot{\gamma}^{\alpha'}$ is shear strain rate on the slip system α', ξ_∞ is the resistance saturation value, $h^{\alpha\alpha'}$ is the slip hardening matrix (including both self- and latent hardening), and a is a material-dependent exponent.

The elastic and plastic flow constants used for the SS 316, PSB, coatings, and buffer layers are summarized in Tables 2 and 3. The stress–strain behaviors produced by the elastic and plastic constants are shown in Figure 3. The elastic constants of both the Cr coating and the SS 316 substrate (including both the PSBs and the matrix) were obtained from the open literature [36–38]. To obtain the plastic flow constants of SS 316, the stress–strain response of a "virtual single crystal"—which was an average of many tensile tests (>50) on single crystals of randomized orientations under the isostrain assumption—was fitted to an experimental curve [39–41]. This technique, also referred to as the "material point" simulation, is a standardized practice to establish flow constants for crystal plasticity simulations [30,42,43].

Table 2. Anisotropic elastic and plastic material constants used for SS316 substrate (including PSB).

Material	C_{11} (GPa)	C_{12} (GPa)	C_{44} (GPa)	$\dot{\gamma}_0$ (1/sec)	n	ξ_0 (MPa)	ξ_∞ (MPa)	a	h_0 (MPa)
SS 316 matrix	207.0	133.0	117.0	0.001	20.0	115.0	430.0	2.25	50.0
PSB primary	207.0	133.0	117.0	0.001	50.0	40.0 [1]	400.0 [1]	2.25	50.0

[1] These values are for the primary slip system in the PSB. The slip activities in the secondary slip systems were suppressed by using much higher slip resistances (at least 1000 times higher).

Table 3. Isotropic elastic and plastic material constants used for the buffer layers and coatings.

Material	C_{11} (GPa)	C_{12} (GPa)	$\dot{\gamma}_0$ (1/sec)	M	n	ξ_0 (MPa)	ξ_∞ (MPa)	a	h_0 (MPa)
Buffer	20.0	10.0	0.1	3	5	3.0	6.3	2.5	7.5
Coating	279.0	115.0	0.001	30	20	20.0 [1]	100.0 [1]	2.5	18.0

[1] Although the ξ_0 and ξ_∞ appear to be quite low, when combined with the M and n, they produced a true yield strength of 750 MPa and a strength of 2.5 GPa at 0.5 strain (see Figure 3).

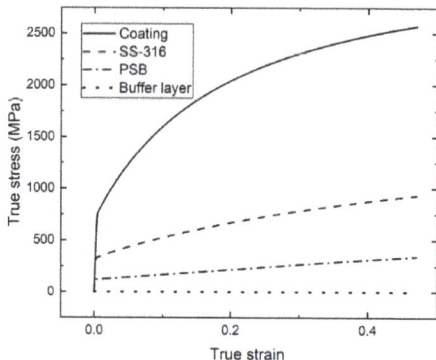

Figure 3. The stress–strain responses of the four material types considered in this study.

PSBs within the substrate were modeled as different materials with identical elastic constants and crystallographic orientations but lower shear resistances (~100 MPa in the PSBs compared to ~300 MPa in the matrix) on the primary slip system. As discussed in the introduction, due to the wider dislocation channels within the PSB compared to the matrix, the PSBs have substantially lower critical resolved shear stress (CRSS). The applied overall strains are therefore localized within the PSBs. For pure Cu, Ni, and Ag, the respective CRSSs are ~30, 50, and 20 MPa [44]. As for SS316, the CRSS of PSBs is not known to the authors' best knowledge and must be assumed. Considering the solid solution strengthening effect in SS316, the highest known CRSSs among the three aforementioned elemental metals, i.e., 50 MPa of Ni, was used. Assuming a Schmid factor of 0.5 (which applies for the current model geometries), the corresponding tensile yield strength is 100 MPa. As will be shown in Section 3, this choice of the CRSS appears to be sufficient to capture the strain localization within the PSBs. The slip activity on the secondary systems was completely suppressed by applying much higher critical resolved shear stresses (~40 GPa). The plastic flow constants of the coating and the buffer layer were calibrated so that they reproduce yield strengths of 750 and ~0 MPa, respectively.

Note that the crystallographic orientation of the substrate must be defined carefully, so that the primary slip system in the PSB experiences maximum shear stress under the tensile loading applied (see Figure 2b). In other words, the primary slip direction and slip plane should both be 45° off the loading axis. A cubic grid with the characteristic size ~0.09 × ~0.09 × ~0.09 μm has been chosen for all the models, leading to 2 Fourier points (FPs) along the y-direction, 400 to 440 (FP) along the x-direction, and 30 to 135 (FP) along the z-direction. The one-FP per voxel configuration is comparable to the one-integration point, C3D8R finite element type (according to ABAQUS), which was emulated in the model setup by DAMASK [45].

2.2. Post-Processing

The results were visualized using the open-source software package Paraview [46]. In all of the visualizations, the buffer layers have been removed to avoid any confusion. All the results visualized correspond to a specific point in the loading history, i.e., when the average shear strain in the primary slip system in the PSBs equals ~1%. Therefore, PSB volume fractions of 50%~11% correspond to overall shear plastic strain ranges of approximately $\Delta\gamma_{pl} = 1\%~0.2\%$. Due to the presence of the coatings, the shear strain on the active slip system within the PSBs may vary depending on the distance to the coating–substrate interface. To assess the suppressive effect of coatings on the operations of PSBs, the plastic strain within the PSBs were plotted as a function of the x-coordinate. For this purpose, the PSBs were sliced into discrete bins of equal thickness along the x-direction. Average shear strain on the primary slip system was then obtained for each bin.

3. Results and Discussion

The results of the simulations were analyzed with respect to four characteristics of the coating–PSB interaction—i.e., (1) the suppressive effects of coatings on the operation of PSBs, (2) the normal stress developed in the coating layer along the loading direction, (3) the interfacial normal stress developed perpendicular to the interface, and (4) the interfacial shear stress developed parallel to the interface. For instance, typical results, including the distributions of shear strain, as well as shear and normal stresses of three selected simulations, are shown in Figure 4 for three cases: (a) $t_{coat} = 0.5$ μm, $d_{PSB} = 1$ μm, and $\Delta\gamma_{pl} = 0.01$, (b) $t_{coat} = 0.5$ μm, $d_{PSB} = 8$ μm, and $\Delta\gamma_{pl} = 0.002$, and (c) $t_{coat} = 2$ μm, $d_{PSB} = 1$ μm, and $\Delta\gamma_{pl} = 0.01$, respectively. Note that the d_{PSB} and $\Delta\gamma_{pl}$ parameters are coupled, as discussed in the previous section. Apparently, all these characteristics appear to be influenced by both the coating thickness and overall loading amplitude (reflected in the density of PSBs). In what follows, we analyze and discuss these influences from the four aforementioned aspects.

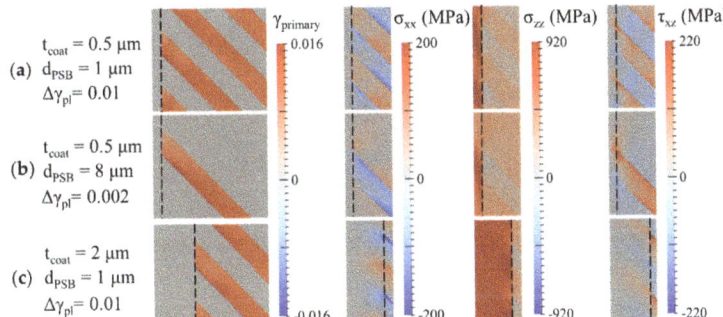

Figure 4. Contour plots of shear strains within the PSBs, as well as normal and shear stresses near the coating–substrate interfaces (marked using the black dashed lines) for four cases of simulations: (a) $t_{coat} = 0.5$ μm, $d_{PSB} = 1$ μm, and $\Delta\gamma_{pl} = 0.01$, (b) $t_{coat} = 0.5$ μm, $d_{PSB} = 8$ μm, and $\Delta\gamma_{pl} = 0.002$, and (c) $t_{coat} = 2$ μm, $d_{PSB} = 1$ μm, and $\Delta\gamma_{pl} = 0.01$.

3.1. Suppressive Effects of the Coating on the Operation of PSBs (Distribution of Plastic Shear Strain in the PSBs)

Comparing the first column of Figure 4a,c, it is apparent that the thicker coating ($t_{coat} = 2$ μm) inhibited the slip activities ($\gamma_{primary}$) within the PSB, especially at locations near the coating–substrate interfaces. As expected, this suppression effect was attenuated with the increase in the strain amplitude applied (compare Figure 4a,b, which represented higher and lower applied strain amplitudes, respectively). To quantitatively compare the suppressive effect of the coatings, the plastic shear strain on the primary slip system within the PSBs are plotted in Figure 5 as a function of the distance from the coating–substrate interface along the x-direction. The plots were generated using the binning analysis, described in Section 2.2. Each panel of the figure corresponds to a different PSB spacing (d_{PSB}) and, accordingly, a different applied plastic strain range ($\Delta\gamma_{pl}$). The plots were obtained at the point of loading history which corresponds to an average of ~1% plastic shear strain over the entire PSBs.

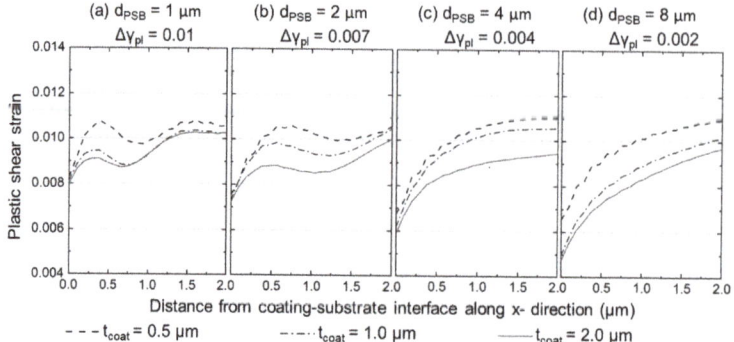

Figure 5. Distribution of shear strain along x on the primary slip system within the PSBs for various PSB spacings and equivalent applied plastic strain ranges: (a) $d_{PSB} = 1$ μm and $\Delta\gamma_{pl} = 0.01$, (b) $d_{PSB} = 2$ μm and $\Delta\gamma_{pl} = 0.007$, (c) $d_{PSB} = 4$ μm and $\Delta\gamma_{pl} = 0.004$, and (d) $d_{PSB} = 8$ μm and $\Delta\gamma_{pl} = 0.002$.

The qualitative observations made in Figure 4 are echoed in Figure 5. Coatings with increasing thickness had stronger suppression effects on the operation of PSBs. This is reflected by the consistently lower shear strain magnitudes in PSBs that interacted with thicker coatings. Comparing Figure 5a–d), the suppression effects were attenuated significantly with the increasing applied strain amplitude. This is concomitant with the increasing plastic shear strain in the PSBs near the coating–substrate interfaces as the loading amplitude increases. For instance, at a lower applied strain range (such as at 0.002 for the case of $d_{PSB} = 8$ μm as shown in Figure 5d), the coatings had a profound effect suppressing the plastic shear strains near the coating–substrate interface. With a coating thickness of 0.5 μm, the shear strain within the distance of 0.5 μm from the interface was lower than the prescribed 1% magnitude. With a coating thickness of 2 μm, this distance extended beyond 2 μm. However, at a higher applied plastic shear strain range of 0.01, this distance reduced to 0.13 μm for the thinner coating and ~1.3 μm for the thicker coating, shown in Figure 5a. The attenuation at higher strain amplitudes (smaller d_{PSB}) can be ascribed to the close coupling of neighboring PSB–substrate interactions, which induced larger elastic deflections within the coatings. As will be demonstrated in the later sections, this coupling also influences the stresses at/near the coating–substrate interfaces.

3.2. Fracture Tendency within the Coatings (Distribution of σ_{zz} in the Coatings)

Hard coatings, especially ceramic ones, may fracture from pre-existing flaws due to tensile stresses parallel to the coating layer. As shown in the third column of Figure 4a,b, the tensile stress (σ_{zz}) in the coating appear to rise at the locations where PSBs intersect with the coating for specimens at a small coating thickness of $t_{coat} = 0.5$ μm. In addition, due to the load transfer from the substrates to the coatings, σ_{zz} is discontinuous across the interface. Due to the superposition of the neighboring PSB–coating interactions, the maximum σ_{zz} may be higher at a smaller PSB spacing (higher applied strain ranges). As an example, the profiles of σ_{zz} in the coating near the interface along the z-direction (i.e., values on the dashed lines shown in Figure 4) are plotted in Figure 6 for the 0.5-μm thick coating.

The observations made from Figure 4 are confirmed in the σ_{zz} profiles shown in Figure 6, noting the values of the stress within the blue shades (locations on the interfaces where PSBs intersect with the coatings). At higher applied strain amplitudes (such as the 0.01 plastic strain range reflected by $d_{PSB} = 1$ μm, Figure 6a), both the overall and the peak values of the σ_{zz} are higher compared to lower applied strain amplitudes, indicating a higher tendency to develop tensile fracture, as expected. Figure 7a shows the variation of the peak σ_{zz} values as a function of d_{PSB} (reflecting the applied strain amplitude $\Delta\gamma_{pl}$), demonstrating a similar decreasing trend of the peak σ_{zz} with increasing d_{PSB} for all t_{coat} values.

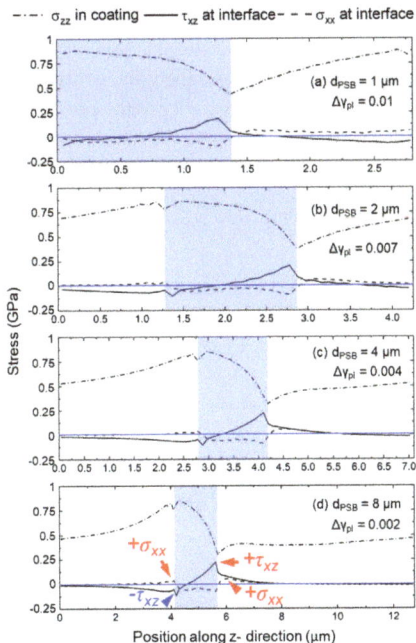

Figure 6. Stress profiles, including the σ_{zz} in the coating near the interface, σ_{xx} at the interface, and τ_{xz} at the interface, of the computational cells with 0.5-μm coating thickness. The blue shades indicate the locations where the PSBs intersect with the coatings. Note that the horizontal axes are not of the same scale, which led to their different appearance in thickness. The four panels respectively show data for (**a**) $d_{PSB} = 1$ μm and $\Delta\gamma_{pl} = 0.01$, (**b**) $d_{PSB} = 2$ μm and $\Delta\gamma_{pl} = 0.007$, (**c**) $d_{PSB} = 4$ μm and $\Delta\gamma_{pl} = 0.004$, and (**d**) $d_{PSB} = 8$ μm and $\Delta\gamma_{pl} = 0.002$.

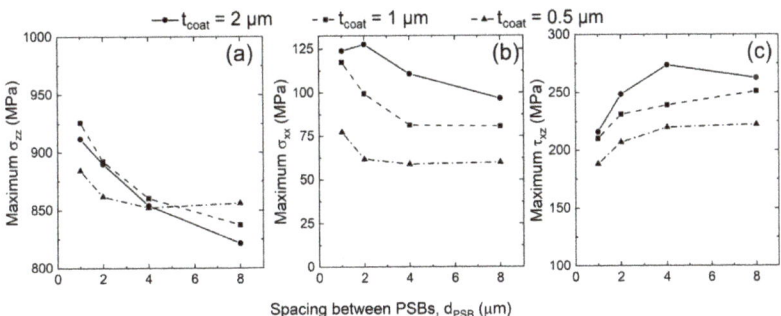

Figure 7. The maximum stresses, including (**a**) σ_{zz} in the coating at the interface, as well as (**b**) σ_{xx} and (**c**) τ_{xz} at the interface, as functions of the increasing PSB spacing, d_{PSB}. Note that the applied plastic shear strain range is inversely related to d_{PSB}—i.e., $d_{PSB} = 1$ μm corresponds to $\Delta\gamma_{pl} = 0.01$, $d_{PSB} = 2$ μm corresponds to $\Delta\gamma_{pl} = 0.007$, $d_{PSB} = 4$ μm corresponds to $\Delta\gamma_{pl} = 0.004$, and $d_{PSB} = 8$ μm corresponds to $\Delta\gamma_{pl} = 0.002$.

Comparing the third column of Figure 4c with that of the Figure 4a,b, it is interesting to note that when the thickness of the coating is large, the distribution of the σ_{zz} within the coating is less influenced by the presence of the PSBs. For instance, Figure 8 shows the profiles of σ_{zz} in the coating along the z-direction at both the surface and the interface. In both coating thicknesses shown, the profiles of σ_{zz} in the coatings at the interface were nearly identical (see the thick and the thin dashed lines). However, for the case of a thin coating ($t_{coat} = 0.5$ μm), significant variations in the stress can be observed on the surface (thin solid line). However, when the coating is thick ($t_{coat} = 2$ μm), σ_{zz} is approximately invariant at the surface (thick solid line). The combined observations made in the third column of Figure 4 and in Figure 8 imply that a coating's sensitivity to the presence of a potential surface flaw in the coating is different for different coating thicknesses. For instance, a thinner coating only experiences higher tensile stresses near the PSBs, a flaw at other locations may still be relatively safe and may not lead to early onset of fracture. On the other hand, the tensile stress on the surface of thicker coatings is uniform which makes thick coatings more susceptible to tensile fracture from surface flaws.

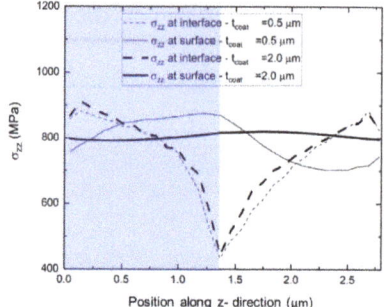

Figure 8. The profiles of σ_{zz} along the z-direction on both sides of the coating for two simulations: (1) $t_{coat} = 0.5$ μm, $d_{PSB} = 1$ μm and (2) $t_{coat} = 2$ μm, $d_{PSB} = 1$ μm. The corresponding stress and strain contours of these two simulations have been shown in Figure 4a,c. The blue shade indicates the location where the PSBs intersect with the coating.

To further investigate the variation of σ_{zz} at different locations within the coating, its standard deviation (SD) along the z-direction for all model geometries was calculated. Figure 9 shows the SD of σ_{zz} at interface and the coating surface with respect to the ratio of d_{PSB} to the coating thickness t_{coat} (i.e., $\lambda = d_{PSB}/t_{coat}$). It is evident that, as the ratio λ increases (i.e., thickness of the coating decreases with respect to the PSB spacing), the variation of σ_{zz} along the z-direction at the coating surface significantly increases. On the other hand, the variation of σ_{zz} at the interface is always quite significant and is not affected by λ. This agrees with the observations made earlier in Figure 8.

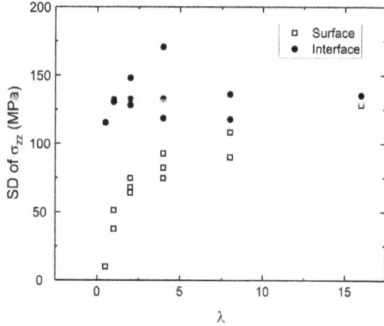

Figure 9. The standard deviation (SD) of σ_{zz} at two locations—i.e., at the coating–substrate interface and coating surface, versus the ratio $\lambda = d_{PSB}/t_{coat}$.

3.3. Delamination Tendency at the Coating–Substrate Interface

The tendency for coating–substrate delamination was assessed by evaluating the interfacial stresses σ_{xx} and τ_{xz}, which were perpendicular and parallel to the interfaces, respectively. As shown in

Figure 4, due to the model setup of a perfectly bonded interface, both σ_{xx} and τ_{xz} stress components were continuous across the interface. Similar to the behavior of σ_{zz}, σ_{xx} appeared to slightly increase when the loading amplitude increased (i.e., when d_{PSB} decreased). Interestingly, the magnitude of τ_{xz} showed an opposite trend—it appeared to decrease when the loading amplitude increased (i.e., when d_{PSB} decreased). This can be seen by comparing Figure 4a, 4b in the second and the fourth columns. The opposite trends observed here may be ascribed to the "symmetries" in the σ_{xx} and τ_{xz} values on both sides of the location where the PSBs intersect with the coating.

In Figure 6, the profiles of σ_{xx} and τ_{xz} at the interface along the z-direction (i.e., values on the dashed lines shown in Figure 4) are plotted for the 0.5-μm coating thickness. At the PSB spacing of 8 μm, the interfacial stresses induced by an individual PSB is clear (see Figure 6d). The sign of the τ_{xz} component is opposite at the locations left and right of the intersection between the PSB and coating (see the red and blue arrows marking the $\pm\tau_{xz}$). On the other hand, the sign of the σ_{xx} is the same on the left and right of this intersection—i.e., both values are positive (see the red arrows marking the $+\sigma_{xx}$ in Figure 6d). As a result, when the density of the PSB increases, the superposition of the stress fields from neighboring PSB–coating interactions increases the maximum magnitude of σ_{xx} and reduces the maximum magnitude of the τ_{xz}. The variation of the maximum σ_{xx} and τ_{xz} stresses are discernable from Figure 6 by comparing the four panels. In addition, the magnitude of both stress components also increased with increasing coating thickness (compare Figure 4a and Figure 4c in the second and fourth columns). The σ_{xx} and τ_{xz} appeared to somewhat signify the "suppressive" action of the coatings—as a function of the thickness—on the operations of the PSBs (compare Figures 4 and 5). In other words, the suppressive effects tend to be stronger when the values of these stresses increase.

For a clearer comparison, the peak values of these stresses for all the simulations have been obtained and plotted in Figure 7b,c. The substantial increases in the peak σ_{xx} and τ_{xz} values due to the increase in the coating thickness are also evident. The combined action of the two stresses may be responsible for the delamination of the coating–substrate interface. This is in line with the critical plane approach put forth by Fatemi and Socie [47], which stated that the planes with large plastic shear strain and large normal stress tend to initiate fatigue cracks.

Figure 5 has shown that the suppression imposed by the coating on the operation of the PSBs is more effective at larger coating thicknesses. Nevertheless, excessively thick coatings are associated with the risk of early crack initiation due to coating tensile fracture (with pre-existing surface flaws) and coating–substrate delamination (due to the combined action of both tensile and shear stresses at the interface). Therefore, thicker coatings are only preferred if a coating with higher fracture toughness as well as an ideal coating–substrate adhesion can be achieved. Otherwise, thicker coatings may be detrimental to the fatigue performance of the coated parts.

On the other hand, while the applied plastic shear strain range is expected to monotonically affect the tendency of tensile fracture in the coating (i.e., higher amplitude leads to easier fracture), its effect on coating–substrate delamination may be more complex. Since the increase in the applied strain range decreases τ_{xz} but increases σ_{xx}, there may exist an intermediate plastic shear strain amplitude that favors delamination the most, assuming that the interfacial delamination is driven by the combined action of normal and shear stresses.

4. Conclusions

Crystal plasticity-based simulations were used to investigate the suppressive effects of coatings on the operation of surface PSBs. The impact of both coating thickness and the applied strain amplitude (reflected by a variable density of PSBs) on the suppressive effects were evaluated. Four characteristics of the PSB–coating interactions—including (1) the suppressive effects of coatings on the operation of PSBs, (2) the normal stress developed in the coating layer along the loading direction, (3) the interfacial normal stress developed perpendicular to the interface, and (4) the interfacial shear stress developed parallel to the interface—were investigated.

The following conclusions can be drawn from this research:

(1) Assuming a perfect coating–substrate adhesion and the absence of coating fracture, thicker coatings offered better suppression against the plastic shear deformation of the PSBs.
(2) The suppression effects of the coatings were attenuated at higher applied plastic shear strains (higher PSB densities).
(3) The distribution of the normal stress parallel to the loading direction in thinner coatings was highly heterogeneous and was strongly affected by the PSBs. However, the distribution of this stress was much more uniform for thicker coatings.
(4) The interfacial shear stress parallel to the loading direction and the interfacial normal stress perpendicular to the interface increased significantly with increasing coating thickness, which can potentially result in delamination.
(5) The peak values of the stresses mentioned in Conclusions (3) and (4) varied differently with increasing applied strains due to the superposition of the stress fields caused by the neighboring PSB–coating interactions.

In general, although thicker coatings may be beneficial to the fatigue performance of the coated parts, excessive coating thickness can lead to early crack initiation due to coating tensile fracture and coating–substrate delamination. The beneficial effect of the coatings on fatigue performance therefore is limited by the fracture toughness of the coating, and the adhesion strength of the coating–substrate interfaces. Therefore, the enhancement in the fatigue performance of a coating–substrate system hinges upon the careful selection of the correct coating–substrate material combination as well as the appropriate coating thickness.

Author Contributions: Conceptualization, S.S., J.W., and N.S.; methodology, S.S.; validation, M.S.D. and S.S.; formal analysis, M.S.D., S.S., J.W., and N.S.; investigation, M.S.D., S.S., J.W., and N.S.; resources, S.S., N.S., and J.W.; data curation, M.S.D. and S.S.; writing—original draft preparation, S.S.; writing—review and editing, M.S.D., S.S., J.W., and N.S.; visualization, M.S.D. and S.S.; supervision, S.S. and N.S.; project administration, S.S., J.W., and N.S.; funding acquisition, S.S., J.W., and N.S. All authors have read and agreed to the published version of the manuscript.

Funding: M.S.D., S.S., and N.S. acknowledge the support by the U.S. Department of Energy, Office of Science, Office of Basic Energy Sciences, under Award Number DE-SC0019378. J.W. acknowledges the support by the Department of Energy (DOE) Office of Nuclear Energy and Nuclear Energy University Program through Award No. DE-NEUP-18-15703.

Acknowledgments: The authors acknowledge the helpful discussions with Wenjin Meng, Michael M. Khonsari, and Shengmin Guo at Louisiana State University.

Conflicts of Interest: The authors declare no conflict of interest. Disclaimer: This report was prepared as an account of work sponsored by an agency of the United States Government. Neither the United States Government nor any agency thereof, nor any of their employees, makes any warranty, express or implied, or assumes any legal liability or responsibility for the accuracy, completeness, or usefulness of any information, apparatus, product, or process disclosed, or represents that its use would not infringe privately owned rights. Reference herein to any specific commercial product, process, or service by trade name, trademark, manufacturer, or otherwise does not necessarily constitute or imply its endorsement, recommendation, or favoring by the United States Government or any agency thereof. The views and opinions of authors expressed herein do not necessarily state or reflect those of the United States Government or any agency FA R&D Special TC NOVEMBER 2017- FF Page 4 of 12 thereof.

References

1. Ewing, J.A.; Humfrey, J.C.W. The Fracture of Metals under Repeated Alternations of Stress. *Philos. Trans. R. Soc. A* **1903**, *200*, 241–250. [CrossRef]
2. Stephens, R.I.; Fatemi, A.; Stephens, R.R.; Fuchs, H.O. *Metal Fatigue in Engineering*, 2nd ed.; John Wiley & Sons: Hoboken, NJ, USA, 2000; ISBN 978-0-471-51059-8.
3. Thompson, N.; Wadsworth, N.; Louat, N. Xi. The origin of fatigue fracture in copper. *Philos. Mag.* **1956**, *1*, 113–126. [CrossRef]
4. Dodaran, M.; Khonsari, M.M.; Shao, S. Critical operating stress of persistent slip bands in Cu. *Comput. Mater. Sci.* **2019**, *165*, 114–120. [CrossRef]
5. Kuhlmann-Wilsdorf, D.; Laird, C. Dislocation behavior in fatigue. *Mater. Sci. Eng.* **1977**, *27*, 137–156. [CrossRef]

6. Kuhlmann-Wilsdorf, D. Theory of plastic deformation—Properties of low energy dislocation structures. *Mater. Sci. Eng. A* **1989**, *113*, 1–41. [CrossRef]
7. Neumann, P. Low energy dislocation configurations: A possible key to the understanding of fatigue. *Mater. Sci. Eng.* **1986**, *81*, 465–475. [CrossRef]
8. Mughrabi, H. The cyclic hardening and saturation behaviour of copper single crystals. *Mater. Sci. Eng.* **1978**, *33*, 207–223. [CrossRef]
9. Hancock, J.; Grosskreutz, J. Mechanisms of fatigue hardening in copper single crystals. *Acta Metall.* **1969**, *17*, 77–97. [CrossRef]
10. Castelluccio, G.M.; Musinski, W.D.; McDowell, D.L. Computational micromechanics of fatigue of microstructures in the HCF–VHCF regimes. *Int. J. Fatigue* **2016**, *93*, 387–396. [CrossRef]
11. Mughrabi, H.; Wang, R.; Differt, K.; Essmann, U. Fatigue Crack Initiation by Cyclic Slip Irreversibilities in High-Cycle Fatigue. In *Fatigue Mechanisms: Advances in Quantitative Measurement of Physical Damage*; ASTM International: West Conshohocken, PA, USA, 1983; Volume 5, pp. 35–41.
12. Mughrabi, H. Cyclic Slip Irreversibilities and the Evolution of Fatigue Damage. *Metall. Mater. Trans. A* **2009**, *40*, 1257–1279. [CrossRef]
13. Man, J.; Obrtlík, K.; Polák, J. Extrusions and intrusions in fatigued metals. Part 1. State of the art and history†. *Philos. Mag.* **2009**, *89*, 1295–1336. [CrossRef]
14. Man, J.; Klapetek, P.; Man, O.; Weidner†, A.; Obrtlík, K.; Polák, J. Extrusions and intrusions in fatigued metals. Part 2. AFM and EBSD study of the early growth of extrusions and intrusions in 316L steel fatigued at room temperature. *Philos. Mag.* **2009**, *89*, 1337–1372. [CrossRef]
15. Basinski, Z.S.; Pascual, R.; Basinski, S.J. Low amplitude fatigue of copper single crystals—I. The role of the surface in fatigue failure. *Acta Metall.* **1983**, *31*, 591–602. [CrossRef]
16. Soyama, H.; Chighizola, C.R.; Hill, M.R. Effect of compressive residual stress introduced by cavitation peening and shot peening on the improvement of fatigue strength of stainless steel. *J. Mater. Process. Technol.* **2021**, *288*, 116877. [CrossRef]
17. Lee, C.M.; Chu, J.P.; Chang, W.Z.; Lee, J.W.; Jang, J.S.C.; Liaw, P.K. Fatigue property improvements of Ti-6Al-4V by thin film coatings of metallic glass and TiN: A comparison study. *Thin Solid Films* **2014**, *561*, 33–37. [CrossRef]
18. Bagherifard, S.; Fernández Pariente, I.; Ghelichi, R.; Guagliano, M. Fatigue properties of nanocrystallized surfaces obtained by high energy shot peening. *Procedia Eng.* **2010**, *2*, 1683–1690. [CrossRef]
19. Zhang, B.; Haghshenas, A.; Zhang, X.; Zhao, J.; Shao, S.; Khonsari, M.M.; Guo, S.; Meng, W.J. On the failure mechanisms of Cr-coated 316 stainless steel in bending fatigue tests. *Int. J. Fatigue* **2020**, *139*, 105733. [CrossRef]
20. Yu, C.-C.; Chu, J.P.; Jia, H.; Shen, Y.-L.; Gao, Y.; Liaw, P.K.; Yokoyama, Y. Influence of thin-film metallic glass coating on fatigue behavior of bulk metallic glass: Experiments and finite element modeling. *Mater. Sci. Eng. A* **2017**, *692*, 146–155. [CrossRef]
21. Tsai, P.H.; Li, T.H.; Hsu, K.T.; Ke, J.H.; Jang, J.S.C.; Chu, J.P. Coating thickness effect of metallic glass thin film on the fatigue-properties improvement of 7075 aluminum alloy. *Thin Solid Films* **2019**, *677*, 68–72. [CrossRef]
22. Su, Y.L.; Yao, S.H.; Wei, C.S.; Kao, W.H.; Wu, C.T. Influence of single- and multilayer TiN films on the axial tension and fatigue performance of AISI 1045 steel. *Thin Solid Films* **1999**, *338*, 177–184. [CrossRef]
23. Zhou, Y.; Rao, G.B.; Wang, J.Q.; Zhang, B.; Yu, Z.M.; Ke, W.; Han, E.H. Influence of Ti/TiN bilayered and multilayered films on the axial fatigue performance of Ti46Al8Nb alloy. *Thin Solid Films* **2011**, *519*, 2207–2212. [CrossRef]
24. Wang, Y.M.; Zhang, P.F.; Guo, L.X.; Ouyang, J.H.; Zhou, Y.; Jia, D.C. Effect of microarc oxidation coating on fatigue performance of Ti–Al–Zr alloy. *Appl. Surf. Sci.* **2009**, *255*, 8616–8623. [CrossRef]
25. Du, D.; Liu, D.; Ye, Z.; Zhang, X.; Li, F.; Zhou, Z.; Yu, L. Fretting wear and fretting fatigue behaviors of diamond-like carbon and graphite-like carbon films deposited on Ti-6Al-4V alloy. *Appl. Surf. Sci.* **2014**, *313*, 462–469. [CrossRef]
26. Stoudt, M.; Ricker, R.; Cammarata, R. The influence of a multilayered metallic coating on fatigue crack nucleation. *Int. J. Fatigue* **2001**, *23*, 215–223. [CrossRef]
27. Akebono, H.; Komotori, J.; Suzuki, H. The Effect of coating thickness on fatigue properties of steel thermally sprayed with ni -based self-fluxing alloy. *Int. J. Mod. Phys. B* **2006**, *20*, 3599–3604. [CrossRef]

28. Sangid, M.D.; Maier, H.J.; Sehitoglu, H. A physically based fatigue model for prediction of crack initiation from persistent slip bands in polycrystals. *Acta Mater.* **2011**, *59*, 328–341. [CrossRef]
29. Sangid, M.D. The physics of fatigue crack initiation. *Int. J. Fatigue* **2013**, *57*, 58–72. [CrossRef]
30. Roters, F.; Diehl, M.; Shanthraj, P.; Eisenlohr, P.; Reuber, C.; Wong, S.L.; Maiti, T.; Ebrahimi, A.; Hochrainer, T.; Fabritius, H.-O.; et al. DAMASK—The Düsseldorf Advanced Material Simulation Kit for modeling multi-physics crystal plasticity, thermal, and damage phenomena from the single crystal up to the component scale. *Comput. Mater. Sci.* **2019**, *158*, 420–478. [CrossRef]
31. Eisenlohr, P.; Diehl, M.; Lebensohn, R.A.; Roters, F. A spectral method solution to crystal elasto-viscoplasticity at finite strains. *Int. J. Plast.* **2013**, *46*, 37–53. [CrossRef]
32. Shanthraj, P.; Eisenlohr, P.; Diehl, M.; Roters, F. Numerically robust spectral methods for crystal plasticity simulations of heterogeneous materials. *Int. J. Plast.* **2015**, *66*, 31–45. [CrossRef]
33. Pokharel, R.; Lind, J.; Kanjarla, A.K.; Lebensohn, R.A.; Li, S.F.; Kenesei, P.; Suter, R.M.; Rollett, A.D. Polycrystal Plasticity: Comparison Between Grain—Scale Observations of Deformation and Simulations. *Annu. Rev. Condens. Matter Phys.* **2014**, *5*, 317–346. [CrossRef]
34. Dodaran, M.; Wang, J.; Chen, Y.; Meng, W.J.; Shao, S. Energetic, structural and mechanical properties of terraced interfaces. *Acta Mater.* **2019**, *171*. [CrossRef]
35. Lavenstein, S.; Gu, Y.; Madisetti, D.; El-Awady, J.A. The heterogeneity of persistent slip band nucleation and evolution in metals at the micrometer scale. *Science* **2020**, *370*, eabb2690. [CrossRef]
36. Palmer, S.B.; Lee, E.W. The elastic constants of chromium. *Philos. Mag. A J. Theor. Exp. Appl. Phys.* **1971**, *24*, 311–318. [CrossRef]
37. Dodaran, M.; Ettefagh, A.H.; Guo, S.M.; Khonsari, M.M.; Meng, W.J.; Shamsaei, N.; Shao, S. Effect of alloying elements on the γ′ antiphase boundary energy in Ni-base superalloys. *Intermetallics* **2020**, *117*. [CrossRef]
38. Bolef, D.I.; de Klerk, J. Anomalies in the Elastic Constants and Thermal Expansion of Chromium Single Crystals. *Phys. Rev.* **1963**, *129*, 1063–1067. [CrossRef]
39. Kweon, H.D.; Kim, J.W.; Song, O.; Oh, D. Determination of true stress-strain curve of type 304 and 316 stainless steels using a typical tensile test and finite element analysis. *Nucl. Eng. Technol.* **2020**. [CrossRef]
40. Zhang, X.; Mu, Y.; Dodaran, M.; Shao, S.; Moldovan, D.; Meng, W.J. Mechanical failure of CrN/Cu/CrN interfacial regions under tensile loading. *Acta Mater.* **2018**, *160*, 1–13. [CrossRef]
41. Zhang, B.; Dodaran, M.; Ahmed, S.; Shao, S.; Meng, W.J.; Juul, K.J.; Nielsen, K.L. Grain-size affected mechanical response and deformation behavior in microscale reverse extrusion. *Materialia* **2019**, *6*. [CrossRef]
42. Marin, E.B. *On the Formulation of a Crystal Plasticity Model*; Sandia National Laboratories: Albuquerque, NM, USA; Livermore, CA, USA, 2006.
43. Lienert, U.; Han, T.-S.; Almer, J.; Dawson, P.R.; Leffers, T.; Margulies, L.; Nielsen, S.F.; Poulsen, H.F.; Schmidt, S. Investigating the effect of grain interaction during plastic deformation of copper. *Acta Mater.* **2004**, *52*, 4461–4467. [CrossRef]
44. Tu, S.-T.; Zhang, X.-C. Fatigue Crack Initiation Mechanisms. In *Reference Module in Materials Science and Materials Engineering*; Elsevier: Amsterdam, The Netherlands, 2016.
45. Diehl, M. *A Spectral Method Using Fast Fourier Transform To Solve Elastoviscoplastic Mechanical Boundary Value Problems*; TU München: München, Germany, 2010.
46. Ahrens, J.; Geveci, B.; Law, C. ParaView: An End-User Tool for Large-Data Visualization. In *Visualization Handbook*; Elsevier: Amsterdam, The Netherlands, 2005; pp. 717–731.
47. Fatemi, A.; Socie, D.F. A critical plane approach to multiaxial fatigue damage including out-of-phase loading. *Fatigue Fract. Eng. Mater. Struct.* **1988**, *11*, 149–165. [CrossRef]

Publisher's Note: MDPI stays neutral with regard to jurisdictional claims in published maps and institutional affiliations.

© 2020 by the authors. Licensee MDPI, Basel, Switzerland. This article is an open access article distributed under the terms and conditions of the Creative Commons Attribution (CC BY) license (http://creativecommons.org/licenses/by/4.0/).

Article

The Effect of Strain Rate on the Deformation Processes of NC Gold with Small Grain Size

Jialin Liu [1], Xiaofeng Fan [1,*], Yunfeng Shi [2], David J. Singh [3] and Weitao Zheng [1,4]

[1] Key Laboratory of Automobile Materials, Ministry of Education, and College of Materials Science and Engineering, Jilin University, Changchun 130012, China; skpliujialin@126.com
[2] Department of Material Science and Engineering, Rensselaer Polytechnic Institute, Troy, NY 12180, USA; shiy2@rip.edu
[3] Department of Physics and Astronomy, University of Missouri, Columbia, MO 65211-7010, USA; singhdj@missouri.edu
[4] State Key Laboratory of Automotive Simulation and Control, Jilin University, Changchun 130012, China; wtzheng@jlu.edu.cn
* Correspondence: xffan@jlu.edu.cn; Tel.: +86-1594-301-3495

Received: 13 August 2020; Accepted: 17 September 2020; Published: 24 September 2020

Abstract: The strength of nanocrystalline (NC) metal has been found to be sensitive to strain rate. Here, by molecular dynamics simulation, we explore the strain rate effects on apparent Young's modulus, flow stress and grain growth of NC gold with small size. The simulation results indicate that the apparent Young's modulus of NC gold decreases with the decrease of strain rate, especially for strain rates above 1 ns^{-1}. The rearrangement of atoms near grain boundaries is a response to the decrease of apparent Young's modulus. Indeed, the flow stress is also sensitive to the strain rate and decreases following the strain rate's decrease. This can be found from the change of strain rate sensitivity and activation volume with the strain rate. Temperature has little effect on the activation volume of NC gold with small grain size, but has an obvious effect on that of relatively large grain size (such as 18 nm) under low strain rate (0.01 ns^{-1}). Finally, grain growth in the deformation process is found to be sensitive to strain rate and the critical size for grain growth increases following the decrease of strain rate.

Keywords: strain rate; molecular dynamics simulation; strain rate sensitivity; activation volume; grain growth

1. Introduction

Nanocrystalline (NC) materials, especially NC metals and alloys, have attracted much attention due to their novel properties, such as improved wear resistance, high yield and high fracture strength [1–3]. It is well known that with the Hall–Petch rule the yield stress is increased following the grain size decrease from millimeter to submicron in coarse grained metals [4,5]. Interestingly, the range of application of the Hall–Petch rule about the yield stress and grain size can be expanded to nanoscale in NC metals such that the hardness and yield stress can increase 5–10 times, compared with their partners of coarse grain [6]. However, the micromechanism of the deformation processes in both regimes is considered to be different. In coarse grained metals, dislocations are generated from intragranular sources and they are stored and rearranged by the interaction between dislocation–dislocation. In NC metals, grain boundary mediated deformation is considered to control the strengthening [6–11].

The plastic deformation of NC metals may be related to many factors. Using grain size as the sole parameter to characterize its mechanical properties may be overly simplified, and thus sometimes give rise to uncertainties [12]. It has been revealed that the Hall–Petch rule breaks down when grain size decreases down to some critical size. To understand these novel results in experiments, even controversial findings [13,14], computer simulations at atomic level, especially molecular dynamics (MD), are expected to offer key insights. Indeed, MD is very helpful in understanding the deformation processes including plastic and elastic deformation, since it can provide real-time behavior and uncover the transient responses which are difficult to detect in experiments [15]. A lot of work has been taken to explore the critical size quantitatively and it is considered that the strength decreases generally as grain size decreases down to 20–10 nm. For example, the MD simulation showed that the critical size of NC Cu was about 10–15 nm [9].

It is well known that the mechanical response is always rate-sensitive [16–18]. For example, in the stainless steel, it was found that high strain rates (such as 10^4–10^5 s^{-1}) could produce twin bundles with high density and nanoscale thickness [19]. Under high strain rates, such as pulsed shocking wave loading, it was found the shear and tensile strengths in metals could have very high values [18]. It is also noticed that the process of plastic deformation in NC metals is very sensitive to the loading rate [20,21]. In NC metals, the strain rate sensitivity (m) is an order higher than that of coarse grain [16,22,23]. It is considered that the large value of m is related to the interaction between grain boundaries (GBs) and dislocations in the plastic process. Another parameter, activation volume (V^*) which is related to the m, is considered to affect the rate-controlling mechanism. The V^* of NC metal is about two orders smaller than that of coarse grain [24,25]. Recently, many studies have found the m and V^* are very helpful to quantify the deformation mechanism in NC metals further [10,11,17,26,27]. Experimental measuring [27] indicated the m of NC gold with a grain size of 30 nm was 0.01 under the strain rate above 10^{-4} s^{-1}. Asaro et al. [17] showed theoretically that the value of m increased with the decrease of average grain size in NC metals, while the V^* was increased with the increase of average grain size. Wang et al. found that the V^* decreased with the increase of temperature in the NC Ni experiment with a grain size of 15 nm [10].

MD with its inherent constraints makes the time scale of simulation limited. The dynamics of the system is probed over just a few nanoseconds. Even with the quick development of computational techniques, the time period for the dynamics of a system with intermediate size (about 10^6 atoms) can be probed to be about 10^3 ns and thus the deformation of the system is simulated under very high strain rates, such as the typically used 1 ns^{-1} for deformation processes, corresponding to the strain of 0.1 in 0.1 ns. Through this limit of simulation time, we can modulate the strain rate to explore the deformation processes to provide some insights into the atomic mechanism. Some simulation works about strain rate on NC metals with small grain size have been performed [11,15,28–30]. For example, in simulations of NC copper with grain sizes of 2.1–11.5 nm, it was found that grain coarsening was closely related to strain rate [31]. The grain size of grain coarsening increased with the decrease of strain rate. The simulation on 2D NC copper with a grain size of 9 nm [26] showed Young's modulus was kept almost constant at strain rates below 5×10^5 s^{-1}. When the strain rate was more than 5×10^5 s^{-1}, Young's modulus increased with the increase of strain rate. However, there is still a lot of work needing to be undertaken. For instance, the mechanism of the change of elastic modulus with the strain rate isn't fully understood. The temperature effect combined with the strain rate on the deformation processes needs to be explored further. The effect of strain rate on grain growth also needs to be studied in depth.

In this work, we use MD simulations to study the effect of strain rate on the mechanical behaviors and deformation mechanisms in NC gold with small grain size. Two models of NC gold are constructed. One is with an average grain size of 6 nm, and the other is with 18 nm. For the deformation under applied tensile strain, the strain rate is modulated from 0.01 per ns to 10 per ns. Two temperatures including 300 K and 800 K are adopted to combine the change of strain rate to explore the deformation mechanism. The simulation results clearly show strain rate effects on Young's modulus, flow stress and grain growth of NC gold in tensile deformation and also reveal the atomic mechanism to some extent by combing the results from known experiments.

2. Computational Approach and Models

All the simulations were performed with standard molecular dynamics simulation (MD) methods which were implemented in large-scale atomic/molecular massively parallel simulator (LAMMPS) [32]. The atomic interaction (V_p) between gold atoms was constructed under the frame of embedded-atom method (EAM) potential and composed by a pairwise potential and a many-body embedding energy, as indicated by the formula,

$$V_p = 1/2 \sum\nolimits_{i,j\ (i \neq j)} U_{i,j}(r_{i,j}) + \sum\nolimits_{i,j\ (i \neq j)} E_i \rho_j(r_{i,j}), \tag{1}$$

where $U_{i,j}(r_{i,j})$ was the pairwise potential for the atom at r_i and that at r_j, E_i was the embedding energy of atom at r_i from the contributions of nearby atoms whose density was described by the item of $\sum_{j\ (i \neq j)} \rho_j(r_{i,j})$. Here we used the parameters from Foiles et al. [33] to parameterize the EAM of gold. This potential can predict the mechanical properties of gold well. The simulated lattice constant of gold is 0.4078 nm and the stable lattice is fcc. The calculated Young's modulus of polycrystalline gold with this potential is about 78 GPa and is consistent to the value from experiments.

The NC gold models were constructed with the Voronoi method which was popularly used for the building of atomic models for polycrystalline systems. Here the Voronoi method used was implemented in the program Atomsk [34]. We constructed the models of NC gold with average grain sizes of 6 nm and 18 nm, respectively. The periodic boundary conditions were applied for the three directions. To account for statistical effects, there are more than 15 Voronoi grains in each model. Grain orientations are random and the expected mean values (6 nm and 18 nm in two models) around distribution of grain size are used, as shown in Figure 1a,b.

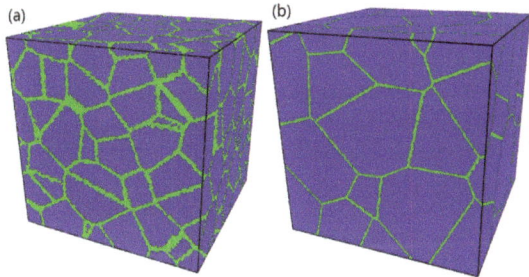

Figure 1. Atomic configurations of nanocrystalline gold samples with mean grain sizes of (a) 6 nm and (b) 18 nm. Blue and green represent grain interiors with fcc lattice and atoms at grain boundaries, respectively.

It is known that the Voronoi approach is just the method of geometric construction with atoms in discrete lattice site. In atomic model of NC, it is popular that the atoms of GBs are unstable. In order to obtain the reasonable atomic configurations at GBs, we needed to relax the unfavorable atomic configurations at GBs. Thus, before the simulations of mechanical properties were performed, the atomic models of NC were annealed at room temperature for 100 ps. All the annealing processes were carried out with an isothermal–isobaric (NPT) ensemble. Then, the NC gold samples were subjected to uniaxial tension tests along the x-direction with NPT ensemble. To check the effect of strain rate, the strain rate was modulated from 100 ns^{-1} to 0.01 ns^{-1}. The pressure in y- and z-direction was kept at zero in the process of uniaxial tension. The time interval for the step of Newton equation of motion was 1 fs. Besides the method of strain–stress, we considered the second method (method 2) to measure the Young's modulus. In this method, the model of NC is firstly stretched quickly with the strain of 1% along x direction under a strain rate of more than 10 ns^{-1}. Then the strain in x direction is kept to be 1% and the length of the sample in y and z directions can be changed under zero pressure, while the internal coordinates of atoms can relax and we measure the change of stress in x direction by following the increase of time (which implies the decrease of strain rate).

In order to visualize and analyze the simulated atomic structures, we used the visualization tool OVITO [35]. Here, the dislocation types and dislocation densities were identified with the method of dislocation extraction algorithm (DXA) [36]. The conventional common neighbor analysis (CNA) [37] was designed to characterize the local structural environment by the atomic pattern matching algorithm, which could detect and classify grain interiors (fcc), stacking faults, GBs and surfaces atoms. The atomic-level strains were analyzed on the basis of the displacement of atoms between the two nearby configurations in the process of tensile strain.

To analyze the grain growth in the process of strain, the change of grain size of each grain in the atomic structure of NC was calculated with the rule as described below. The core of each grain was firstly detected by checking the atoms with their nearest neighbors who had the fcc lattice. The cluster analysis on these cores was applied to distinguish each grain. Then the grain size was calculated from the number of atoms in each cluster with a grain skin of 0.816 nm thickness. The calculated grain size with this method is consistent with that from the Voronoi method in the initial configuration of the sample.

3. Results and Discussion

3.1. The Effect of Strain Rate on the Young's Modulus

The behavior of grain size dependence of mechanical properties in the elastic region is very different from that in the plastic region. In the plastic region, the flow stress and/or yield stress increase and then decrease by following the grain size decrease from hundreds of nanometers to several nanometers, as indicated by the Hall–Petch rule and inverse Hall–Petch relation. For an example, as the observation in the simulation of Cu and Au NC systems, the max value of flow stress appeared at the grain sizes of 10–15 nm under a strain rate of 0.1 ns^{-1} and 10–18 nm under 1 ns^{-1} [9,38], respectively. However, the max value of elasticity modulus, such as Young's modulus, appears in polycrystalline systems. For instance, the Young's modulus of Au polycrystalline is about 78 GPa from the simulation [39]. From our simulation, the Young's moduli of grain sizes of 6 nm and 18 nm were about 39.54 and 51.67 GPa at a strain rate of one per ns, respectively. The early experimental measurements in different materials, such as Fe and Pd, also indicated that the Young's modulus of NC was smaller than that of corresponding polycrystalline [40,41]. The low value of Young's modulus in NC is considered to be due to GBs. It is also possible that it is related to the pores and cracks which are unavoidable in experimental samples [40,41]. Based on the model of crystalline grains with grain-boundary fixed phase, the effective Young's modulus can be analyzed to decrease following the decrease of grain size by fixing the boundary thickness [42].

How do the GBs affect the Young's modulus? Here we expect to modulate the strain rate to check the response of Young's modulus. The stress–strain curves of grain size of 6 nm in the elastic region for five different strain rates are shown in Figure 2a (stress–strain curves under larger strain in Figure S2). We uniformly take the slope of the strain from 0.1% to 1% on the stress–strain curve as the Young's modulus. The Young's modulus of grain sizes of 6 nm increases following the increase of strain rate. The value changes from 30.90 to 58.20 GPa as the strain rate is from 10^7 s^{-1} to 10^{10} s^{-1}. The change of modulus for grain size of 18 nm has a similar rule, though the value is larger than that of a grain size of 6 nm at fixed strain rate in Figure 2b. We checked the Young modulus of single crystalline gold under the different strain rate, such as tensile strain along [100] and [111] directions. It is clear that the Young's modulus doesn't change following the strain rate in single crystalline. This can be understood from the elastic theory based on change of potential. The modulus is the mechanical response to the small deviation of atoms in lattice sites from equilibrium positions under applied strain, and thus has nothing to do with strain rate. Therefore, the reason is GBs which are a response to the change of Young's modulus with strain rate and grain size.

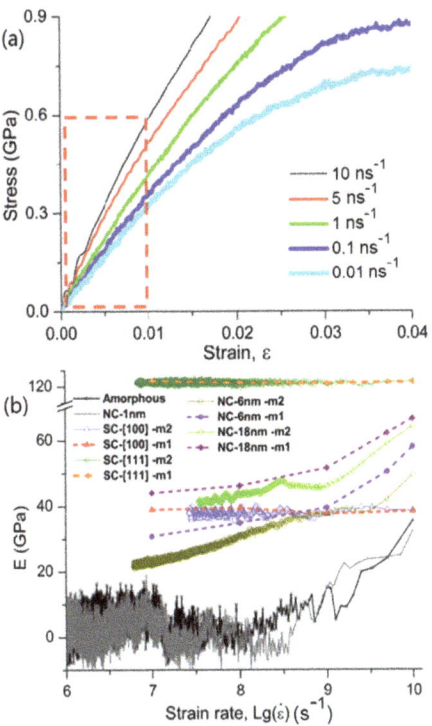

Figure 2. (a) Stress-strain curves for nanocrystalline gold simulations with grain size of 6 nm under different strain rates from 0.01 ns^{-1} to 10 ns^{-1}; and (b) Young's modulus as the function of strain rate for amorphous gold with two models (Amorphous, NC-1 nm), nanocrystalline gold with a grain size of 6 nm (NC-6 nm-m1 from strain–stress method, NC-6 nm-m2 from method 2 mentioned in the text), and grain size of 18 nm (NC-18 nm-m1, NC-18 nm-m2) and single crystal under strain long [100] direction (SC-[100]-m1, SC-[100]-m2) and [111] directions (SC-[111]-m1, SC-[111]-m2).

In the beginning of the deformation process under small strain less than 1%, NC's response to applied strain being rate sensitive implies that the system is non-elastic in the strict traditional view of elastic deformation of single crystalline. However, in this region of deformation with strain less than 1%, there is no generation of dislocations and stacking faults. The "plastic stage" is considered to begin by following the increase of dislocation density and/or stacking faults from zero. In Figure 3, the strain is typically more than 3% for the case of NC gold. Here we can call this region with strain less than 1% as the quasi-elastic region. The obtained Young's modulus is called as apparent Young's modulus (AYM) to distinguish the traditional view about elastic deformation and Young's modulus. In Figure S3, the loading and unloading processes are performed under strain rates of 10 ns^{-1} and 1 ns^{-1} for the sample with grain size of 6 nm. We have considered two cases, including one loading with a max strain of 0.4% and the other with a max strain of 4%. Under the smaller loading strain (0.4%), the structure can be very close to the initial state after the unloading process. For the larger loading strain (such as 4%), it is clear that the structure cannot restitute to the initial state after the unloading. This is because of the generation of dislocations under larger strain. From this view, the quasi-elastic region under the small strain is reasonable.

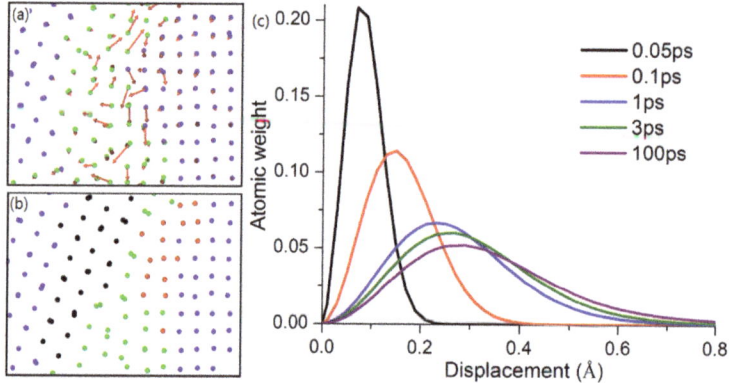

Figure 3. (a) Atomic configuration of grain size of 6 nm after a strain of 1% in 0.1 ps (blue and green represent grain interiors with fcc and atoms at grain boundaries, respectively); (b) atomic configuration after the structural relaxation of 850 ps under a strain of 1% whose initial structure is the structure in (a); and (c) distribution of atomic weight as a function of atomic displacement for grain size of 6 nm under stain of 1% (the structure in (a) is the initial structure) at different relaxation times. In (a), the red arrows in each atom represent the displacement size of atoms from the initial structure to the configuration after the relaxation of 850 ps under a strain of 1%. In (b), the black represents the atoms of grain interiors with fcc whose initial positions are disordered and belong to grain boundaries, and the red represents the atoms of grain boundaries whose initial positions are ordered with fcc and belong to grain interiors in (a).

We have found that the value at higher strain rate is larger. Is it possible that under higher strain rate, the stretching process is too fast and thus the response of atoms at GBs is dull? Thus the GBs become stiff and the effective Young's modulus from the contribution of GBs is larger. In order to consider the response of GBs to small strain applied, we have considered the second method to measure the AYM. We tested this method in single crystal Au and the results were consistent to that from the strain–stress test under fixed strain rate, as shown in Figure 2b.

We check the evolution of stress on NC models over time. The results have a similar rule to that from strain–stress curves for the different strain rate, though the measured stress is a little lower than that from strain–stress. For example, for the model of grain size of 6 nm, the AYM decreases quickly with the decrease of strain rate down to 2.5 ns^{-1}, and then does not obviously change under the strain rate of 2.5 ns^{-1}–0.5 ns^{-1}. For the strain rate less than 0.5 ns^{-1}, the AYM has a weak decrease trend and is difficult to converge. Similarly, for a grain size of 18 nm, the AYM remains constant under the strain rate of 1 ns^{-1}–0.4 ns^{-1}. Then, it also has a trend of decrease with the decrease of strain rate further, but it seems that the value of AYM in larger grain size (18 nm) is easier to converge than that in smaller size (6 nm). It may be understood that the contribution of GBs becomes weaker following the increase of grain size. We have proposed two amorphous models of gold (Supplementary Materials Figure S1), including the structure with a grain size of 1 nm (NC-1 nm in Figure 2b, atomic fraction of GBs is 88.3%) and one typical amorphous structure (Amorphous in Figure 2b). Under a strain of 1%, the AYM decreased by following the strain rate decreasing down to 0.5 per ns for both models. Then it began to oscillate around a small value (3.19 GPa). From these results, we can confirm that the decrease of AYM in nanoscale is basically due to GBs and not from others, such as pores. Grain boundaries become soft with low effective Young's modulus under the low strain rate. Thus, it is the time-dependent deformation mechanism related to GBs affecting the AYM under different strain rates.

We analyzed the evolution of atomic structures after the applied strain. As shown in Figure 3a,b, after the deformation of 1%, the atomic structures are relaxed for 0.1 ps and 850 ps (corresponded with a strain rate of 0.012 ns^{-1}) for the model of grain size of 6 nm. The red arrow in each atom in Figure 3a represents the displacement of atoms from the configuration at 0.1 ps to that at 850 ps. We noticed that the atomic displacements at GBs are much larger than that in the grain. In Figure 3c, we show the distribution of atomic weight as the function of atomic displacement at different times. Before the time of 1 ps (strain rate of 10 ns^{-1}), the change of distribution is very large and indicates the system is unstable and tries to response to the applied strain. After 100 ps (strain rate of 0.1 ns^{-1}), the change of distribution is almost indistinguishable. This indicates that the strain rate being set to 1 ns^{-1}–0.1 ns^{-1} in the usual simulation of deformation process is reasonable. It is known that the distance between the nearest neighbor atoms in Au lattice is 0.286 nm. We can see in Figure 3c that the atomic displacements are larger than the 1% of this value. This is due to the thermal movement of atoms and local larger displacement at GBs (Figure 3a). Due to the larger displacement, we found that the configuration of GBs had been changed (Figure 3b) even at the small strain of 1% with small strain rate. The arrangement of some atoms at GBs became ordered with fcc lattice and some near GBs became disordered. The time-dependent mechanism includes the rearrangements of GBs (in Figure 3) probably even in the elastic regime. Thus, at the very low strain rate in experiments, the AYM of NC is smaller than that of polycrystalline due to the special response of GBs.

3.2. The Effect of Strain Rate on Flow Stress

At the usual strain state of 1 ns^{-1}, for the strain–stress test of the model with the grain size of 18 nm, the deformation enters the plastic region after the elastic region, as shown in Figure 4a. The max stress is reached at a strain of about 4% and then the stress enters the plateau region in which the stress is called as flow stress, by following the increase of strain. From the beginning of plastic deformation (at a strain of about 3%), the dislocation density increases and becomes to be constant after entering the stable region of flow stress at a strain of about 14%.

In Figure S2, we show the strain–stress curves of grain sizes of 6 nm and 18 nm. We can find that the flow stress decreases with the decrease of strain rate, whatever the size of grains. In Figure 4b, we plot the flow stress as the function of strain rate. It is noticed that the flow stress increases rapidly with the increase of strain rate when the strain rate is more than 1 ns^{-1}, while the flow stress decreases slowly with the strain rate decrease down to 1 ns^{-1}. It seems there is a critical strain rate above which the mechanical properties will be affected obviously by strain rate. This may be found from the change of dislocation density with strain and strain rate. As shown in Figure 4d, the dislocation density of grain size of 6 nm changes following the strain for different strain rates. For the strain rates of 1 ns^{-1} and 0.1 ns^{-1}, the dislocation densities don't have an obvious difference.

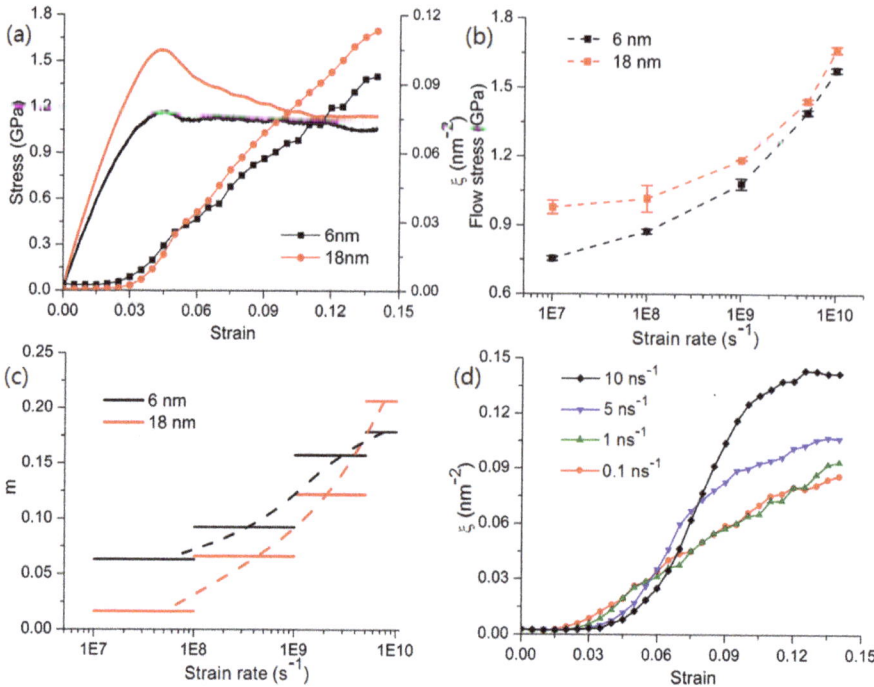

Figure 4. (a) Stress and dislocation density as functions of tensile strain for nanocrystalline gold of 6 nm grain size and 18 nm grain size at a strain rate of 1 ns^{-1}; (b) flow stress; (c) strain rate sensitivity as functions of strain rate for nanocrystalline gold; and (d) dislocation densities as a function of tensile strain for grain size of 6 nm under different strain rate.

There is an important parameter called as strain rate sensitivity (*m*) which can very helpful to quantify the deformation mechanism. It is defined as [14,17],

$$m = \frac{\sqrt{3}kT}{V^*\sigma}, \tag{2}$$

where *k*, *T* and *σ* are Boltzmann constant, absolute temperature and flow stress, respectively. In the formula, the parameter activation volume (*V**) can be expressed as,

$$V^* = \sqrt{3}kT\left(\frac{\partial \ln \dot{\varepsilon}}{\partial \sigma}\right), \tag{3}$$

where $\dot{\varepsilon}$ is the strain rate. Thus, from the relation of flow stress and strain rate, we can obtain the average value of *m* in a range of strain rate, such as 1 ns^{-1}–0.1 ns^{-1}.

The calculated strain rate sensitivities are shown in Figure 3c. The value of *m* decreases with the decrease of strain rate. The *m* value of 18 nm NC gold is less than that of 6 nm at the same range of strain rate. When the strain rate is below 0.1 per ns, the strain rate sensitivities of 6 nm and 18 nm NC gold are 0.063 and 0.016, respectively. This implies that the strain rate sensitivity decreases with the increase of grain size. This result is consistent with many experiments [12,14,16,22,43], the *m* of NC gold with a grain size of 30 nm is 0.01 [27]. It is reasonable to compare the strain rate sensitivity for the simulation under the strain rate lower than 0.1 per ns with the experimental value. The simulation of NC copper [28] also indicates a critical strain rate and the strain rate sensitivity decreases with the increase of grain size below the critical strain rate of 0.1 ns^{-1}. This is consistent with the observation about strain rate sensitivity in our NC gold.

In Figure 4d, we can find that at high strain rates, the activation of dislocation occurs under relatively large strain. In Figure 5, we show the atomic structures of 6 nm NC gold under a strain of 4% with the strain rate of 0.1 ns^{-1} and 10 ns^{-1}. It is clear that the stacking faults are easy to form with dislocation nucleated near GBs under lower strain rates (such as 0.1 ns^{-1}). This also results in the larger localized shear strain appearing at GBs. Thus, the deformation is easy to appear at GBs under low strain rate, and this is consistent to the observation at elastic region under small strain. At the stable flow stress region, the dislocation density is relatively high under larger strain rate (Figure 4d). This indicates that the NC gold is more prone to dislocation movement under higher strain rate (Figure 5c,g, atomic structures at the strain of 10% with strain rate of 0.1 ns^{-1} and 10 ns^{-1}). While at the lower strain rate, GBs are more prone to responding to the applied large strain, as the distributions of atomic shear strain in Figure 5d,h. The high dislocation density under larger strain rates is consistent to the higher stress observed in strain–stress curve.

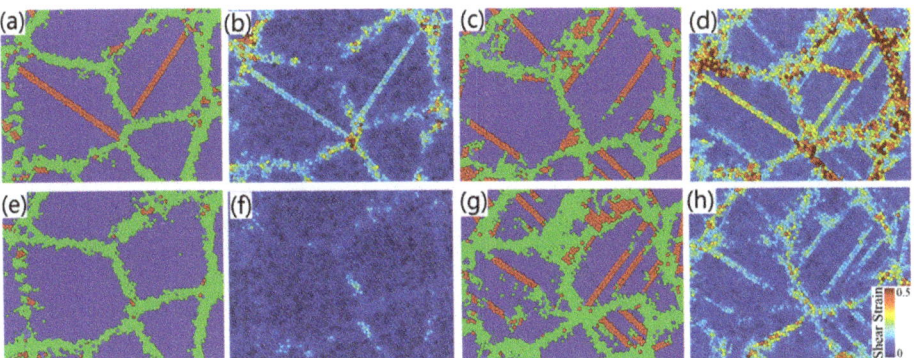

Figure 5. Atomic configurations and distribution of local shear strain of grain size of 6 nm at tensile strain of 4% under the strain rate of (**a,b**) 0.1 ns^{-1} and (**e,f**) 10 ns^{-1}, and that at tensile strain 10% under the strain rate of (**c,d**) 0.1 ns^{-1} and (**g,h**) 10 ns^{-1}. In (**a,c,e,g**), blue, red and green represent grain interiors with fcc, stacking faults with hcp, and atoms at grain boundaries, respectively. In (**b,d,f,h**), the change of color from blue to red indicates the increase of atomic local shear strain.

As we know, under the assistance of GBs, the dislocations under local shear stress become easy to nucleate in NC metals. From the view of plastic flow activated thermally, the shear deformation rate for overcoming the barrier to dislocation motion is related to the activation volume by the relation [6],

$$\dot{\gamma} \propto \exp[(-\Delta F + \tau_e^* V^*)/kT], \tag{4}$$

where ΔF is the change of Helmholtz free energy and τ_e^* is the thermal component of total stress. The item of $\tau_e^* V^*$ is the contribution of thermally activated stress to reduce the energy barrier. Thus the V^* is related to the deformation mechanism [44–46]. Here we check the change of V^* by modulating the temperature and strain rate.

As shown in Figure 6a, from the strain–stress curves, the formation of structures at high temperatures (800 K) is easier in the plastic region under low strain and thus with lower flow stress, compared to the case of low temperature (300 K). In Figure 6b, the flow stress at 800 K is shown as the function of strain rate for grain sizes of 6 nm and 18 nm. It is clear that the flow stress decreases continuously following the strain rate decrease in the range of our test. It is considered that the GBs slipping are activated at 800 K for both cases of 6 nm NC and 18 nm NC.

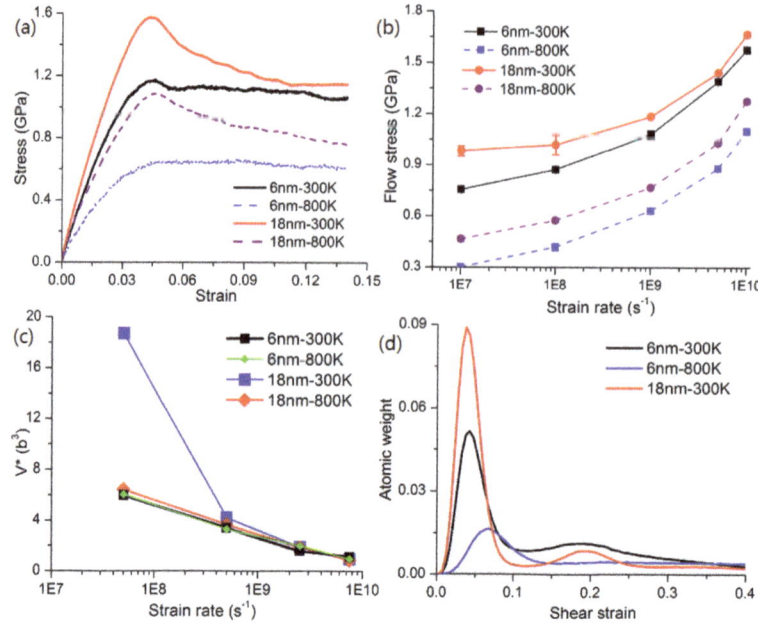

Figure 6. (a) Stress–strain curves for nanocrystalline gold with grain sizes of 6 nm and 18 nm with a strain rate of 1 ns^{-1} with 300 K and 800 K, (b) flow stress and (c) activation volume as functions of strain rate for grain sizes of 6 nm and 18 nm with 300 K and 800 K, and (d) the distribution of atomic weight as a function of atomic shear strain for 18 nm grain size with 300 K and 6 nm grain size with 600 K and 600 K at tensile strain 7.5% under the strain rates of 0.01 ns^{-1}.

As mentioned above, the m below the strain rate of 1 ns^{-1} is closely dependent on the grain size. From the relation between m and V^*, the activation volume should also be closely dependent on the grain size at the low strain rate of less than 1 ns^{-1}. In Figure 6c, we plot the activation volume as the function of strain rate for different cases. It is noticed that in the range of 0.01 ns^{-1}–0.1 ns^{-1}, the activation volumes of 18 nm NC and 6 nm NC at 800 K are similar to that of 6 nm NC at 300 K and about 6.06 b^3. However, the activation volume of 18 nm NC at 300 K is about 18.69 b^3. The activation volumes of NC Ni and Cu from experimental strain rate tests [10,14,47] are about 10–20 b^3. This is consistent to the case of 18 nm NC Au at 300 K. Thus the dislocation pile-up against GBs is the main formation mechanism for grain size 18 nm at 300 K. From the distribution of atomic weight as the function of local shear strain at the applied strain of 7.5% and strain rate of 0.01 per ns in Figure 6d, there is a second peak with large local shear strain. This is an indicator which implies GBs are important sources to nucleate the dislocations and emit the stacking faults into the grain interior. We can see in Figure 6c that temperature has little effect on the activation volume of grain size 6 nm, which implies the main deformation mechanism does not change for both temperatures (300 and 800 K). In Figure 6d, there is just one broadened peak and no second peak appears in the distribution of atomic weight for grain size of 6 nm at 800 K. Thus, it implies that the response of GBs slipping may be the main mechanism of deformation. At 300 K, though there is a second peak at large shear strain, its width is very large and implies that GBs is not only for dislocations pile-up but also the relative slipping between grains. Therefore, at high temperatures and in the case of small grain sizes, the GBs slipping are the main deformation mechanism and thus this system is with low flow stress.

3.3. The Effect of Strain Rate on Grain Growth

In NC metals, the large amounts of GBs due to the small grain size make them unstable. The grain growth (GG) in NC with small grain size is popular at annealing processes, even in rapid annealing [48], such as laser annealing. Besides the GG in high temperatures, it is also possible for GG to appear at low temperatures [49], especially at the deformation processes [50]. Thus, it has an influence on the mechanical properties of NC. In the processes of deformation, the local shear strain induced by applied tensile strain will facilitate the grain coalescence [38]. In Figure S4, the initial atomic structure and that under the strain of 4.5% with strain rate of 5 ns^{-1} for NC gold with grain size of 3.8 nm are shown. As per the circles in Figure S4a,b, the configurations of GBs are modulated due to the realignment of atoms at GBs under the local stress and thus the size of the grain changes, accompanied by the stacking faults emission through grain interior.

We have noticed that the atomic realignment at GBs leads to the change of grain size in the processes of tensile strain. We have analyzed statistically the distribution of grain size at different strains by the methods mentioned above. We used the NC gold with average grain sizes of 3.8 nm and 4.5 nm under applied strain with a strain rate of 5 ns^{-1} as examples. We checked the change of the ten largest grains in the simulated cell. It was found that for the sample of average grain of 3.8 nm under the strain of 4.5%, the sizes of the three largest grains increase from 5.67, 5.07 and 4.79 nm in initial the structure to 5.83, 5.58 and 5.34 nm, respectively. Clearly, the three largest grains have been grown significantly relative to the initial structure, accompanied by decrease of other smaller grains, like the Ostwald ripening. Interestingly, in the NC gold of 4.5 nm with strain rate of 5 ns^{-1}, it isn't found that the grains grow up.

In Figure 7a–c, we show the atomic configurations of 6 nm NC gold under the tensile strain of 7% with the strain rate of 0.1 ns^{-1} and 1 ns^{-1}. It can find that the grain grows up under a strain rate of 0.1 ns^{-1} and doesn't grow under a strain rate of 1 ns^{-1}, as indicated by the circles in Figure 7b,c. Thus, it is proposed that there is a critical grain size for each strain rate. Under a fixed strain rate, it is possible to make the GG appear when grain size is less than the corresponding critical grain size. We analyzed the critical grain size by the statistical method of grain size mentioned above for the cases of different initial grain sizes under different strain rates. The details are listed in Figures S5–S8. In Figure 7d, we show the relation of critical grain size and strain rate. We found that the critical grain size became large, by following the decrease of strain rate. By the extrapolation, we can propose the critical grain size is about 25 nm under the strain rate of 10^{-4} s^{-1} (which is the regular strain rate used in experiments) if the main mechanism of GG isn't changed. This is consistent with the previous observations in experiments about other NC metals [51–55]. For example, in NC Pt thin film, the GG appears with grain sizes for a dozen, even tens of, nanometers during the tensile deformation test under a strain rate of 3×10^{-5} s^{-1}.

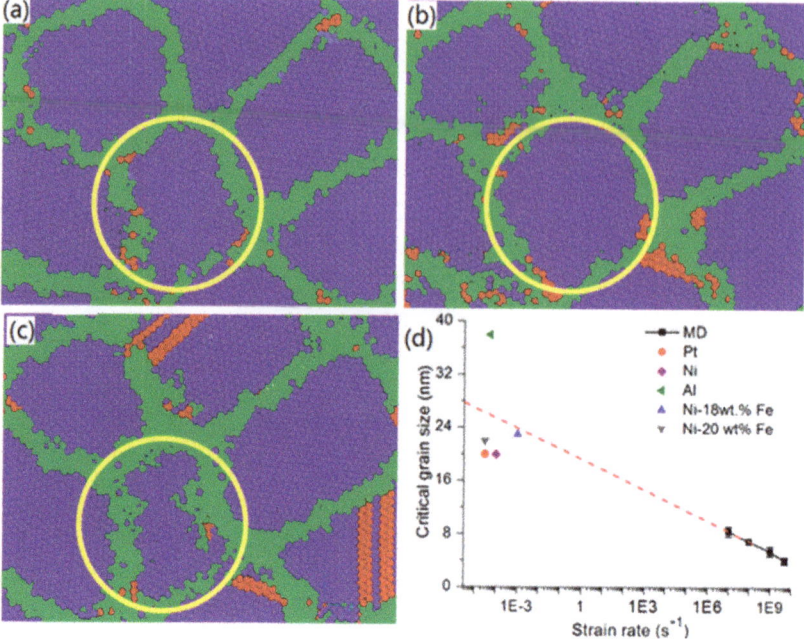

Figure 7. (a) The initial structures of grain size of 6 nm, atomic configurations of it at tensile strain of 7% under the stain rate of (b) 0.1 ns^{-1} and (c) 1 ns^{-1}, and (d) critical grain size for grain growth as a function of strain rate. In (a–c), blue, red and green represent grain interiors with fcc, stacking faults with hcp, and atoms at grain boundaries, respectively. In (d) experimental results from [46–50] are provided for comparison.

For the GG during the tensile strain, the change of GBs' configuration is an important way as mentioned above. One of the important mechanisms is grain rotation, in which the dependence of GB's energy on misorientation between two nearby grains is the driving force. The local shear stress will rotate the grain to form low energetic GBs, as indicated by the previous simulations [38,56]. Generally, the GB migration and grain rotation derived by the local large stress near GBs with the assistance of dislocation take the main rule. Thus, the growth up of single grain or/and the coalescence between grains to form a larger grain appear in the tensile deformation processes. As an example, the coalescence between grains in 7 nm NC gold under tensile strain is shown in Figure S9.

4. Conclusions

We have explored the effect of strain rate on the elasticity, plastic deformation and grain growth of NC gold with small grain size by using molecular dynamics simulations. By considering the models of NC gold with grain sizes of 6 nm and 18 nm, we found that that the apparent Young's modulus of NC gold decreases by following the strain rate decreases. By comparing with that of single crystalline and amorphous gold, it was found that this couldbe attributed to the decrease of effective Young's modulus from GBs with the decrease of strain rate. It was noticed that under a low strain rate of less than 0.5 per ns, the value of apparent Young's modulus of amorphous gold oscillates around 3.19 GPa. The GBs had similar atomic arrangements to amorphous structure and thus small effective Young's modulus under low strain rate. Thus, by following the decrease of grain size the apparent Young's modulus of NC decreased due to the contribution of larger atomic fraction of GBs. The low apparent Young's modulus at low strain was due to the atomic rearrangement around GBs even under small applied tensile strain.

For the response of NC gold to strain rate, flow stress has similar laws to apparent Young's modulus. Following the decrease of strain rates of less than 1 per ns, the decrease of flow stress became slow, especially for larger grain size (such as 18 nm). From the relation between flow stress and strain rate, we found that strain rate sensitivity decreased and activation volume increased with the increase of strain rate. The larger the grain size, the quicker the decrease of strain rate sensitivity. Thus, under lower strain rate, the strain rate sensitivity of larger grain size was smaller. When the strain rate was below 0.1 per ns, the strain rate sensitivities of 18 nm NC gold was just 0.016. Temperature had little effect on the strain rate sensitivity and thus activation volume of 6 nm NC gold. However, there was an obvious effect on NC gold with a grain size of 18 nm. We found that the activation volume of 18 nm NC at 300 K was about 18.69 b^3 under a strain rate of 0.01 ns^{-1}. This indicates that the larger flow stress of 18 nm NC is due to the dislocation pile-up against GBs as the main deformation mechanism under applied strain. For the smaller grain size (such as 6 nm) and high temperature, the smaller activation volume is due to GB slipping and/or change of atomic configuration near GBs. We found that grain growth in NC was related to strain rate in the process of tensile strain. There is a critical grain size about grain growth for each strain rate and the critical grain size increases by following the decrease of strain rate. We propose the critical size for grain growth is about 25 nm under the strain state used in the usual experiments.

Supplementary Materials: The following are available online at http://www.mdpi.com/2073-4352/10/10/858/s1, Figure S1: both modes of amorphous gold, Figure S2: Strain-stress curves of 6 nm and 18 nm grain sizes under different strain rates, Figure S3: The loading and unloading processes for the case of grain size of 6 nm, Figure S4: Atomic configurations of 3.8 nm grain size under strains, Figures S5–S8: analysis of critical size for grain growth, Figure S9: atomic configuration of 7 nm grain size.

Author Contributions: Conceptualization, X.F. and W.Z.; methodology, X.F.; software, J.L.; validation, X.F., J.L., Y.S., and D.J.S.; formal analysis, X.F., J.L.; investigation, X.F., J.L.; resources, X.F.; data curation, X.F., J.L.; writing—original draft preparation, J.L.; writing—review and editing, X.F., Y.S., D.J.S., W.Z.; visualization, J.L., X.F.; supervision, X.F.; project administration, X.F.; funding acquisition, X.F., W.Z. All authors have read and agreed to the published version of the manuscript.

Funding: This research was funded by the National Key R&D Program of China, grant number 2016YFA0200400, and the National Natural Science Foundation of China, grant number 51627805. The APC was funded by the National Key R&D Program of China.

Conflicts of Interest: The authors declare no conflict of interest.

References

1. Youssef, K.M.; Scattergood, R.O.; Murty, K.L. Ultrahigh strength and high ductility of bulk nanocrystalline copper. *Appl. Phys. Lett.* **2005**, *87*, 091904. [CrossRef]
2. Zhang, M.; Yang, B.; Chu, J.; Nieh, T.G. Hardness enhancement in nanocrystalline tantalum thin films. *Scr. Mater.* **2006**, *54*, 1227–1230. [CrossRef]
3. Schuh, C.A.; Nieh, T.G.; Yamasaki, T. Hall–Petch breakdown manifested in abrasive wear resistance of nanocrystalline nickel. *Scr. Mater.* **2002**, *46*, 735–740. [CrossRef]
4. Pande, C.S.; Cooper, K.P. Nanomechanics of Hall–Petch relationship in nanocrystalline materials. *Prog. Mater Sci.* **2009**, *54*, 689–706. [CrossRef]
5. Armstrong, R.W. 60 years of Hall-Petch: Past to present nano-scale connections. *Mater. Trans.* **2014**, *55*, 2–12. [CrossRef]
6. Dao, M.; Lu, L.; Asaro, R.J.; Hosson, J.T.M.D.; Ma, E. Toward a quantitative understanding of mechanical behavior of nanocrystalline metals. *Acta Mater.* **2007**, *55*, 4041–4065. [CrossRef]
7. Yamakov, V.; Wolf, D.; Phillpot, S.R.; Mukherjee, A.K.; Gleiter, H. Deformation-mechanism map for nanocrystalline metals by molecular-dynamics simulation. *Nat. Mater.* **2004**, *3*, 43–47. [CrossRef]
8. Li, H.; Choo, H.; Ren, Y.; Saleh, T.A.; Lienert, U.; Liaw, P.K.; Ebrahimi, F. Strain-dependent deformation behavior in nanocrystalline metals. *Phys. Rev. Lett.* **2008**, *101*, 015502. [CrossRef] [PubMed]
9. Schiøtz, J.; Jacobsen, K.W. A maximum in the strength of nanocrystalline copper. *Science* **2003**, *301*, 1357–1359. [CrossRef] [PubMed]

10. Wang, Y.M.; Hamza, A.V.; Ma, E. Temperature-dependent strain rate sensitivity and activation volume of nanocrystalline Ni. *Acta Mater.* **2006**, *54*, 2715–2726. [CrossRef]
11. Renk, O.; Maier-Kiener, V.; Issa, I.; Li, J.H.; Kiener, D.; Pippan, R. Anneal hardening and elevated temperature strain rate sensitivity of nanostructured metals: Their relation to intergranular dislocation accommodation. *Acta Mater.* **2019**, *165*, 409–419. [CrossRef]
12. Huang, P.; Wang, F.; Xu, M.; Xu, K.W.; Lu, T.J. Dependence of strain rate sensitivity upon deformed microstructures in nanocrystalline Cu. *Acta Mater.* **2010**, *58*, 5196–5205. [CrossRef]
13. Carlton, C.E.; Ferreira, P.J. What is behind the inverse Hall–Petch effect in nanocrystalline materials? *Acta Mater.* **2007**, *55*, 3749–3756. [CrossRef]
14. Chen, J.; Lu, L.; Lu, K. Hardness and strain rate sensitivity of nanocrystalline Cu. *Scr. Mater.* **2006**, *54*, 1913–1918. [CrossRef]
15. Schiøtz, J.; Vegge, T.; Tolla, F.D.D.; Jacobsen, K.W. Atomic-scale simulations of the mechanical deformation of nanocrystalline metals. *Phys. Rev. B* **1999**, *60*, 11971. [CrossRef]
16. Wei, Q.; Cheng, S.; Ramesh, K.T.; Ma, E. Effect of nanocrystalline and ultrafine grain sizes on the strain rate sensitivity and activation volume: Fcc versus bcc metals. *Mater. Sci. Eng. A* **2004**, *381*, 71–79. [CrossRef]
17. Asaro, R.J.; Suresh, S. Mechanistic models for the activation volume and rate sensitivity in metals with nanocrystalline grains and nano-scale twins. *Acta Mater.* **2005**, *53*, 3369–3382. [CrossRef]
18. Kanel, G.I.; Zaretsky, E.B.; Razorenov, S.V.; Ashitkov, S.I.; Fortov, V.E. Unusual plasticity and strength of metals at ultra-short load durations. *Phys. Uspekhi* **2017**, *60*, 490. [CrossRef]
19. Chen, A.Y.; Ruan, H.H.; Wang, J.; Chan, H.L.; Wang, Q.; Li, Q.; Lu, J. The influence of strain rate on the microstructure transition of 304 stainless steel. *Acta Mater.* **2011**, *59*, 3697–3709. [CrossRef]
20. Jonnalagadda, K.N.; Chasiotis, I.; Yagnamurthy, S.; Lambros, J.; Pulskamp, J.; Polcawich, R.; Dubey, M. Experimental investigation of strain rate dependence of nanocrystalline Pt films. *Exp. Mech.* **2010**, *50*, 25–35. [CrossRef]
21. Karanjgaokar, N.J.; Oh, C.-S.; Lambros, J.; Chasiotis, I. Inelastic deformation of nanocrystalline Au thin films as a function of temperature and strain rate. *Acta Mater.* **2012**, *60*, 5352–5361. [CrossRef]
22. Lu, L.; Li, S.X.; Lu, K. An abnormal strain rate effect on tensile behavior in nanocrystalline copper. *Scr. Mater.* **2001**, *45*, 1163–1169. [CrossRef]
23. Schwaiger, R.; Moser, B.; Dao, M.; Chollacoop, N.; Suresh, S. Some critical experiments on the strain-rate sensitivity of nanocrystalline nickel. *Acta Mater.* **2003**, *51*, 5159–5172.
24. Lu, L.; Schwaiger, R.; Shan, Z.W.; Dao, M.; Lu, K.; Suresh, S. Nano-sized twins induce high rate sensitivity of flow stress in pure copper. *Acta Mater.* **2005**, *53*, 2169–2179.
25. Lu, L.; Shen, Y.; Chen, X.; Qian, L.; Lu, K. Ultrahigh strength and high electrical conductivity in copper. *Science* **2004**, *304*, 422–426. [CrossRef] [PubMed]
26. Rida, A.; Micoulaut, M.; Rouhaud, E.; Makke, A. Understanding the strain rate sensitivity of nanocrystalline copper using molecular dynamics simulations. *Comp. Mater. Sci.* **2020**, *172*, 109294.
27. Jonnalagadda, K.; Karanjgaokar, N.; Chasiotis, I.; Chee, J.; Peroulis, D. Strain rate sensitivity of nanocrystalline Au films at room temperature. *Acta Mater.* **2010**, *58*, 4674–4684.
28. Zhang, T.; Zhou, K.; Chen, Z.Q. Strain rate effect on plastic deformation of nanocrystalline copper investigated by molecular dynamics. *Mater. Sci. Eng. A* **2015**, *648*, 23–30.
29. Yaghoobi, M.; Voyiadjis, G.Z. The effects of temperature and strain rate in fcc and bcc metals during extreme deformation rates. *Acta Mater.* **2018**, *151*, 1–10. [CrossRef]
30. Rupert, T. Strain localization in a nanocrystalline metal: Atomic mechanisms and the effect of testing conditions. *J. Appl. Phys.* **2013**, *114*, 033527.
31. Zhou, K.; Liu, B.; Yao, Y.; Zhong, K. Grain coarsening in nanocrystalline copper with very small grain size during tensile deformation. *Mater. Sci. Eng. A* **2014**, *595*, 118–123. [CrossRef]
32. Plimpton, S. Fast parallel algorithms for short-range molecular dynamics. *J. Comput. Phys.* **1995**, *117*, 1–19. [CrossRef]
33. Foiles, S.M.; Baskes, M.I.; Daw, M.S. Embedded-atom-method functions for the fcc metals Cu, Ag, Au, Ni, Pd, Pt, and their alloys. *Phys. Rev. B* **1986**, *33*, 7983. [CrossRef] [PubMed]
34. Hirel, P. Atomsk: A tool for manipulating and converting atomic data files. *Comput. Phys. Commun.* **2015**, *197*, 212–219. [CrossRef]

35. Stukowski, A. Visualization and analysis of atomistic simulation data with OVITO—The Open Visualization Tool. *Model. Simul. Mater. Sci. Eng.* **2009**, *18*, 015012. [CrossRef]
36. Stukowski, A.; Bulato, V.V.; Arsenlis, A. Automated identification and indexing of dislocations in crystal interfaces. *Model. Simul. Mater. Sci. Eng.* **2012**, *20*, 085007. [CrossRef]
37. Stukowski, A. Computational analysis methods in atomistic modeling of crystals. *JOM* **2014**, *66*, 399–407. [CrossRef]
38. Liu, J.; Fan, X.; Zheng, W.; Singh, D.J.; Shi, Y. Nanocrystalline gold with small size: Inverse Hall–Petch between mixed regime and super-soft regime. *Philos. Mag.* **2020**, *100*, 2335–2351. [CrossRef]
39. Liu, J.; Fan, X.; Shi, Y.; Singh, D.J.; Zheng, W. Nanopores in nanocrystalline gold. *Materialia* **2019**, *5*, 100195. [CrossRef]
40. Fougere, G.E.; Riester, L.; Ferber, M.; Weertman, J.R.; Siegel, R.W. Young's modulus of nanocrystalline Fe measured by nanoindentation. *Mater. Sci. Eng. A* **1995**, *204*, 1–6. [CrossRef]
41. Sanders, P.G.; Eastman, J.A.; Weertman, J.R. Elastic and tensile behavior of nanocrystalline copper and palladium. *Acta Mater.* **1997**, *45*, 4019–4025. [CrossRef]
42. Sharma, P.; Ganti, S. On the grain-size-dependent elastic modulus of nanocrystalline materials with and without grain-boundary sliding. *J. Mater. Res.* **2003**, *18*, 1823–1826. [CrossRef]
43. Wang, Y.M.; Ma, E. Strain hardening, strain rate sensitivity, and ductility of nanostructured metals. *Mater. Sci. Eng. A* **2004**, *375*, 46–52. [CrossRef]
44. Petegem, S.V.; Brandstetter, S.; Schmitt, B.; Swygenhoven, H.V. Creep in nanocrystalline Ni during X-ray diffraction. *Scr. Mater.* **2009**, *60*, 297–300.
45. Kottada, R.S.; Chokshi, A.H. Low temperature compressive creep in electrodeposited nanocrystalline nickel. *Scr. Mater.* **2005**, *53*, 887–892. [CrossRef]
46. Chokshi, A.H. Unusual stress and grain size dependence for creep in nanocrystalline materials. *Scr. Mater.* **2009**, *61*, 96–99. [CrossRef]
47. Ma, E. Watching the nanograins roll. *Science* **2004**, *305*, 623–624. [CrossRef] [PubMed]
48. Liu, J.; Fan, X.; Shi, Y.; Singh, D.J.; Zheng, W. Melting of nanocrystalline gold. *J. Phys. Chem. C* **2018**, *123*, 907–914. [CrossRef]
49. Rösner, H.; Markmann, J.; Weissmüller, J. Deformation twinning in nanocrystalline Pd. *Philos. Mag. Lett.* **2004**, *84*, 321–334. [CrossRef]
50. Haslam, A.J.; Moldovan, D.; Yamakov, V.; Wolf, D.; Phillpot, S.R.; Gleiter, H. Stress-enhanced grain growth in a nanocrystalline material by molecular-dynamics simulation. *Acta Mater.* **2003**, *51*, 2097–2112. [CrossRef]
51. Fan, G.J.; Fu, L.F.; Choo, H.; Liaw, P.K.; Browning, N.D. Uniaxial tensile plastic deformation and grain growth of bulk nanocrystalline alloys. *Acta Mater.* **2006**, *54*, 4781–4792.
52. Sharon, J.A.; Su, P.-C.; Prinz, F.B.; Hemker, K.J. Stress-driven grain growth in nanocrystalline Pt thin films. *Scr. Mater.* **2011**, *64*, 25–28.
53. Gianola, D.S.; Petegem, S.V.; Legros, M.; Brandstetter, S.; Swygenhoven, H.V.; Hemker, K.J. Stress-assisted discontinuous grain growth and its effect on the deformation behavior of nanocrystalline aluminum thin films. *Acta Mater.* **2006**, *54*, 2253–2263.
54. Wang, Y.B.; Li, B.Q.; Sui, M.L. Deformation-induced grain rotation and growth in nanocrystalline Ni. *Appl. Phys. Lett.* **2008**, *92*, 011903.
55. Wang, Y.B.; Ho, J.C.; Liao, X.Z.; Li, H.Q.; Ringer, S.P.; Zhu, Y.T. Mechanism of grain growth during severe plastic deformation of a nanocrystalline Ni–Fe alloy. *Appl. Phys. Lett.* **2009**, *94*, 011908.
56. Sansoz, F.; Dupont, V. Grain growth behavior at absolute zero during nanocrystalline metal indentation. *Appl. Phys. Lett.* **2006**, *89*, 111901.

© 2020 by the authors. Licensee MDPI, Basel, Switzerland. This article is an open access article distributed under the terms and conditions of the Creative Commons Attribution (CC BY) license (http://creativecommons.org/licenses/by/4.0/).

Article

On the Size Effect of Strain Rate Sensitivity and Activation Volume for Face-Centered Cubic Materials: A Scaling Law

Xiazi Xiao [1], Hao Liu [2] and Long Yu [3],*

[1] Department of Mechanics, School of Civil Engineering, Central South University, Changsha 410075, China; xxz2017@csu.edu.cn
[2] Department of Applied Physics, School of Physics and Electronics, Hunan University, Changsha 410082, China; haoliu@hnu.edu.cn
[3] State Key Laboratory for Turbulence and Complex System, Department of Mechanics and Engineering Science, College of Engineering, Peking University, Beijing 100871, China
* Correspondence: yulong123@pku.edu.cn; Tel.: +86-13126976377

Received: 7 September 2020; Accepted: 2 October 2020; Published: 3 October 2020

Abstract: In a recent experimental study of indentation creep, the strain rate sensitivity (SRS) and activation volume v^* have been noticed to be dependent on the indentation depth or loading force for face-centered cubic materials. Although several possible interpretations have been proposed, the fundamental mechanism is still not well addressed. In this work, a scaling law is proposed for the indentation depth or loading force-dependent SRS. Moreover, v^* is indicated to scale with hardness H by the relation $\partial \ln(v^*/b^3)/\partial \ln H = -2$ with the Burgers vector b. We show that this size effect of SRS and activation volume can mainly be ascribed to the evolution of geometrically necessary dislocations during the creep process. By comparing the theoretical results with different sets of reported experimental data, the proposed law is verified and a good agreement is achieved.

Keywords: indentation creep; size effect; strain rate sensitivity; activation volume; geometrically necessary dislocations

1. Introduction

Over recent decades, instrumented indentation tests have been recognized as an effective tool for probing the thermally activated deformation of metallic materials [1–3]. Typical experimental methods for the study of indentation creep contain the constant load and hold (CLH) test [4], constant strain rate (CSR) test [5], constant loading rate (CLR) test [6], strain rate jump (SRJ) test [7], etc. Being different from the conventional creep tests, the strain rate sensitivity (SRS) and activation volume (the two critical rate sensitive parameters) measured by these creep tests have been noticed to exhibit an obvious size effect [2,7–9]. The comprehension of these size-dependent parameters is critically essential for the interpretation of the fundamental creep deformation mechanisms [1,10–12].

So far, there exist two types of size effect as informed from the tests of indentation creep, including the interface-dominant creep size effect [7,9,11,13–16] and indentation depth- or force-related creep size effect [2,8,17–23]. For the former, previous literature has indicated that both the SRS and activation volume are affected by the intrinsic microstructures like grain and twin boundaries at the micro- or nano-scale [24,25]. For face-centered cubic (FCC) materials, enhanced SRS with decreasing grain size has been observed for nanocrystalline gold [14], copper [13] and nickel [11]. As for nano-twinned materials, a similar scaling relation has also been noticed for the dependence of SRS on the twin thickness [26,27]. It is, therefore, realized that there exists an intrinsic length scale for the SRS and activation volume, which is determined by the grain size and twin width of nanostructured materials [7,9,11,13–15] or the film thickness of nanocrystalline films [28–30]. In order to interpret the thermally activated mechanisms for this length scale, Asaro and Suresh [26] proposed an analytical model by considering the emission of partial dislocations from grain and twin boundaries. Following this idea, a non-homogeneous nucleation model was later developed that can rationalize the size-dependent SRS and activation volume for nanocrystals and nano-twinned materials [31].

Besides the influence of intrinsic microstructures, there exists another form of size effect when addressing the indentation creep of single crystals and polycrystals with large grain size, i.e., the SRS decreases or the activation volume increases with increasing indentation depth or loading force, and this phenomenon has been widely observed in the CLH [19,21,22,32,33], CSR [8,23,34,35] and SRJ [36–38] test. For example, the size effect of indentation creep has been studied for polycrystalline pure aluminum through CLH tests at room temperature, which exhibits an obvious decreasing tendency of the SRS with increasing loading force even after the correction of thermal drift effects [22]. Similarly, the SRS of both annealed and 80% cold-worked 70/30 brass has been noticed to decrease with increasing indentation depth when performed under CSR tests [35]. Moreover, when applying SRJ tests on sintered silver nanoparticles, the SRS decreases from 0.04 to 0.024 with the increase in indentation depth from 1100 nm to 1700 nm [37]. Therefore, it is anticipated that there exist some different mechanisms for the depth- or force-related creep size effect of single or polycrystals, when compared with the interface-dominant creep size effect of nanocrystals or nano-twinned polycrystals.

In recent years, several possible explanations have been proposed for addressing the depth- or force-related creep size effect, including the consideration of the free surface effect [21], thermal drift [2,19] and the evolution of geometrically necessary dislocations (GNDs) [2,8,39,40]. Sadeghilaridjani et al. [21] attribute this creep size effect to the high diffusion and mobility of dislocations near the sample surface, which result in a comparatively high SRS at shallow indents. However, even when the indentation depth extends 100 nm so that the influence of the free surface can be ignored, the size effect can still be observed in brass [35] and Al alloys [23]. Another possible explanation is considered to be the influence of thermal drift [2,19]. It is believed that the measurement error could exceed 100% when the indentation displacement rate gets close to the thermal drift rate [2]. However, even if the thermal drift is artificially inhibited or corrected during the indentation creep tests, the creep size effect still exists, especially at shallow indentation depths [23]. Actually, it is interesting to note that the depth- or force-related creep size effect seems to follow a similar evolution tendency as the hardness–force (or depth) relation of polycrystalline aluminum and alpha brass [8]. For the latter, it is with the well-known indentation size effect that the hardness decreases with increasing indentation depth due to the influence of GNDs [41]. Consequently, the fundamental mechanisms addressing the creep size effect are believed to originate from the thermally activated interaction between GNDs, which could become comparatively difficult as the density of GNDs becomes high at shallow indents [8,39,40].

In this work, we intend to propose a mechanistic model scaling the depth- or loading force-dependent SRS and activation volume of FCC materials, as corresponding theoretical analyses addressing this creep size effect have been seldomly reported in the literature. The outline of this paper is given as follows: in Section 2, the theoretical model is proposed in detail for the creep size effect. In Section 3, the experimental data of alpha brass, aluminum and austenitic steel are considered to verify the rationality and accuracy of the model results. Finally, we close with a brief conclusion in Section 4.

2. Theoretical Model for the Creep Size Effect

To begin with, the creep process under indentation tests is considered to be accommodated with the evolution of dislocation microstructures beneath the indenter tip. At the onset of creep deformation, dislocation loops with Burgers vectors normal to the surface plane are generated to address the geometrical shape change at the contact surface [41]. Then, the existing GNDs are forced to move radially from the inner creep region (close to the indenter tip) to the boundary between the creep and elastic regions. For simplicity, we assume that the creep deformation can be discretized into numbers of sequential activation events. During each activation event, the indenter tip moves downwards by a distance of b, and forces the i-th dislocation loop ($1 \leq i \leq N$ with N being the number of dislocation loops) to sweep a distance of s. Here, b is the magnitude of the Burgers vector and s is the spacing between dislocation loops [41], as illustrated in Figure 1.

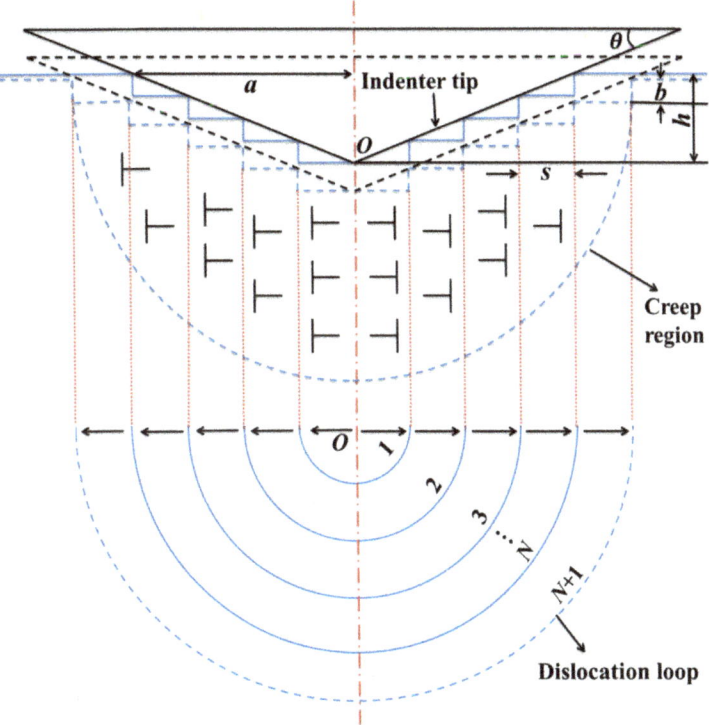

Figure 1. Schematic of indentation creep with the evolution of geometrically necessary dislocations (GNDs), which are performed by a conical indentation. When $\tan\theta = \sqrt{\pi/24.5} = 0.358$ (θ is the angle between the indenter and sample), the model is also applicable to the Berkovich indentation following the self-similar principle [41]. The creep process is discretized into the expansion of circular dislocation loops. During each creep activation, the indenter tip moves forward by a length of b, and a new dislocation loop is generated from the indenter tip that forces existing dislocation loops to creep radially with a distance of s. Thereinto, b and s are, respectively, the magnitude of the Burgers vector and the spacing between individual dislocation loops.

2.1. Size Effect of the Activation Volume

Considering that the total length of GNDs is $\lambda = \pi h a/b$ [41] and the number of dislocation loops equals $N = h/b$, the average length of the dislocation segment within the creep region yields

$$l^*_G = \frac{\lambda}{N} = \pi a = \pi h \cot\theta \tag{1}$$

where a and h are, respectively, the contact radius and indentation depth. θ is the angle between the indenter and sample. In addition, the activation distance between each dislocation loop becomes

$$d^*_G = s = \frac{ba}{h} \tag{2}$$

When one further considers the obstacle-determined dislocation plasticity [24,42], the activation volume determined by GNDs can then be expressed as

$$v_G^* = bl_G^* d_G^* = \pi b^2 \cot^2 \theta h = \frac{3\pi b}{2} \frac{1}{\rho_G} \quad (3)$$

where $\rho_G = 3\tan^2\theta/(2bh)$ is the depth-dependent GND density, as defined by Nix and Gao [41]. Equation (3) indicates that v_G^* increases proportionally with h but varies inversely with ρ_G. A similar evolution tendency has already been observed for swaged and annealed copper where the activation volume increases almost linearly with h [18]. Given that the activation distance is a constant, as indicated by Equation (2), the variation of v_G^* is then realized to be determined by the evolution of l_G^*.

As the creep process goes on, the density of GNDs gradually decreases due to the expansion of the creep region with increasing indentation depth. Although the transition of pre-existing statistically stored dislocations (SSDs) to GNDs might be possible by the cross-slip mechanism [43], its influence on the decrease in GND density may not be obvious, as the dominant type of dislocation is the edge dislocation for FCC materials considered in this work. Then, the thermal activation of SSDs tends to dominate the creep deformation. For crystalline materials with large grain size, the average segment length and activation distance of SSDs can be, respectively, estimated as $l_S^* \sim k/\sqrt{\rho_S}$ and $d_S^* \sim 1/\sqrt{\rho_S}$ with ρ_S being the density of SSDs [24]. Then, to be consistent with Equation (3), the activation volume related to SSDs could be taken as

$$v_S^* = bl_S^* d_S^* = \frac{3\pi b}{2} \frac{1}{\rho_S} \quad (4)$$

where $\rho_S = 3\tan^2\theta/(2bh^*)$ and $k = 3\pi/2$. Thereinto, h^* is a characteristic length related to the bulk hardness [41]. It is indicated by Equation (4) that v_S^* is independent of h but characterizes the intrinsic creeping properties of materials without size effect.

When simultaneously addressing the contribution of GNDs and SSDs, one may consider the relation $1/v^* = 1/v_G^* + 1/v_S^*$ [27], and then the general expression of the activation volume yields

$$v^* = \frac{3\pi b}{2} \frac{1}{\rho_S + \rho_G} \quad (5)$$

which can be reduced to Equation (3) when there exists an obvious indentation size effect (i.e., $\rho_G \gg \rho_S$), or be degraded into Equation (4) when $\rho_G \to 0$ at deep indents. Given the expressions of ρ_G and ρ_S as mentioned above, the activation volume can be recast as

$$v^* = \frac{\pi b^2 h^* H_0^2}{\tan^2\theta} \frac{1}{H^2} \quad (6)$$

and

$$\ln\left(\frac{v^*}{b^3}\right) = \ln k_2 - 2\ln H \quad (7)$$

where $H = H_0\sqrt{1+h^*/h}$ is the depth-dependent indentation hardness [41], and $k_2 = \pi h^* H_0^2/(b\tan^2\theta)$ is a constant related to the bulk hardness H_0 and characteristic length h^* [44]. As indicated by Equation (7), $\ln(v^*/b^3)$ scales linearly with $\ln H$, which is consistent with the experimental observations for most FCC materials, like aluminum, silver and nickel [40]. In addition, the decrease in v^* with increasing H can be ascribed to the accumulation of dislocations at shallow indents that leads to a small activation area swept out by gliding dislocations during the thermal activation event [30].

2.2. Size Effect of the Strain Rate Sensitivity

Next, following the Taylor relation, the shear strength of FCC materials is determined by the dislocation density, i.e.,

$$\tau = \mu b\alpha\sqrt{\rho_S + \rho_G} \quad (8)$$

where μ is the shear modulus and α is the dislocation strength coefficient. The lattice friction τ_0 is ignored in the expression of τ as it is usually very small for FCC materials [45]. By further considering the von Mises flow rule [46] and Tabor's factor [47], the hardness can be given as

$$H = 3\sqrt{3}\tau = 3\sqrt{3}\mu b\alpha \sqrt{\rho_S + \rho_G} \tag{9}$$

Submitting Equations (5) and (9) into the definition of SRS, it yields

$$m = \frac{3\sqrt{3}k_B T}{v^* H} = \frac{2k_B T}{3\pi\mu\alpha b^2} \sqrt{\rho_S + \rho_G} \tag{10}$$

where k_B and T are the Boltzmann constant and testing temperature, respectively. Recalling the expressions of ρ_S and ρ_G as mentioned above, one can further have

$$m = \frac{2k_B T}{3\pi\mu\alpha b^2} \sqrt{\frac{3\tan^2\theta}{2bh^*} + \frac{3\tan^2\theta}{2bh}} = m_0 \sqrt{1 + \frac{h^*}{h}} \tag{11}$$

where $m_0 = k_B T \tan\theta/(\pi\mu\alpha b^2)\sqrt{2/(3bh^*)}$ is the SRS without size effect for bulk materials that depends on the density of SSDs through h^*. Equations (10) and (11) indicate that both GNDs and SSDs contribute to the SRS measured by indentation creep tests. However, the variation of SRS with respect to the indentation depth is determined by the contribution of GNDs.

When one further takes the relation between the loading force P and h^2 in a proportional form [48], i.e., $P = Kh^2$ and $P^* = K(h^*)^2$, where K is a proportionality factor and P^* is the characteristic loading force corresponding to h^*, then the expression of Equation (11) can be recast as

$$m = m_0 \sqrt{1 + \sqrt{\frac{P^*}{P}}} \tag{12}$$

It is interesting to note that Equations (11) and (12) offer a characteristic form for the depth dependence or loading force dependence of the SRS so that the square of the SRS scales linearly with the reciprocal of the indentation depth or of the square root of the loading force. When the raw experimental data of polycrystals are drawn in this way, a straight line is anticipated so that the intercept informs the value of m_0 and the slope yields h^* or P^*. In order to verify this proposed scaling law, four different sets of experimental data are considered in the following, including annealed and work-hardened alpha brass [8,49], annealed aluminum [8] under CLH tests, austenitic steel [38] under SRJ tests and annealed alpha brass [35] under CSR tests.

3. Comparison between Theoretical Results and Experimental Data

We firstly present the $m - P$ relationships of work-hardened mechanically polished alpha brass [49] and annealed aluminum [8], which are compared between the experimental data (black dots) and theoretical results (red lines), as illustrated in Figure 2. The creep tests are performed at room temperature with the loading force ranging from 10^{-4} to 10^0 N, and the type of indenter is a Berkovich indenter [8,49]. The theoretical results are predicted by Equation (12) with m_0 and P^* calibrated by comparison with the converted experimental data (see the inset of Figure 2). The model parameters are listed in Table 1. An excellent agreement is observed where the SRS decreases with the increase in P until m_0 is reached. This decreasing tendency is ascribed to the variation of creep mechanisms, which range from the dislocation–dislocation interaction to the dislocation–solute interaction, as the former becomes dominant at a high stress level [50]. Moreover, the fitted values of m_0, i.e., 0.016 and 0.007, for alpha brass and aluminum are, respectively, close to the literature values of 0.018 obtained by CSR tests for copper [19] and 0.01 obtained by CLH tests for aluminum [40].

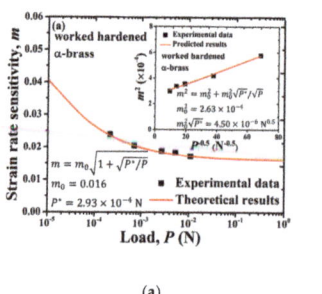 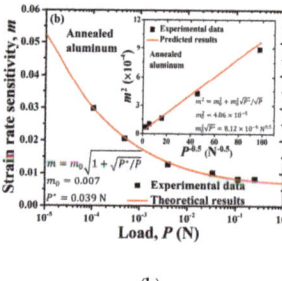

Figure 2. Strain rate sensitivity m-loading force P relationships compared between the experimental data (black dots) and theoretical results (red lines) of (**a**) work-hardened mechanically polished alpha brass [49] and (**b**) annealed aluminum [8]. The inset figure illustrates the calibration of model parameters by comparison with experimental data.

Table 1. Model parameters for work-hardened alpha brass [49], annealed aluminum [8], austenitic steels [38] and annealed alpha brass [35].

Parameter	Work-Hardened Alpha Brass	Annealed Aluminum	Parameter	Austenitic Steels	Annealed Alpha Brass
m_0	0.016	0.007	m_0	0.006	0.0023
P^* (N)	2.94×10^{-4}	0.039	h^* (nm)	2023	6097

The proposed $m - P$ relation can also be expressed in a similar form as the classic Nix–Gao model [41] so that the SRS decreases with increasing indentation depth. In order to verify the $m - h$ relation, as indicated by Equation (8), the experimental data of austenitic steel (SRJ test) [38] and annealed alpha brass (CSR test) [35] are considered. A Berkovich indenter is applied for these creep tests. In Figure 3, the comparison between theoretical results and experimental data is illustrated so that a reasonable agreement is observed for both materials. In this case, $m_0 = 0.006$ for austenitic steel is close to 0.0066 for 310 stainless steel [51] and $m_0 = 0.0023$ matches well with 0.002 obtained from [49] for annealed alpha brass. However, one may note that the experimentally measured value of m for the alpha brass gradually deviates from the theoretically predicted line. This is believed to originate from the effect of thermal drift during the indentation creep tests. According to the analysis of [2], the SRS without thermal drift correction $\overline{m}$ can be approximated as $\overline{m} \equiv m(1 + \lambda/\dot{h})^2$ with λ as the thermal drift rate (of the order of $\pm 10^{-2}$ nm·s^{-1}) and $\dot{h}$ as the penetration rate. With increasing indentation depth, $\dot{h}$ decreases from an initial high value down to the absolute value of λ. Therefore, it is rational to observe that $\overline{m} < m$ with the increase in h when $\lambda < 0$, as indicated in Figure 3b.

Besides the characteristic $m - h$ and $m - P$ relationships as proposed above, the normalized activation volume with respect to the indentation hardness also follows a scaling law, as expressed in Equation (7). To verify this conclusion, the experimental data of copper obtained from [30,52], as well as the data of alpha brass and aluminum taken from [8], are plotted in Figure 4. Correspondingly, the scaling relation is illustrated by solid lines with the slope $\partial \ln(v^*/b^3)/\partial \ln H = -2$. It seems that the creep data follow this scaling law well, and similar linear relationships between $\ln(v^*/b^3)$ and $\ln H$ have also been noticed in the experimental data of some other FCC materials, but with the slope ranging from −1 to −3 [40]. This discrepancy may originate from ignoring the effect of strain hardening induced by the dislocation–dislocation interaction, and might also come from the influence induced by thermal drift through $v^* = k_B T/(m\tau)$, as the simulation work of [19] has captured an obvious variation of SRS with increasing thermal drift rate.

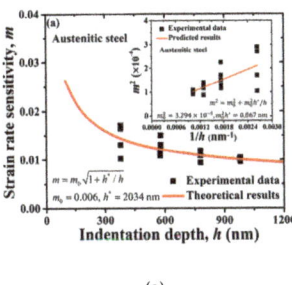 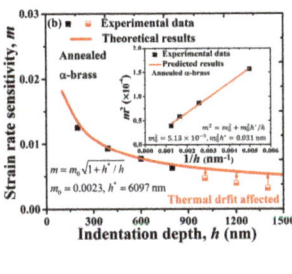

Figure 3. Strain rate sensitivity m-indentation depth h relationships compared between the experimental data (black dots) and theoretical results (red lines) of (**a**) austenitic steel [38] and (**b**) annealed alpha brass [35]. The inset figure illustrates the calibration of model parameters by comparison with experimental data.

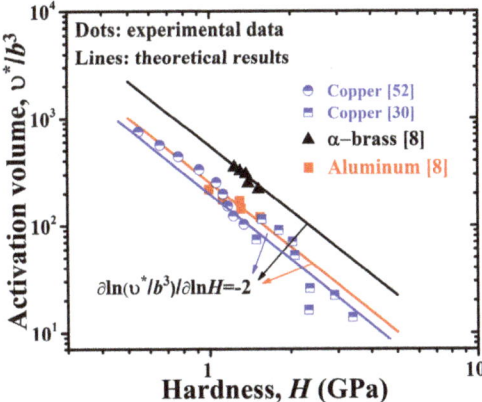

Figure 4. Normalized activation volume v^*/b^3–hardness H relationships compared between experimental data (dots) and theoretical results (lines) for copper [30,52], alpha brass [8] and aluminum [8].

4. Conclusions

To sum up, a scaling law is proposed in this work to address the indentation depth- or loading force-dependent SRS for FCC materials, i.e., $m = m_0 \sqrt{1 + h^*/h}$ or $m = m_0 \sqrt{1 + \sqrt{P^*/P}}$. In addition, the activation volume is found to scale with the hardness by the relation $\partial \ln(v^*/b^3)/\partial \ln H = -2$. The model is deduced by the consideration of a dislocation-dominant mechanism so that the mutual interaction of GNDs at shallow indents plays a critical role in determining this scaling law. Moreover, the proposed law has been verified by comparison with different sets of experimental data.

Author Contributions: Conceptualization, X.X. and L.Y.; writing—original draft preparation, X.X.; writing—review and editing, H.L. and L.Y.; supervision, X.X.; project administration, X.X.; funding acquisition, H.L. and X.X. All authors have read and agreed to the published version of the manuscript.

Funding: This work is supported by the National Nature Science Foundation of China (NSFC) under Contract No. 11802344, 11872379 and 11805061, and the Natural Science Foundation of Hunan Province, China (Grant No. 2019JJ50809 and 2019JJ50072). Hao Liu thanks the Fundamental Research Funds for the Central Universities.

Conflicts of Interest: The funders had no role in the design of the study; in the collection, analyses, or interpretation of data; in the writing of the manuscript, or in the decision to publish the results.

References

1. Maier-Kiener, V.; Durst, K. Advanced Nanoindentation Testing for Studying Strain-Rate Sensitivity and Activation Volume. *JOM* **2017**, *69*, 2246–2255. [CrossRef] [PubMed]
2. Alkorta, J.; Martinez-Esnaola, J.M.; Sevillano, J.G. Critical examination of strain-rate sensitivity measurement by nanoindentation methods: Application to severely deformed niobium. *Acta Mater.* **2008**, *56*, 884–893. [CrossRef]
3. Gibson, J.S.K.L.; Schroeders, S.; Zehnder, C.; Korte-Kerzel, S. On extracting mechanical properties from nanoindentation at temperatures up to 1000 °C. *Extreme Mech. Lett.* **2017**, *17*, 43–49. [CrossRef]
4. Mayo, M.J.; Siegel, R.W.; Narayanasamy, A.; Nix, W.D. Mechanical properties of nanophase TiO_2 as determined by nanoindentation. *J. Mater. Res.* **1990**, *5*, 1073–1082. [CrossRef]
5. Lucas, B.N.; Oliver, W.C. Indentation power-law creep of high-purity indium. *Met. Mater. Trans. A* **1999**, *30*, 601–610. [CrossRef]
6. Mayo, M.; Nix, W. A micro-indentation study of superplasticity in Pb, Sn, and Sn-38 wt% Pb. *Acta Met.* **1988**, *36*, 2183–2192. [CrossRef]
7. Maier, V.; Durst, K.; Mueller, J.; Backes, B.; Hoeppel, H.W.; Goeken, M. Nanoindentation strain-rate jump tests for determining the local strain-rate sensitivity in nanocrystalline Ni and ultrafine-grained Al. *J. Mater. Res.* **2011**, *26*, 1421–1430. [CrossRef]
8. Elmustafa, A.A.; Stone, D.S. Nanoindentation and the indentation size effect: Kinetics of deformation and strain gradient plasticity. *J. Mech. Phys. Solids* **2003**, *51*, 357–381. [CrossRef]
9. Zhao, J.; Wang, F.; Huang, P.; Lu, T.J.; Xu, K.W. Depth dependent strain rate sensitivity and inverse indentation size effect of hardness in body-centered cubic nanocrystalline metals. *Mater. Sci. Eng. A* **2014**, *615*, 87–91. [CrossRef]
10. Niu, J.J.; Zhang, J.Y.; Liu, G.; Zhang, P.; Lei, S.Y.; Zhang, G.J.; Sun, J. Size-dependent deformation mechanisms and strain-rate sensitivity in nanostructured Cu/X (X = Cr, Zr) multilayer films. *Acta Mater.* **2012**, *60*, 3677–3689. [CrossRef]
11. Wei, Q. Strain rate effects in the ultrafine grain and nanocrystalline regimes-influence on some constitutive responses. *J. Mater. Res.* **2007**, *42*, 1709–1727. [CrossRef]
12. Bai, Z.; Fan, Y. Abnormal Strain Rate Sensitivity Driven by a Unit Dislocation-Obstacle Interaction in bcc Fe. *Phys. Rev. Lett.* **2018**, *120*, 125504. [CrossRef] [PubMed]
13. Chen, J.; Lu, L.; Lu, K. Hardness and strain rate sensitivity of nanocrystalline Cu. *Scr. Mater.* **2006**, *54*, 1913–1918. [CrossRef]
14. Nyakiti, L.O.; Jankowski, A.F. Characterization of Strain-Rate Sensitivity and Grain Boundary Structure in Nanocrystalline Gold-Copper Alloys. *Met. Mater. Trans. A* **2010**, *41*, 838–847. [CrossRef]
15. Jia, D.; Ramesh, K.T.; Ma, E. Effects of nanocrystalline and ultrafine grain sizes on constitutive behavior and shear bands in iron. *Acta Mater.* **2003**, *51*, 3495–3509. [CrossRef]
16. Li, S.; Wang, F.; Zhang, L.F.; Huang, P. Indentation depth dependence of nanoscale Al/W multilayers on strain rate sensitivity. *Mater. Sci. Eng. A* **2019**, *767*, 138438. [CrossRef]
17. Bhakhri, V.; Klassen, R.J. The depth dependence of the indentation creep of polycrystalline gold at 300 K. *Scr. Mater.* **2006**, *55*, 395–398. [CrossRef]
18. Klassen, R.J.; Diak, B.J.; Saimoto, S. Origin of the depth dependence of the apparent activation volume in polycrystalline 99.999% Cu determined by displacement rate change micro-indentation. *Mater. Sci. Eng. A* **2004**, *387*, 297–301. [CrossRef]
19. Peykov, D.; Martin, E.; Chromik, R.R.; Gauvin, R.; Trudeau, M. Evaluation of strain rate sensitivity by constant load nanoindentation. *J. Mater. Res.* **2012**, *47*, 7189–7200. [CrossRef]
20. Liu, W.; Wang, Y.; Ma, Y.; Huang, Y.; Tang, Y.; Cheng, F. Indentation size effect of stress exponent and hardness in homogeneous duplex eutectic 80Au/20Sn. *Mater. Lett.* **2014**, *120*, 151–154. [CrossRef]
21. Sadeghilaridjani, M.; Muskeri, S.; Hasannaeimi, V.; Pole, M.; Mukherjee, S.; Hassannaeimi, V. Strain rate sensitivity of a novel refractory high entropy alloy: Intrinsic versus extrinsic effects. *Mater. Sci. Eng. A* **2019**, *766*, 138326. [CrossRef]
22. Li, H.; Ngan, A.H.W. Size effects of nanoindentation creep. *J. Mater. Res.* **2004**, *19*, 513–522. [CrossRef]
23. Haghshenas, M.; Wang, L.; Klassen, R.J. Depth dependence and strain rate sensitivity of indentation stress of 6061 aluminium alloy. *Mater. Sci. Technol.* **2012**, *28*, 1135–1140. [CrossRef]

24. Wei, Q.; Cheng, S.; Ramesh, K.T.; Ma, E. Effect of nanocrystalline and ultrafine grain sizes on the strain rate sensitivity and activation volume: Fcc versus bcc metals. *Mater. Sci. Eng. A* **2004**, *381*, 71–79. [CrossRef]
25. Liu, Y.; Hay, J.; Wang, H.; Zhang, X. A new method for reliable determination of strain-rate sensitivity of low-dimensional metallic materials by using nanoindentation. *Scr. Mater.* **2014**, *77*, 5–8. [CrossRef]
26. Asaro, R.J.; Suresh, S. Mechanistic models for the activation volume and rate sensitivity in metals with nanocrystalline grains and nano-scale twins. *Acta Mater.* **2005**, *53*, 3369–3382. [CrossRef]
27. Lu, L.; Dao, M.; Zhu, T.; Li, J. Size dependence of rate-controlling deformation mechanisms in nanotwinned copper. *Scr. Mater.* **2009**, *60*, 1062–1066. [CrossRef]
28. Cao, Z.H.; Li, P.Y.; Lu, H.M.; Huang, Y.L.; Zhou, Y.C.; Meng, X.K. Indentation size effects on the creep behavior of nanocrystalline tetragonal Ta films. *Scr. Mater.* **2009**, *60*, 415–418. [CrossRef]
29. Cao, Z.H.; Huang, Y.L.; Meng, X.K. Size-dependent rate sensitivity and plasticity of nanocrystalline Ru films. *Scr. Mater.* **2010**, *63*, 993–996. [CrossRef]
30. Mohammed, Y.S.; Stone, D.S.; Elmustafa, A.A. Strain Rate Sensitivity of the Nanoindentation Creep of Ag, Cu, and Ni Thin Films. *JOM* **2019**, *71*, 3734–3743. [CrossRef]
31. Gu, P.; Dao, M.; Asaro, R.J.; Suresh, S. A unified mechanistic model for size-dependent deformation in nanocrystalline and nanotwinned metals. *Acta Mater.* **2011**, *59*, 6861–6868. [CrossRef]
32. Haghshenas, M.; Khalili, A.; Ranganathan, N. On room-temperature nanoindentation response of an Al-Li-Cu alloy. *Mater. Sci. Eng. A* **2016**, *676*, 20–27. [CrossRef]
33. Sadeghilaridjani, M.; Mukherjee, S. High-temperature nano-indentation creep behavior of multi-principal element alloys under static and dynamic loads. *Metals* **2020**, *10*, 250. [CrossRef]
34. Ma, Z.S.; Long, S.G.; Zhou, Y.C.; Pan, Y. Indentation scale dependence of tip-in creep behavior in Ni thin films. *Scr. Mater.* **2008**, *59*, 195–198. [CrossRef]
35. Haghshenas, M.; Klassen, R.J. Assessment of the depth dependence of the indentation stress during constant strain rate nanoindentation of 70/30 brass. *Mater. Sci. Eng. A* **2013**, *572*, 91–97. [CrossRef]
36. Maier, V.; Schunk, C.; Goeken, M.; Durst, K. Microstructure-dependent deformation behaviour of bcc-metals—indentation size effect and strain rate sensitivity. *Philos. Mag.* **2015**, *95*, 1766–1779. [CrossRef]
37. Long, X.; Tang, W.; Feng, Y.; Chang, C.; Keer, L.M.; Yao, Y. Strain rate sensitivity of sintered silver nanoparticles using rate-jump indentation. *Int. J. Mech. Sci.* **2018**, *140*, 60–67. [CrossRef]
38. Kasada, R.; Konishi, S.; Hamaguchi, D.; Ando, M.; Tanigawa, H. Evaluation of strain-rate sensitivity of ion-irradiated austenitic steel using strain-rate jump nanoindentation tests. *Fusion Eng. Des.* **2016**, *109*, 1507–1510. [CrossRef]
39. Liu, Y.; Liu, W.; Yu, L.; Chen, L.; Sui, H.; Duan, H. Hardening and Creep of Ion Irradiated CLAM Steel by Nanoindentation. *Crystals* **2020**, *10*, 44. [CrossRef]
40. Stegall, D.E.; Elmustafa, A.A. The Contribution of Dislocation Density and Velocity to the Strain Rate and Size Effect using Transient Indentation Methods and Activation Volume Analysis. *Met. Mater. Trans. A* **2018**, *49*, 4649–4658. [CrossRef]
41. Nix, W.D.; Gao, H.J. Indentation size effects in crystalline materials: A law for strain gradient plasticity. *J. Mech. Phys. Solids* **1998**, *46*, 411–425. [CrossRef]
42. Kobrinsky, M.J.; Thompson, C.V. Activation volume for inelastic deformation in polycrystalline Ag thin films. *Acta Mater.* **2000**, *48*, 625–633. [CrossRef]
43. Lee, S.; Meza, L.; Greer, J.R. Cryogenic nanoindentation size effect in [001]-oriented face-centered cubic and body-centered cubic single crystals. *Appl. Phys. Lett.* **2013**, *103*, 101906. [CrossRef]
44. Xiao, X.; Chen, Q.; Yang, H.; Duan, H.; Qu, J. A mechanistic model for depth-dependent hardness of ion irradiated metals. *J. Nucl. Mater.* **2017**, *485*, 80–89. [CrossRef]
45. Xiao, X.; Song, D.; Xue, J.; Chu, H.; Duan, H. A self-consistent plasticity theory for modeling the thermo-mechanical properties of irradiated FCC metallic polycrystals. *J. Mech. Phys. Solids* **2015**, *78*, 1–16. [CrossRef]
46. Gurson, A.L. Continuum theory of ductile rupture by void nucleation and growth: Part I—Yield criteria and flow rules for porous ductile media. *J. Eng. Mater. Technol.* **1977**, *99*, 2–15. [CrossRef]
47. Tabor, D. A simple theory of static and dynamic hardness. *Proc. R. Soc. Lond. Ser. A* **1948**, *192*, 247–274.
48. Cheng, Y.T.; Cheng, C.M. Scaling, dimensional analysis, and indentation measurements. *Mater. Sci. Eng. R* **2004**, *44*, 91–149. [CrossRef]

49. Elmustafa, A.A.; Stone, D.S. Size-dependent hardness in annealed and work hardened at-brass and aluminum polycrystalline materials using activation volume analysis. *Mater. Lett.* **2003**, *57*, 1072–1078. [CrossRef]
50. Butt, M.Z.; Feltham, P. Work-hardening of polycrystalline copper and alpha brasses. *Met. Sci.* **1984**, *18*, 123–126. [CrossRef]
51. Lin, M.R.; Wagoner, R.H. Effect of temperature, strain, and strain rate on the tensile flow-stress of I.F. steel and strainless-steel Type-310. *Scr. Met.* **1986**, *20*, 143–148. [CrossRef]
52. Zehetbauer, M.; Seumer, V. Cold work-hardening in stage-IV and stage-V of F.C.C. metals—I. Experiments and indentation. *Acta Met. Mater.* **1993**, *41*, 577–588. [CrossRef]

© 2020 by the authors. Licensee MDPI, Basel, Switzerland. This article is an open access article distributed under the terms and conditions of the Creative Commons Attribution (CC BY) license (http://creativecommons.org/licenses/by/4.0/).

Article

Influence of the Rake Angle on Nanocutting of Fe Single Crystals: A Molecular-Dynamics Study

Iyad Alabd Alhafez and Herbert M. Urbassek *

Physics Department and Research Center OPTIMAS, University Kaiserslautern, Erwin-Schrödinger-Straße, D-67663 Kaiserslautern, Germany; alhafez@rhrk.uni-kl.de
* Correspondence: urbassek@rhrk.uni-kl.de

Received: 13 May 2020; Accepted: 13 June 2020; Published: 17 June 2020

Abstract: Using molecular dynamics simulation, we study the cutting of an Fe single crystal using tools with various rake angles α. We focus on the (110)[001] cut system, since here, the crystal plasticity is governed by a simple mechanism for not too strongly negative rake angles. In this case, the evolution of the chip is driven by the generation of edge dislocations with the Burgers vector $b = \frac{1}{2}[111]$, such that a fixed shear angle of $\phi = 54.7°$ is established. It is independent of the rake angle of the tool. The chip form is rectangular, and the chip thickness agrees with the theoretical result calculated for this shear angle from the law of mass conservation. We find that the force angle χ between the direction of the force and the cutting direction is independent of the rake angle; however, it does not obey the predictions of macroscopic cutting theories, nor the correlations observed in experiments of (polycrystalline) cutting of mild steel. Only for (strongly) negative rake angles, the mechanism of plasticity changes, leading to a complex chip shape or even suppressing the formation of a chip. In these cases, the force angle strongly increases while the friction angle tends to zero.

Keywords: molecular dynamics; nanocutting; iron; dislocations; cutting theory

1. Introduction

While in applications, cutting processes are usually performed on poly-crystalline materials, from a materials-science point of view, it is also interesting to study the cutting of single crystals. In micro- and nano-cutting [1], the cutting depth may be smaller than the grain size of the material, so that crystal plasticity effects need to be taken into account.

However, also from the point of view of machining mechanics, this topic is interesting. The theory of cutting is largely based on the concept of a single shear plane responsible for the cut and the geometry of the chip formed [2,3]. For cutting polycrystalline materials, it is known that this concept is an idealization (see for instance the critical review by Astakhov [4]), and several extensions of the basic framework have been formulated [5–7]. The application of these concepts to nanocutting has been recently reviewed by Fang and Xu [1]. However, in the cutting of single crystals, plasticity will as a rule be based on the dislocation slip. If the cut system—that is, the surface orientation and the cut direction—is carefully chosen, only a single slip system will be activated, and thus, the shear plane is determined by crystallography.

Molecular dynamics (MD) simulation has been repeatedly used for studying machining processes such as nanoindentation [8,9] or the scratching of surfaces [10,11]. Furthermore, cutting has been simulated as well for fcc [12–18] as for bcc [19–22] metals and also for metallic glasses [23], ceramics [24–31], and composites [32]. MD is based on solving Newton's equations of motion for the atoms that make up the workpiece such that this simulation method allows obtaining atomistic insight into the plasticity processes relevant for machining at the nanoscale.

The analysis of MD simulations of the cutting process has often been based for the macroscopic concept of a "stagnation point" that divides the material moving upwards to form the chip from the material that moves downward into the material interior below the tool [1,33]. In this scenario, in particular the effect of tool edge curvature has been investigated [34,35]. Furthermore, a minimum cutting depth and a minimum rake angle (of the order of −65° to −70°) have been postulated; for shallower cuts or blunter rakes, the material will be buried beneath the tool, and no chip is formed [1,33]. However, these ideas do not appear to have been combined up to now with a crystal plasticity analysis of the dislocation processes responsible for material separation and chip formation.

In the present paper, we will use MD simulation to study the application of machining theories to a well studied cut system, the (110)[001] cutting system of bcc Fe, since here, it was shown that indeed a single dislocation slip system is dominant in chip formation [21,22]. By varying the rake angle of the tool, we can vary the magnitude and orientation of the forces during cutting and analyze to what extent traditional machining theories describe the cutting of this system. While a few MD simulations studied the effect of the rake angle previously [33,36,37], they did not analyze the results obtained for dislocation plasticity in terms of available cutting theories. It is the objective of the present study to identify the influence of the rake angle of the cutting tool on the cutting of single crystals in terms of the dislocation plasticity.

A further motivation for studying the effect of the rake angle on metal cutting is given by the fact that in nanomachining, the geometry of the tool edge is not well characterized in the nanometer scale. On this scale, often the tool edge may be idealized by a geometry with a negative rake angle, even though the macroscopic form is sharp, possessing a positive rake angle. This issue provides a further motivation to study the influence of the rake angle on machining processes.

2. Simulation Method

In this study, we focus on the (110)[001] cutting system of bcc Fe; this means that the top surface had a (110) orientation, and the cutting direction was along [001]. The tool edge was along the [$\bar{1}$10] direction. We showed previously [21,22] that this cutting system allows for a simple quasi-two-dimensional understanding of the cutting process, since the major dislocation systems involved in chip formation—those with the Burgers vector $b = \frac{1}{2}[111]$—have their dislocation lines aligned with the tool edge. In Appendix A, we illustrate the complexities that can arise for other cut systems.

The substrate was single crystalline with a (110) surface, while the cutting direction was along [001]. It was built from 1.72×10^6 Fe atoms, extended 611 Å along the cutting direction, and had a height of 404 Å and a thickness of 81 Å. Fe atoms interact with each other according to the Mendelev potential [38], which is known to describe the elastic and plastic properties of Fe, as well as the surface energies relevant for cleavage processes, faithfully [39–41].

The simulation system is displayed schematically in Figure 1. In order to prevent the substrate from any translation or rotation during the machining process, its bottom and the left boundary contained two fixed atom layers. The next two atom layers were thermostatted at a low temperature, <1 K, to ease detection of dislocations and other defects in the system. In the direction along the tool edge (the [$\bar{1}$10] direction), periodic boundary conditions were applied, while the top and the right-hand side boundaries were free.

The tool was built from between (44–166) $\times 10^3$ carbon atoms; it was rigid, i.e., no C atom could move with respect to the others. It is known from nanoindentation studies [42] that the atomistic surface structure of the tool may influence the nucleation of dislocations; we believe, however, that this effect lost its importance during the long cutting simulations performed in the present study. It had a clearance angle of 5° and a rounded edge with a curvature radius of $r = 10$ Å. The rake angle, α, of the tool was varied in this study between +45° and −45°. Fe and C atoms interact purely repulsively with each other; their interaction is described by a Lennard–Jones potential [43] that was cut off at its minimum at 4.2 Å [20]. In all simulations presented here, the cutting depth amounted to $d = 50$ Å.

The tool moved with a velocity of 20 m/s, and the total cutting length was 100 Å. This length might be considered as too small to provide steady-state cutting values, in particular for the negative rake angles; we therefore check this issue in Appendix B.

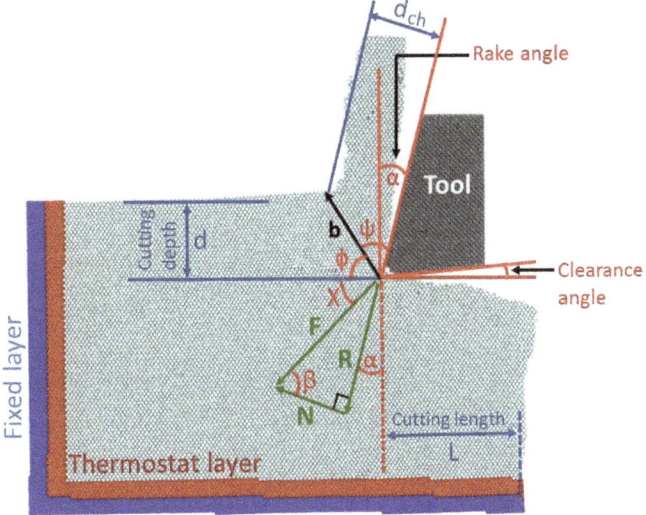

Figure 1. Schematics of the simulation setup showing important quantities characterizing the geometry of the cutting and of the chip. Quantities are defined in the text.

The open-source code LAMMPS [44] was used to perform the simulations. Dislocations were identified using the Crystal Analysis Tool (CAT) [45–47]. The atomistic configurations were visualized using the free software OVITO [48], while dislocations and surfaces were rendered by ParaView [49].

We note that previously, we simulated this cutting system with a rake angle of $\alpha = 15°$ [21,22]. Since the results were quite close to our new data for $\alpha = 0°$, we do not show snapshots for this case, but use the quantitative results of the forces and chip thickness in our discussion.

3. Results

3.1. Plastic Deformation

Let us first discuss cutting for $\alpha = 45°$ and $0°$, which proceeded via the same plastic mechanism. Figure 2 shows the dislocations that formed during the cutting process. The dominant mechanism was the formation of edge dislocation with the Burgers vector $b = \frac{1}{2}[111]$, which moved upward in the [111] direction. Their motion transported atoms upwards at the front of the tool out of the surface and thus formed the chip. The motion could effectively be described as the shear of the material with the shear plane given by the glide plane of the dislocation system; this shear plane is conventionally denoted as the primary shear plane (PSZ). The shear angle ϕ was hence given by the angle between the Burgers vector of the dislocations and the cutting direction [001]; it amounted to $\phi = 54.7°$. We emphasize that Figure 2 demonstrates that all plasticity occurring during the cut was confined to a single shear plane; this cutting system therefore provided a textbook example of the so-called single shear plane cutting model in machining theories [2,3,5–7,50,51]. A previous analysis of this cut system for the rake angle of $\alpha = 15°$ gave the same conclusions on the plastic behavior during cutting [21,22].

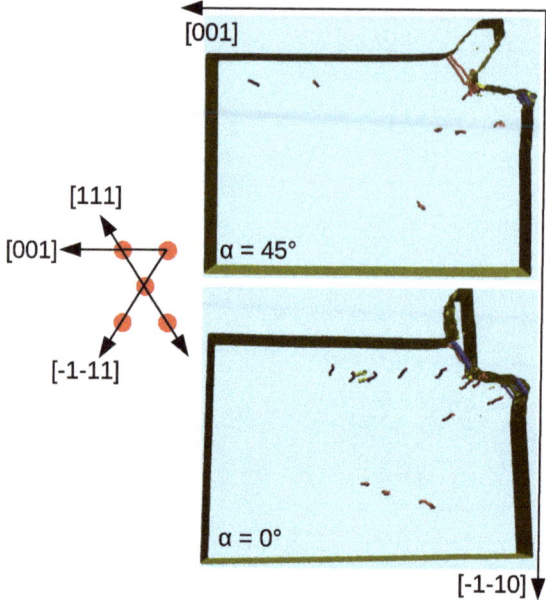

Figure 2. Side view of the dislocations generated in the Fe crystal after a cutting length of 100 Å with a cutting tool of rake angle 45° and 0°. On the left-hand side, an atomistic sketch shows a 2D bcc unit cell in the plotted plane that explains the slip directions. Yellow: deformed surface including unidentified defects. Color code for dislocations classified by the Burgers vector b: red: $b = \langle 100 \rangle$; orange: $b = \frac{1}{2} \langle 111 \rangle$; blue: $b = \frac{1}{p} \langle 111 \rangle$ partial with $p = 3, 6,$ or 12.

Besides the $b = \frac{1}{2}[111]$ dislocations that are immediately responsible for chip formation, Figure 2 also shows the presence of several other dislocation systems. All of them are edge dislocations with their dislocation lines oriented parallel to the tool edge. Among these are $b = [001]$ edge dislocations, which move along the cutting direction, [001], to the left, parallel to the surface, as well as $b = \frac{1}{2}[\bar{1}\bar{1}1]$ dislocations, which move into the substrate interior. Both of these glide systems transport material away from the cut zone, but do not contribute to chip formation.

For a rake angle of $\alpha = -22.5°$, plasticity is illustrated in Figure 3 for various cutting lengths L. Here, the action of the $b = \frac{1}{2}[111]$ dislocations, close to the rake face, as barely visible, since the angle between the glide direction and the rake face was quite small, only 12.8°, such that dislocations had only little time (and space) for moving towards the surface, before the rake surface covered them up. Due to the small activity of dislocations in the [111] direction, the upper surface started bending elastically upwards. This bending generated a large number of partial dislocations (in the following abbreviated as "partials") flowing in the $[\bar{1}\bar{1}1]$ direction; they created a step on the surface at the positions where the partials were emitted (seen at a cutting length $L = 31$ Å). These partials had Burgers vectors of the form $b = \frac{1}{p} \langle 111 \rangle$ with $p = 3, 6,$ or 12; they formed twinning boundaries, such that the volume enclosed by these partials was by nanotwins. After this nanotwin grew to span almost the entire simulation volume ($L = 60$ Å), the process of the generation of twinning partials in the elastically bent material in front of the tool repeated itself ($L = 100$ Å).

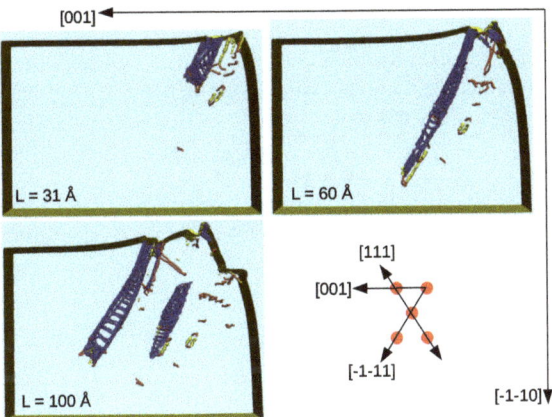

Figure 3. As in Figure 2, but for a rake angle of $-22.5°$ and various cutting lengths L.

Finally, Figure 4 shows the plastic activity when cutting with a rake angle of $\alpha = -45°$. Emission of $b = \frac{1}{2}[111]$ dislocations was completely suppressed as their glide direction lied within the rake tool. Now, a large amount of $b = \frac{1}{p}[\bar{1}\bar{1}\bar{1}]$ partial dislocations were activated, which transported the material backward. These latter dislocations led to a deformation of the frontal (right-hand side) surface, where the tool entered. As a consequence, the cutting process did not lead to chip formation on the top surface, but rather to a deformation of the frontal surface.

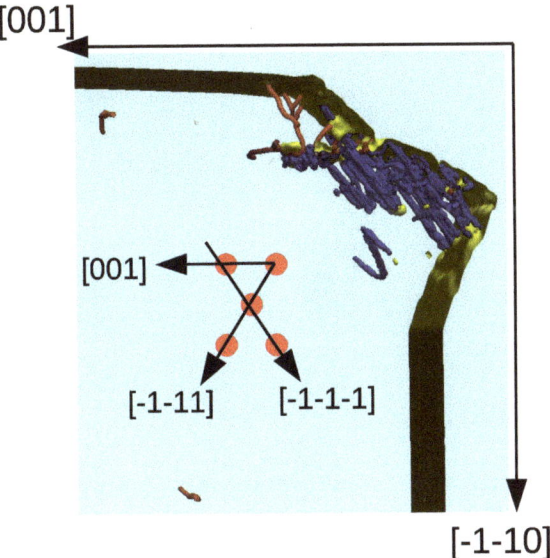

Figure 4. As in Figure 2, but for a rake angle of $-45°$ at the end of the cut, $L = 100$ Å.

We concluded that for $\alpha \geq 0°$, a single dislocation system—with the Burgers vector $b = \frac{1}{2}[111]$—was responsible for cutting. For a positive rake angle, hence, cutting could be well described by a single shear angle, $\phi = 54.7°$. Even for $\alpha = -22.5°$, this system still contributed to cutting, but was no longer dominant. For $\alpha = -45°$, the system was completely deactivated.

3.2. Angles

For the analysis of the simulation results, we introduced several angles; see Figure 1. The shear angle ϕ is the angle between the PSZ and the original surface plane. Analogously, we define ψ as the angle between the rake face and the PSZ; it is:

$$\psi = \frac{\pi}{2} + \alpha - \phi. \tag{1}$$

The force angle χ is the angle between the total force and the original surface plane. It satisfies the relation:

$$\chi = \beta - \alpha \tag{2}$$

with the friction angle β. This friction angle is given by the components of the force parallel (R) and perpendicular (N) to the rake face by:

$$\tan \beta = \frac{R}{N}. \tag{3}$$

3.3. Chip Thickness

Figure 5 assembles the chip shapes obtained by the cutting process. For $\alpha = 45°$ and $0°$ (and also for $15°$; see [21,22]), the chip had a roughly rectangular cross-section with a constant thickness d_{ch}. This feature was easily explained by the action of the single dislocation system forming the chip. Even for $\alpha = -22.5°$, a small chip immediately adjacent to the rake face could be discerned, which was formed by the same dislocation system. In addition, for this rake angle, the other plastic processes discussed above contributed to forming a very broad chip extending a considerable distance in front of the tool. For $\alpha = -45°$, no chip was formed.

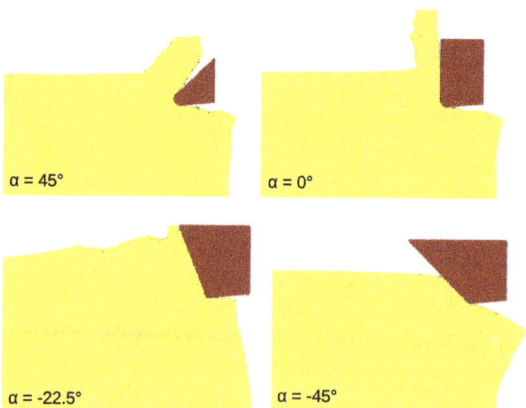

Figure 5. Side views of the chips formed during cutting.

We determined the chip thickness d_{ch} as the extension of the chip perpendicular to the rake face (see Figure 1) and measured it at the base of the chip close to the PSZ; the values are assembled in Table 1. For the complex chip form of $\alpha = -22.5°$, we only considered the part of the chip adjacent to the rake face. The decline of d_{ch} with decreasing α illustrated in Figure 6 could be rationalized as follows.

Table 1. Force and chip characteristics for various rake angles α: cutting force, F_c; thrust force, F_t; force components perpendicular, N, and parallel, F, to the rake surface of the tool; friction angle, $β$; and chip thickness, d_{ch}. Force values are averaged over the last 30 Å of cutting. For the negative rake angles, data were also obtained for a cutting length of 200 Å see Appendix B; in the rows marked by (*), averages over the last 100 Å of cutting are presented.

α	F_c (μN)	F_t (μN)	R (μN)	N (μN)	$β$	d_{ch} (Å)
45°	0.41	0.12	0.37	0.21	60.4°	70.3
15°	0.48	0.13	0.25	0.43	31.0°	39.9
0°	0.69	0.18	0.18	0.69	14.6°	38.0
−22.5°	0.99	0.48	0.06	1.10	2.9°	19.6
−22.5° (*)	1.01	0.48	0.06	1.12	3.1°	20.0
−45°	0.60	0.66	0.04	0.89	2.3°	0
−45° (*)	0.56	0.60	0.03	0.82	2.1°	0

From mass conservation, the chip thickness, d_{ch}, is related to the cutting depth, d, by:

$$d_{ch} = d \frac{\sin ψ}{\sin φ'} \qquad (4)$$

This relation may simply be obtained by noting that the length of the PSZ, b, obeys both $d = b \sin φ$ and $d_{ch} = b \sin ψ$; see Figure 1. Using Equation (1), Equation (4) thus predicts that $d_{ch} = 0$ for $α = −35°$ and $d_{ch} = d$ for $α = 2φ − 90° = 19.5°$, since then, $ψ = φ$; see Figure 1.

Figure 6 compares our simulation results with the simple theory of Equation (4). Since only the shear angle enters this theory, the fair agreement obtained shows that indeed, the single shear plane model can describe the cutting process well.

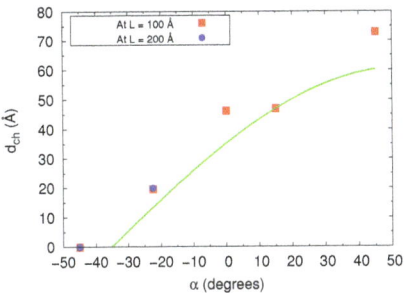

Figure 6. Chip thickness, d_{ch}, as a function of the rake angle α. Simulation results (squares) are compared to the theoretical prediction, Equation (4). For the negative rake angles, data were also obtained for a cutting length of 200 Å (circles); see Appendix B.

3.4. Forces

The force on the tool was readily measured in an MD simulation. Figure 7 shows the evolution of the force with cutting length; the components of the force in the cutting direction, F_c, and normal to the original surface, F_t, which are denoted as cutting and thrust forces, respectively, are displayed. After the tool contacted the substrate (at Cutting Length 0), the forces started increasing. The increase was, however, not monotonous, since the generation of dislocations led to force drops. Towards the end of the cut, a sort of steady-state appeared to develop—with the possible exception of the thrust force at rake angle $α = −45°$. We assemble the averages of the forces over the last 30 Å of the cut in Table 1.

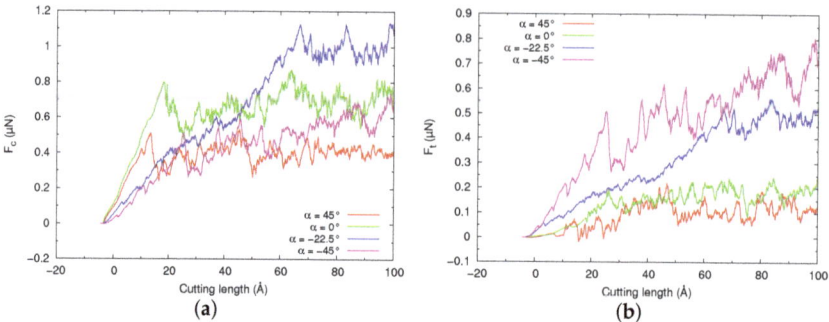

Figure 7. (a) Cutting force, F_c, and (b) thrust force, F_t, as a function of the cutting length L.

The thrust forces increased with decreasing α showing that a stronger perpendicular force was needed to keep a wider tool at its prescribed depth. With decreasing α, also the cutting force increased; this increase illustrated the fact that larger forces were needed to shear the material and form the chip. Here, the case of $\alpha = -45°$ was exceptional, as the cutting force was quite small, even below the values of $\alpha = 0°$. This could be explained by the fact that in this case, no chip was formed (see Section 3.3), and we hence, did not have a case of real "cutting" here.

3.5. Force Angle

From the forces determined in the simulation, we could calculate the friction angle; see Table 1. It allowed us to calculate the force angle; we display it in Figure 8. It showed that in the regime of chip formation, $\alpha \geq 0$, the force angle was constant, $\chi = 15.3° \pm 0.7°$, and only increased for negative α, where no chips were formed. This feature—constant χ for constant ϕ–is important as the relation between these two angles is an important ingredient of available cutting theories.

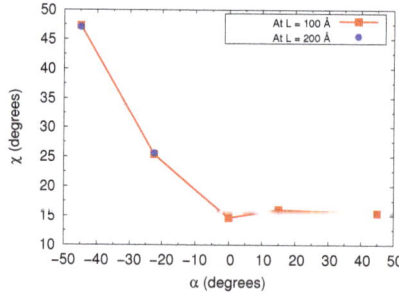

Figure 8. Force angle, χ, as a function of the rake angle α. For the negative rake angles, data were also obtained for a cutting length of 200 Å (circles); see Appendix B.

Mechanical theories of cutting result in a linear relationship between shear angle ϕ and force angle χ [6],

$$\phi = c_1 - c_2 \chi. \tag{5}$$

From theoretical arguments, the law, Equation (5), is derived with $c_1 = 45°$ and $c_2 = 1$ or 0.5, depending on whether it is assumed that shear should occur in the direction of maximum shear stress or the power needed for cutting is minimized, respectively [2,3,5–7,50,51]. In particular, the so-called Merchant's law [2] with $c_1 = 45°$ and $c_2 = 0.5$ forms the basis of textbook treatments of cutting [51].

On the other hand, from experimental data compiled for mild steel by Pugh [52], coefficients of $c_1 = 32°$ and $c_2 = 0.44$ can be read off (Figure 2) [6]. A shear angle of $\phi = 54.7°$ would hence correspond to a force angle of $\chi = -52°$; this corresponds to a force pointing out of the surface, which is totally unphysical. On the other hand, our measured force angle of $\chi = 15.3°$ corresponded to $\phi = 25.3°$, which was in direct contrast to the shear observed in the simulation.

Our results also lay strongly outside the simple theoretical dependencies $c_1 = 45°$ and $c_2 = 1$ or 0.5, which would predict a force angle of $-9.7°$ ($c_2 = 1$) and $-19.4°$ $c_2 = 0.5$. Again, the large shear angle would require a negative force angle.

We note that MD simulation of cutting an isotropic material (metallic glass) corroborated Merchant's law [2]), i.e., Equation (5) with $c_1 = 45°$ and $c_2 = 0.5$ [53].

We concluded that our results of single crystal cutting showed a fixed force angle for a fixed shear angle, in agreement with available cutting theories. However, the value of χ was incompatible with these theories.

4. Summary

We used MD simulation to study the cutting of an Fe single crystal using tools with various rake angles. Focusing on the (110)[001] cut system, we obtained the following findings.

1. For the (110)[001] cut system, at not too strongly negative rake angles, the crystal plasticity was governed by a simple mechanism: the evolution of the chip was driven by the generation of edge dislocations with the Burgers vector $b = \frac{1}{2}[111]$. These fixed the shear angle to $\phi = 54.7°$, independently of the rake angle of the tool.
2. For positive α, the thickness of the cut chip corresponded well to the law predicted by mass conservation, Equation (4).
3. While macroscopic cutting of (polycrystalline) iron is governed by a linear relationship between the shear angle ϕ and the force angle χ, in our system, the shear angle was fixed, and also, the force angle was fixed, at $\chi \cong 15°$. The relation between ϕ and χ was, however, outside that found for macroscopic cutting, Equation (5), of macroscopic (polycrystalline) mild steels and also of other data found experimentally for metal cutting.
4. The relation observed between ϕ and χ was even far away from relations derived theoretically, which were shown to hold true in MD simulations of isotropic materials (metallic glasses) [53].
5. The chip form was simple—a rectangular shape, showing negligible curvature—as long as a single dislocation glide mechanism governed plasticity. This changed for (strongly) negative rake angles.
6. Only for (strongly) negative rake angles, the mechanism of plasticity changed, leading to a complex chip shape or even suppressing the formation of a chip. In these cases, the force angle strongly increased while the friction angle tended to zero.

The dependencies were simple as long as cutting was dictated by a single dislocation glide mechanism. For other cut systems, the situation became more complex and could no longer be described in the framework of the single shear cutting model. An example is given in Appendix A.

In future work, it will be interesting to investigate how the cutting depth d and the curvature radius r of the tool edge influence our conclusions. In applications, these two lengths will be larger than the values adopted in the present study, but may have a similar relation, d/r, as the one adopted in the present work, where we used $d/r = 5$. As long as a single crystalline grain is cut, one could therefore presume that our ideas are relevant; but this hypothesis needs to be further investigated. In particular, experimental verification of our prediction of the effect of the rake angle on the cutting of single crystals would be highly appreciated.

Author Contributions: I.A.A. performed the simulations and analyzed the results. I.A.A. and H.M.U. designed the work, discussed the results, and wrote the manuscript. All authors read and agreed to the published version of the manuscript.

Funding: This research was funded by the Deutsche Forschungsgemeinschaft (DFG, German Research Foundation), Project Number 172116086–SFB 926.

Acknowledgments: Access to the computational resources provided by the compute cluster "Elwetritsch" of the University of Kaiserslautern is appreciated.

Conflicts of Interest: The authors declare no conflict of interest.

Appendix A. The (100)[011] Cut System: Twinning

For general cut surfaces and cut directions, more than one dislocation slip system—and hence shear plane—contributes to chip formation [22], and the analysis provided here for the (110)[001] system will not hold. We illustrate the complexities arising with the example of the (100)[011] system.

In this system, the slip direction responsible for chip formation, [111], made an angle of $\phi = 35.3°$ to the surface. Figure A1 illustrates the deformation occurring under cutting for three rake angles, $\alpha = 45°, 0°$, and $-45°$. Perfect dislocations with the Burgers vector $b = \frac{1}{2}\langle 111 \rangle$ were activated, which glided to the top surface. In addition, and more pronouncedly, twinning partials $b = \frac{1}{p}\langle 111 \rangle$ showed up. The generation of a twinning boundary is most clearly seen in the figure for the rake angle of $0°$, where it is marked by a dense net of partial $b = \frac{1}{6}[111]$ dislocations. The slip direction $[\bar{1}11]$ was only activated for $\alpha = -45°$.

The twinning partials led to the generation of twin boundaries, which are highlighted in Figure A2. The largest twin was generated for $\alpha = -0°$. Twinned surfaces were recognizable by their smooth surfaces, while the $b = \frac{1}{2}\langle 111 \rangle$ dislocations arriving at the surface led to surface roughening. Note that for this cutting system, even the $\alpha = -45°$ tool led to chip formation, since now the shear angle $\phi = 35.3°$ was smaller, enabling dislocation glide to the surface even for strongly negative rake angles.

The twinning occurring in front of the tool was equivalent to several shear planes being active temporarily during the cutting process and thus complicated any simple analysis in terms of the single shear plane model.

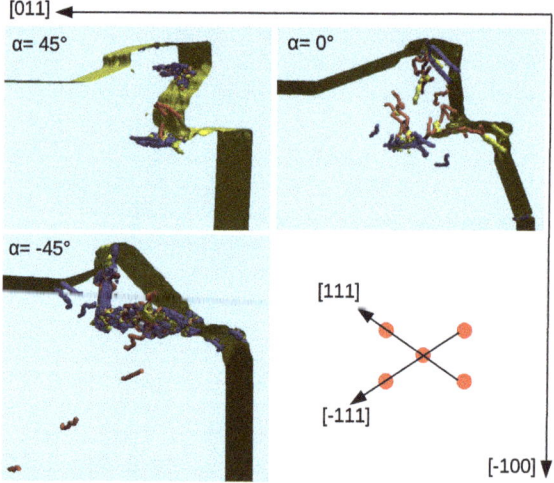

Figure A1. Side view of the dislocations generated in the Fe (100)[011] cutting system at cutting length 100 Å with a cutting tool of rake angles 45°, 0°, and −45°. Dislocations and slip directions are denoted as in Figure 2.

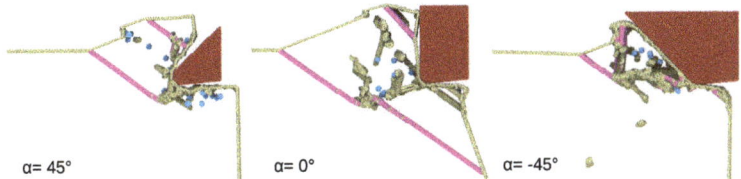

Figure A2. Same cutting systems as in Figure A1, but highlighting twinning boundaries (pink). Vacancies are shown in blue, while surfaces and dislocations are colored gray.

Appendix B. Extended Cutting Length

It might be speculated that a cutting length of $L = 100$ Å is not sufficient to obtain steady-state cutting conditions, in particular for the negative rake angles. We therefore increased the cutting length in these cases to 200 Å. The evolution of the cutting force and of the thrust force shown in Figure A3 demonstrated that the simulation results indeed stabilized at $L = 100$ Å. The data for cutting lengths between 100 and 200 Å showed, apart from fluctuations caused by the discontinuous emission of dislocations, the same average behavior. We assemble the averages over the last 100 Å in these extended runs in Table 1, which shows that deviations of ≤ 10 % showed up between the 100 Å run and the new extended run. As a consequence, also the friction angle and the force angle showed only minor deviations from the shorter run, cf. Figure 8, and the conclusions stated in the main part of the text remain valid.

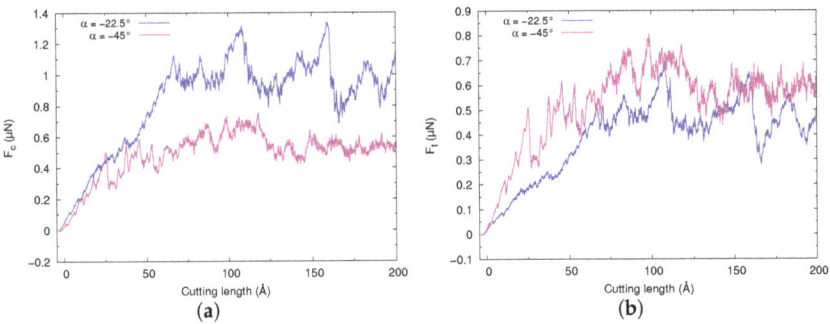

Figure A3. (a) Cutting force, F_c, and (b) thrust force, F_t, as a function of the cutting length L for an extended simulation up to 200 Å for the negative rake angles, $\alpha = -22.5°$ and $-45°$.

The chip shapes also (Figure A4) showed no new features compared to the shorter runs shown in Figure 5. In particular, for the most negative rake angle of $\alpha = -45°$, no chip was formed on the top surface. The chip for the $\alpha = -22.5°$ rake angle grew in size, but its structure consisting of several sub-peaks remained intact. Note that the very first peak (close to the rake face), which was created by the glide of 1/2[111] dislocations, retained its identity. Its width did not change (see Table 1), which indicated that this feature did not change any more after $L = 100$ Å, in close agreement with the result obtained after cutting of 100 Å. Our conclusion that the chip thickness fits again nicely with the theoretical prediction, Equation (4), was hence unchanged.

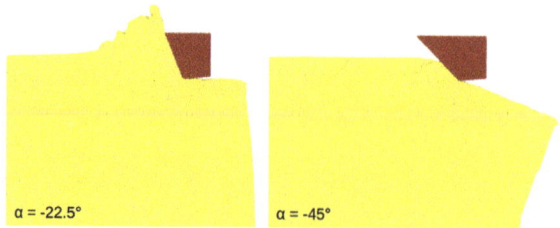

Figure A4. Side views of the chips formed during cutting for an extended simulation up to 200 Å for the negative rake angles, $\alpha = -22.5°$ and $-45°$.

Finally, we display in Figure A5 the dislocations generated after cutting the crystals for the extended cutting length of $L = 200$ Å. For the rake angle of $\alpha = -22.5°$, again, the large number of $[\bar{1}\bar{1}1]$ partials dominated the picture; these were responsible for the substructure of the chip containing several peaks. Indeed, when comparing Figure A5a with Figure 3, a new dislocation emission center at the chip front emerged, which induced a new peak at the front of the chip. For the rake angle of $\alpha = -45°$, dislocation emission in $[\bar{1}\bar{1}\bar{1}]$ directions proceeded and was responsible for the increased deformation of the frontal (right-hand side) surface, which now extended almost over the entire simulated crystal face in the $[\bar{1}\bar{1}0]$ direction. The tangle of dislocations beneath the front part of the tool increased under extended cutting. Due to dislocation reactions, for both rake angles shown in Figure A5, vacancies were created, which are visualized as small yellow spheres.

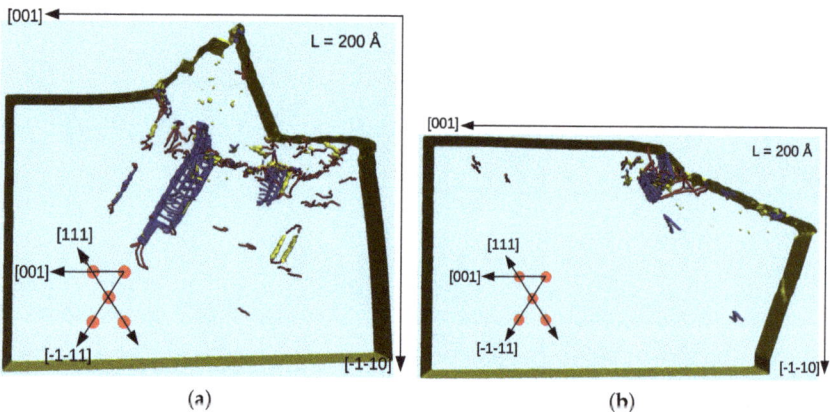

Figure A5. As in Figure 2, but for an extended cutting length, $L = 200$ Å. Snapshots are for a rake angle of (a) $\alpha = -22.5°$ and of (b) $-45°$.

We concluded that after increasing the cutting length from $L = 100$ to 200 Å, the dislocation generation proceeded in analogy to the shorter cutting length, and no qualitatively new features showed up.

References

1. Fang, F.; Xu, F. Recent Advances in Micro/Nano-cutting: Effect of Tool Edge and Material Properties. *Nanomanuf. Metrol.* **2018**, *1*, 4–31. [CrossRef]
2. Merchant, M.E. Mechanics of the Metal Cutting Process. I. Orthogonal Cutting and a Type 2 Chip. *J. Appl. Phys.* **1945**, *16*, 267–275. [CrossRef]

3. Merchant, M.E. Mechanics of the Metal Cutting Process. II. Plasticity Conditions in Orthogonal Cutting. *J. Appl. Phys.* **1945**, *16*, 318–324. [CrossRef]
4. Astakhov, V.P. On the inadequacy of the single shear plane model of chip formation. *Int. J. Mech. Sci.* **2005**, *47*, 1649–1672. [CrossRef]
5. Oxley, P.L.B. A strain-hardening solution for the "shear angle" in orthogonal metal cutting. *Int. J. Mech. Sci.* **1961**, *3*, 68–79. [CrossRef]
6. Atkins, A.G. Modelling metal cutting using modern ductile fracture mechanics: quantitative explanations for some longstanding problems. *Int. J. Mech. Sci.* **2003**, *45*, 373–396. [CrossRef]
7. Atkins, A.G. Toughness and cutting: A new way of simultaneously determining ductile fracture toughness and strength. *Eng. Fract. Mech.* **2005**, *72*, 849–860. [CrossRef]
8. Ruestes, C.J.; Bringa, E.M.; Gao, Y.; Urbassek, H.M. Molecular dynamics modeling of nanoindentation. In *Applied Nanoindentation in Advanced Materials*; Tiwari, A., Natarajan, S., Eds.; Wiley: Chichester, UK, 2017; Chapter 14, pp. 313–345. [CrossRef]
9. Ruestes, C.J.; Alabd Alhafez, I.; Urbassek, H.M. Atomistic Studies of Nanoindentation–A Review of Recent Advances. *Crystals* **2017**, *7*, 293. [CrossRef]
10. Gao, Y.; Brodyanski, A.; Kopnarski, M.; Urbassek, H.M. Nanoscratching of iron: A molecular dynamics study of the influence of surface orientation and scratching direction. *Comput. Mater. Sci.* **2015**, *103*, 77–89. [CrossRef]
11. Alabd Alhafez, I.; Brodyanski, A.; Kopnarski, M.; Urbassek, H.M. Influence of Tip Geometry on Nanoscratching. *Tribol. Lett.* **2017**, *65*, 26. [CrossRef]
12. Komanduri, R.; Chandrasekaran, N.; Raff, L.M. MD simulation of exit failure in nanometric cutting. *Mat. Sci. Eng. A* **2001**, *311*, 1. [CrossRef]
13. Fang, T.H.; Chang, W.J.; Weng, C.I. Nanoindentation and nanomachining characteristics of gold and platinum thin films. *Mat. Sci. Eng. A* **2006**, *430*, 332–340. [CrossRef]
14. Pei, Q.X.; Lu, C.; Fang, F.Z.; Wu, H. Nanometric cutting of copper: A molecular dynamics study. *Comput. Mater. Sci.* **2006**, *37*, 434–441. [CrossRef]
15. Chu, C.Y.; Tan, C.M. Deformation analysis of nanocutting using atomistic model. *Int. J. Solids Struct.* **2009**, *46*, 1807–1814. [CrossRef]
16. Pen, H.M.; Liang, Y.C.; Luo, X.C.; Bai, Q.S.; Goel, S.; Ritchie, J.M. Multiscale simulation of nanometric cutting of single crystal copper and its experimental validation. *Comput. Mater. Sci.* **2011**, *50*, 3431–3441. [CrossRef]
17. Romero, P.A.; Anciaux, G.; Molinari, A.; Molinari, J.F. Friction at the tool-chip interface during orthogonal nanometric machining. *Model. Simul. Mater. Sci. Eng.* **2012**, *20*, 055007. [CrossRef]
18. Zhang, L.; Zhao, H.; Dai, L.; Yang, Y.; Du, X.; Tang, P.; Zhang, L. Molecular dynamics simulation of deformation accumulation in repeated nanometric cutting on single crystal copper. *RCS Adv.* **2015**, *5*, 12678. [CrossRef]
19. Narulkar, R.; Bukkapatnam, S.; Raff, L.M.; Komanduri, R. Graphitization as a precursor to wear of diamond in machining pure iron: A molecular dynamics investigation. *Comput. Mater. Sci.* **2009**, *45*, 358–366. [CrossRef]
20. Gao, Y.; Urbassek, H.M. Evolution of plasticity in nanometric cutting of Fe single crystals. *Appl. Surf. Sci.* **2014**, *317*, 6–10. [CrossRef]
21. Alabd Alhafez, I.; Gao, Y.; Urbassek, H.M. Nanocutting: A comparative molecular-dynamics study of fcc, bcc, and hcp metals. *Curr. Nanosci.* **2017**, *13*, 40–47. [CrossRef]
22. Alabd Alhafez, I.; Urbassek, H.M. Orientation dependence in nanocutting of Fe single crystals: A molecular-dynamics study. *Comput. Mater. Sci.* **2018**, *143*, 286–294. [CrossRef]
23. Avila, K.E.; Küchemann, S.; Alabd Alhafez, I.; Urbassek, H.M. An atomistic study of shear-band formation during cutting of metallic glasses. *J. Appl. Phys.* **2020**, *127*, 115101. [CrossRef]
24. Komanduri, R.; Chandrasekaran, N.; Raff, L.M. Molecular dynamics simulation of the nanometric cutting of silicon. *Philos. Mag. B* **2001**, *81*, 1989–2019. [CrossRef]
25. Han, X.S.; Lin, B.; Yu, S.Y.; Wang, S.X. Investigation of tool geometry in nanometric cutting by molecular dynamics simulation. *J. Mater. Process. Tech.* **2002**, *129*, 105–108. [CrossRef]
26. Fang, F.Z.; Wu, H.; Liu, Y.C. Modelling and experimental investigation on nanometric cutting of monocrystalline silicon. *Int. J. Mach. Tool Manu.* **2005**, *45*, 1681–1686. [CrossRef]

27. Cai, M.B.; Li, X.P.; Rahman, M. Study of the temperature and stress in nanoscale ductile mode cutting of silicon using molecular dynamics simulation. *J. Mater. Process. Technol.* **2007**, *192–193*, 607–612. [CrossRef]
28. Lai, M.; Zhang, X.; Fang, F.; Wang, Y.; Feng, M.; Tian, W. Study on nanometric cutting of germanium by molecular dynamics simulation. *Nanoscale Res. Lett.* **2013**, *8*, 13. [CrossRef]
29. Goel, S.; Stukowski, A.; Luo, X.; Agrawal, A.; Reuben, R.L. Anisotropy of single crystal 3C-SiC during nanometric cutting. *Model. Simul. Mater. Sci. Eng.* **2013**, *21*, 065004. [CrossRef]
30. Goel, S.; Kovalchenko, A.; Stukowski, A.; Cross, G. Influence of microstructure on the cutting behaviour of silicon. *Acta Mater.* **2016**, *105*, 464–478. [CrossRef]
31. Xu, F.; Fang, F.; Zhang, X. Study on surface generation in nano-cutting by large-scale molecular dynamics simulation. *Int. J. Adv. Manuf. Technol.* **2019**, *104*, 4325–4329. [CrossRef]
32. Vardanyan, V.H.; Zhang, Z.; Alabd Alhafez, I.; Urbassek, H.M. Cutting of Al/Si bilayer systems: Molecular dynamics study of twinning, phase transformation, and cracking. *Int. J. Adv. Manuf. Technol.* **2020**, *107*, 1297–1307. [CrossRef]
33. Lai, M.; Zhang, X.D.; Fang, F.Z. Study on critical rake angle in nanometric cutting. *Appl. Phys. A* **2012**, *108*, 809–818. [CrossRef]
34. Hosseini, S.V.; Vahdati, M. Modeling the effect of tool edge radius on contact zone in nanomachining. *Comput. Mater. Sci.* **2012**, *65*, 29. [CrossRef]
35. Xu, F.; Wang, J.; Fang, F.; Zhang, X. A study on the tool edge geometry effect on nano-cutting. *Int. J. Adv. Manuf. Technol.* **2017**, *91*, 2787–2797. [CrossRef]
36. Komanduri, R.; Chandrasekaran, N.; Raff, L.M. Some aspects of machining with negative-rake tools simulating grinding: A molecular dynamics simulation approach. *Philos. Mag. B* **1999**, *79*, 955. [CrossRef]
37. Dai, H.; Du, H.; Chen, J.; Chen, G. Investigation of tool geometry in nanoscale cutting single crystal copper by molecular dynamics simulation. *Proc. Inst. Mech. Eng. Part J J. Eng. Tribol.* **2019**, *233*, 1208–1220. [CrossRef]
38. Mendelev, M.I.; Han, S.; Srolovitz, D.J.; Ackland, G.J.; Sun, D.Y.; Asta, M. Development of new interatomic potentials appropriate for crystalline and liquid iron. *Philos. Mag.* **2003**, *83*, 3977–3994. [CrossRef]
39. Malerba, L.; Marinica, M.C.; Anento, N.; Björkas, C.; Nguyen, H.; Domain, C.; Djurabekova, F.; Olsson, P.; Nordlund, K.; Serra, A.; et al. Comparison of empirical interatomic potentials for iron applied to radiation damage studies. *J. Nucl. Mater.* **2010**, *406*, 19–38. [CrossRef]
40. Gunkelmann, N.; Bringa, E.M.; Kang, K.; Ackland, G.J.; Ruestes, C.J.; Urbassek, H.M. Polycrystalline iron under compression: Plasticity and phase transitions. *Phys. Rev. B* **2012**, *86*, 144111. [CrossRef]
41. Haghighat, S.M.H.; von Pezold, J.; Race, C.P.; Körmann, F.; Friak, M.; Neugebauer, J.; Raabe, D. Influence of the dislocation core on the glide of the $\frac{1}{2}\langle 111\rangle\{110\}$ edge dislocation in bcc-iron: An embedded atom method study. *Comput. Mater. Sci.* **2014**, *87*, 274–282. [CrossRef]
42. Wagner, R.J.; Ma, L.; Tavazza, F.; Levine, L.E. Dislocation nucleation during nanoindentation of aluminum. *J. Appl. Phys.* **2008**, *104*, 114311. [CrossRef]
43. Banerjee, S.; Naha, S.; Puri, I.K. Molecular simulation of the carbon nanotube growth mode during catalytic synthesis. *Appl. Phys. Lett.* **2008**, *92*, 233121. [CrossRef]
44. Plimpton, S. Fast Parallel Algorithms for Short-Range Molecular Dynamics. *J. Comput. Phys.* **1995**, *117*, 1–19. Available online: http://lammps.sandia.gov/ (accessed on 13 May 2020). [CrossRef]
45. Stukowski, A.; Bulatov, V.V.; Arsenlis, A. Automated identification and indexing of dislocations in crystal interfaces. *Model. Simul. Mater. Sci. Eng.* **2012**, *20*, 085007. [CrossRef]
46. Stukowski, A. Structure identification methods for atomistic simulations of crystalline materials. *Model. Simul. Mater. Sci. Eng.* **2012**, *20*, 045021. [CrossRef]
47. Stukowski, A.; Arsenlis, A. On the elastic-plastic decomposition of crystal deformation at the atomic scale. *Model. Simul. Mater. Sci. Eng.* **2012**, *20*, 035012. [CrossRef]
48. Stukowski, A. Visualization and analysis of atomistic simulation data with OVITO –the Open Visualization Tool. *Model. Simul. Mater. Sci. Eng.* **2010**, *18*, 015012. Available online: http://www.ovito.org/ (accessed on 13 May 2020). [CrossRef]
49. Henderson, A. *ParaView Guide, A Parallel Visualization Application*; Kitware Inc.: Clifton Park, NY, USA, 2007. Available online: http://www.paraview.org (accessed on 13 May 2020).
50. Krystof, J.; Schallbroch, H. *Grundlagen der Zerspanung*; Berichte über betriebswissenschaftliche Arbeiten; VDI-Verlag: Berlin, Germany, 1939; Volume 12.

51. Klocke, F.; König, W. *Fertigungsverfahren*, 8th ed.; Drehen, Fräsen, Bohren; Springer: Berlin, Germany, 2008; Volume 1.
52. Pugh, H.L.D. Mechanics of the cutting process. In *Proceedings of the Conference on Technology of Engineering Manufacture*; The Institute of Mechanical Engineers: London, UK, 1958; pp. 237–254.
53. Avila, K.E.; Vardanyan, V.H.; Alabd Alhafez, I.; Zimmermann, M.; Kirsch, B.; Urbassek, H.M. Applicability of cutting theory to nanocutting of metallic glasses: Atomistic simulation. **2020**, submitted.

© 2020 by the authors. Licensee MDPI, Basel, Switzerland. This article is an open access article distributed under the terms and conditions of the Creative Commons Attribution (CC BY) license (http://creativecommons.org/licenses/by/4.0/).

Article

Transformation of SnS Nanocompisites to Sn and S Nanoparticles during Lithiation

Haokun Deng [1,*], Thapanee Sarakonsri [2], Tao Huang [3], Aishui Yu [3] and Katerina Aifantis [4]

1. EVE Energy Co., Ltd., Huizhou 516006, China
2. Department of Chemistry, Chiang Mai University, Chiang Mai 50200, Thailand; scchi017@chiangmai.ac.th
3. Institute of New Energy, Fudan University, Shanghai 200438, China; huangt@fudan.edu.cn (T.H.); asyu@fudan.edu.cn (A.Y.)
4. Mechanical and Aerospace Engineering, University of Florida, Gainesville, FL 36211, USA; kaifantis@ufl.edu
* Correspondence: 039749@evebattery.com

Abstract: SnS nanomaterials have a high initial capacity of 1000 mAh g^{-1}; however, this cannot be retained throughout electrochemical cycling. The present study provides insight into this capacity decay by examining the effect that Li intercalation has on SnS "nanoflowers" attached on carbon substrates' such as artificial graphite. Scanning and transmission electron microscopy reveal that lithiation of such materials disrupts their initial morphology and produces free-standing Sn and SnS nanoparticles that dissolve in the electrolyte and disperse uniformly over the entire electrode surface. As a result, the SnS is rendered inactive after initial cycling and contributes to the formation of the solid electrolyte interface layer, resulting in continuous capacity decay during long term cycling. This is the first study that illustrates the morphological effects that the conversion mechanism has on SnS anodes. In order to fully utilize SnS materials, it is necessary to isolate them from the electrolyte by fully encapsulating them in a matrix.

Keywords: anode; tin sulfide; lithium ion battery; conversion reaction; nanoflower

1. Introduction

The present study aims to understand the capacity degradation that occurs in SnS nanomaterials in terms of the morphological changes they undergo during Li-ion insertion. Sn has attracted significant attention for its use as an anode in Li-ion batteries, since it allows for a theoretical specific capacity that is over 990 mAh g^{-1} [1]. This is approximately three times the capacity of commercially used graphitic anodes (372 mAh g^{-1}) [2]. Despite its promising electrochemical performance, upon the formation of Li alloys, Sn suffers from severe volume expansions as high as 298%, which result in fracture of the Sn and delamination from the electrode. This leads to a loss of material that can host Li, and a rapid capacity decay occurs after several charge–discharge cycles. In order to alleviate high volume expansions, Sn is rarely used in its pure form, but is rather used as an oxide (SnO$_2$) or as an alloy (such as SnSb and SnS) [1,3–5]. To further increase mechanical stability, the size of the Sn-based particles is reduced to the nanoscale [6,7] and they are embedded or attached on a carbon-based matrix, since it can buffer and constrain the volume expansion.

The maximum initial capacity of Sn or SnO$_2$ cannot exceed 990 mAh g^{-1}; however, that of SnS anodes is greater than 1000 mAh g^{-1}. Despite this, the initial capacity of SnS anodes significantly decreases during cycling, as is the case for all Sn-based materials. For example, SnS nanoparticles (5–6 nm) embedded in a carbon matrix [5] experienced a capacity loss of ~500 mAh g^{-1} after 40 cycles, while SnS single phase alloy anodes in the form of microflakes showed an initial capacity of 1085 mAh g^{-1}, which dropped to ~550 mAh g^{-1} after 45 cycles [8].

The capacity drop observed during the first cycle has been attributed to the conversion reaction that gives rise to the formation of Li$_2$S [5,9], which forms irreversibly upon initial

lithiation. The resulting microstructure is, therefore, Sn particles in a Li_2S matrix, and with continuous lithiation/delithiation, Li_xSn alloys form reversibly. The mechanisms that give rise to the continuous capacity fade over long term cycling have not been explored.

A detailed examination of the microstructural changes that occur during the conversion and alloying reactions of SnS electrodes has not been performed. However, based on the severe fracture observed for Sn/C and SnO_2/C nanocomposites during cycling [10], it is of interest to examine whether mechanical instability can be correlated to the capacity decay of SnS. Hence, the present article focuses on the long term cycling performance of SnS/C nanomaterials and captures the effect that conversion and alloying reactions have on the microstructure by performing transmission and scanning electron microscopy before and after cycling. The particular microstructure considered is that of a flower-like nanoscale SnS alloy attached to graphite. The motivation for using such structures is that their thin petal structure is not dense and could accommodate the volume expansion of Sn. Such flower-type active materials have not been cycled before, and although similar SnS microstructures have recently been fabricated [11], they were significantly larger and their "flower petals" were much denser.

In order to examine the effect of the carbon substrate in SnS/C, both artificial graphite and microcarbon microbeads were used as the matrix, while the SnS content was varied by allowing the precursor compounds in solution to be either 10% or 20% by weight. Although the nanoflower-like SnS materials used here have not been examined as electrodes before, the purpose of this article is not to propose a promising SnS/C electrode (since SnS cannot compare to the promising high-capacity Si-based anodes), but to provide insight as to why SnS materials cannot retain their initial high capacity of over 1000 mAh g^{-1}, and interpret their capacity fade in terms of their microstructural changes.

2. Materials and Methods

2.1. Fabrication

The SnS/C nanopowders were fabricated using either artificial graphite (AG) or microcarbon microbeads (MCMB) as the matrix, as both have very good electrochemical properties [12]. Both MCMB and AG were purchased, rather than being synthesized in the lab.

The synthesis of the composite SnS/C materials was carried out using electroless deposition, similar to the processes reported in [13]. $SnCl_2 \cdot 2H_2O$ (99.9%, Fluka, Charlotte, NC, USA) and NH_2CSNH_2 (99.0%, Merck, Kenilworth, NJ, USA) were used as the metal ion precursors, while AG or MCMB were used as the carbon precursors. The metal ion precursors were refluxed with the carbon in 100mL ethylene glycol (99.9%, J.T. Baker, Radnor, PA, USA) at 200 °C for 8 h. The precipitate obtained was collected by centrifugation and washed using ethanol. Finally, the product was dried at 70 °C for 2 h to obtain the final powders.

In order to evaluate the effect of SnS content on the electrochemical performance, two different amounts (10% or 20%) of SnS were considered. For the case of 10%Sn10%S/C (abbreviated as 10SnS/C), the solution contained the inorganic materials in the proportion of 0.1 g Sn atoms and 0.1 g S atoms for every 0.8 g C atoms. Similarly, for the 20%Sn20%S/C (abbreviated as 20SnS/C), the solution contained 0.2 g Sn atoms and 0.2 g S atoms for every 0.6 g of C atoms.

2.2. Electrochemical Cycling

Electrochemical tests were performed using a CR2016-type coin cell that employed lithium foil as the counter electrode. The working electrode was fabricated by creating a slurry from 85% active material (which was the newly fabricated SnS/C), 5% super P carbon black, and 10% polyvinylidene fluoride (PVDF), and then coating it on a copper foil, which was the current collector.

The coin cells were assembled in an argon-filled glove box (Mikarouna, Superstar 1220/750/900) with 1 M $LiPF_6$ solution with ethylene carbonate–diethyl carbonate (EC–

DEC = 1:1, v/v) as the electrolyte and Celgard 2300 as the separator. The galvanostatic charge–discharge tests were performed in a battery test system (Land CT2001A, Wuhan Jinnuo Electronic Co., Ltd., China) at a constant current density of 50 mA g^{-1} in the potential range from 0 to 2.5 V. Finally, electrochemical impedance spectroscopy (EIS) of the Sn/C anodes was conducted through a potentiostat/galvanostat system (Autolab PGSTAT302N, Riverview, FL, USA).

2.3. Microstructure Analysis

The morphology and element analysis of the SnS/C materials before and after cycling was examined using a Hitachi S-4800 (Hitachi, Chiyoda City, Japan) field emission scanning electron microscope (FESEM) and a Philips CM200-FEG (Philips, Amsterdam, The Netherland) transmission electron microscope (TEM); both were equipped with electron dispersive spectroscopy (EDS). After cycling, the cells were opened in a fume hood and the working electrode was dipped in a 1:1 by volume mixture of ethylene carbonate (EC) and dimethyl carbonate (DMC) for 30 s to remove electrolyte residue, and was then dried using argon. Powders were then scraped onto (i) a piece of conductive carbon tape that was placed on a standard FESEM sample holder, and (ii) a TEM carbon lacy grid, which was then placed in the TEM. X-ray diffraction (XRD) was performed using a Philips X'Pert MPD diffractometer (Philips, Amsterdam, The Netherland).

3. Results and Discussion

In this section, we present the microstructure and electrochemical performance of the SnS/C nanocomposites with two different SnS contents: 10SnS/C and 20SnS/C. Two C matrices were used, either artificial graphite (AG) or multicarbon–microbeats (MCMB). Both of these C matrices are considered promising for use in Li-ion batteries, and, thus, both were examined. A total of four samples were fabricated and tested: 10SnS/AG, 10SnS/MCMB, 20SnS/AG, and 20SnS/MCMB. First, the morphology of the new SnS/C nanocomposites is presented using FESEM and TEM, while the phases present are identified by employing selected area electron diffraction (SAED) and X-ray diffraction patterns. Then, long-term electrochemical cycling results are shown and the morphology is re-examined.

3.1. Microstructure Prior Electrochemical Cycling

In Figure 1 the FESEM images show that for all nanocomposites fabricated, the SnS particles had an open (flower-like) structure and were attached on the carbon powder (AG or MCMB) surfaces.

To determine the average particle size, 50 areas were imaged for each sample. In addition, EDS was also performed in order to determine the at% of the Sn and S. Table 1 presents the averaged semiquantitative EDS analysis, which indicates that the Sn and S at% was higher when MCMB was used as the precursor. The 10SnS/AG sample contained the lowest at% of active SnS material, as well as the smallest average particle size.

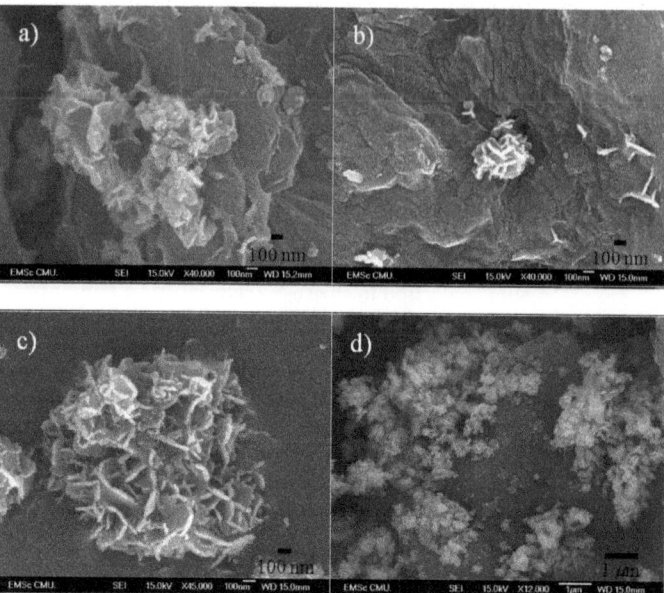

Figure 1. The field emission scanning electron microscope (FESEM) images of the new as prepared samples, before lithiation: (**a**) the 10SnS/artificial graphite (AG) composite; (**b**) the 10SnS/multicarbon–microbeats (MCMB) composite; (**c**) the 20SnS/AG composite; (**d**) the 20SnS/MCMB composite.

Table 1. FESEM–electron dispersive spectroscopy (EDS) analysis conducted before cycling to determine the content and particle size of Sn and S.

Sample	Particle Size (nm)	Elements (% atom)		
		Sn	C	S
10SnS/AG	254.33	3.04	95.48	1.48
10SnS/MCMB	494.35	3.5	94.17	2.34
20SnS/AG	411.34	3.71	92.16	4.13
20SnS/MCMB	559.91	5.34	89.09	5.57

Performing TEM on the as prepared powders, as seen in Figure 2, indicated that the SnS particles consisted of multiple nanofiber/nanorod-like petals. Selected-area electron diffraction (SAED) conducted on the nanofibers indicated the presence of an SnS phase (JCPDS card #39-354 for SnS).

The formation of the SnS alloy was further verified by X-ray diffraction (XRD). In Figure 3, it is illustrated that all as prepared samples contained only C (JCPDS card #41-1487 for C) and SnS, while pure Sn or S phases were not present, which is consistent with the SAED results of Figure 2. Furthermore, using the Panalytical X' Pert Plus software, the crystal size within the SnS flower-like particles before cycling was found to be ~10 nm.

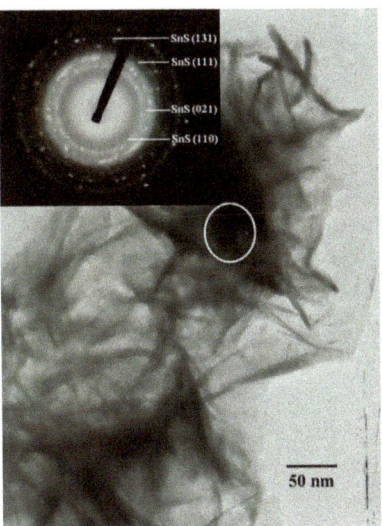

Figure 2. Transmission electron microscope (TEM) image and selected-area electron diffraction (SAED) pattern of the SnS structure of the as prepared 10SnS/AG sample. In all prepared nanocomposites, the SnS particles had this microstructure.

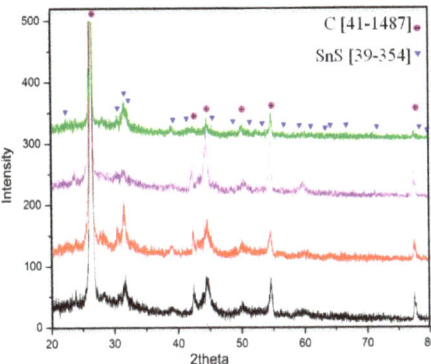

Figure 3. X-ray diffraction (XRD) of the SnS/carbon composites before cycling. **Green**: 20SnS/AG, **purple**: 20SnS/AG, **red**: 20SnS/MCMB, **black**: 10SnS/MCMB.

3.2. Electrochemical Performance

Figure 4a presents the capacity of all samples when cycled against Li for 200 cycles. It is seen that in all cases the capacity faded continuously with cycling. The lower SnS content electrodes (10SnS/AG and 10SnS/MCMB) had an initial capacity of ~630 mAh g^{-1} that reduced to ~260 mAh g^{-1} after 120 cycles. This is approximately the capacity of the carbon bases (AG and MCMB) used to support the SnS. The higher SnS content samples (20SnS/AG and 20SnS/MCMB) showed a higher initial capacity of over 700 mAh g^{-1}, but a greater capacity decrease during cycling, which reached 100 mAh g^{-1} after 120 cycles.

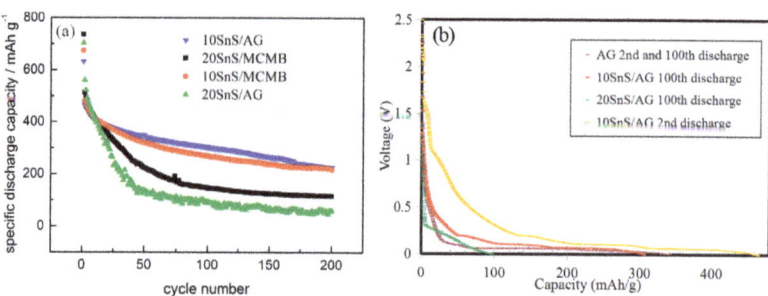

Figure 4. (a) Cycling performance of the different SnS/C samples; (b) discharge voltage capacity profiles for the 100th discharge of the 10SnS/AG, 20SnS/AG, and AG, and for the 2nd discharge of the 10SnS/AG.

The capacity fade shown in Figure 4a is consistent with the decay observed in [5,8], where after 45–50 cycles the initial capacity (~1000 mAh g^{-1}) of SnS/C materials had decreased to half its original value. However, this study goes beyond previous studies, as it is the first to perform long term cycling for SnS/C materials. Particularly, Figure 4a indicates that, for low SnS contents (10%), the capacity drops to that of the graphite base, indicating that SnS may not be playing the role of an active material after initial cycling, while for higher SnS contents (20%), the capacity is much lower than that of both AG and MCMB after 200 cycles.

To better illustrate that, after long term cycling, the capacity of the SnS/C drops to that of pure carbon, the discharge curves at the 100th cycle for pure AG, 10SnS/AG, and 20SnS/AG are plotted together in Figure 4b. It can be observed that all three curves are in close proximity, indicating that the SnS ceased to partake in the reaction with continuous cycling. For comparison purposes, the 2nd discharge curves for the 10SnS/AG and pure AG anodes are also shown in Figure 4b, where it is clear that SnS contributes to the voltage capacity, as it is above the curves obtained after 100 cycles. It should be noted that the 2nd and 100th discharge curves coincide for pure AG and no distinction can be made between them in Figure 4b, which is consistent with the long term electrochemical stability of graphite. Further examination of Figure 4b indicates that the degradation of the voltage capacity curve during the 100th cycle is more severe for the 20SnS/AG material than for the 10SnS/AG material, which is consistent with the capacity fade seen in Figure 4a.

3.3. Microstructure after Cycling

To interpret the capacity fading observed in Figure 4, XRD and electron microscopy were performed on the anodes after the 200 cycles were completed. The XRD characterization of the SnS/C anodes after 200 cycles is shown in Figure 5, indicating that the Sn phase, in addition to SnS, was present, after cycling. Using the Panalytical X' Pert Plus software, the grain size of the SnS alloy was estimated to be ~10 nm, which is the same as that of prior cycling, suggesting that crystal size was not affected by lithiation.

Figure 5 illustrates that pure Sn metal formed during the lithiation and delithiation process. This can occur because initial lithiation gives rise to the irreversible conversion reaction SnS + 2Li+ 2e$^-$ →Sn + Li$_2$S, by which Sn is dispersed in a Li$_2$S matrix. Further lithiation forms Li–Sn alloys, which is a reversible process during the charge/discharge cycle. The lack of the Li$_2$S phase in the XRD of Figure 5 may be explained by the following three possibilities: (i) Li$_2$S is soluble in the electrolyte [14], (ii) it may be amorphous, or (iii) the crystal size is too small to be detected by XRD.

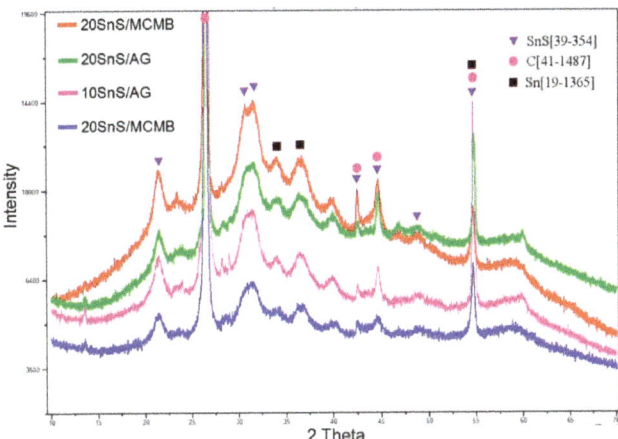

Figure 5. XRD of the SnS/C composites after 200 cycles.

Although XRD has not been performed on cycled SnS anodes, these observations are consistent with those of XRD studies of SnS_2 anodes [15], which reported the formation of metallic Sn during lithiation, as both conversion and alloying mechanisms occurred during the lithiation of SnS_2. Furthermore, our conclusions are in accordance with detailed XRD studies for the first Li-ion insertion (but not deinsertion) in Sb_2S_3 electrodes [16], which showed that Sb_2S_3 underwent both a conversion reaction (forming a pure Sb phase) and an alloying reaction; however, neither Li–Sb nor Li–S was observed in XRD, which was attributed to them either being amorphous or having a very small crystal size.

As no other studies have examined the phases present upon complete Li deinsertion of SnS electrodes, high resolution TEM (HRTEM) was conducted on the cycled materials, as shown in Figure 6.

Comparing Figure 6 with the initial microstructure depicted in Figure 2 reveals that after cycling, the flower-like structure of the SnS particles was "destroyed". The initial SnS nanofibers decomposed into nanoparticles that were 2–10 nm. Similarly, as the XRD indicated, the SAED patterns in Figure 6 documented the existence of pure Sn and SnS phases; however, no C rings were observed, even though the area considered contained a lighter phase region. This indicated that the light contrast region surrounding the nanoparticles (in Figure 6) was not the initial carbon particle substrate, but solid electrolyte interface (SEI) or electrolyte residue. This is verified by the representative EDS spectrum in Figure 6 that illustrates that the light shaded area of Figure 6c a contained high contents of F, O, C, and P. This C would be from the binder and carbon additive of the electrode.

The TEM analysis thus illustrates that, during the Li insertion and deinsertion process, the SnS detached from the C substrate and dissolved into the electrolyte, getting dispersed within the inactive SEI layer. This is consistent with the continuous capacity fade, as the SnS that detached from the carbon could not store Li ions. Therefore, even though the addition of SnS seemed to benefit the performance of the AG and MCMB (as the initial capacity was high), it was actually a drawback, as with continuous cycling it dissolved in the electrolyte and resulted in a lower capacity than carbon materials.

To examine the extent by which this inactive SnS/Sn SEI layer formed, FESEM was performed on the cycled electrodes, as this method could depict a wider area than TEM. For all samples, the same microstructure was observed. Representative images are shown in Figure 7 for the 10SnS/AG and in Figure 8 for the 20SnS/MCMB. It can be seen that the initial flower-like structure was disrupted and large solid particles were present on the carbon surfaces. To better understand the composition throughout the anode, EDS mapping (Figures 7b–g and 8b–g) was also performed. It is of particular interest to see that the micron size particles in Figures 7a and 8a were not comprised of SnS, but had a high C and F content, indicating that it was part of the SEI. These results are in accordance with the TEM images, since the EDS mapping illustrated that, during Li insertion and deinsertion, the flower-like SnS particles reduced into nanoparticles covering the anode surface, as Figures 7g–f and 8g–f illistrate.

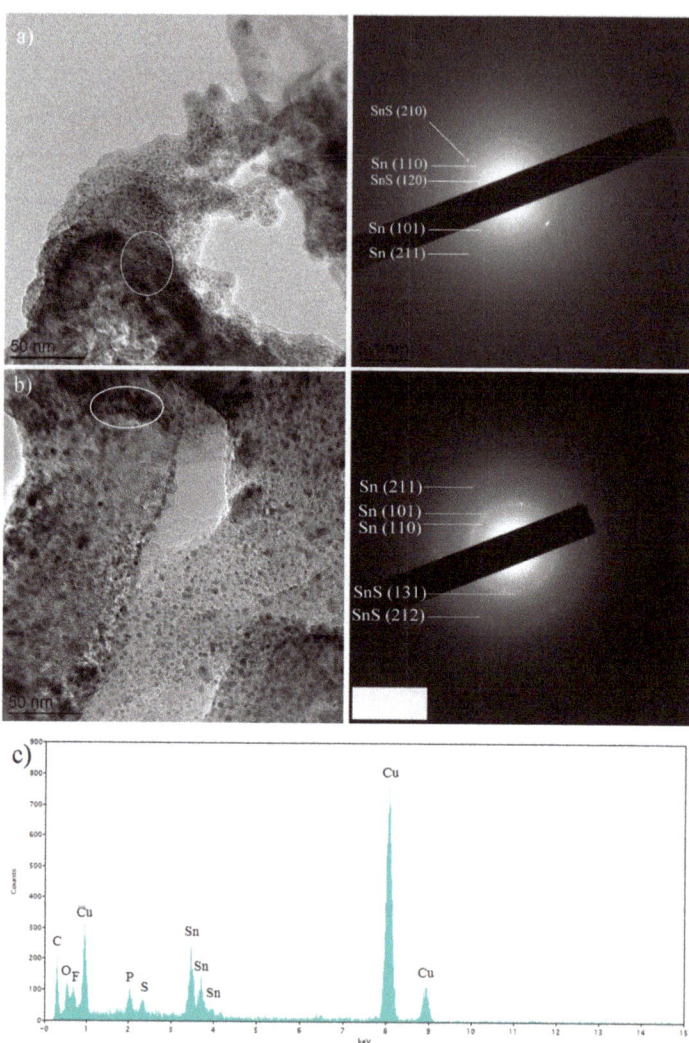

Figure 6. High resolution TEM (HRTEM) images with SAED and EDS results of the SnS/C composite anodes after 200 cycles. (a) 10SnS/AG, (b) 10SnS/MCMB, and (c) the results of EDS on 10SnS/AG.

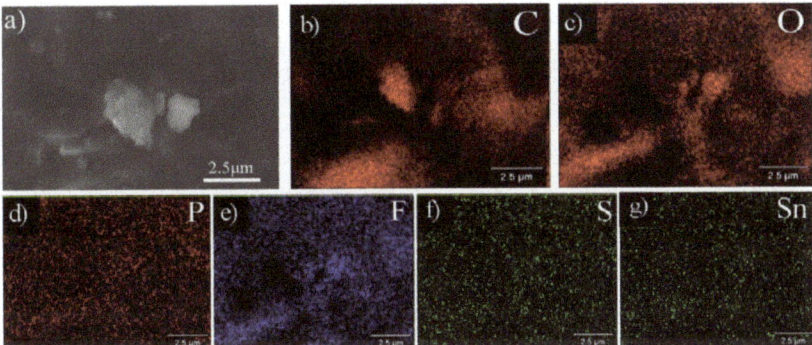

Figure 7. FESEM images of (**a**) 10SnS/AG, (**b**–**g**) EDS mapping of the area in (**a**). The element being mapped is indicated in each image.

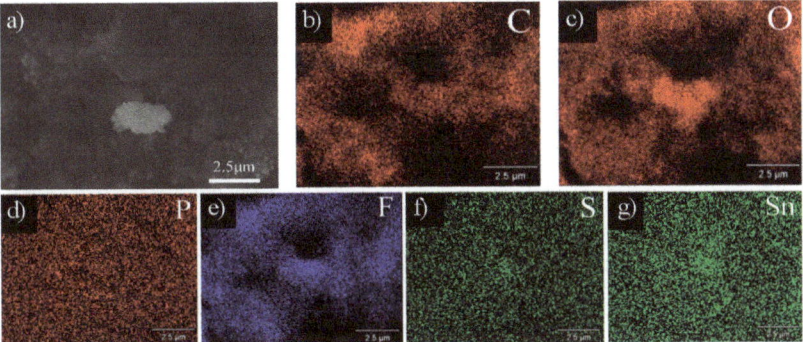

Figure 8. FESEM images of (**a**) 20SnS/AG, (**b**–**g**) EDS mapping of the area in (**a**). The element being mapped is indicated in each image.

Table 2 summarizes the EDS semiquantitative element analysis inside the micron particle of Figure 7a and outside the adjacent matrix, which again supports the conclusion that the SnS was distributed throughout the anode surface.

Table 2. EDS analysis after cycling of the area inside the particles and outside the adjacent matrix.

Sample	Area	% Elements (% atom)					
		Sn	S	C	O	F	P
10SnS/AG	Inside particle	1.25	0.66	37.79	37.73	22.07	0.50
	Adjacent matrix	1.75	1.52	45.45	29.60	19.40	2.28

The high concentrations of F, O, C, and P (as seen in the EDS of Figures 7 and 8), together with the observed free-standing SnS particles (as seen in the TEM images of Figure 6), suggest that the electrode surface was coated by an electrochemically inactive layer, which was comprised of Sn and SnS nanoparticles distributed throughout the SEI layer. Such an electrochemically inactive coating increases internal resistance, and the Sn trapped within could not host Li ions, explaining the capacity and voltage loss seen in Figure 4.

The anodes that contained higher contents of Sn and S (20SnS/AG and 20SnS/MCMB) would have a higher Sn and S content in the inactive layer, which would result in a higher internal resistance and greater capacity loss (as seen in Figure 4), as compared to that in the lower content SnS samples. Therefore, even though the addition of SnS on the C resulted in an initial increase in the capacity, the Sn and S dispersed into the electrolyte over continuous cycling, becoming part of the SEI layer and increasing the internal resistance of the cell. This conjecture was tested by comparing the impedance spectroscopy before and after 20 cycles for the 10SnS/AG and 20SnS/AG electrodes. In Figure 9 it can be seen that the impedance (i.e., interface/internal resistance) was higher for the 20SnS/AG sample, which supports the concept that SnS becomes chemically inactive, and the greater its content, the more negative its effects on electrochemical performance (as seen in Figure 4).

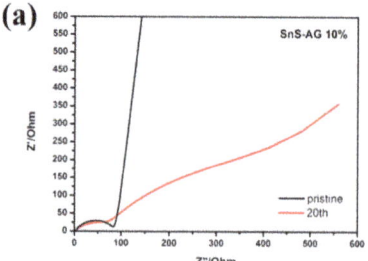 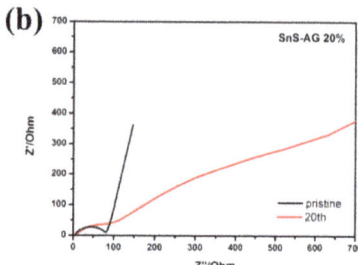

Figure 9. Electron impedance spectroscopy (EIS) diagrams for (**a**) 10SnS/AG and (**b**) 20SnS/AG, before and after 20 cycles.

Hence, the better performance noted for the 10SnS/AG materials was due to the lower Sn and S content that could dissolve into the solid electrolyte interface layer and increase internal resistance. These results show that the conversion reaction and alloying that SnS undergoes during cycling result in the dispersion of Sn particles within the electrolyte and SEI layer. Therefore, even though the capacity is initially higher, in the long term, the addition of SnS nanoflowers on graphite increases the internal resistance and results in a capacity that is even lower than that of the carbon substrate. In order to produce efficient SnS/C nanocomposites, the SnS must be fully isolated from the electrolyte using a protective coating.

4. Conclusions

The present article provided significant insight into the underlying mechanisms that give rise to the capacity fade of SnS materials during cycling. Although it has been suggested that SnS undergoes a conversion mechanism upon lithiation, its morphological consequences have not been examined. The microscopy observations herein indicate that lithiation of SnS/C leads to the formation of free-standing Sn and SnS nanoparticles that are dispersed throughout the electrolyte/SEI layer on the electrode surface. As a result, the Sn and SnS become inactive and isolated from the current collector, leading to continuous capacity fade. The microstructural observations are in accordance with the electrochemical cycling results, which indicate that, after 120 cycles, the capacity had dropped to that of the pure carbon matrices, suggesting the SnS was not participating in the reactions. Furthermore, EIS showed that increasing the SnS content increased the interface resistance, as the content of Sn and SnS in the inactive layer increased. Hence, in order to commercialize SnS electrodes, which have the ability to offer a capacity greater than other Sn-based anodes, it is necessary to inhibit such dissolution of Sn and SnS into the electrolyte. Approaches similar to those followed for S–C electrodes (for Li–S batteries), which "trap" S in highly porous carbon substrates, may be appropriate [1]. However, the goal of the present study was to provide insight into the capacity fade of SnS, not fabricate a promising new anode.

Author Contributions: Conceptualization: K.A., T.S. and H.D.; methodology, K.A., T.S., A.Y. and H.D.; validation, K.A., H.D., T.S., A.Y. and T.H.; data curation, K.A., H.D., A.Y., T.S. and T.H.; writing—review and editing, K.A, H.D., A.Y. and T.S.; All authors have read and agreed to the published version of the manuscript.

Funding: This research was initially funded by K.A., A.Y., T.S., T.H., grant number ERC-MINATRAN 211166.

Institutional Review Board Statement: Not applicable.

Informed Consent Statement: Not applicable.

Acknowledgments: This work began with the support of ERC-MINATRAN 211166 (K.A., A.Y., T.S., T.H.). Many of the present results were included in the PhD thesis of H.D.

Conflicts of Interest: The authors declare no conflict of interest.

References

1. Wang, Z.; Tian, W.; Li, X. Synthesis and Electrochemistry Properties of Sn-Sb Ultrafine Particles as Anode of Lithium-ion Batteries. *J. Alloys Compd.* **2007**, *439*, 350–354. [CrossRef]
2. Fong, R.; von Sacken, U.; Dahn, J.R. Studies of Lithium Intercalation into Carbons Using Nonaqueous Electrochemical Cells. *J. Electrochem. Soc.* **1990**, *137*, 2009–2013. [CrossRef]
3. Wang, K.; He, X.; Ren, J.; Wang, L.; Jiang, C.; Wan, C. Preparation of Sn_2Sb alloy encapsulated carbon microsphere anode materials for Li-ion batteries by carbothermal reduction of the oxides. *Electrochim. Acta* **2006**, *52*, 1221–1225. [CrossRef]
4. Li, Y.; Tu, J.P.; Huang, X.H.; Wu, H.M.; Yuan, Y.F. Nanoscale SnS with and without carbon-coating as an anode material for lithium ion batteries. *Electrochim. Acta* **2006**, *52*, 1383–1389. [CrossRef]
5. Li, Y.; Tu, J.P.; Huang, X.H.; Wu, H.M.; Yuan, Y.F. Net-like SnS/carbon nanocomposite film anode material for lithium ion batteries. *Electrochem. Commun.* **2007**, *9*, 49–53. [CrossRef]
6. Besenhard, J.O.; Yang, J.; Winter, M. Will advanced lithium-alloy anodes have a chance in lithium-ion batteries? *J. Power Sources* **1997**, *68*, 87–90. [CrossRef]
7. Armstrong, R.W. Crystal Engineering for Mechanical Strength at Nano-Scale Dimensions. *Crystals* **2017**, *7*, 315. [CrossRef]
8. Kumar, G.G.; Reddy, K.; Nahm, K.S.; Angulakshmi, N.; Stephan, A.M. Manuel Stephan, Synthesis and electrochemical properties of SnS as possible anode material for lithium batteries. *J. Phys. Chem. Solids* **2012**, *73*, 1187–1190. [CrossRef]
9. Gou, X.; Chen, J.; Shen, P.W. Synthesis, characterization and application of SnS_x (x = 1, 2) nanoparticles. *Mat. Chem. Phys.* **2005**, *93*, 557–566. [CrossRef]
10. Aifantis, K.E.; Haycock, M.; Sanders, P.; Hackney, S.A. Fracture of nanostructured Sn/C anodes during Li-insertion. *Mat. Sci. Eng. A* **2011**, *529*, 55–61. [CrossRef]
11. Cai, W.; Hu, J.; Zhao, T.; Yang, H.; Wang, J.; Xiang, W. Synthesis and characterization of nanoplate-based SnS microflowers via a simple solvothermal process with biomolecule assistance. *Adv. Powder Technol.* **2012**, *23*, 850–854. [CrossRef]
12. Aifantis, K.E.; Huang, T.; Hackney, S.A.; Sarakonsri, T.; Yu, A. Capacity fade in Sn-C nanopowder anodes due to fracture. *J. Power Sources* **2012**, *197*, 246–252. [CrossRef] [PubMed]
13. Adpakpang, K.; Sarakonsri, T.; Aifantis, K.E.; Hackney, S.A. Morphological study of SnSb/graphitecomposites influenced by different ratio of Sn:Sb. *Rev. Adv. Mater. Sci.* **2012**, *32*, 12–18.
14. Liang, C.; Dudney, N.J.; Howe, J.Y. Hierarchically structured sulfur/carbon nanocomposite material for high-energy lithium battery. *Chem. Mater.* **2009**, *21*, 4724–4730. [CrossRef]
15. Brousse, T.; Lee, S.M.; Pasquereau, L.; Defives, D.; Schleich, D.M. Composite negative electrodes for lithium ion cells. *Solid State Ionics* **1998**, *115*, 51–56. [CrossRef]
16. Denis, Y.W.Y.; Hoster, H.E.; Batabyal, S.K. Bulk antimony sulfide with excellent cycle stability as next-generation anode for lithium-ion batteries. *Sci. Rep.* **2014**, *4*, 4562–4567.

MDPI
St. Alban-Anlage 66
4052 Basel
Switzerland
Tel. +41 61 683 77 34
Fax +41 61 302 89 18
www.mdpi.com

Crystals Editorial Office
E-mail: crystals@mdpi.com
www.mdpi.com/journal/crystals

www.ingramcontent.com/pod-product-compliance
Lightning Source LLC
LaVergne TN
LVHW070209100526
838202LV00015B/2023